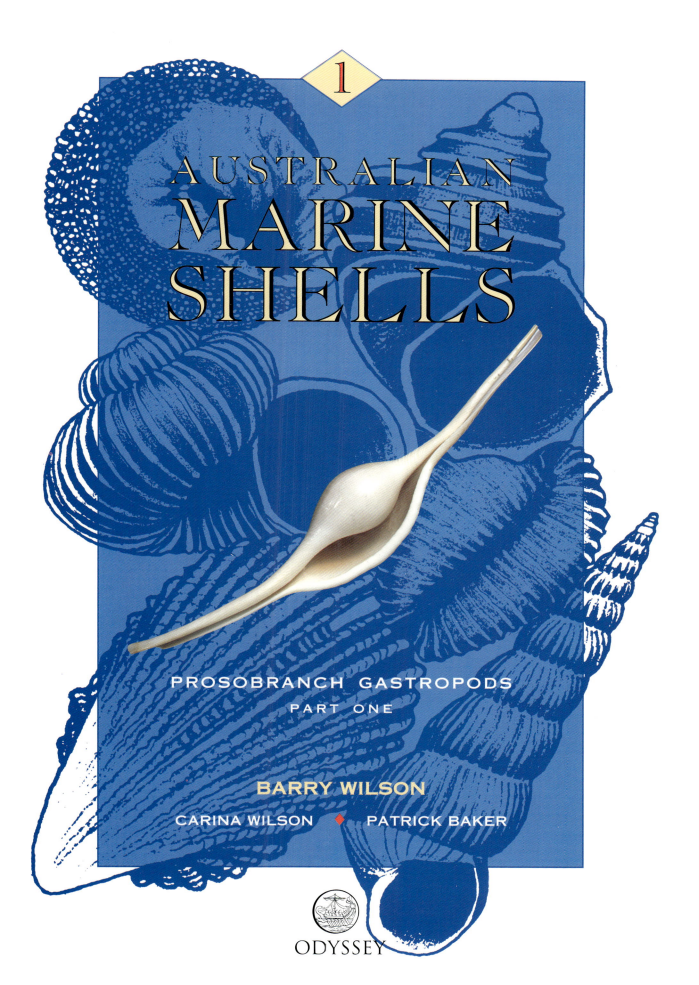

AUSTRALIAN MARINE SHELLS

PROSOBRANCH GASTROPODS

PART ONE

BARRY WILSON

CARINA WILSON ◆ PATRICK BAKER

ODYSSEY

Publisher
Odyssey Publishing
4 Saint Ives Loop
Kallaroo
Western Australia 6025
Telephone and facsimile (09) 307 1469

USA Representative
John D. Jackson
11558 Rolling Hills Drive
El Cajon, CA 92020
USA
Telephone (619) 579 8405
Facsimile   (619) 579 7901

Author
Dr. Barry Wilson

Illustrations
Carina Wilson

Photography
Patrick Baker

Design
Robyn Mundy
Mundy Design
Leederville, Western Australia

Colour Laboratory
Churchill Color Laboratories & Computer Graphics
Subiaco, Western Australia

Negative Preparation
Quadrascan Graphics Pty. Ltd.
West Perth, Western Australia

Printing
Kaleidoscope Print and Design
West Perth, Western Australia

Finishing
M & M Binders
Mount Waverley, Australia

First published in Australia in 1993

© 1993 Odyssey Publishing

(Odyssey Publishing Co is a joint venture between Murex Consultants
Pty Ltd and Odyssey Ventures Pty Ltd)

National Library of Australia Cataloguing-in-Publication data

Wilson, B. R. (Barry Robert), 1935-
Australian marine shells. Volume 1

Bibliography.
Includes index.
ISBN 0 646 15226 2 (v. 1).
ISBN 0 646 15225 4 (set).

1. Mollusks - Australia. I. Wilson, Carina, 1960- . II. Baker, Patrick,
1943- . III. Title. IV. Title : Prosobranch gastropods.

594.994

Preface

MANY TIMES during the four years that I have been preparing this study I have asked myself "Why am I doing this?" Why spend countless days on the endless curatorial and secretarial tasks associated with such a project, which kept getting bigger instead of finished? It was not as if I had empty time to occupy, and some aspects of the matter could be said to be in conflict with the conservation ethics that have been the basic principles of my professional life in recent years. Yet there were several reasons.

In the first place I am by training a taxonomist. Although a disgracefully messy individual in many ways, and hopelessly disorganised, I share the taxonomists' delight in finding order in the seemingly chaotic world of nature. There is something immensely satisfying in discovering new kinds of shells, finding names for them and resolving as best you can the often confusing details of variation between one kind and another. Even more satisfying is the task of putting this information together in concise detail in a book. Producing a book of this scope demands patience and perseverance but it is very rewarding.

Secondly, shells are such beautiful things. Just their variety is a joy. Some shells are simply exquisite in their symmetry of form, sculpture and colour patterns and the animals that make them are among the most interesting of earth's creatures with lives more diverse than almost any other group.

Thirdly, while this book is intended to serve a useful purpose in helping people identify species of marine gastropod for pleasure or for study purposes, I hope it will also convey to its readers some of the delight and satisfaction I get myself from studying these beautiful creatures.

Then there is the question of conservation. It has sometimes been said to me that collecting shells is immoral because it involves taking the lives of small creatures simply to gratify the human pleasure of ownership. I agree with that sentiment to some extent. Having witnessed wanton destruction of habitat by thoughtless collectors seeking rare cowries among living corals, and the systematic rape of reef-flats by greedy collectors taking every shell in sight, dead or alive, I have to admit that among the shell collector fraternity there are those who lack any environmental ethic or even any basic common sense. Yet that is a matter of degree and there is another side to the story.

Of far greater danger to the survival of any mollusc species and the marine environment generally are the massive destruction of coastal habitats by encroaching urban and industrial development, the pollution of oceanic waters and the deadly eutrophication of coastal bays and estuaries. Two things are needed to contain the current rate of marine habitat destruction - knowledge and political will.

Too often we find that very little is known about the animals and plants in threatened habitats and hardly anything at all about the basic ecological principles involved in designing appropriate marine environmental management programs. Identification of the species and knowledge of the ecological processes involved are prerequisites for wise environmental protection and management.

The political will necessary for emplacement of appropriately balanced development projects and environmental management programs comes from a well informed, knowledgeable and environmentally conscious community. Politicians hear and respond (eventually) to what people are demanding. The loudest and most effective demands come from people who know and care about the natural world. It seems to me that shell collecting is one of those healthy pursuits that fosters growth of the environmental ethic.

It is crass overkill and commercialisation that I find offensive. Most collectors are careful to take only what is needed to fill a gap in a series and avoid destructive collecting methods. Such people are driven by a thirst for knowledge rather than for objects. With time they become defenders of the earth, not destroyers of it. My books are meant for them.

Barry Wilson

Acknowledgments

During the early stages of this project significant support was received from the Western Australian Department of Conservation and Land Management. The Western Australian Museum generously allowed us to use the museum's photographic studios and library facilities. Also of critical importance was assistance provided by several museums and their staff who provided so many of the specimens used for the illustrations, most notably the Western Australian Museum, The Australian Museum, the South Australian Museum, the Museum of Victoria, and the British Museum (Natural History). This usually involved museum staff in many hours of work preparing the loans. The support of all these government institutions and their staff is gratefully acknowledged.

Specimens for the illustrations were also drawn heavily from the private collection of Hugh Morrison. In fact the book would not have been possible without the generous assistance of many private collectors who made specimens available and provided information. Among those who contributed were Mollie Bowman, Barbara Collins, Andrew Edinger, Tony Gabelish, David Greenacre, Glad Hansen, Joy Hewitt, Alan Limpus, Max Marrow, Luigi Raybaudi, Joe Rinkens, Lorraine Rutherford, Trevor Sutcliffe and Thora Whitehead.

Many colleagues freely gave information and advice about the families of their particular expertise. These helpful people are acknowledged in the relevant sections. Alan Limpus and Ian Loch provided assistance by telephone, facsimile and letter throughout the project.

Walter Sage and Carole and Jules Hertz reviewed the manuscript and made many helpful suggestions.

CONTENTS

INTRODUCTION

HE shores and continental shelves surrounding the Australian continent are the home of an extraordinarily large number of marine molluscs. Australia undoubtedly has the richest fauna of marine molluscs of any nation in the world. There are several reasons for this. The island continent sits astride the Tropic of Capricorn, its northern coast within the tropical zone facing the verdant coasts of South-East Asia and its southern coast in the temperate zone, separated from Antarctica by the cold waters of the Southern Ocean. The north coast features coral reef and mangrove habitats while the cooler south coast is characterised by rocky shores and sandy beaches, with high wave action and a dense cover of seagrasses or brown algae in the sublittoral zone.

This climatic and habitat diversity accounts for much of the species richness of the Australian coastal marine fauna. There are also historical factors. The marine fauna of the north coast has had very different origins from that of the south coast.

After a little more than 200 years of study in the Linnaean tradition, the Australian species of marine molluscs remain incompletely known. The families of larger and more conspicuous species have been fairly well catalogued but many species of the lesser families remain to be discovered and described. Consequently there is no clear idea of how large the Australian fauna of marine molluscs actually is but it seems likely to be in the order of 10,000 species.

This volume is the first of two which gather together current information about Australian marine prosobranch molluscs and describe and illustrate their shells as an aid to identification for the purposes of business or pleasure. They will provide information to guide all those who collect shells simply for the satisfaction of gathering beautiful things. Most importantly it should bear testimony to the beauty of shells and the living creatures which make them. I hope that the illustrations will make people aware of the incredible variety of molluscs and think about the lives they lead.

The Subclass Prosobranchia is the largest group of the Class Gastropoda and contains most of the coiled shells found in the sea and along the shore. The Subclass contains seven Orders of which six are marine or include marine representatives. In this volume, five of the six marine Orders are covered. The sixth marine Order, the Neogastropoda, including the predatory murex, olive, volute, cone and terebra shells, is the subject of volume two, which is to follow.

Although an effort has been made to include all the more conspicuous species, this study is far from a complete inventory. Nor is it the last word on the nomenclature of the included species. There is still much to be done on the taxonomy of the Australian marine fauna. Nevertheless, the information here should be an advance on what was available before and useful to amateur collectors, students of natural history, and environmental scientists who need to identify Australian molluscs for whatever purpose.

Shell collecting

In the early years of shell collecting, interest focused almost entirely on the shells as beautiful objects and on giving them names. In our century the rate of new discoveries has slowed and attention has shifted from describing new kinds to the study of the living animals and their place in the ecosystems of our planet. In our time of changing climate and alarming impact by human activity on the environment that supports us, it has become urgent that we understand how our ecosystems function. While scientists carry out their investigations using increasingly sophisticated techniques and technology, they still need to have knowledge of the species that comprise living communities of plants and animals. The work of the earlier collectors and classifiers provides that basis, incomplete though it is. Yet there is still a need for inventory work on our flora and fauna, and amateur collectors will continue to play a significant role in this for some time to come.

It is also a fact that for the results of our scientists' discoveries to be put to effective use in realising resource potentials and resolving environmental problems, there must be a higher level of public understanding and support. Shell collecting is one of those healthy natural history pursuits which enriches people's lives and helps build public understanding and appreciation of the natural world.

Hints on how and where to collect shells

Some collectors search for their specimens in dealers' catalogues and showcases. That isn't half as much fun as finding your own in the field. Molluscs live in a vast range of different ecological niches. Learning where to find them in the wild adds a very different dimension to the joys of shell collecting.

The simplest way to find shells is to search for them among the flotsam and jetsam thrown up on the shore by waves and currents. After a little experience you will learn where the best places are. Beaches adjacent to intertidal and sublittoral habitats rich in living animals are obviously the places to find empty ("dead") shells in most abundance. But even there, some corners accumulate things from the sea more frequently than others.

Beach-collected dead shells are, in many ways, preferable. Live-collected specimens pose a problem because it is necessary to remove the body, which would otherwise soon become smelly. There is also the matter that taking live molluscs for their shells requires the sacrifice of the animals. Some collectors believe that this is wrong in principle. The other side of this argument, however, is that dead shells have often been rolled around in the sand for some time and have become worn and broken. The shells of living animals are more likely to be in good condition. The collector must examine his or her own conscience and decide whether or not taking live animals for their shells is acceptable. (It is also necessary to examine the law in the collecting locality in case collecting live shells is illegal.)

If the decision is that collecting living specimens is okay, then there are many ways to go about it. For most collectors the obvious way is to visit the shore when the tide is low and access is possible to the animals in their living habitats. Reference to published local tide charts will assist in selecting the right times.

Rocky or coral shores at low tide are perhaps the most productive of all sites for collecting molluscs. One soon learns to find them in their hiding places in crevices and gutters. Many species spend the daylight hours, during low tide periods at least, hiding under stones which have space beneath them and may be found by turning the stones on their edge so that the undersides may be examined. But remember to put the stones back the way they were - many creatures live there and some lay their eggs in these places. If the stones are left upturned most will die and their habitat will be destroyed. The number of such hiding places and egg-laying sites is limited on most shores and too much and too frequent disturbance of this type will soon make the area uninhabitable, just as destruction of hollow limbs in a forest makes them useless as breeding areas for hollow-nesting birds.

Many rocky shore and coral reef molluscs actually live in pockets of sand in the tide pools. They may be found by sieving the sand. Others live hidden among the fronds of seagrasses or algae growing on the rocks. They can be found by washing the weeds in dishes of fresh water but this collecting technique is habitat destructive and is not recommended.

For the brave at heart, perhaps the best time to collect living molluscs on rocky shores is during low tides at night. Many of the gastropods leave their hiding places at night and go searching for food. This is true of cowries, for example. They may be found then without disturbing the habitat.

Intertidal sand flats may also be very productive collecting areas for molluscs. This is actually the domain of the bivalves but there are many gastropods too. However, it takes an experienced eye to find many of the sand-flat species as most bury themselves when the sand is exposed to the air. Some make characteristic trails in the sand which give away their location.

Dredging and trawling are methods of collecting bottom-living animals from the sublittoral zone. A dredge is a box-like structure made of metal which is dragged along the sea bed behind a boat. There are many different designs for different situations. A typical box dredge has a curved blade on both wide sides of the opening; they dig into the sand or mud as the dredge is pulled along. The side of the box may have runners so that the back end is lifted slightly off the bottom, allowing finer sand to filter through the mesh sides. In effect the dredge acts as a sieve dragged through the substrate. Shells and other objects are retained in finer mesh in the back of the box.

A trawl does a similar job but is a net, not a box, designed so that the mouth of the net is held open as it is dragged along the bottom. The rope around the opening of the net is fitted with lead weights so that it digs into the sand, and the back end (the "cod end") is lined with fine mesh to retain small objects. A beam trawl has a transverse beam across the top of the net at the front holding the mouth open. This type of trawl has the disadvantage that it does not work if it falls to the bottom upside down and this may be hard to avoid. An Agassiz trawl has a transverse beam across the middle of the opening with a "lead line" both top and bottom so that it doesn't matter which way down it goes. An otter trawl has two boards at the front which spread the net open when it is under tow. This is the type of trawl the prawn boats use and it is very effective; but it takes skill to set so that it opens properly.

The most difficult thing about dredging and trawling is to find the right length of tow rope for the depth of operation. If the rope is too long the dredge or trawl will anchor the boat. If it is too short the front blade or lead line will not dig or scrape the bottom. A general rule is that the tow rope should be two and a half times the depth, but this will vary according to the equipment used and the bottom type. Another problem is that the dredge or trawl may get "hung up" on rocks or other objects on the sea floor. Careful selection of the area of operation (i.e. avoiding rocky bottom) is necessary. It is also desirable to have a "break link" on the connecting chain of one side so that if a hang-up does occur it will snap and allow the equipment to tip sideways or upside down and be rescued.

Another problem with dredging and trawling is that they are very hard work. It is possible to use a small dredge or trawl behind a small motorised or sailing boat and to pull it up by hand. This can be very effective in shallow water. However, most dredging and trawling are done using larger vessels equipped with a winch and davits for lifting the full (hopefully) dredge or trawl aboard. In many coastal areas dredging and trawling for specimens are illegal or require a special permit.

For many years the only means available to marine scientists and naturalists for sampling the creatures of

Above - *Cypraea reevei* (p. 176) with mantle extended. In an underwater cave off Fremantle, WA.

Above - *Cypraea venusta* (p. 198). Female sitting on an egg-mass laid in an old *Scutus* shell. Marmion Marine Park, WA. February, 1993.

Left - *Natica stellata* (p. 217). These are very active animals with a large foot.

Photos: Barry Wilson

the sea floor were dredging and trawling. At the best of times it is rather like a penny dip. You do not know what you catch until the contents are tipped onto the sorting table. Such collecting gives little information about the habitats and life styles of the animals caught. The advent of scuba and snorkel diving in recent years has revolutionised marine collecting. Collectors are no longer restricted to the intertidal zone to get first-hand experience of the habitat. We may now explore the sublittoral habitats ourselves, just as we do in the intertidal zone. This is one reason for the host of new species found and described in recent years. Divers can seek out animals in their hiding places on the sea bed where no dredge or trawl could find them.

Conservation

Although there is a lack of scientific evidence, there is strong anecdotal evidence that heavy collecting pressure can deplete populations of marine shells. Populations of cowry shells, for example, can be stripped from an intertidal reef if there is easy access to it. However, most marine gastropods are not confined to the intertidal zone. If left alone for a while, damaged populations there will soon recover by means of immigration from elsewhere. With planktonic veliger larvae most species' populations are likely to receive new recruits from afar brought by the ocean currents.

The most vulnerable species are those with direct development which do not have a capacity for planktonic larval dispersal. Once these are locally extinct they may never recover. The volutes are of particular concern in this regard. The Queensland *Cymbiola* species, for example, may easily be wiped out on local intertidal sand cays. The *Zoila* cowries of the south-west are also vulnerable to local extinction as a result of over-collecting by divers. Although it is very

unlikely that a whole species could be collected to total extinction, this can certainly happen to local genetic forms.

For conservation of species and genetic variations the problem is not so much over-collecting but the habitat damage which may be caused by too many people disturbing the environment. Turning stones and breaking coral to look for shells is like cutting down the trees in a forest to catch the monkeys. Not only is that generation of shells removed, but the habitat is rendered unsuitable for the new generation which might otherwise have replaced them.

There are some golden rules for responsible collectors:

◆ Ensure minimal disturbance to habitat.

◆ Do not break corals to look for shells.

◆ Carefully replace overturned rocks to the original position.

◆ Take only sufficient specimens to represent the species in your collection.

◆ Do not remove juveniles or shells with eggs.

◆ Leave damaged shells to breed.

Regulations

Some species of marine mollusc are totally protected in Australia. These are all the species of the giant clam family *Tridacnidae*, the helmet shell *Cassis cornuta* and the triton *Charonia tritonis*. There are size and bag limits for some others that are fished commercially. Those regulations vary from state to state and apply to amateur collectors as well as to commercial fishermen.

Commercial collecting of shells for sale is not permitted in Australia without a professional fisherman's licence. Permits are also required by amateur collectors in marine parks. In Queensland's Great Barrier Reef Marine Park shell collecting is permitted only in General Use 'A' and 'B' zones. Similar restrictions apply in Western Australia's Ningaloo Marine Park and elsewhere. There are also limits on how many specimens may be taken in some areas. Collectors should contact the local offices of the park management agencies for details.

There are also regulations governing the export overseas of marine shells. It is necessary to provide a certified list and obtain a Customs export permit before taking or sending Australian shells out of the country. The Customs authorities have made special arrangements with the state natural history museums and shell clubs which can assist in this matter.

Some may find these regulations and restrictions irksome. However, most collectors will agree that we all have a responsibility to ensure that our descendants do not inherit from us an impoverished molluscan fauna because of our greediness and disregard for basic environmental care.

Caring for a shell collection

If a living specimen is taken the first problem facing the collector is what to do about the body. Most marine gastropods die out of water within a few days. If only the shell specimen is required the body must be removed. There are several ways to do this.

Specimens with sturdy, uncoloured shells may be boiled until the adductor muscles fastening the body to the inner surface detach and the remains may be shaken or picked out with a curved wire or pin. However, heat can cause polished shells like cowries or olives to crack. It also fades many colour pigments.

Another method is to allow the body to rot until it can be shaken or picked out. There are two dangers with this method. If the tissues dry it can be almost impossible to get the body out. Secondly, if there is too much moisture acids may develop during the decay process and acidic fluids may spill out, seriously damaging polished surfaces. This can absolutely ruin shells like cowries. Rotting is best done by burying the specimens in clean beach sand for a few days. (But do not use garden soil because it also contains acids.) The dampness prevents the tissues from drying. The acidic fluid problem can be avoided by keeping the sand damp and burying the specimens upside down so that any fluids do not spill out. When the specimens are "ripe" the meat may be washed out under a high-pressure jet of water. Although effective, this method will not please other members of your family - the smell can be terrible.

A non-smelly variation of the sand method is to repeatedly freeze and thaw the specimens until the body can be washed out. This has advantages for travellers who have access to a household freezer. Alternatively, a few hours in dilute alcohol will often allow the body to be removed.

There are some things not to do. Never rot your specimens in water. Acids and gases will develop and ruin the specimens after a very short time. And if you put them in a jar with a lid screwed on, you (and all the folk around) will have a very unpleasant surprise when you remove the lid. There is a popular myth that shells may be "put on an ant heap" so that the ants may do the dirty work for you. This method is a failure. In the first place most ants do not eat meat and in any case the body of the molluscs will become dry and permanent long before meat-eating ants could remove it.

Many greedy collectors make the mistake of collecting more specimens than they can process. The result too often is that part of the sample is thrown away.

For those who need to keep the body as well as the shell, there is no better way than freezing the specimens. But eventually it will be necessary either to remove the body or preserve it in fluid. The commonly available formalin solution is not to be used for shelled molluscs unless it is very carefully buffered. Formalin is an acid and will have disastrous effects on fine sculpture and polished surfaces. The formalin solution may be buffered with borax powder or lime but getting the concentration correct is a job for an expert. It is much better for most collectors to use a 70% alcohol solution.

If the shell is not important and the specimen is intended for study of the animal, especially if histological sections are to be made from the tissues, it is best to fix them for a few hours in a 10% formalin solution and then change to 70% alcohol.

Most molluscs withdraw into their shells when they die. To overcome this you can anaesthetise them first. Several chemicals may be used but the most commonly effective one for gastropods is magnesium chloride. Seventy-three grams of this compound dissolved in a litre of fresh water makes a solution which has the same osmotic pressure as sea water. Most gastropods placed in this solution will hardly notice the difference until they become incapable of contracting their muscles. For the more sensitive species you can add the stock solution to sea water in stages until it is full strength. After anaesthetisation the animals may be preserved with the body extended.

With many kinds of gastropod shell the job is not over even when the animal's body is removed. The outside of the shell is often covered with algae, calcareous growths and plain dirt. If the collector wants to have "clean" specimens they can be soaked in dilute household bleach for a few hours. This will usually loosen up the unwanted material so that it may be washed or scrubbed off. This treatment can be overdone, however, to the extent that the shell surface is damaged.

Some collectors also like to remove the periostracum (a natural furry outer layer) from those species which possess one, so that the true shell surface is exposed. Household bleach and a little elbow grease will remove that as well. But remember that the periostracum is a natural part of the shell and it should be retained if the specimen is intended for scientific purposes.

The other part which is often discarded but should be retained is the operculum. If the collector is serious about the scientific value of his or her specimens, care should be taken to keep the operculums. They can be glued to cotton wool and put back into their place in the aperture once the animal is removed. But make sure that the operculums are put back into the right shells.

Once shells are clean and dry, a little light oil (baby oil is very good) brushed gently onto the surface will help preserve the colours.

For shell collections to have any scientific value they must be accurately "provenanced", that is, there must be a reliable locality for each specimen. It is also desirable for the name of the collector, the date of collection, and habitat data to be recorded. This information can be written on small paper labels and tucked into the aperture of gastropods. If the specimens are too small for this, they should be kept in vials or plastic bags together with their labels. The value of a collection is greatly enhanced by being well labelled. In fact a collection of shells without labels is scientifically worthless.

Cypraea (Zoila) friendii, on a sponge 20m, off Dunsborough, WA. (Photo: Barry Wilson)

References

The following general books on shells and shell collecting have relevance to Australia. Books and articles on specific issues and families will be found in the relevant sections.

Abbott, R. T. (1972). *Kingdom of the Seashell.* Rutledge Books, New York.

Abbott, R. T. & Dance, S. P. (1982). *Compendium of Seashells.* E. P. Dutton, New York.

Allan, J. (1959). *Australian Shells.* Georgian House, Melbourne.

Cernohorsky, W. O. (1967). *Marine Shells of the Pacific* **1**. Pacific Publications, Sydney.

Cernohorsky, W. O. (1972). *Marine Shells of the Pacific* **2**. Pacific Publications, Sydney.

Cernohorsky, W. O. (1978). *Tropical Pacific Marine Shells.* Pacific Publications, Sydney.

Coleman, N. (1975). *What Shell is That?* Paul Hamlyn, Sydney.

Coleman, N. (1976). *Shell Collecting in Australia.* A. H. & A. W. Reed, Sydney.

Coleman, N. (1981). *Shells Alive.* Rigby, Sydney.

Cotton, B. C. (1959). *South Australian Mollusca - Archaeogastropoda.* Government Printer, Adelaide.

Cotton, B. C. (1961). *South Australian Mollusca - Pelecypoda.* Government Printer, Adelaide.

Cotton, B. C. & Godfrey, F. K. (1940). *The Molluscs of South Australia, Part 2, Scaphopoda, Cephalopoda, Aplacophora and Crepipoda.* Government Printer, Adelaide.

Habe, T. (1964). *Shells of the Western Pacific* **2**. Hoikusha Publishing Co., Osaka.

Hinton, A. (1972). *Shells of New Guinea and the Central Indo-Pacific.* Robert Brown and Associates, Port Moresby.

Hinton, A. (1978). *Guide to Australian Shells.* Robert Brown and Associates, Port Moresby.

Kay, E. Alison. (1979). *Hawaiian Marine Shells. Reef and Shore Fauna of Hawaii, Section* **4**: *Mollusca.* Bishop Museum Press, Honolulu.

Kira, T. (1962). *Shells of the Western Pacific* **1**: Hoikusha Publishing Co., Osaka.

Macpherson, J. H. & Gabriel, C. J. (1962). *Marine Molluscs of Victoria.* Melbourne University Press. (*Victorian National Museum Handbook No. 2.*)

May, W. L. (1958). *An Illustrated Index of Tasmanian Shells* (revised edition by J. H. Macpherson). Government Printer, Hobart.

McMichael, D. F. (1960). *Shells of the Australian Sea Shore.* Jacaranda Press, Brisbane.

Powell, A. W. B. (1979). *New Zealand Mollusca.* Collins, Auckland.

Rippingale, O. H. & McMichael, D. F. (1961). *Queensland and Great Barrier Reef Shells.* Jacaranda Press, Brisbane.

Short, J. W. & Potter, D. G. (1987). *Shells of Queensland and the Great Barrier Reef - Marine Gastropods.* Robert Brown & Associates, Bathurst.

Springsteen, F. J. & Leobrera, F. M. (1986). *Shells of the Philippines.* Carfel Seashell Museum, Manila.

Wells, F. E., Bryce, C. W., Clark, J.E. & Hanson, G.M. (1990). *Christmas Shells: The Marine Shells of Christmas I. (Indian Ocean).* Christmas I. Natural History Association, Christmas I., Australia.

Wells, F. E. & Bryce, C. W. (1985). *Seashells of Western Australia*. Western Australian Museum, Perth. (Revised 1990.)

Wilson, B. R. & Gillett, K. (1971). *Australian Shells*. A. H. & A. W. Reed, Sydney. (Revised 1974.)

Wilson, B. R. & Gillett, K. (1979). *Field Guide to Australian Shells*. A. H. & A. W. Reed, Sydney.

Shell magazines and newsletters

Several of the shell clubs and societies regularly publish magazines of interest to the Australian collector and naturalist. Many of them include technical as well as general articles.

Australian Shell News. Newsletter of the Malacological Society of Australia. (Editorship rotates among the Australian natural history museums.)

Hawaiian Shell News. Published by the Hawaiian Malacological Society, P.O. Box 22130 Honolulu 96823.

Keppel Bay Tidings. Published by the Keppel Bay Shell Club, Rockhampton.

La Conchiglia. Obtainable from: C. Federici, 1-00147 Rome.

The Festivus. Published by the San Diego Shell Club. 3883 Mt Blackburn Ave. San Diego, CA 92111.

World Shells. Obtainable from World Shells, P.O. Box 561, Rome 00187.

Xenophora. Bulletin de l'Association Française de Conchyliologie, Paris.

Technical journals on molluscs

Archiv für Molluskenkunde. Journal of the German Malacological Society. Senekenberganlage 25 6000 Frankfurt, AM MAIN 1.

Haliotis. The journal of the Société Française Malacologie. Obtainable from: Laboratoire de Malacologie, Museum National d'Histoire Naturelle, 55 Rue de Buffon, 75005, Paris, France.

Journal of the Malacological Society of Australia. Obtainable from: Honorary Secretary, Malacological Society of Australia, c/- Dept. of Zoology, University of Queensland, St. Lucia, Queensland 4067.

Journal of Molluscan Studies. This is the journal of the Malacological Society of London. Obtainable from: Journal Subscriptions, Oxford University Press, Pinkhill House, Southfield Rd., Eynsham, Oxford, OX 8 IJJ, UK.

Malacologia. Obtainable from: Dept. of Malacology, Academy of Natural Sciences of Philadelphia, 1900 Benjamin Franklin Parkway, Philadelphia, PA., 19103-1195, USA.

Malacological Review. Obtainable as for *Malacologia*.

Monographs of Marine Mollusca. Published by American Malacologists, Inc., P.O. Box 7279, Silver Spring, MD, 20907, USA.

The Nautilus. Published by the Trophon Corporation. Obtainable from: The Nautilus, P.O. Box 3430, Silver Spring, MD, 2091, USA.

The Veliger. Published by the California Malacozoological Society, Inc. Obtainable from: The Veliger,

Santa Barbara Museum of Natural History, 2559 Puesto del Sol Road, Santa Barbara, CA 93105, USA.

Venus. Published by the Malacological Society of Japan. Obtainable from: c/- National Science Museum (Natural History Institute) 23-1, Hyakun-incho-3, Shinjukuku, Tokyo 169, Japan.

Remarks about molluscs

The "success" of any group of animals could be judged in several quite different ways. Groups which have survived long periods of time with little obvious change in their body forms are undoubtedly well adapted to the environmental niches they have occupied and therefore could be said to be highly successful. Success could also mean that the group has a large number of living individuals, a large number of species, or has rapidly evolved into many different habitats and has great diversity of body forms and habits.

Usually when a group is said to be successful, abundance, diversity and persistence through time are all implied. In all senses the Mollusca is recognised as one of the most successful groups of animals. The abundance and diversity of molluscs is surpassed at present only by the arthropods (i.e. crustaceans, insects, spiders, and their relatives). Even the vertebrates, which certainly have reached the peak of complex organisation, do not equal the molluscs in terms of abundance, number of kinds, or antiquity.

Fossil molluscs occur in some of the oldest known fossil-bearing rocks. It seems that the early molluscan body pattern was very adaptable and evolution has successfully modified it to fit many of the kinds of habitats the earth's surface has to offer.

Marine molluscs are found in most abundance and diversity in shallow seas near shore, but some occur in the abyssal deeps. There are some that spend their whole lives swimming in mid-water of the oceans or floating at the surface, drifting at the mercy of the ocean currents. Although the majority spend their adult lives on the sea floor, most have larval stages which drift in the plankton. Gastropod and bivalve molluscs have also invaded the fresh waters of inland rivers and lakes. Air-breathing pulmonate gastropods are a very diverse group that has successfully colonised land, including the driest deserts.

As you might expect with such a large range of habitats, molluscs are extremely varied in their life styles. Compare the garden snails which eat your lettuces with the giant predatory squids of the ocean deeps. Or, a delicate, soft-bodied, beautifully coloured nudibranch with a crusty old oyster cemented to a rock on the seashore. Yet, varied as they are, all these creatures share features which confirm their relationships and stamp them as being molluscs. Not all of them have shells but they are all fascinating animals and they play very significant roles in the planet's ecosystems.

Classification

Scholars of the 18th century devised a system of classification of plants and animals based on similarities of structure. They also established conventions for naming the classification units based on the ancient Latin and Greek languages. Although somewhat modified and expanded these systems and conventions are followed today, and comprise the scientific discipline known as taxonomy.

Species and subspecies

The basic unit of zoological classification is the species. We may think of the species as a *kind* of animal, or a group of individuals which can or could interbreed and share their genetic inheritance.

In most cases it is easy to recognise the members of a species because they share genes and have many visible characteristics which distinguish them from members of other species. But the process of gene inheritance from one generation to the next is dynamic. Some species are extremely variable in their visible characteristics and it is difficult to tell whether one or more species is involved without laboratory experimentation. In other species there may be little variation in those genes which express themselves in visible characteristics but large variation in those governing physiological or behavioural characteristics. Sometimes different species may look the same but lead very different lives in one way or another, to the extent that they are unable to interbreed.

There are, of course, many millions of species of plants and animals living on our world and it is not possible to experiment with them all. In the majority of cases we must rely on visible and measurable characteristics to determine the taxonomic limits of species. Nowadays there are intricate statistical methods which can assist, but in the absence of sophisticated laboratory work, distinguishing between species often remains a matter of guesswork and judgement. This is the reason why taxonomists disagree so often. Species which look different may be the same and species which look alike may be different.

The subspecies is an even more ambiguous category. It is used for discrete forms of a species which are geographically separated from each other. It is assumed that members of subspecies could interbreed if there were the opportunity. Their differences are because isolated populations have evolved some genetic peculiarities, but not to the point where they have become fixed on separate evolutionary paths.

Some collectors still use the subspecies category for extreme forms within an interbreeding population but the practice has no place in modern taxonomy. Even within a local population there may be discrete forms resulting from certain gene combinations but which nevertheless interbreed. Distinct forms within a breeding population can also result as individual adaptations to different environmental conditions. In none of these situations is the subspecies considered to be a valid taxonomic category. Naming forms within a breeding population may be convenient to those interested in the various forms as collectable items, but it serves no scientific purpose and can create terrible confusion in the taxonomic literature. For convenience such genetic variants can be given "form" names, but this should not be confused with the formal procedure for naming species and subspecies.

Genera

Species which have had a common ancestor will usually have many similarities and be recognisably different as a group from species of different origin. Similar species believed to have had the same evolutionary origin are grouped together in genera. In practice the genus is often an arbitrary category, its limits depending on the whims of the taxonomist. There are principles which most taxonomists follow and some effective statistical analysis methods; but even so, generic classification is often subjective. Some taxonomists like to use broad generic groups while others, the "splitters", prefer to arrange species in smaller groups, thus emphasising the differences among them rather than the similarities. Consequentially there is enormous taxonomic confusion in groups like the cones and cowries that have been studied and classified by workers who have taken different approaches to the process.

The use of subgeneric units is a compromise. Subgenera are useful taxonomic categories that acknowledge that there are distinctive groups of species within a genus, while at the same time preserving the information that they are all related. In this book broad generic units are usually used, with smaller groups ranked as subgenera. In many cases other contemporary authors rank the subgenera used here as full genera.

Higher categories

Still on the basis of similarities and presumed common ancestry, species and genera are classified in higher categories of increasing rank, viz. families, orders, classes, phyla and kingdoms. Often these classification units are further broken up into super- and sub-categories. Some phyla have additional categories.

Nomenclature - the rules and conventions for naming shells

As well as classifying the range of species, genera and higher categories, the various groups in the system must be named. This is not as easy as it may sound for there are many ways in which the whole system might become hopelessly confused, for example when different names are used for the same units, or the same names are used for different units. To prevent this happening, taxonomists have established a set of rules and conventions which everyone is supposed to follow. A body known as the International Commission for Zoological Nomenclature has been established which sets the rules and adjudicates in the event of disputes.

Names

The "scientific" names of plants and animals are always given in Latin or Greek form. This is a convention established in the early days when scholars wrote their works in those languages. This may be annoying to laymen and modern scholars without knowledge of the classical language. But it sets standards, and it is surprising how quickly you become used to it. Many vernacular names are the anglicised forms of Latin or Greek names; for example, tiger cowry is the vernacular name for *Cypraea tigris*. Even proper nouns in English (or other modern languages) can be easily turned into scientific names by changing the end to Latin or Greek form; Friend's cowry, for example, becomes *Cypraea friendii*.

In western tradition individual people have two essential parts to their name, i.e. a parental name and a given name. In a similar way species have names in two essential parts, a generic name followed by a specific name, e.g. *Cypraea tigris*. This is known as the binominal system and it was introduced by Linnaeus in his classic work *Systema Naturae* published in 1758, which marks the beginning of modern nomenclature. Strictly speaking the full binominal form should be used, but it is customary to use an abbreviated form, e.g. *C. tigris,* once the generic name has been established in any passage of writing.

When a subgeneric name is used it is written in brackets after the generic name, e.g. *Cypraea (Zoila) friendii*. This too can be abbreviated as *C. (Zoila) friendii* or *C. (Z.) friendii*. However, if the subgeneric name is not relevant to the context it may be ignored, e.g. *C. friendii*. For the sake of simplicity, in this book subgeneric names are not used in combination although species are listed beneath subgeneric headings within the generic groupings.

When a subspecies name is used it follows the species name, e.g. *Cypraea (Zoila) friendii thersites* or in an abbreviated form *C. (Z.) f. thersites* or *C. f. thersites.*

Another convention is that the author of the species or subspecies name, and the year of the description, are regarded as being part of the name defining the group under discussion, e.g. *Cypraea tigris* Linnaeus, 1758. This is necessary because the species referred to by Linnaeus in his 1758 publication may not be the same thing as, say, *Cypraea tigris* Linnaeus, 1772 or *Cypraea tigris* Smith, 1990. Also, if the generic name being used in a binominal combination is not the same as the one used by the original author of the species, then the author's name and the date are enclosed in brackets. For example, an Australian limpet was first named *Patella peroni* Blainville, 1825; but since the original description it has been re-classified into another genus and is now referred to as *Patellanax peroni* (Blainville, 1825).

When generic or subgeneric names are used by themselves for the first time in any passage of writing,

it is also customary to add the author's name and the date of description to make it clear which genus or subgenus is being referred to, e.g. *Cypraea* Linnaeus, 1758 or *Zoila* Jousseaume, 1884.

Note that generic, subgeneric, species and subspecies names are written in italics. Generic and subgeneric names always begin with a capital letter but species and subspecies do not.

Type specimens

Sometimes an author of a "new" species or subspecies has a mixed sample upon which the name is based and he does not recognise that more than one entity is involved. To guard against the confusion that may come about because of this the author nominates one specimen as the *"type"*, or better still the *"holotype"*. Other specimens in the sample are known as the *"paratypes"*. If later workers discover that the original material included more than one species, then the original name belongs to the species represented by the holotype.

In the old days many taxonomists did not nominate types. In such a case a subsequent author may identify the original specimens as the *"syntypes"* (if there is more than one). In the context of a revision of the group, a subsequent author may select one of the original specimens as a *"lectotype"*, which henceforth has the status of a holotype, and the others as *"paralectotypes"*.

Type species

When describing a "new" genus or subgenus an author is required to nominate one species which he regards as typical of the group, as he understands it, as the *"type species"*. If there is only one species associated

with the new generic name in the original description then it automatically becomes the type species. If the author of the new name listed several species under that name, but did not nominate one as the type species, then the first one which appears after the new name automatically becomes the type species. These procedures are necessary to avoid any confusion which might otherwise come about later if it is discovered that several species associated with the generic name do not in fact all belong to one genus. The generic name must always be assigned to the type species.

The laws of priority

Very often, especially in the early days, taxonomists gave names to species and genera (and subspecies and subgenera) unaware that they had already been named by earlier authors. In such cases the internationally accepted rule is that the earliest name has *priority*. There are even cases where the one author has given two or more names to the same species in the same publication. In that situation the name that appears first has what is known as *"page priority"*. Different names given to the same entity are known as *"synonyms"*.

The reverse of this situation occurs when the same name is given to two or more different species of a genus. These names are known as *"homonyms"*. To resolve conflicts of this kind the same laws of priority apply - the first use of the name is accepted and new names must be found for the others.

Many species have long synonymy lists, which is not surprising given the relatively poor communications between taxonomists in years gone by and varying levels of competence and access to adequate material for study. Much of the work of modern taxonomists is devoted to sorting out the nomenclatural confusion created by earlier workers. The International Laws of Nomenclature are designed to give a solid basis for this work.

Cypraea (Zoila) rosselli (p. 197). The "golden form" from Shark Bay, WA. (Photo: Barry Wilson)

Classification of the Mollusca

The Phylum **Mollusca** is one of the largest and most diverse phyla in the animal kingdom. It contains seven Classes:

Monoplacophora	-	primitive limpets
Aplacophora	-	worm-like solengastres
Polyplacophora	-	chitons
Gastropoda	-	snails and slugs
Bivalvia	-	bivalved cockles, mussels and oysters
Scaphopoda	-	tusk-shells
Cephalopoda	-	squids, cuttles and octopus

The Class **Gastropoda**, part of which is the subject of this volume, has four subclasses:

Prosobranchia

Heterobranchia

Opisthobranchia

Pulmonata

The Subclass **Prosobranchia** contains by far the majority of marine gastropods familiar to shell collectors. Although each of the other three subclasses contain shelled marine species found in Australian waters, this study (volumes one and two) considers only marine prosobranchs.

Classification of the Prosobranchia

Recent advances in knowledge of the comparative anatomy and fossil history of gastropods have led to radical changes in the classification of the Prosobranchia. Another factor has been the discovery of the so-called "hot-vent limpets", i.e. limpets living around hydrothermal vents in the ocean deeps, many of them belonging to entirely "new" groups or to groups previously thought to be extinct.

In 1986 a symposium on prosobranch phylogeny was held in Edinburgh, resulting in the publication of a collection of papers on this topic edited by Dr Winston Ponder (1988, Prosobranch Phylogeny, *Malacological Review, Supplement* **4**: 346 pp.). The classification of the prosobranchs used here follows arrangements recommended in those papers and in the earlier publications referred to in their bibliographies (see reference list). However, the debate is ongoing and disagreements about methodology remain. We may expect further changes as new information is brought forward by the research workers in this field.

For more than half a century it has been customary to recognise three prosobranch Orders, i.e. Archaeogastropoda, Mesogastropoda and Neogastropoda. These groups represent grades of complexity which were supposed to relate to levels of evolutionary development, as the names imply, from ancient (primitive) to advanced (specialised). Molluscan taxonomists have now reached consensus, more or less, that this system is much too simplistic and does not truly represent the evolutionary history of these molluscs. There has certainly not been a simple archaeogastropod-mesogastropod-neogastropod sequence of evolution.

In particular it has been realised that the living and fossil families traditionally grouped in the Archaeogastropoda are in fact a mixed lot. Early in the history of the class there were many divergent families which have been regarded as "archaeogastropods". Most of these have become extinct. Those which have survived belong to several quite distinct evolutionary lines and it is not appropriate to classify them together in the same Order.

Resulting from these new data and new ideas, a new classification at ordinal level has been proposed. Also there have been many re-arrangements of the families and subfamilies. Some families previously regarded as prosobranchs have been shifted to a new Subclass, the Heterobranchia. These include the Architectonicidae, Valvatidae, Rissoellidae and Pyramidellidae, which therefore are not included in this study.

The following arrangement is used here for the living prosobranch species, derived from the symposium proceedings referred to above. Note that the Archaeogastropoda (equivalent to Vetigastropoda of some recent authors) has been limited to exclude the docoglossan limpets, nerites and cocculinids, which are all now given separate ordinal rank. The Superorder Caenogastropoda combines the old Mesogastropoda and Neogastropoda. The Order Mesogastropoda has been discarded; most of the families previously placed there are now grouped in the Order Neotaenioglossa while two small families are separated in the new Order Architaenioglossa (neither of which has marine representatives). The old Neogastropoda remains but is substantially reclassified internally.

Subclass **PROSOBRANCHIA**

    Order **DOCOGLOSSA**

    Order **COCCULINIFORMIA**

    Order **NERITIMORPHA**

    Order **ARCHAEOGASTROPODA**

        Superfamily **Pleurotomarioidea**

        Superfamily **Fissurelloidea**

        Superfamily **Trochoidea**

Superorder **CAENOGASTROPODA**

    Order **ARCHITAENIOGLOSSA**

        Superfamily **Cyclophoroidea**

        Superfamily **Ampullarioidea**

    Order **NEOTAENIOGLOSSA**

    Suborder **Discopoda**

        Superfamily **Loxonematoidea**

        Superfamily **Cerithioidea**

        Superfamily **Littorinoidea**

        Superfamily **Cingulopsoidea**

        Superfamily **Truncatelloidea**

        Superfamily **Stromboidea**

        Superfamily **Vanikoroidea**

        Superfamily **Calyptraeoidea**

        Superfamily **Xenophoroidea**

        Superfamily **Vermetoidea**

        Superfamily **Cypraeoidea**

        Superfamily **Lamellarioidea**

        Superfamily **Naticoidea**

        Superfamily **Tonnoidea**

    Suborder **Heteropoda**

        Superfamily **Carinarioidea**

    Suborder **Ptenoglossa**

        Superfamily **Triphoroidea**

        Superfamily **Janthinoidea**

        Superfamily **Eulimoidea**

    Order **NEOGASTROPODA**

        Superfamily **Muricoidea**

        Superfamily **Cancellarioidea**

        Superfamily **Conoidea**

References

Bieler, R. (1990). Haszprunar's "clado-evolutionary" classification of the Gastropoda - a critique. *Malacologia* **31** (2): 371-380.

Haszprunar, G. (1985). The Heterobranchia - a new concept in the phylogeny of the higher Gastropoda. *Zeitschrift für Phylogenetische Systematik und Evolutionsforschung* **23** (1): 15-37.

Haszprunar, G. (1988). A preliminary phylogenetic analysis of the streptoneurous gastropods. *Malacological Review, Supplement* **4**: 7-16.

Hickmann, C. S. (1988). Archaeogastropod evolution, phylogeny and systematics: a re-evaluation. *Malacological Review, Supplement* **4**: 17-34.

Moore, R.C. (*ed.*) (1960). *Treatise on Invertebrate Paleontology*. Part 1. Mollusca 1. University of Kansas Press and Geological Society of America Inc. 351 pp.

Ponder, W. F. & Warén, A. (1988). Appendix. Classification of the Caenogastropoda and Heterostropha - a list of the family-group names and higher taxa. *Malacological Review, Supplement* **4**: 288-326.

Vaught, K. C. (1989). *A Classification of the Living Mollusca*. American Malacologists Inc. Melbourne, Florida. 1-X11+1-195

Some hints on using this book

Identification of specimens

Although the reader will find many different kinds of information in this book its main use will probably be as an aid to identification of specimens. Most users will of course refer to the illustrations and, to a lesser extent, the descriptions. But it must always be remembered that most species are quite variable in shell form and colour and that neither the illustrations nor the descriptions can represent the full range of variation. The experienced conchologist aquires "an eye" for distinguishing between species but it takes time to learn the skills.

Special terms

It is not possible to avoid a host of technical terms in describing shells. To assist the reader a comprehensive glossary is given at the end of the book.

Format of the descriptions

The families within each order are arranged according to the customary concepts of their evolutionary relationships. Within each family and subfamily the genera are listed in alphabetical order, written in bold capitals. The generic name itself is followed by the author's name and the year of description, all on the same line. Within each genus the species are listed in alphabetical order, written in bold lower case. The species name is followed on the same line by its author's name and the year of description.

When there are subgenera within a genus they are presented in the same way as the genera, i.e. written in capitals, but not in bold type, and again in alphabetical order. For convenience's sake subgeneric names are not given in brackets between the generic and specific names, but the reader may assume that all the species appearing below a subgeneric name belong to that subgeneric group within the genus whose name appeared before it.

The name of the type species of a genus or subgenus is written in italics immediately below the generic or subgeneric name. The description which follows is that of the genus or subgenus.

Below the species descriptions the shell dimensions are given. This is usually the shell height (often referred to as the length) measured from the tip of the spire to the anterior end. See diagram at the end of the book. However, for globular or flattened shells it may be the width, or sometimes both height and width. The dimension is usually the maximum size of adults. In this work dimensions are given in metric units. For readers' convenience, a conversion table between imperial and metric units is given at the end of the book.

The species dimension is followed by a geographic distribution statement. First the world-wide distribution is given, followed by a statement about the species' distribution in Australia. In the latter case the distribution is usually stated as between two well-known localities in a clockwise direction around the coast. In some cases where the exact limits of distribution are unknown, generalised terms such as "southern Qld" are used.

After the species' distribution statement there may be remarks about habitat, nomenclature, comparisons with other species and synonymy. The given synonymies are not always complete. For many species there are very long synonymies and it is impractical to list more than those which have been used in the Australian context. Also, the author has not had access to a comprehensive library of the older literature and some nomenclatural information was not readily available, so there is some inconsistency in the treatment given to the synonyms. Where information has been available it has usually been given.

There is also some inconsistency in the extent to which the families are treated in this book. Families such as the cowries have been thoroughly studied in Australia and a comprehensive account of these is given. However, there have been no reviews of many families and, since it is beyond the scope of a work like this to carry out original taxonomic studies, the treatment here of such groups is cursory.

Prosobranch Gastropods

PHYLUM **MOLLUSCA**
CLASS **GASTROPODA**
SUBCLASS **PROSOBRANCHIA**

Prosobranch gastropods originated early in the history of the Mollusca. Their basic body plan has proved extremely adaptable and the group persists today in a very wide range of marine, freshwater and terrestrial habitats. It is believed that the other gastropod subclasses evolved from prosobranch ancestors.

Although a coiled shell is a feature of the gastropods it is neither unique nor universal in the class. Docoglossan limpets do not have coiled shells but have retained the ancestral shell form. Some other modern prosobranchs have reverted to limpet-like shells, while in many of the opisthobranch and pulmonate gastropods there is no shell at all.

Prosobranchs are distinguished from the other subclasses by features of their anatomy and the possession of an operculum which closes the aperture of the shell. The majority of modern prosobranchs are marine although there is a large number of terrestrial forms. The modern representatives include detrital gatherers, herbivores, carnivores and parasites. In most forms sexes are separate but protandry is common. Their reproductive strategies range from simple mass spawning to complex mating procedures and production of protective egg masses. Typically there is a planktonic (planktotrophic) veliger larval stage but in many the larvae develop entirely within egg capsules and the young hatch as crawling, shelled snails.

According to contemporary classification there are seven orders, six of which are marine or include marine species. This volume includes the Australian marine docoglossans, cocculinids, nerites, archaeogastropods and neotaenioglossans. Volume two will deal with the sixth marine order, the neogastropods.

Charonia tritonis

SUPERFAMILY

PATELLOIDEA

THE conical "limpet" shell form has evolved several times during the history of the gastropods but it found its greatest success in the docoglossan superfamily Patelloidea. These molluscs dominate rocky shores in many parts of the world. They are browsers on algae growing on the rocks.

For many years most authors included these limpets in the Archaeogastropoda, but strong arguments have recently been presented for their separation as a distinct Order, preserving many very ancient gastropod features but derived quite separately from other groups (Golikov & Starobogatov, 1975, *Malacologia* **15**: 185-232). Lindberg (1988, in Ponder & Warén) preferred the name Patellogastropoda to Docoglossa.

There are five living families, Acmaeidae, Patellidae, Lepetidae, Nacellidae and Lottiidae. The simple, bilaterally symmetrical, conical shell is common to them all and they share many characters in their simple body anatomy. They also all have a characteristic radula, known as the docoglossan type, in which the central tooth is small or absent and the tips of the denticles are specially hardened for scraping algal particles from hard rocks.

The Acmaeidae and Patellidae are strongly represented in Australian waters and some of their species are among the most conspicuous creatures on our rocky shores. Most of the species live in the intertidal zone, where they are extremely abundant.

The adductor muscles of docoglossan limpets begin as a pair, with the left and right members equal. As the animal grows the muscles grow backwards, meeting at the rear and forming a U-shape with a corresponding muscle scar on the inner shell surface. This scar encloses a spoon-shaped central area known as the spatula, which has an anterior constriction and may be distinctively coloured.

FAMILY **ACMAEIDAE**

Acmaeid Limpets

These mostly small limpets retain one true gill, the left one. The ancestral right gill has been lost in this family. The gill lies free in the shallow, anterior mantle cavity except at its base and it has filaments along both sides of its axis (bipectinate). When fully extended the gill may project beyond the margin of the shell. The family is further characterised by the lack of central teeth on the radula. Each half tooth row has two lateral teeth and 0-2 marginals.

Acmaeid limpets may be found on most rocky shores around Australia, especially in the south. The taxonomy of the group remains badly confused, partly because the shells of most species tend to be very variable and there is little to distinguish one from another. Collectors and biologists alike will have difficulty identifying them. The present account draws heavily from a study by Winston Ponder & R.G. Creese (1980 - see references) of three of the six genera. However, much work needs to be done on the anatomy and natural history of the species before the taxonomy can be finally settled.

The classification used here is rather simplified, following Ponder and Creese. Unfortunately the largest groups, *Notoacmea* and *Patelloida*, have few differentiating shell characters and classification of the species is not possible on the shells alone. The serious student must resort to the laborious task of examining radulae in order to determine generic position.

Species of *Notoacmea*, *Patelloida* and *Collisella* are found at most levels of the intertidal zones, as well as in the shallow sublittoral. Most of them are rock dwellers. *Patelloida nigrosulcata* is unique in living on the shells of other rocky shore gastropods. *Asteracmea* species are minute and are dredged from deeper water; nothing is known of their life histories. The two species of *Naccula* are tiny fragile creatures which live on seagrasses in shallow water.

Altogether there are at least twenty-eight Australian species of the family, representing five genera.

References

Macpherson, J.H. (1955). Preliminary revision of the families Patellidae and Acmaeidae in Australia. *Proceedings Royal Society Victoria* **67**: 229-256.

Oliver, W.R.B. (1926). Australasian Patelloididae. *Transactions N.Z. Institute* **56**: 547-582.

Ponder, W.F. & Creese, R.G. (1980). A revision of the Australian species of *Notoacmea*, *Collisella* and *Patelloida* (Mollusca: Gastropoda: Acmaeidae). *Journal Malacological Society Australia* **4** (4): 167-208.

▲ GENUS **ASTERACMEA** Oliver, 1926
Helcioniscus illibratus Verco

Minute shells, translucent; ovate to elongate-ovate; apex subcentral or anterior; surface smooth or radially striate. Most species with radial pink bands. Radula with two lateral and two marginal teeth.

Southern Australia. There are five species assigned to this genus. The additional species are *A. axiaerata* (Verco, 1912) from southern WA; *crebristriata* (Verco, 1904) from western SA to Fremantle, WA; *roseoradiata* (Verco, 1912) from SA to Cape Naturaliste, WA; *stowae* (Verco, 1906) from SA to Fremantle, WA.

Asteracmea illibrata (Verco, 1906)

Minute but rather solid; conical with high profile and a blunt, almost central apex; anterior slope almost straight, posterior slope convex; radial sculpture lacking. Apex pink, surface brown with four radial white bands which divide near the margin.

2.6 mm. SA to Bunbury, WA. Dredged from deep water.

▲ GENUS **COLLISELLA** Dall, 1871
Acmaea cassis Eschscholtz (= *pelta* Eschscholtz)

Shells small, thin, ovate, usually narrowing anteriorly, conical; apex subcentral to anterior. Radula with one marginal tooth.

Cosmopolitan. Distinguished from *Notoacmea* by the presence of marginal teeth in the radula.

Collisella mixta (Reeve, 1855)

Pl. 1; fig. 7 a-b   Ovate; often high in profile, with the apex at about the anterior third position, anterior slope straight, posterior slope convex; smooth or with strong, irregular, radial ribs. Grey-brown to yellow, with a reticulate pattern of darker brown rays, which often form a conspicuous "Maltese Cross"; interior white with brown blotches, margin brown.

2 cm. Western Vic. to Streaky Bay, SA. Uncommon in Tas. Lives on exposed rock platforms in the middle and upper intertidal zones.

Collisella onychitis (Menke, 1843)

Pl. 1; figs. 2 a-b; 3 a-b   Ovate, narrowing anteriorly, low-conical; usually heavily eroded but sculpture sometimes remaining, consisting of about 20 rounded radial ribs. Exterior white with dark brown radial rays or uniformly dark chocolate brown; interior outer part white or brown with a continuous or interrupted black marginal band; spatula light brown, muscle scar white.

2.5 cm. Ceduna, SA to Quobba, WA. Very common on exposed rocky shores in the middle and upper intertidal zone. The back is usually heavily eroded. A common form which has a uniformly black margin is tentatively assigned to this species (figs. 3 a-b).

Collisella septiformis (Quoy & Gaimard, 1834)

Pl. 1; figs. 8 a-c Ovate, moderately low in profile, often narrowing anteriorly, apex pointed, subcentral; sculptured with strong radial ribs, primaries and secondaries more or less alternating, crossed by concentric striae making the surface scaly. Exterior fawn with dark grey radials, mottling and speckling; interior white, black at the margin; spatula white, brown or bluish.

2 cm. Western SA to Cape Naturaliste, WA. There has been doubt about the identity and locality of this species but the specimens illustrated here from the type locality seem to match the types well. Moderately common in the upper intertidal zone on limestone rocks of open ocean shores. Placement of this species in *Collisella* is provisional until the radula details are known.

▲ GENUS **NACCULA** Iredale, 1924
Nacella parva Angas (= *punctata* Quoy & Gaimard)

Shell small, elongate, ovate, depressed, glassy; apex pointed, recurved, projecting posteriorly; anterior slope concave, posterior slope convex; radial sculpture lacking. Radula with two laterals but lacking marginals.

Southern Australia. There are two named species currently assigned to this genus. The radula is very similar to that of *Notoacmea* but because of the very different shells *Naccula* is regarded as a distinct genus.

Naccula compressa (Verco, 1906)

Elongate, narrowly elliptical, sides straight, apex over-hanging the posterior end.

1.25 mm. SA to King George I., WA. Lives on strands of seagrass *Posidonia*. See Cotton (1959, fig. 210).

Naccula punctata (Quoy & Gaimard, 1834)

Ends rather squared, apex sub-marginal. White or buff, dotted with red; interior white.

6 mm. Bass St. to Fremantle, WA. Lives on seagrasses in the sublittoral zone. Cotton (1959, fig. 29) records it from *Zostera*, while Macpherson and Gabriel record it (as *N. parva* (Angas, 1876) which is a synonym) from *Cymodocea antarctica*.

▲ GENUS **NOTOACMEA** Iredale, 1915
Patelloida pileopsis Quoy & Gaimard

Shell characters as for *Collisella*. Radula with no marginal teeth.

Southern Australia and New Zealand. Some recent authors have treated this group as a subgenus of *Collisella*. Synonyms: *Parvacmea* Iredale, 1915 (*daedala* Suter); *Conacmea* Oliver, 1926 (*parviconoidea* Suter); *Subacmea* Oliver, 1926 (*scopulina* Oliver).

Notoacmea alta Oliver, 1926

Thin, depressed to elevated; apex position variable; anterior slope straight or slightly convex; surface nearly smooth, sculptured with fine and close radial and concentric striae. Exterior dark grey to black, sometimes with irregular white radial markings; interior black, or at least black around the margins.

8.5 mm. Southern NSW to Spencer Gulf, SA including the east coast of Tas. Lives in sheltered areas in the mid-littoral, often on the shells of other molluscs. See Ponder & Creese, Pl. 1, figs. 1-5. Synonym: *corrosa* Oliver, 1926.

Notoacmea conoidea (Quoy & Gaimard, 1834)

Elevated conical, sometimes height and length almost equal; apex subcentral; sculpture very weak. White to grey-brown, with narrow, dark brown radial lines; interior white, brown-spotted at the margin.

1.4 cm. Esperance to Fremantle, WA. Under stones in the shallow sublittoral zone of protected areas. Resembles *N. insignis* but much taller and lacks the conspicuous cross of that species. See Ponder & Creese, Pl. 1, figs. 6-11.

Notoacmea corrodenda (May, 1920)

Ovate, rather low in profile, apex at about the anterior third position; anterior slope slightly concave, posterior slope slightly convex; sculptured with low narrow radial ribs. Ribs white or cream, interspaces dark-brown; interior off-white, grey or pale brown, brown spots around the margin; spatula usually dark brown with a white border.

1.8 cm. Western Vic. and Tas. On rocks; lower littoral zone. See Ponder & Creese, Pl. 1, figs. 12-14.

Notoacmea flammea (Quoy & Gaimard, 1834)

Ovate, thin, apex at about the anterior third; anterior slope slightly concave, posterior slope slightly convex; with or without a few low, wide radial ribs, whole surface covered with fine, minutely granular radial striae. Exterior cream to black, sometimes cream, orange or rose rays alternate with black rays, often with white flecks; exterior colour and pattern show through on the interior; spatula greenish brown, poorly defined.

1.7 cm. Central NSW to Fremantle, WA. Lives in protected areas on or under boulders in the lower littoral. Synonyms: *scabrilirata* Angas, 1865; *subundulata* Angas, 1865; *diminuta* Iredale, 1924. See Ponder & Creese, Pl. 2, figs. 1-18.

Notoacmea mayi (May, 1923)

Pl. 1; fig. 5 a-b Thick, profile low, apex at the anterior margin; surface lacking radial sculpture. Uniform light brown to grey, sometimes mottled with darker grey or brown; interior dark, with a grey-white band near the margin; spatula white, brown or black.

2.5 cm. Bass St. Lives on rock faces in the upper intertidal zone. *N. petterdi* may occur in the same habitats but is easily distinguished by its conical shape.

Notoacmea petterdi (Tenison Woods, 1876)

Pl. 1; fig. 6 a-b Elliptical, with a moderate conical profile; apex subcentral; sculptured with low, rounded, smooth, radial ribs. Exterior light brown with darker radial bands; interior light brown; external pattern showing through at the margin; spatula mid-brown.

2.3 cm. Southern Qld to Bass St. Lives on vertical rock faces high in the intertidal zone.

▲ GENUS PATELLOIDA Quoy & Gaimard, 1834
P. rugosa Quoy & Gaimard

Ovate; conical, apex subcentral to anterior; usually radially ribbed. Radula with two marginal teeth.

Cosmopolitan. Oliver (1926) treated *Collisellina* Dall as a subgenus of *Patelloida* and introduced *Chiazacmea* as a distinct genus but Ponder & Creese (1980) synonymised both names with *Patelloida*. Synonyms: *Collisellina* Dall, 1871 (*saccharina* Linnaeus); *Radiacmea* Iredale, 1915 (*cingulata* Hutton); *Chiazacmea* Oliver, 1926 (*flammea* auct. = *insignis* Menke); *Actinoleuca* Oliver, 1926 (*campbelli* Filhol).

Patelloida alticostata (Angas, 1865)

Pl. 1; fig. 14 a-b Solid, ovate; sculpture of 12-30 strong, sometimes angular, radial ribs; margin strongly crenulate. Exterior grey with prominent crescent-shaped brown marks in the spaces between the ribs; interior porcellaneous, white or brown, margin spotted with black; spatula brown.

4 cm. Northern NSW around southern Australia, including Tas., to Kalbarri, WA. Very common in the lower intertidal zone of rocky shores exposed to strong wave action. Iredale, (1924) introduced two subspecies names, *P. a. antelia* and *P. a. complanata,* for NSW specimens but these can be regarded as synonyms.

Patelloida bellatula (Iredale, 1929)

Elongate-ovate, thick, low in profile, apex subcentral; sculpture irregular, consisting of prominent radial ribs, the whole surface finely corded. Exterior and interior white except for dark brown lines between the ribs and a brown-blotched spatula.

1.6 cm. North Qld and Papua New Guinea. Found in the lower intertidal and sublittoral zones: on coral rubble, beach rock and dead shells. See Ponder & Creese, Pl. 5, figs. 7-9.

Patelloida cryptalirata (Macpherson, 1955)

Elliptical to ovate, thin, low in profile, apex at the anterior third; anterior and posterior slopes straight or slightly convex; surface smooth. Cream or light brown, with a reticulate pattern of fine red-brown lines, sometimes with a few wide radial bands or a broad cross; interior white or brown with a bluish marginal band and brown spatula.

1.2 cm. Broome WA to southern Qld. Intertidal rock crevices in more or less sheltered areas. Synonym: *ater* Macpherson, 1955. See Ponder & Creese, Pl. 5, figs. 10-15.

Patelloida heteromorpha (Oliver, 1926)

Elongate-ovate, moderately thick, profile low, apex subcentral; margin often irregular; finely radially corded. Exterior white to pale brown with several dark radial bands which may coalesce; interior margin black, or white with black stripes; spatula pale-bluish to dark brown, sometimes spotted.

1.9 cm. Qld. On rocks in the mid and upper intertidal zone, usually in silty situations. Distinguished from *P. cryptalirata* by its heavier, larger shell, darker interior, more central apex, and radial sculpture. See Ponder & Creese, Pl. 5, figs. 16-18.

Patelloida insignis (Menke, 1843)

Pl. 1; fig. 15 a-c Ovate, conical, apex subcentral, anterior slopes straight or slightly convex; finely radially striate and sometimes with weak ribs. White, grey or fawn, with irregular blotches of brown and usually a conspicuous brown "Maltese Cross"; interior white, brown external markings showing through at the margins; spatula brown, poorly defined.

3 cm. Bass St. to about Geraldton, WA. Under stones in the lower intertidal and shallow sublittoral zones in sheltered situations. Synonyms: *crucis* Tenison Woods, 1876; *mixta* Oliver, 1926.

Patelloida latistrigata (Angas, 1865)

Pl. 1; fig. 13 Ovate to elongate-ovate, usually high in profile apex subcentral, located at the anterior third position; back usually eroded but otherwise sculptured with irregular rounded ribs, more ribs in northern specimens than in southern ones. Exterior brown, darker brown between the ribs; interior off-white, grey to pale brown, margin with black spots; spatula brown, densely speckled or streaked with blue or black and usually with a white outline.

1.8 cm. Coolangatta, Qld to Tas. and westwards to eastern SA. Lives in the middle and upper intertidal zone of rocky shores where there is strong wave action. Synonyms: *marmorata* Tenison Woods, 1876; *submarmorata* Pilsbry, 1891. The Pilsbry name was apparently intended to give subspecies status to the northern populations which have more ribs; but the populations intergrade, and there seems little to be gained from maintaining the distinction.

Patelloida mimula (Iredale, 1924)

Ovate, high and conical in profile with straight slopes, apex subcentral; exterior smooth. Brown or greenish, with irregular dark brown radial marks which sometimes form a "Maltese Cross"; interior cream or pale brown, with the external pattern visible near the margin; spatula white, brown or mottled, usually with a white outline.

2 cm. Northern WA to Qld and south to eastern Vic. Lives in sheltered bays and estuaries, usually associated with the rock-oyster *Saccostrea*. The subspecies *Chiazacmea flammea queenslandiae* Oliver, 1926 and *Patelloida flammea marrowi* Christiaens, 1975 belong in the complex synonymy of this species. Where *P. mimula* and *P. mufria* occur together in south-eastern Australia they can be readily distinguished by the colour of the foot, which is orange in the former and grey in the latter. See Ponder & Creese, Pl. 6, figs. 7-16.

Patelloida mufria (Hedley, 1915)

Ovate to elongate-ovate, usually high and conical in profile; apex subcentral, anterior slope straight, posterior slope straight or slightly convex; sculptured with wide, smooth radial ribs which may be weak or strong. White, flecked with brown, and with brown radials, sometimes forming a "Maltese Cross"; interior white but showing the external pattern near the margin; spatula brown or white.

1 cm. NSW to southern WA. Lives in the lower littoral and shallow sublittoral zones in sheltered to moderately exposed situations, often on shells of other gastropods. Very like *P. insignis* in shell characters but the radulae of the two species are different. The two species are very hard to distinguish in the areas where they both occur. Synonym: *insignis cavilla* Iredale, 1924. See Ponder & Creese, Pl.7, figs. 5-13.

Patelloida nigrosulcata (Reeve, 1855)

Pl. 1; fig. 4 a-b   Elongate-ovate, usually low in profile with the lower margin concave to fit the convexity of the host shell; apex at the anterior third position; back often eroded, otherwise sculptured with low, wide, flat radial ribs with narrow interspaces; margins finely scalloped. Cream to grey, sometimes with brown radials between the ribs; interior white, usually with irregular light tan blotches and brown marginal spots; spatula light brown, poorly defined.

4 cm. Esperance to Geraldton, WA. Lives attached to other shells, particularly *Patella laticostata* and *Haliotis roei*, in the lower littoral and shallow sublittoral zones. Synonym: *patellavecta* Verco, 1912.

Patelloida profunda (Deshayes, 1863)

Pl. 1; fig. 10   Ovate, conical in profile, apex subcentral, slopes straight; sculptured with many fine, unequal radial ribs. Dull white or yellowish, with irregular purple-brown markings; interior white, often lined with pink; spatula poorly developed, brown or white.

2 cm. This species has a wide distribution in the southern Indian Ocean. The figured southern Australian form is the subspecies *P. p. calamus* (Crosse & Fischer, 1864) which occurs from NSW to southern WA. Christiaens, 1975 named another subspecies, *P. p. ivani* from Dampier, WA with strong radial ribs but the status of this name needs further study.

Patelloida saccharina (Linnaeus, 1758)

Pl. 1; fig. 1 a-c   Elongate-ovate, sometimes narrowing anteriorly; sculpture of about 18-20 strong raised radial ribs which are produced at the margin; margin strongly crenulate. Exterior greyish, usually dark grey or brown between the ribs, sometimes with inverted V-shaped lines close to the margin; interior white; spatula olive-green, yellow or white profusely peppered with brown spots, margin rimmed or spotted with black.

5 cm. Indo-West Pacific; North West Cape, WA to southern Qld. Lives in the middle and lower intertidal zones of rocky shores in exposed situations. The Australian form is recognised as a distinct subspecies *P. s. stella* (Lesson, 1830). Synonym: *paropsis* Iredale, 1929.

Patelloida victoriana (Singleton, 1937)

Pl. 1; fig. 9   Ovate, moderately high in profile; apex at the anterior third position, slopes convex; surface with many fine, irregular radial cords, margin finely crenulate. Dark greenish brown, apex usually eroded; interior white or grey, margin brown.

3 cm. Bass St. to Spencer Gulf, SA. Lives among algae in the lower sublittoral zone of exposed rocky shores.

FAMILY **PATELLIDAE**

Although the shells of this family closely resemble those of acmaeid limpets, close examination of the soft parts reveals significant differences. Most importantly there is no gill in the shallow anterior mantle cavity. Instead, patellids have a row or cordon of respiratory filaments along a shallow groove between the mantle and the foot on each side of the body. These respiratory organs are "secondary gills", evolved independently and not from the ancestral gills of the mantle cavity. The radula is also different in that it usually possesses a median central tooth in each transverse row, although the tooth is very small or absent in some species. It is flanked on each side by a pair of centrals, then a large multi-cuspid lateral, and finally three slender and weak marginals which may be fused to form a single plate.

Patellid limpets are gregarious spawners. Their veliger larvae are short-lived, usually remaining in the plankton for only a few days before settlement. The initial larval shell has one simple coil but it is replaced by the new conical one after settlement. The English *Patella vulgata* is said to be a protandrous hermaphrodite; that is, it begins life as a male but changes sex later. Some patellid limpets have been shown to maintain individual home territories, returning to stick fast to the same spot after foraging for food.

Two genera are recognised in the Australian fauna, *Patella* (with four species) and *Cellana* (with eight species). Both have many species in other parts of the world as well. They are all inhabitants of the intertidal or shallow sublittoral zones of rocky shores. Although their shells are quite variable, the species are much easier to identify than the acmaeid limpets.

For a more detailed account of the family the reader could consult a review by A.B. Powell (1973, *Indo-Pacific Mollusca* **3** (15): 75-205, pls. 60-182).

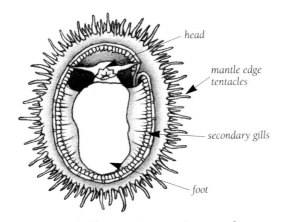

Underside of a patellid limpet showing the row of secondary gills around the mantle cavity.

▲ GENUS **CELLANA** Adams, 1869
C. cernica Adams

Shell similar to that of *Patella* but interior strongly glazed and iridescent; gill cordon interrupted by the head, epipodial fringe absent; radular median central tooth vestigial or absent, long curved centrals alternate with similar laterals; marginals weak.

Indo-West Pacific.

Cellana analogia Iredale, 1940

Pl. 1; fig. 17 Broadly ovate, only moderately elevated in profile; sculpture of numerous strong carinate radials, primaries and secondaries alternating, interspaces deep, whole surface made scabrous by concentric ridges; margin crenulate. Dull white, interior grey-white with short brown radial lines near the margin; spatula yellow to fawn, sometimes covered with white callus.

4 cm. Endemic to Lord Howe I.

Cellana conciliata Iredale, 1940

Pl. 1; fig. 16 a-d Broadly ovate, profile low, apex subcentral; sculpture delicate, consisting of numerous, evenly sized radial riblets; margin very weakly scalloped. Greenish or bluish grey, sometimes radially banded with brown, with chestnut lines in the interspaces; interior silvery blue to cream or white, faintly rayed with bluish grey near the margin; spatula fawn to yellow-brown, often clouded with white callus.

4.5 cm. Central Qld. Resembles *C. tramoserica* but distinguished by its very fine and dense radial sculpture.

Cellana howensis Iredale, 1940

Pl. 1; fig. 11 Ovate, moderately elevated in profile, apex at the anterior quarter position; sculpture of many flattened radial ribs of varying width which are themselves finely grooved; surface weakly granulose in the upper part, smooth toward the finely crenulate margin. Exterior greenish grey to buff, interspaces lined in dark brown; interior yellow-brown to orange-brown, external radial lines showing through at the margins; spatula dark red-brown but usually clouded by callus.

3.5 cm. Endemic to Lord Howe I.

Cellana radiata (Born, 1778)

Pl. 1; fig. 19 a-c Roundly ovate, apex low and subcentral; with 11-14 primary ribs alternating with secondaries, the whole surface finely corded; margin corrugated. Exterior cream, usually with brown streaks or blotches in the interspaces between the ribs; spatula brown or white with brown edges, surrounded by a yellow zone, white with alternating brown blotches at the margin.

4 cm. Indo-West Pacific; Geraldton, WA to Qld. Common on rocks in the mid-intertidal zone. A number of subspecies are recognised. According to Powell (1973) the Qld form belongs to the nominate subspecies while the population on the Kimberley coast of WA belongs to the subspecies *C. radiata orientalis* (Pilsbry, 1891). The latter is said to have stronger primary ribs than the nominate form. However, both forms occur in the western Pacific and the distinction is dubious. Synonyms of the nominate form are: *rota* Gmelin, 1791; *reynaudi* Deshayes, 1832; *petalata* Reeve, 1844; *aster, luzonica, nimbus* and *scalata* all of Reeve, 1855; *travanconica* Preston, 1911; *bombayana* and *ceylanica* Smith, 1911. Synonyms of the *orientalis* form or subspecies are: *hombroni* Dautzenberg & Bouge, 1933; *eudora* Iredale, 1940.

Cellana solida (Blainville, 1825)

Pl. 1; fig. 22 a-c   Broadly ovate; margin deeply scalloped; with about 26 strong, rounded, radial ribs. Exterior dull grey to buff, often with radial brown rays; spatula bluish or slate-grey, muscle scars yellow; margin orange with the reddish brown radial rays of the exterior showing through boldly.

8 cm. Vic., Tas. and SA. Easily distinguished from *Cellana tramoserica* by the fewer, bolder, rounded radial ribs, more strongly scalloped margin, and the bright orange margin in adults. Found in the same habitats as *C. tramoserica* on the south-east coast where the two species' geographic ranges overlap. Synonyms: *rubraurantica* Blainville, 1825; *limbata* Philippi, 1849.

Cellana testudinaria (Linnaeus, 1758)

Pl. 1; fig. 24   Round or broadly ovate, profile low; apex subcentral; sculpture weak, consisting of low, rounded smooth radial ribs; margin smooth. Exterior greenish brown, with dark brown radial rays which may join concentrically to form zigzag patterns or bold chevron marks; interior bluish silver, with a continuous brown margin; spatula large, elongate, grey-white to yellow-brown.

9 cm. Central Indo-West Pacific; Qld. Synonyms: *patera* Röding, 1798; *rumphii* Blainville, 1825; *insignis* Dunker, 1866; *discrepans* Pilsbry, 1891; *mestayerae* Suter, 1906.

Cellana tramoserica (Holten, 1802)

Pl. 1; fig. 23   Broadly ovate; with about 36 strong sub-carinate radial ribs; margin finely crenulate. Colour variable, exterior usually light green, yellow or brown, some ribs darker fawn, reddish brown rays between the ribs; interior nacreous golden, green or orange, with dark brown marks around the margin; spatula fawn, greenish or white.

6 cm. Southern Qld to SA. Very common in the upper and mid intertidal zone of rocky shores where there is strong wave action. See remarks on *C. solida*. Synonyms: *variegata* Blainville, 1825; *jacksoniensis* Lesson, 1831; *variegata ariel* Iredale, 1924.

Cellana turbator Iredale, 1940

Ovate, rather conical in profile; with a bold sculpture of about 25 nodulose primary ribs, and a single secondary rib almost as strong in each interspace; margin weakly crenulate. Exterior greenish white, speckled with brown; interior cream or pinkish or silvery white, with brown radials usually prominent toward the margin; spatula chestnut.

2 cm. Known only from Caloundra, Qld. This small species may be recognised by its coarsely nodulose radial sculpture, according to Powell. However, in view of the restricted range the status of the species needs review. See Powell, Pl. 146, figs. 5-7; Pl. 148, fig. 4.

▲ GENUS **PATELLA** Linnaeus, 1758
P. vulgata Linnaeus

Round or elliptical; apex subcentral; usually radially sculptured; gill cordon complete; radula with four or five central teeth, median tooth present, vestigial or absent, lateral tooth large and pluricuspid.

Cosmopolitan. There are many species in this wide-spread genus but few on the coasts of North and South America. Six subgenera are recognised; apparently all the Australian species belong to the *Scutellastra*.

▲ SUBGENUS **SCUTELLASTRA** H. & A. Adams, 1854
P. plicata Born (= *barbara* Linnaeus)

Rather massive shells with an opaque, non-iridescent porcellaneous interior; median central radular tooth well developed.

Indo-West Pacific, South Africa and Southern Australia. Synonyms: *Patellanax* Iredale, 1924 (*squamifera* Reeve = *peronii*); *Penepatella* Iredale, 1929 (*inquisitor* Iredale = *flexuosa*).

Patella chapmani Tenison Woods, 1876

Pl. 1; fig. 18 a-b   Irregularly star-shaped, with eight prominent radial ribs which project strongly at the margin; surface with many weakly scaled radial cords, the centre one down each of the ribs being stronger than the rest. Buff, flecked with brown, sometimes with reddish brown lines down the crests of the ribs; interior white, spatula not clearly defined.

3 cm. Northern NSW to Dongara, WA. The outline is very variable in this small species which lives on rocks in the lower intertidal zone. A strongly star-shaped form from NSW was named *P. perplexa* by Pilsbry, 1891. Another synonym is *P. alba* Tenison Woods, 1876.

Patella flexuosa Quoy & Gaimard, 1834

Pl. 1; fig. 20 a-c   Elongate-ovate, apex nearly central, low; sculptured with 7-9 strong radial ribs that project strongly at the margin, and scaly to spinose radial cords over the whole surface. Exterior white, sometimes speckled or radially lined with brown; interior white, spatula yellow, orange-brown or grey.

4 cm. Indo-West Pacific; North West Cape, WA to southern Qld. A very common and variable species on rocky and coral-rock substrates in the intertidal and shallow sub-littoral zones of northern Australia. The Australian form is the nominate subspecies; *P. f. optima* Pilsbry, 1927 is found in southern Japan. Synonyms: *stellaeformis* Reeve, 1842; *inquisitor* Iredale, 1929.

Patella laticostata Blainville, 1825

Pl. 1; fig. 21 a-b   Narrowly ovate, frequently high in profile; with numerous irregular radial ribs; margin finely crenulate with groups of 2-4 elongate denticles around its inner surface. Exterior greenish grey with brown radial rays; interior brown or fawn, muscle scar often dark brown, margin boldly marked with squarish dark brown spots.

11 cm. Esperance to Shark Bay, WA. This is the largest Australian limpet. It lives in the lower intertidal zone of rocky shores exposed to heavy wave action. Synonyms: *neglecta* Gray, 1826; *rustica* Menke, 1843; *zebra* Reeve, 1854.

Patella peronii Blainville, 1825

Pl. 1; fig. 12   Ovate; sculpture variable, usually with about 16 rounded primary radial ribs and 4-6 radial threads in the interspaces, sometimes with about 24 strong, angular and nodulose primaries and a few subsidiary threads; margin crenulate or smooth. Exterior brown or dark grey with pale primary ribs; interior white, spatula yellow, margin often rimmed black.

5 cm. NSW to Shark Bay,WA. Synonyms: *diemenensis* Philippi, 1848; *ustulata*, *aculeata* and *squamifera* all of Reeve, 1855; *tasmanica* Tenison Woods, 1876; *hepatica* Verco, 1906.

SUPERFAMILY

NERITOIDEA

THERE are several families in this group. Until recently they were regarded as odd archaeogastropods, but they did not sit comfortably there and have been reclassified as a separate order with an independent origin from an early prosobranch stock.

The group is mainly tropical and includes marine, estuarine and freshwater species. The shells are strong with a wide aperture, a broad columella, and a thick, calcareous operculum which has a peg-like basal tooth.

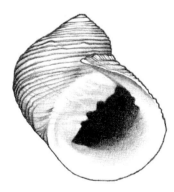

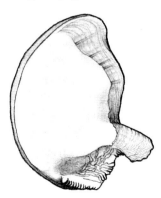

Nertia undosa (above), and the inner surface of the operculum showing the peg (below).

FAMILY **NERITOPSIDAE**

Neritopsids

Although this family of marine nerites has a long fossil history and several extinct genera, there is only one surviving genus. Little is known of its natural history. The shells are thinner than in the Neritidae, usually pustulose or nodulose, the inner walls are not resorbed, and the massive operculum is trapeziform.

▲ GENUS **NERITOPSIS** Grateloup, 1832
N. moniliformis Grateloup

Spire elevated, body whorl globose; sculptured with axial ribs and spiral cords; aperture round, inner lip moderately thick, smooth.

Indo-West Pacific and tropical Western Atlantic. There is only one species in the Indo-West Pacific region. Synonym: *Radula* Gray, 1842 (*radula* Linnaeus).

Neritopsis radula (Linnaeus, 1758)

Pl. 2; fig. 13 a-b   Sculpture of regular spiral ribs bearing crowded, rounded granules, with axial cords crossing the interspaces; aperture wall striate; columella concave, curving in a regular arc to meet the outer lip, with a single, weak, squared tooth anteriorly. Cream or off-white; interior white.

2 cm. Indo-West Pacific; Ashmore Reef, NT to north Qld. Though widespread this species is never common. It appears to inhabit the sublittoral zone and is rarely found alive.

FAMILY **NERITIDAE**

Nerites

The true nerites include marine, estuarine and freshwater species. The marine species live in the intertidal zone, often quite high up where they are exposed to the air and sun for long periods. Their tightly fitting operculae help prevent water loss in this situation.

Shell form is variable within the family. In the marine *Nerita* the shells are thick and globular to depressed-turbinate but in some freshwater genera they are very depressed and limpet-like or almost planately coiled. Characteristically the inner walls of the shells are resorbed in adults. The columellar lip is calloused and wide, often forming a flat shelf or "deck" and usually prominently toothed. The deck may be smooth, pustulose, lirate or wrinkled, characters which are useful for classification purposes. An umbilicus is lacking.

Nerites are herbivorous, browsing on algae growing on the rocks or logs on which they live. Females lay eggs in tough oval or dome-shaped capsules which they attach to the rocks or to the shells of their neighbours. There is usually a planktotrophic veliger larval stage, although development is said to be direct in some species (Anderson, 1962, *Proceedings Linnean Society NSW* **87**: 62-68).

Three subfamilies are recognised, two of them represented in the Australian fauna.

SUBFAMILY **NERITINAE**

Turbinate, globular to planate; operculum with a rib on the inner side and an apophysis.

This subfamily has a long fossil history beginning in the Triassic. There are several extinct genera and five living genera, three of which are represented in the Australian fauna.

▲ GENUS **NERITA** Linnaeus, 1758
N. peloronta Linnaeus

Turbinate, globular or depressed, oblique; surface smooth or spirally ribbed; septum ("deck") well developed, usually sculptured.

Cosmopolitan on tropical and subtropical shores. Inhabit the intertidal zone of rocky shores or, in the case of *N. balteata*, mangrove trees. Several subgeneric names have been proposed but some Australian species do not fit easily into any of them. Although a subgeneric classification is not used here, for reference purposes the subgeneric names are as follows: *Nerita* s.s. has a flat columellar deck with two marginal teeth and two teeth on the thick outer lip; *Ritena* Gray, 1858 (*plicata*) has a relatively high spire, a costate deck with several marginal teeth and a toothed outer lip; *Theliostyla* Mörch, 1852 (*albicilla*) has a low spire and a granulate deck with a finely dentate or sinuate margin; *Melanerita* Martens, 1889 (*atramentosa*) has a smooth or finely striate outer surface and smooth deck.

Nerita albicilla Linnaeus, 1758

Pl. 2; fig. 12 a-c   Almost planispiral; spire small, low, close to the postero-ventral margin; deck wide, calloused, pustulose; outer surface with a few spiral cords; operculum finely granulose. Black or grey with spiral bands and fine, irregular flecks, lip edge often brown; deck green, yellow or white.

2.5 cm wide. Indo-West Pacific; Shark Bay, WA to northern NSW. There are many synonyms, including: *imperfecta* Röding, 1798; *venusta* Dunker, 1844; *ustulata* Sowerby, 1883.

Nerita atramentosa Reeve, 1855

Pl. 2; fig. 5 a-b   Subglobose, spire depressed; surface dull, smooth or sometimes finely striate; outer lip finely serrate along its inner margin, with a rounded nodule posteriorly; deck narrow, smooth or weakly lirate, with several small teeth on the edge. Black; deck and interior white; columella black or grey.

3 cm wide. Southern Qld to Pt. Cloates, WA; also occurs in New Zealand. This is the only nerite on the temperate shores of southern Australia. Synonyms include: *atrata* Gmelin, 1791; *saturata* Hutton, 1884; *melanotragus* Smith, 1884.

Nerita balteata Reeve, 1855

Pl. 2; fig. 9 a-b   Globose, spire low, relatively thin-shelled; outer surface sculptured with numerous spiral ribs; outer lip sharp-edged and weakly toothed within; deck smooth but with several weak teeth centrally on the inner margin; operculum granulose. Light brown with darker spiral lines; columellar deck yellow or orange; operculum grey.

4 cm wide. Indo-West Pacific; Shark Bay, WA to southern Qld. Lives on the trunks and branches of mangrove trees. Commonly known as *N. lineata* Gmelin, 1791 but that name had been used previously for a different species. Although the matter has not been properley settled, it seems that Reeve's is the earliest available replacement name.

Nerita chamaeleon Linnaeus, 1758

Pl. 2; fig. 6 a-e   Broad, spire low; outer surface roughly sculptured with rugose spiral ribs, sometimes with secondary ribs between them; outer lip sharp-edged and weakly toothed within; columellar deck narrow, with a few weak pustules and lirae and 2-4 small marginal teeth centrally; operculum finely granulose. Colour variable, yellow, red, orange grey, white or black, sometimes uniformly coloured but often banded, maculated or spotted with grey, black or purple; interior and columellar deck white; operculum grey.

2 cm wide. Indo-West Pacific; North West Cape, WA to southern Qld. Compare with *N. squamulata*. Synonym: *bizonalis* Lamarck, 1816.

Nerita costata Gmelin, 1791

Pl. 2; fig. 1 a-b   Turbinate, spire relatively high; outer surface strongly sculptured with 12-15 broad, rough, spiral ribs; outer lip sharp-edged and fluted, inner side strongly toothed with one particularly large tooth at the posterior end; columellar deck transversely lirate posteriorly, strongly pustulose anteriorly, with 4-5 large marginal teeth; operculum minutely granulose. Dull black, sometimes with yellow lines between the ribs; interior white; operculum green or grey.

3.5 cm high. Indo-West Pacific; Qld. Less common than most nerite species of our region. Synonyms: *grossa* Born, 1780 (not Linnaeus, 1758); *scabricosta* de Lessert, 1841.

Nerita grossa Linnaeus, 1758

Pl. 2; fig. 2 a-b   Turbinate, spire relatively high; outer surface sculptured with 10-12 prominent, rounded spiral ribs, interspaces deep and concave; outer lip with strong internal nodules, the two posterior nodules quite prominent; columellar deck wide but thinly calloused, roughly lirate, margin with four strong, nodulose teeth. Cream, interspaces darker, ribs spotted with greenish brown; columella green-tinted.

4 cm wide. Indo-West Pacific; North Qld.

Nerita planospira Anton, 1839

Pl. 2; fig. 11 a-b   Wide, spire low, shoulder subangulate; outer lip flaring, turning inward at the suture and partly covering the spire; outer surface sculptured with numerous, rather rough spiral cords; columellar deck narrow, usually smooth, with four small, widely spaced marginal denticles; operculum smooth. Cream or grey with reddish brown or grey spiral bands and maculations; aperture white, deck white or pale yellow but with a purple-black blotch; operculum green.

3 cm wide. Indo-West Pacific; north Qld. Synonyms: *atropurpurea* Récluz, 1841; *angularis* Hombron & Jaquinot, 1854.

Nerita plicata Linnaeus, 1758

Pl. 2; fig. 8 a-b   Turbinate, globose, spire low; outer surface sculptured with coarse spiral ribs; outer lip thick and strongly toothed; columellar deck calloused, wrinkled, rounded, with four strong, squared teeth on the margin; operculum smooth. White or yellow, sometimes spotted with grey; outer lip and deck white, deep interior yellow; operculum fawn.

3 cm high. Indo-West Pacific; Abrolhos, WA to northern NSW. Synonyms: *lactaria* Linnaeus, 1758; *eburnea* Röding, 1798; *otaitensis* Lesson, 1831; *versicolor* Quoy & Gaimard, 1834.

Nerita polita Linnaeus, 1758

Pl. 2; fig. 10 a-b   Subglobose, diameter greater than height, spire flat; surface shiny but sculptured with growth lines; columellar deck smooth but with several weak, rounded, marginal teeth at the centre; operculum smooth except for a finely radially striate zone near the margin. Cream, brown, pink or green, with irregular black or brown markings; aperture orange, columellar deck red or dark orange; operculum black, grey or fawn.

3 cm wide. Indo-West Pacific; North West Cape, WA to northern NSW. This is a very common and attractive shell. Synonyms: *bifasciata* and *litterata* both of Gmelin, 1791; *doreyana* Quoy & Gaimard, 1834; *vitiensis* Hombron & Jaquinot, 1854.

Nerita reticulata Karsten, 1789

Pl. 2; fig. 3 a-c Width greater than height, spire elevated, penultimate whorl exposed; sculptured with numerous scaly spiral cords; outer lip flaring, lirate within; columellar deck smooth but with about four small marginal denticles; operculum finely and evenly granulose. Colour variable, cream, grey, yellow or orange, with obscure spiral bands and rows of spots; aperture and columellar deck white or yellow, with a rusty red blotch on the deck; operculum grey.

2 cm wide. Indo-West Pacific; Broome, WA to north Qld. Resembles *N. chamaeleon* but the spire is more elevated and the columellar blotch is characteristic. Synonym: *signata* Macleay in Lamarck, 1822.

Nerita squamulata Le Guillou, 1841

Pl. 2; fig. 4 a-e Width greater than height, spire low; sculptured with scabrous spiral cords and axial threads; outer lip flaring, toothed, lirate within; columellar deck concave, rugose, nodulose, with 4-6 small marginal denticles; operculum finely granulose. Brown, orange or grey, with darker maculations and spiral bands; interior pale green or white; operculum greenish.

3 cm wide. Indo-West Pacific; Exmouth Gulf, WA to north Qld. The sculpture is more scabrous and the spire more depressed than in *N. chamaeleon* but the two species are very much alike.

Nerita undata Linnaeus, 1758

Pl. 2; fig. 7 a-c Turbinate with a moderately high spire; sculptured with numerous fine spiral threads; fine teeth present along the inner edge of the outer lip plus one or two more prominent teeth posteriorly; columellar deck strongly wrinkled, margin with three or more teeth centrally; operculum finely pustulose. Light brown, yellow, grey or purple-black, with green or black spots or patches; aperture white, often stained with yellow, especially on the columellar deck; operculum grey.

4 cm high. Indo-West Pacific; Geraldton, WA to southern Qld. Synonyms include: *undulata* and *quadricolor* both of Gmelin, 1791; *striata* Burrow, 1815; *novaeguineae* Lesson, 1831; *chrysostoma, papuana* and *flammulata* all of Récluz, 1841; *leguillouana* and *longii* Récluz, 1842; *tongaensis* Hombron & Jaquinot, 1854; *savieana* and *oleagina* both of Reeve, 1855.

▲ GENUS **NERITINA** Lamarck, 1816
Nerita pulligera Linnaeus

Like *Nerita* but smaller, thinner-shelled, with a thin and sharp-edged outer lip and smooth or weakly toothed columellar lip; operculum with a marginal peg like that of *Nerita*.

Cosmopolitan in tropical and subtropical areas. Mainly estuarine but with some marine and freshwater species. There appear to be several species in northern Australian estuaries but there has been no review of them. The species illustrated here are identified only provisionally and classified following the system used in the *Treatise on Invertebrate Paleontology* (Moore, *ed*. 1964).

▲ SUBGENUS **DOSTIA** Gray, 1847
Nerita crepidularia Lamarck = *violacea* Gmelin

Shell depressed, almost planispiral, aperture wide, outer lip entire; columellar deck wide, shield-shaped, with an arcuate and finely toothed edge.

Indo-West Pacific.

Neritina violacea (Gmelin, 1791)

Pl. 2; fig. 18 a-b; 19 a-d Laterally elongate-ovate, oblique, apex overhanging the posterior margin. Cream, grey, pale orange or brown, sometimes with darker axial streaks; aperture and columellar deck orange, shiny.

1.5 cm wide. Darwin, NT to Central Qld. Specimens from Qld appear close to the Indo-West Pacific species *N. violacea* but are much smaller and more oblique than PNG and Philippine specimens seen by this author. Specimens of *N. violacea* from PNG are illustrated here (figs. 18 a, b) for comparison.

▲ SUBGENUS **PROVITTOIDA** Baker, 1923
Nerita smithi Wood

Shell ovate, glossy smooth, spire tall and pointed; aperture hemispherical; columella thick but not forming a shield, inner margin almost straight, obscurely noduled.

Indo-West Pacific.

Neritina cf. zigzag (Lamarck, 1822)

Pl. 2; fig. 14 a-b; 15 a-d Spire high, body whorl oblique exposing much of the penultimate whorl. Green or yellow with broad or narrow wavy axial lines; columella yellow.

1.3 cm high. Central Indo-West Pacific. The specimens illustrated here from a tidal creek near Darwin appear to establish the presence of the species in northern Australia although the locality neeeds confirmation. Specimens from PNG are included for reference (figs. 14 a, b).

▲ SUBGENUS **VITTOIDA** Baker, 1923
Nerita variegata Lesson

Shell like that of *Provittoida* but with the edge of the columella more strongly toothed.

Indo-West Pacific.

Neritina cf. **variegata** (Lesson, 1831)

Pl. 2; fig. 16 a-b   Ovate, spire moderately low and usually eroded; columella smooth but with numerous fine teeth along the margin and a prominent central tooth. Colour extremely variable, usually shades of green or yellow and with tent-shaped marks or spiral or axial lines; aperture bluish white.

1.5 cm high. Central Indo-West Pacific; Kimberley, WA to north Qld. Found in tidal creeks. The identity of the illustrated specimens is provisional.

▲ GENUS **THEODOXUS** Montfort, 1810
T. lutetianus Montfort = *Nerita fluviatilis* Linnaeus

Obliquely turbinate, spire low to moderately low, surface smooth; columellar deck smooth, with a smooth or finely dentate margin; operculum smooth, peg lacking.

Cosmopolitan in tropical areas. Estuarine and freshwater inhabitants. There are several subgenera, at least one of which occurs in Northern Australia.

▲ SUBGENUS **PICTONERITINA** Iredale, 1936
Neritina oualaniensis Lesson

With the characters of the type species.

Indo-West Pacific.

Theodoxus oualaniensis (Lesson, 1831)

Pl. 2; fig. 17 a-i   Ovate, thin-shelled, spire low, surface glossy smooth; columellar deck weakly arched, smooth except for one large and several small teeth along the margin, teeth irregular and sometimes obscure. Colour bright and extremely variable, usually shades of yellow or green, with darker, sometimes black, bands or axial streaks, often with a network pattern of triangular marks.

1 cm high. Indo-West Pacific; north Qld. This colourful little species lives in estuaries and may be extremely abundant in the intertidal zone. It is sometimes placed in *Neritina* but Iredale cited it as the type species of *Pictoneritina*, which is now ranked as a subgenus of *Theodoxus*. The diagnosis of *Pictoneritina* is unsatisfactory; Iredale merely noted that Baker (1923) had included *oualaniensis* in *Vittoclithon* (a West Indian genus), remarking upon "the notable difference in the radula". Apart from this, the stated diagnosis was simply "the columellar dentation is very irregular and obscure and the painting consists of streaks".

SUBFAMILY **SMARAGDIINAE**

Small, obliquely ovate; operculum with ribs and an apophysis.

Three genera are assigned to this subfamily. *Smaragdia* is well known in northern Australia. The type species of the related genus, *Magadis* Melvill & Standen, 1899 (*eumerintha*) was recorded by Charles Hedley in his Queensland list but this author has not seen specimens. The third genus, *Pisulina* Nevill & Nevill, 1869 (*adamsiana*) is from India and Sri Lanka.

▲ GENUS **SMARAGDIA** Issel, 1869
S. viridis Linnaeus

Shells glossy, smooth, spire low; aperture narrow; operculum smooth, with ribs and a peg.

Tropical Indo-West Pacific and Western Atlantic. Species of this genus are marine snails of the shallow sublittoral zone, apparently associated with seagrasses (angiosperms). In addition to *Smaragdia* s.s. there are two described subgenera, both represented in the Australian fauna.

▲ SUBGENUS **SMARAGDISTA** Iredale, 1936
S. tragena Iredale

With the characters of the type species.

Eastern Australia. Monotypic.

Smaragdia tragena (Iredale, 1936)

Globose, spire a little elevated, apical whorl persisting as a small glassy tip; the first teleoconch whorl is wound almost planately, the last descending rapidly; surface smooth and shining; aperture semicircular, outer lip thin; columella straight, with a broad callus and eight small irregular teeth on the margin centrally. White or pinkish, with bands of squarish pink or purple blotches and separate axial marks arranged in series encircling the whorls (sometimes merging near the mouth).

7 mm high. NSW. This is the species which appeared in older Australian lists as "*Neritina rangiana*" but which Iredale considered to be distinct. Iredale's original material was dredged in Sydney Harbour. *N. rangiana* (Recluz, 1842) is a widespread Indo-West Pacific species and further study is needed to ascertain its relationship to the NSW species.

FAMILY **PHENACOLEPADIDAE**

Sugar Limpets

A small family of limpet-like molluscs with bilaterally symmetrical, radially sculptured shells which have a recurved apex, a horse-shoe shaped muscle scar opening anteriorly, and no operculum.

There is a veliger larval stage resembling those of nerites. In describing the gross anatomy of *Cinnalepeta cinnamomea* Hedley (1916, *Proceedings Linnean Society NSW* **41** (4): 707-708, pl. 48, figs 17-19) noted that there is a large anterior mantle cavity containing a single, broad, bipectinate gill, that the males have a large penis, and that there are no epipodial tentacles.

There are two genera in Australia and few species, none of which are commonly collected.

▲ GENUS **CINNALEPETA** Iredale, 1929
Patella cinnamomea Gould

Apex recurved over the posterior margin.

South West Pacific. Iredale (1929, *Memoirs Qld Museum* **9**: 275) introduced the name *Zacalantica* for *Patella linguaviverrae* without diagnosis and, given the posterior apex of that species, this generic name is tentatively placed in the synonymy of *Cinnalepeta*.

Cinnalepeta cinnamomea (Gould, 1846)

Pl. 22; fig. 17 a-b Elongate-ovate; sculptured with many fine and finely granulose radial cords. Exterior brown; interior brown, dark brown within the spatula.

2 cm long. South West Pacific; north Qld to Sydney, NSW. According to Hedley (1916) this animal was found in Sydney Harbour "under large stones in the mud-zone in communities of a dozen or so under the same rock". He described the animal as uniformly crimson. Synonyms: *ferruginea* H & A Adams, 1854; *vagans* Iredale, 1929; *escensa* Iredale, 1929.

Cinnalepeta linguaviverrae (Melvill & Standen, 1899)

Very small, ovate, dorsum convex with the apex turned down and recoiled beyond the posterior margin; sculptured with concentric striae and radial threads forming V-shaped thorns at the intersections; margin finely crenulate. White; interior glossy.

6 mm long. Described from the Torres St. Hedley (1899) believed that this name was a synonym of his species *P. senta* described from Funafuti Atoll. This may be the case.

However, Iredale has argued that *Capulus sagittifer* Gould, 1852 precedes even Hedley's name. A careful re-examination of the original material is needed.

▲ GENUS **PHENACOLEPAS** Pilsbry, 1891
Scutella crenulata Broderip

Round or oval limpets; apex posteriorly recurved; sculpture of radial ribs or cords and concentric striae; operculum lacking.

Circumtropical. There appear to be only minor characters differentiating *Amapileus* Iredale, 1929 (*immeritus* Iredale) and this name is treated here as a synonym. *Phenacolepas reticulata* Thiele, 1909, described from Moreton Bay, southern Qld, is another species of this group not illustrated here.

Phenacolepas calva (Verco, 1906)

Minute, thin, oval, apex subcentral; summit smooth, lower parts sculptured with crowded radial cords crossed and made granulate by concentric striae; margin not crenulate. White.

2 mm high; 2.8 mm long; 1.8 mm wide. Bass St. and Tas. to King George Sound, WA. Dredged in deep water. *P. alboradiata* (Verco, 1906), also dredged off the south coast of SA and WA, differs in that it narrows anteriorly and lacks the radial cords which characterise *P. calva*.

Phenacolepas crenulata (Broderip, 1834)

Pl. 22; fig. 18 a-b Almost round, of variable height, apex recurved and located at the anterior third position; sculptured with many finely nodulose radial cords and concentric striae; interior with a narrow rim near the margin. Exterior cream; interior white.

2.5 cm. Indo-West Pacific; Dampier, WA to Moreton Bay, Qld. Synonym: *mirabilis* Sowerby, 1910.

Phenacolepas immeritus (Iredale, 1929)

Thin, translucent, oval, with about $1^{1}/_{2}$ nuclear whorls which are helically coiled and flattened; surface sculptured with sharp elevated, narrow, radial cords, about 60 at the margin, interspaces wide and minutely concentrically striate; margin smooth, thick, not crenulate.

5 mm high; 10 mm long; 8.5 mm wide. Described from Michaelmas Cay, Qld. Iredale noted that his new species is "apparently related to *P. laevicostalis* Thiele".

SUPERFAMILY

COCCULINOIDEA

THE small limpets of this very small order differ from the docoglossans in having a different type of radula and being hermaphroditic. The radula is like that of the archaeogastropods, i.e. rhipidoglossate with very many brush-like marginal teeth. Haszprunar (1988, *Malacological Review Supp.* 4: 64-84) has described the features of the group, recognising as superfamilies the Cocculinoidea and Lepetelloidea. The latter does not appear to be represented in the Australian fauna.

FAMILY **COCCULINIDAE**

Cap Limpets

This is a small family of minute limpets, more or less bilaterally symmetrical with a recurved apex facing backwards; the protoconch is usually deciduous and the muscle scar is horseshoe-shaped. The animal has a large head and a large plicate gill which points towards the right side and recurves posteriorly at the tip. Eyes are lacking. There is a prominent copulatory organ on the right side. These tiny limpets are found in deep water.

▲ GENUS **NOTOCRATER** Finlay, 1927
Cocculina craticulata Suter

Limpet-like, thin, oval, sides subparallel; apex recurved, located close to the posterior margin; sculpture of concentric and radial grooves; muscle-scar horse-shoe shaped.

Southern Australia and New Zealand. Inhabitants of deep water.

Notocrater meridionalis (Hedley, 1903)

Concentric grooves strong, crossed by fine axial striae; apex smooth; posterior slope almost vertical, anterior slope long and arched. White.

3 mm long; 2 mm wide; 2 mm high. Dredged off central NSW. For an illustration see the original description (*Memoirs Australian Museum* 4: p. 331, fig. 64). A second Australian species, *N. tasmanica* (Pilsbry, 1895), from NSW, Vic., Tas. and SA, is very similar and possibly the same.

▲ GENUS **TECTISUMEN** Finlay, 1927
Cocculina clypidellaeformis Suter

Minute, thin, translucent, saddle-shaped, ends elevated; sculpture of concentric striae only; protoconch minute, globose, polished, usually lost in adults; apex subcentral; muscle-scar distinct, forming a half circle.

Southern Australia and New Zealand. Three Australian species are assigned to this genus.

Tectisumen tasmanica (May, 1919)

Elongate-ovate to subquadrate; sculpture almost non-existent; ends much elevated.

5 mm long; 2.6 mm wide; 2 mm high. East coast of Tas. to SA. The NSW species *T. coercita* Hedley, 1907 is narrower, with less turned up ends. A third species, *T. mayi* Finlay, 1926, also from Tas. and SA, is more ovate and patterned with radial lines of white dots.

SUPERFAMILY

PLEUROTOMARIOIDEA

FAMILY **PLEUROTOMARIIDAE**

Slit Shells

This is a small family with a long fossil history. The shell is rather thin, turbinate or conical, with a slit in the outer lip edge at or close to the middle of the body whorl, behind which there is a closed channel or selenizone ascending around the whorls to the apex. The inner shell layers are nacreous. A horny, multi-spiral operculum is present.

Cosmopolitan. There are several living genera. The species are all deep-water animals.

▲ GENUS **PEROTROCHUS** Fischer, 1885
Pleurotomaria quoyana Fischer & Bernardi

Turbiniform, broader than high; umbilicus lacking; whorls rounded; anal slit long, at or just below mid-whorl; sculpture of fine spiral lines and growth lines.

Cosmopolitan. One species has been collected in Australian waters. The genus was reviewed by Wagner & Coomans (1990, *Gloria Maris* **29** (3): 41-52) who established the identity of the Australian species.

Perotrochus westralis (Whitehead, 1987)

Pl. 3; fig. 1 a-b Thin-shelled; last whorl weakly carinate; iridescence of inner shell layer visible through the thin, almost smooth outer layer. Spire whorls and upper part of the body whorl pale orange-yellow with slightly darker axial flames, base cream; interior with golden nacre extending onto the columella.

10 cm high; 12 cm wide. Southern Japan and Central Indo-West Pacific; recorded from the continental slopes beyond the North West Shelf off Port Hedland, WA.

FAMILY **SCISSURELLIDAE**

Little Slit Shells

The shells of this small family are minute. In form they are turbinate or depressed and ear-shaped, many-whorled and with a wide mouth. As in the pleurotomariids there is an anal fissure in the shell through which digestive and excretory wastes and reproductive products are discharged with excurrent water from the mantle cavity. The anal fissure may be an eye-like hole in the wall of the body whorl or a slit in the outer lip edge. Behind it there is a filled spiral channel (selenizone) representing the position of the fissure at earlier growth stages. The shell is porcelaineous except for a thin internal nacreous layer.

The family is related to the pleurotomariids. The animal has a long snout and the head tentacles are long and ciliated. Most species live on algae in temperate or tropical seas. There are several Australian species belonging to three genera.

▲ GENUS **SCISSURELLA** Orbigny, 1824
S. laevigata Gray

Anal fissure a long slit extending some distance from the lip edge around the middle or upper part of the body whorl.

Cosmopolitan. Two subgenera are represented in the Australian fauna. Synonym: *Schismope* Jeffreys, 1856 (*striatula* Philippi).

▲ SUBGENUS **SCISSURELLA** s.s.

Depressed turbinate; spire small and low; slit on the upper part of the body whorl.

Scissurella cyprina Cotton & Godfrey, 1938

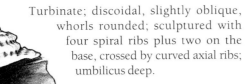

Turbinate; discoidal, slightly oblique, whorls rounded; sculptured with four spiral ribs plus two on the base, crossed by curved axial ribs; umbilicus deep.

2mm. SA. The spiral ribs are stronger than those of *S. ornata*. Drawn from the holotype.

Scissurella ornata May, 1908

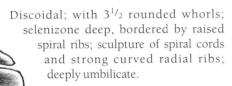

Discoidal; with 3½ rounded whorls; selenizone deep, bordered by raised spiral ribs; sculpture of spiral cords and strong curved radial ribs; deeply umbilicate.

2 mm. NSW to Tas. Lives among the stems of kelps. Drawn from a paratype.

▲ SUBGENUS **ANATOMA** Woodward, 1859
S. crispata Fleming

Globose to conical; spire elevated; anal slit short, located on the middle part of the whorl; selenizone weak.

Cosmopolitan. Synonym: *Schizotrochus* Monterosato, 1877.

Scissurella australis Hedley, 1903

Turbinate with a high conical, gradate spire and deep base, narrowly umbilicate; four whorls, sculptured with a sharp-edged double rib, with the selenizone between, forming an angular peripheral keel, upper surface with oblique-axial, lamellate riblets, base with radial riblets which are most conspicuous near the umbilicus, near the periphery the radials are crossed by numerous spiral threads .

2.5 mm. NSW to Vic. and Tas. For an illustration see the original description (*Memoirs Australian Museum*, 4: p. 330, fig. 63). *S. gunteri* (Cotton & Godfrey, 1933) from SA is less strongly sculptured.

▲ GENUS **INCISURA** Hedley, 1904
Scissurella lytteltonensis Smith

Ear-shaped, last whorl large; spire elevated, sculpture spiral when present; anal slit short, located in the upper or middle part of the whorl; columella concave, reflected over a narrow umbilicus.

Australia and New Zealand. *Incisura* s.s. is a New Zealand group with smooth whorls. New Zealand and Australian species with a spirally sculptured spire belong to *Scissurona*.

▲ SUBGENUS **SCISSURONA** Iredale, 1924
S. rosea Hedley

Early whorls finely spirally ribbed.

New Zealand and southern Australia. The type species is from New Zealand; reported also from Tasmania.

Incisura vincentiana Cotton, 1945

Width greater than height; three whorls; early whorls rounded, spirally threaded; body whorl smooth, oval, flattened above; slit high on the body whorl.

1.2 mm. Vic. to Albany, WA. *I. remota* (Iredale, 1924) from NSW is similar but not so tall-spired. Drawn from the holotype.

▲ GENUS **SINEZONA** Finlay, 1927
S. brevis Hedley

Turbinate; fissure a hole in the body whorl a short distance from the lip edge; sculpture axial or spiral or both.

Cosmopolitan. Australian authors have incorrectly used the generic name *Schismope* Jeffreys, 1856 for this group.

Sinezona atkinsoni (Tenison Woods, 1877)

Globose-tubinate, spire low and blunt; with three whorls, body whorl carinate at the shoulder and periphery, concave between; narrowly umbilicate; anal fissure between spire ridges on the shoulder keel; spirally ribbed below the peripheral keel.

2.5 mm. Southern Qld to southern WA. Also recorded from New Zealand. Found from 0 to at least 200 m. Illustrated by Cotton (1959, fig. 12). Synonyms: *tasmanica* Petterd, 1879; *carinata* Watson, 1886.

Sinezona beddomei (Petterd, 1884)

Depressed-turbinate, spire apex flat; with 3½ whorls and double-ribbed keels at the shoulders, the selenizone and hole between the ribs; strongly axially plicate, especially on the base; umbilicate.

0.75 mm. NSW to Fremantle, WA. Dredged to at least 200 m. Illustrated by Cotton (1959, fig. 13).

Sinezona pulchra (Petterd, 1884)

Depressed-turbinate; selenizone simple; with four whorls, body whorl with two peripheral keels crossed by oblique-axial ribs giving a nodular appearance; deeply umbilicate.

1 mm. Vic., Tas. and SA. Dredged to 300 m. Illustrated by Cotton (1959, fig. 14).

FAMILY **HALIOTIDAE**

Abalone

This is a very distinctive group and there should be no difficulty in recognising its members. The shells are flattened and asymmetrical (ear-shaped), with a low spire and a row of open holes along one side. The holes allow exhalant water from the mantle cavity to be ejected without fouling the head with its sensitive organs. From time to time as the shell grows a new hole is formed in the edge of the lip while the oldest open one is closed. The old closed holes remain visible in a spiral row all the way to the apex.

Within the mantle cavity there are two gills, one on each side. The animals adhere to rocks and have a very wide, flat foot and large, powerful adductor muscles.

Haliotids are herbivorous, feeding on algae. They are spawners, lacking accessory reproductive organs for copulation. Males and females release large quantities of sperm and eggs into the water, where fertilisation takes place. A short larval life in the plankton follows. As usual with animals which reproduce in this way, haliotids tend to be gregarious and abundant in their preferred habitats.

Haliotids have meaty bodies and are excellent eating. Some of the large gregarious southern Australian species are fished commercially. At one time they were called mutton-fish in Australia, but that name has been replaced in recent years by the international name abalone. Species which are fished commercially are *H. ruber,* also known as black-lip abalone (NSW, Vic., Tas.), *H. laevigata,* also known as green-lip abalone (SA and WA) and *H. roei* (WA). The large species *H. conicopora* is also taken sometimes in SA and WA but is not sufficiently abundant to support an industry in its own right.

H. ruber and *H. laevigata* are taken by divers in southern Australia. *H. roei* is very abundant intertidally and it is taken by both commercial and recreational fishermen on rocky shores at low tide as well as by diving.

The tropical abalone are much smaller than their temperate cousins and none of the species are as locally abundant. Consequently they are not fished commercially.

There are strict rules in place now in most states about taking abalone. Collectors should consult local fisheries authorities before collecting specimens to eat or for the collection.

Haliotis scalaris

▲ GENUS **HALIOTIS** Linnaeus, 1758

Haliotis asinina Linnaeus

Depressed, ear-shaped, aperture very wide, spire low and situated more or less posteriorly; with a spiral row of respiratory holes along the left side; columella forming a wide and usually flat ramp around the left side.

Cosmopolitan. There are about 50 species all told, of which 20 occur in Australian waters.

There has been no systematic review of the species of this family and, although it appears that there are several genus-level groups, *Haliotis* is used in the broad sense for all the Australian species. Other available genus-level names could be treated as subgenera but it is impossible to determine the relationships of the species until an anatomical review has been done. For reference purposes the available genus-level names for Australian haliotids are:

Exohaliotis Cotton & Godfrey, 1933 (*cyclobates* Péron);

Marinaurus Iredale, 1927 (*melculus* Iredale);

Notohaliotis Cotton & Godfrey, 1933 (*ruber* Leach);

Ovinotis Cotton, 1943 (*ovina* Gmelin);

Padollus Montfort 1810 (*rubicundus* Montfort)
   = *Neohaliotis* Cotton & Godfrey, 1933 (*scalaris* Leach);

Sanhaliotis Iredale, 1929 (*varia* Linnaeus);

Schismotis Gray, 1856 (*excisa* Gray = *laevigata* Donovan).

Haliotis asinina Linnaeus, 1758

Pl. 4; fig. 3 a-b   Elongate-ovate, thin; spire near the posterior margin; outer surface rounded, smooth except for spiral cords on the early whorls and on the outer side; holes simple, elongate, 5-6 open. Olive green with irregular pale green or yellowish patches.

12 cm. Indo-West Pacific; North West Cape, WA to Capricorn and Bunker Groups, Qld. This beautiful and very active animal is a feature of the coral reefs of our region. It is widespread and common but does not aggregate in large populations. The green animal is large and fleshy and the mantle almost covers the shell in life, quite unlike those of the other Australian species (see p. 300).

Haliotis brazieri Angas, 1869

Pl. 5; fig. 1 a-b   Ovate to spherical; thin; spire relatively high and subcentral; upper surface with a low, broad, smooth, central spiral rib bordered by a few incised spiral lines on each side, outer side with three angular ribs at the bottom and three very faint ribs in the upper zone; holes round, on prominent tubercles, four open. Rose to red, with green patches and irregular white radial lines on the upper surface, outer side green above, red below, both zones crossed by oblique whitish bands.

3 cm. Southern Qld to Jervis Bay, NSW. The species is rare in collections.

Haliotis coccoradiata Reeve, 1846

Pl. 5; fig. 7 a-b   Ovate but narrowing anteriorly, thin; spire low, posterior; outer surface with many fine but rough spiral cords crossed by fine striae; outer side with rough spiral cords in the upper zone but three sharp spiral ridges with smooth interspaces in the lower zone; holes round, simple, 6-7 open. Red-brown with broad, curved, irregular radial cream rays; lower zone of the side green and crossed by wide oblique cream bands.

5 cm. NSW and eastern Vic. The sculpture and colour pattern of the outer side resemble those of the WA *Haliotis semiplicata* but the dorsal sculpture is very different.

Haliotis conicopora Péron, 1816

Pl. 4; fig. 9   Almost spherical, spire subcentral, elevated; area between holes and margin concave; upper surface rather flat, sculptured with irregular, obliquely radiating folds crossed by weak spiral striae, radial growth striae strong; holes round, on elevated tubercles, 6-7 open. Light green or red-brown with greenish patches and broad, curved radiating green rays.

20 cm. Vic. to Fremantle, WA. This large species lives under ledges in the sublittoral zone and is only rarely found intertidally. Synonyms: *cunninghami* Gray, 1826; *gigantea* Menke, 1843; *vixlirata* Cotton, 1943; *granti* Pritchard and Gatliff, 1902.

Haliotis crebrisculpta Sowerby, 1914

Pl. 5; fig. 9 a-b   Ovate, spire posterior; upper surface sculptured with numerous rounded and transversely nodulose spiral cords, and narrow irregular radial folds; outer side with five sharp cords; holes low, oval, about five open. Brownish yellow; internal nacre yellowish.

3 cm. Scott Reef, WA to north Qld. Synonym: *dissona* Iredale, 1929.

Haliotis cyclobates Péron, 1816

Pl. 4; fig. 8 a-b   Almost spherical, spire elevated; upper surface rounded, with weak obliquely radiating folds and rough spiral riblets; with a shallow concave zone on the side below the holes and, between this and the margin, oblique rows of beaded ribs; holes simple, 5-6 open. Brown and green, with curved radiating cream rays.

6 cm. SA to Recherche Archipelago, WA. Lives gregariously in the lower intertidal zone. Synonym: *excavata* Lamarck, 1822.

Haliotis elegans Philippi, 1874

Pl. 4; fig. 2 a-b   Elongate-ovate; spire small, situated posteriorly; upper margin rounded, sculptured with heavy, raised spiral ribs; holes simple, elongate, 8-9 open in juveniles but hole formation irregular in adults. Red-brown or orange with radiating cream rays.

10 cm. Esperance to Geraldton, WA. This species lives in crevices and under stones below low tide level and is hard to find alive.

Haliotis emmae Reeve, 1846

Pl. 4; fig. 6   Ovate, rather thin; spire low to slightly elevated; upper surface with a low, wide central spiral fold and narrow, angular, obliquely radial ridges close to the spire, all crossed by scaly spiral ribs; outer side concave with a nodulose marginal rib; holes on elevated tubercles, 5-6 open. Orange-red with irregular curved cream radial rays.

10 cm. Vic. and Tas. to western SA. Moderately common but not gregarious; under stones in the intertidal and sublittoral zones. Appears to be related to the western *H. scalaris* but the sculpture is much less pronounced.

Haliotis ethologus (Iredale, 1927)

Pl. 5; fig. 5   Almost spherical; upper surface with about 11 flat-topped, spiral ribs, separated by deeply incised and transversely striate grooves, and bearing close-set transverse nodules which may become elevated scales; outer side with a sharp-edged spiral rib at the periphery, concave above this with three ribs similar to those of the upper surface, convex below the periphery with two or three roughly nodulose ribs; holes low, elongate-ovate, about six open. Yellow-brown with reddish blotches.

3 cm. Known only from a few localities in southern Qld.

Haliotis hargravesi Cox, 1869

Pl. 5; fig. 3   Ovate; upper surface with about 11 evenly spaced, narrow, elevated, flat-topped, regularly transversely nodulose spiral ribs and a few wide radial folds; upper outer side slightly concave with deeply excavated, transversely striate, weakly nodulose ribs, lower outer side convex and with four stronger ribs; columellar plate unusually wide; holes small, round, on slightly elevated turrets. Reddish brown with green blotches.

4 cm. Southern Qld and northern NSW. Rare in collections.

Haliotis laevigata Donovan, 1808

Pl. 4; fig. 10   Almost spherical, high and domed, spire not protruding; upper surface and side rounded and smooth except for growth lines; holes small, ovate, simple, 7-8 open. Grey to green.

15 cm. Vic. and Tas. to Cape Naturaliste, WA. Abundant sublittorally where it lives in the open on boulders. Fished commercially and known in the abalone trade as the "green-lip". Synonyms: *albicans* Quoy & Gaimard, 1834; *excisa* Gray, 1856.

Haliotis melculus (Iredale, 1927)

Pl. 5; fig. 4   Ovate, spire elevated; upper surface with a raised central spiral zone, sculptured with numerous narrow,

flat-topped, spiral ribs of irregular size, crossed by sharp, crescent-shaped, transverse ridges, transverse striae in the interspaces; outer side with a sharp-edged peripheral rib, concave above with four ribs like those of the upper surface, convex below with about six roughly nodulose ribs; holes rather large, oval, on slightly elevated tubercles, six open. Fawn with wide, expanding radial rays of orange on the upper surface, wide alternating bands of orange, fawn and green on the upper part of the outer side, and fawn or pale orange bands with thin vertical orange lines on the lower part.

4 cm. Restricted to southern Qld. Cited as the type species of *Marinaurus* by Iredale.

Haliotis ovina Gmelin, 1791

Pl. 4; fig. 5 a-b   Ovate to almost spherical; spire low; upper surface convex, bearing many weak spiral cords and low, rounded, radiating folds; upper part of outer side concave, lower part convex and weakly spirally ribbed; holes round, on slightly elevated tubercles giving the spire a coronate appearance, four holes open. Green or light brown, with broad radiating rays of yellow or white.

6 cm. Indo-West Pacific; Carnarvon, WA to the Capricorn and Bunker Groups, Qld. Synonym: *latilabris* Philippi, 1848. A moderately common species in the tropics.

Haliotis planata Sowerby, 1833

Pl. 5; fig. 6   Ovate, flat; spire small and low; upper surface with low radial folds near the spire and fine spiral cords, outer side with a rather sharp peripheral keel and fine spiral cords; ventral ramp flat and unusually wide; holes simple, about five open. Upper surface reddish brown, outer side with alternating vertical fawn and green bands.

4 cm. Indian Ocean. Recorded from the Rowley Shoals and Cartier I., NT. Lives in coral reef habitats but not yet recorded from Qld.

Haliotis roei Gray, 1826

Pl. 4; fig. 4 a-b   Ovate, spire usually low but may be high in specimens exposed to strong wave action; upper surface with thick, rough, spiral cords of variable width; holes simple, about seven open. Reddish brown, sometimes with green rays.

12 cm. Western Vic. to Shark Bay, WA. Abundant and gregarious in the lower intertidal and sublittoral zones. Fished commercially in WA. Synonyms: *sulcosa* Philippi, 1845; *scabricostata* Menke, 1843.

Haliotis ruber Leach, 1814

Ovate with rounded dorsal surface and low spire, broad radial folds are crossed by many fine, beaded spiral cords; holes on small conical tubercles, 6-7 open; with a spiral concave zone between holes and margin. Red-brown with narrow, greenish curved radiating rays.

16 cm. NSW to SA and Tas. A common commercial species known in the abalone trade as the "black-lip". *H. clathrata* Reeve, 1846 from Qld has a small, depressed, almost circular shell and may be a variant of *H. ruber,* but material is hard to obtain and the matter needs further study. See Wilson & Gillett, plate 8, fig. 1. Synonyms: *naevosa* Martyn, 1784 (not binominal); *gigantea* Péron, 1816; *tubifera* Lamarck, 1822.

Haliotis scalaris Leach, 1814

Pl. 4; fig. 7 a-b Ovate, thin, spire moderately elevated; complexly sculptured with a thick central rib which is itself sculptured with scaly spiral cords, and thin inclined obliquely radial lamellae between the rib and the spire; outer side with a deep channel between the holes and a roughly nodulose marginal rib; holes on high conical tubercles, 4-6 open. Orange-red with curved, radiating cream rays.

10 cm. Western SA to Geraldton, WA. This is one of Australia's most beautiful and unusual shells. It lives beneath stones in the intertidal and sublittoral zones and is common but never abundant. The south-eastern species *H. emmae* is obviously related. Specimens of *H. emmae* from western SA are intermediate in sculpture. Until an analysis is done of the geographical variation in shell sculpture of the eastern *emmae* and western *scalaris* this author prefers to treat them as different species. See also illustration on p. 47. Synonyms: *rubicundus* Gray, 1826; *tricostalis* Menke, 1843.

Haliotis semiplicata Menke, 1843

Pl. 5; fig. 8 a-c Ovate to elongate-ovate, thin, spire low; upper surface rather flat, sculptured with broad, irregular, oblique radial folds, a poorly developed spiral fold centrally, and fine spiral cords; outer side with a concave, finely corded upper part and spirally ridged but otherwise smooth lower part; holes round, on low tubercles, 6-7 open. Olive-green or fawn, often with a central spiral orange band and alternating bands of pink or brown and green on the smooth part of the outer side.

6 cm. Esperance to Geraldton, WA. An uncommon, sublittoral species. It is found under stones. The shell colour varies remarkably. See remarks on the eastern species *Haliotis coccoradiata,* which is similar in many respects. Synonym: *lauta* Reeve, 1846.

Haliotis squamata Reeve, 1846

Pl. 4; fig. 1; Pl. 5; fig. 10 a-b Elongate-ovate, juveniles often more slender than adults; spire small, posterior; upper surface rounded, sculptured with rough growth striae and finely beaded spiral cords; holes simple, oval, 7-8 open. Red-brown with irregular greenish or cream zig-zag or obliquely radial rays, and conspicuous cream patches in a row around the outer margin.

8 cm. Shark Bay, WA to NT. Common under stones in the intertidal and sublittoral zones but not gregarious.

Haliotis varia Linnaeus, 1758

Pl. 5; fig. 2 a-b Elongately ovate, rounded dorsally, spire located posteriorly; upper surface sculptured with irregular radiating folds crossed by low, rounded spiral ribs of varying thickness, some weakly nodulose; outer side convex, upper part weakly corded, lower part with three or four nodulose ribs; holes round, on slightly elevated tubercles, four or five open. Brown or greenish, usually with white or cream patches.

4 cm. Indo-West Pacific; Abrolhos, WA to southern Qld. Common under stones in shallow coral reef or rocky shore habitats. This variable, small, widely distributed species is probably the most common abalone in the tropics of northern Australia and has been named several times. Synonyms: *concinna, semistriata* and *viridis* all of Reeve, 1846; *pustulifera* Pilsbry in Tryon, 1890; *aliena* Iredale, 1929; *howensis* Iredale, 1929.

SUPERFAMILY
FISSURELLOIDEA

FAMILY **FISSURELLIDAE**

Keyhole and Slit Limpets

In their shell form and anatomy the species of this large family preserve many features which may be considered as "primitive" in the scale of gastropod evolution. Generally the animals and shells are bilaterally symmetrical and the mantle cavity is anterior. There are two gills within the mantle cavity, one on each side, and the heart, auricles, kidneys and gonads are paired.

The porcelaineous shells of most fissurellids are conical and limpet-like. They are characterised by an anal notch, slit or hole which serves as the passage for excurrent discharge which is functionally equivalent to the row of respiratory holes in haliotids. There is a horseshoe-shaped muscle scar which has its open side facing anteriorly. Excretory and digestive wastes, and reproductive products, are carried out with the exhalent water. The family takes its scientific and vernacular names from the hole or slit.

There is no operculum. Often the body is too large to fit beneath the shell so that the animal has a slug-like character.

Fissurellids graze on plant or animal tissues growing on firm substrates, using a multi-toothed brush-like radula. Some species appear to associate with sponges or ascidians and may be specialised predators on these organisms. However, little is known of the biology of fissurellids.

There has been no review of the Australian species and a great deal of taxonomic work is needed to clarify relationships and nomenclature. In this account the classification of the *Treatise of Invertebrate Paleontology* is followed, except where otherwise indicated. There are three subfamilies, all of them well represented in the Australian fauna.

SUBFAMILY **DIODORINAE**

The shell perforation is a hole at the apex, bordered internally by a posteriorly truncate callus, muscle scar with hook-shaped ends.

▲ GENUS **DIODORA** Gray, 1821
Patella apertura Montagu

Approximately oval in plan view, depressed or conical in elevation; sculpture cancellate; margin in single plane, crenulate within; apical hole oval, rectangular or sub-triangular.

Cosmopolitan. The group has been divided into subgenera on the shape and position of the apical hole, but this character is inconsistent among the Indo-West Pacific species. These animals are unable to cover the shell with the mantle. They may be found under stones or in crevices or in the intertidal and sublittoral zones. There are many species in Australian waters but their taxonomy is still confused; only a selection of the common ones are described here. Synonyms: *Austroglyphis* Cotton & Godfrey, 1934 (*lincolnensis*); *Elegidion* Iredale, 1924 (*audax*).

Diodora corbicula (Sowerby, 1862)

Rectangularly ovate, rather depressed; hole rectangular-ovate, slightly anterior to centre; sculptured with numerous weak radial ribs and concentric striae. Brown, lighter brown to cream toward the margins, with darker brown radial rays.

3 cm. Qld. Hedley (1913) noted that the types of this species and that of *lanceolata* Sowerby, 1862 appear to be conspecific. In that case *corbicula* takes priority because of its earlier figure number even though both names appear on the same page in the original description.

Diodora galeata (Helbling, 1779)

Pl. 6; fig. 19   Elongate-ovate, slightly narrower anteriorly; apex overhanging the anterior margin; anterior slope short, concave, posterior slope long, convex; hole inclined from horizontal; sculpture of about 50 low, broad, nodulose radials, with crossbars in the interspaces; margin dentate. White.

1.3 cm. Qld.

Diodora granifera (Pease, 1861)

Pl. 6; fig. 20   Ovate, conical; hole elongate, barely on the anterior slope; about 30 rounded, nodulose ribs and finer concentric cords. Cream to greenish, often with brown radial rays.

8 mm. Indo-West Pacific; Quobba, WA to Caloundra, Qld.

Diodora jukesii (Reeve, 1850)

Pl. 6; fig. 25 a-c   Ovate, narrowing anteriorly; hole ovate or quadrate, situated slightly anterior of centre; about 50 irregular, coarsely imbricate, rather angulate radial ribs, crossed by concentric cords which are lamellose in the interspaces. White or cream; interior white, sometimes tinged with rose or with rose rays.

6 cm. Cape Leeuwin, WA to southern Qld. Often with three prominent ribs posteriorly. Intergrades with the southern Australian *D. lincolnensis*. Some northern shells (e.g. fig. 25 c) are identical to southern shells (e.g. fig. 28) and *lincolnensis* may need to be placed in the synonymy of *jukesii*.

Diodora lincolnensis (Cotton, 1930)

Pl. 6; fig. 28   Elongate-ovate, narrower anteriorly; posterior slope convex, anterior slope concave; hole rectangular, slightly in front of centre; sculpture pronounced and intricate, about 20 primary ribs bearing foliose processes; smaller secondary ribs in the interspaces, one in each interspace anteriorly, two or more posteriorly; margin strongly crenulate. White or yellowish with irregular brown-grey blotches in interstices.

4 cm. SA to about Geraldton, WA. See remarks on *D. jukesii*.

Diodora lineata (Sowerby, 1835)

Pl. 6; fig. 27   Elongate-oval, slightly narrowing anteriorly; posterior slope convex, anterior slope concave to straight; hole rectangular or oval, slightly anterior of centre; margin crenulate; sculpture of rough sub-equal, radial ribs, sometimes there are about 20 primary ribs with two or three secondaries between, the radials and interspaces crossed by regular concentric threads or, in juveniles, raised concentric lamellae giving a cancellate appearance. Yellowish white, sometimes blotched with brown or with faint brown radial stripes.

5.5 cm. NSW to Port Phillip Bay, Vic. Synonym: *audax* Iredale, 1924. Thiele (1930) described a similar shell from Shark Bay and named it *D. rugosa*.

Diodora occidua (Cotton, 1930)

Pl. 6; fig. 24   Ovate; rather tall, narrowing slightly anteriorly; hole oblong-ovate, slightly constricted laterally and with a tiny central projection on the anterior margin; sculpture of 40 to 50 coarse, imbricate radial ribs, with three secondary ribs between the primary ribs posteriorly, but laterally and anteriorly the middle intermediate ribs become enlarged so that primary and secondary ribs alternate; concentric lamellae make the sculpture cancellate. Cream, tinged with rose or green.

Dimensions of holotype: length 2.72 cm; width 1.95 cm; height 1.33 cm. The type locality is Shark Bay, WA. In the same year, Thiele described *D. plicifera*, *D. ovalis* and *D. nigropunctata* also from Shark Bay. Comparison of the types is needed before the relationships of these names can be determined. The species also resembles *D. jukesii* Reeve. Until the matter can be resolved *D. occidua* may be treated as a species.

Diodora octogona (Reeve, 1850)

Ovate; apex almost central, anterior and posterior slopes almost straight; hole elongate-ovate; 12-14 primary radial ribs and 2-3 secondary ribs in the interspaces; radials crossed by spiral cords of about equal thickness to the secondary ribs, forming scabrous nodules at the intersections. Brown and white.

1.2 cm. Indo-West Pacific. Recorded from Qld by Hedley (1910). See Kay, 1979, fig. 11 e-g for illustrations.

Diodora quadriradiata (Reeve, 1850)

Small, elongate-ovate; hole elongate, rectangular-ovate, slightly anterior of centre; sculptured with uniform radial ribs crossed by concentric cords forming transverse nodules at the intersections. White with four wide brown radial rays; hole callus bordered with black internally.

12mm. South West Pacific. Recorded from Qld by Hedley (1910). See Cernohorsky, 1978, Pl. 4; fig. 3 for illustrations.

Diodora ruppelli (Sowerby, 1834)

Pl. 6; fig. 18 a-b   Ovate, elevated; margin crenulate; hole tri-cleft; 25-30 equally spaced ribs with interspaces of equal diameter, crossed by smaller scabrous concentric cords. White with triangular black marks.

Length 20 mm; height 13 mm. Indo-West Pacific. Recorded from Qld by Hedley (1910).

Diodora singaporensis (Reeve, 1850)

Pl. 6; fig. 16 a-c   Elongate-ovate to rectangular, often arched, depressed to moderately elevated; hole oval, situated at about the anterior $^1/_3$ position; sculpture of about 80 fine, weakly nodulose radial ribs, crossbars present in the narrow interspaces near the apex but absent near the margins. Fawn, with broad radial rays of brown.

2.5 cm. Indo-West Pacific. Shark Bay WA to southern Qld; moderately common under stones in the intertidal zone.

Diodora ticaonica (Reeve, 1850)

Pl. 6; fig. 17 a-c   Elongate-ovate, elevated, sometimes notably so; hole forward of centre, anterior slope concave, posterior slope convex; hole tending to be tipped forward, elongate, with three round portions like beads on a string; sculpture of 36-40 radial ribs with fluted nodules, high crossbars in the interspaces and deep pits between. Cream, often with four or more broad, grey radial rays.

2.2 cm. Indo-West Pacific. Geraldton, WA to northern NSW.

SUBFAMILY **FISSURELLINAE**

Perforation in the shell is a hole at or near the apex, bordered internally by a rounded callus.

▲ GENUS **AMBLYCHILEPAS** Pilsbry, 1890
Fissurella javanicensis Lamarck

Depressed, saddle-shaped, ovate or oblong-ovate; hole, more or less central, large; shell margins thickened and not crenulate.

This is a peculiarly southern Australian genus. Synonym: *Sophismalepas* Iredale, 1924 (*nigrita* Sowerby).

Amblychilepas compressa Thiele, 1930

Pl. 6; fig. 14   Resembling *A. omicron* but with stronger and more nodulose radial sculpture; anterior end sometimes upturned.

1.2 cm. Esperance to Shark Bay, WA. Cotton's (1959) record of *A. omicron* from south-western Australia probably refers to this species.

Amblychilepas crucis (Beddome, 1882)

Pl. 6; fig. 13   Ovate, with cancellate sculpture. White or yellowish with four red radial rays forming a cross on the back.

1.5 cm. Bass St.

Amblychilepas javanicensis (Lamarck, 1822)

Pl. 6; fig. 4   Ovate to squarish; hole elongate; sculptured with fine, crowded concentric cords; radial sculpture lacking. Whitish with radiating reddish brown rays.

2.5 cm. Southern Qld to Fremantle, WA. Under stones; intertidal and shallow sublittoral. Less common than *A. nigrita*.

Amblychilepas nigrita (Sowerby, 1834)

Pl. 6; fig. 5   Oblong-ovate; hole also oblong-ovate; sculpture of fine, weakly beaded, crowded radial cords and fine concentric threads in the interstices. Whitish, usually with reddish radial rays.

2.5 cm. Southern Qld to Fremantle, WA. Common under stones in the intertidal and shallow sublittoral. The yellow animal is very active. Synonym: *pritchardi* Hedley, 1895.

Amblychilepas oblonga (Menke, 1843)

Pl. 6; fig. 6   Resembles *A. nigrita* but more elongate and with coarser, more granulose radial sculpture. Cream with red or reddish brown radial rays, sometimes almost uniformly rose.

2.8 cm. SA to Geraldton, WA. Cotton (1959) recorded the species from SA localities as far east as the Victorian border.

Amblychilepas omicron (Crosse & Fischer, 1864)

Oblong-ovate, slightly narrower anteriorly; sides concave; with about 24 weak radial ribs (sometimes finer radial cords between), crossed by concentric cords, giving finely granular appearance, sculpture strongest towards the summit. Slate-blue, yellowish or pink, with interrupted radial rays of grey or brown.

13 mm. Vic. to SA.

▲ GENUS **COSMETALEPAS** Iredale, 1924
Fissurella concatenata Crosse & Fischer

Shell thin, depressed, oval; surface with small rounded pits; hole large, oblong, central. Mantle lobes completely cover the shell in life; anterior end of foot bifid.

This is a very distinctive southern Australian genus with only one living species, but there are several Tertiary fossil species in the region.

Cosmetalepas concatenatus (Crosse & Fischer, 1864)

Pl. 6; fig. 15   Surface sculpture consists of irregular, chain-like concentric rings of rounded pits; hole elongate-ovate with slight lateral constrictions giving it a keyhole shape. Animal orange; mantle almost covers the shell.

2.2 cm. NSW to Abrolhos, WA. Cotton (1959) gives the depth range as down to 100 fathoms and the habitat as "under rocks at low tide" and "sometimes taken living in association with a grey Ascidian". This author has collected the species in the open on large simple ascidians growing on rock walls at 5-10 m in the Marmion Marine Park, WA. Because of the rarity of these animals there is uncertainty whether more than one species may be involved.

▲ GENUS **MACROSCHISMA** Sowerby, 1839
M. macroschisma Sowerby (= *hiatula* Swainson)

Shell elongate, saddle-shaped, with an elongate hole close to the posterior margin. The shell covers the anterior end of the animal but the long posterior end of the foot is uncovered.

Indo-West Pacific and southern Australian regions.

When introducing this generic name Sowerby cited as type a species from NSW which he misidentified as the Japanese *macroschisma* Solander. Australian workers have concluded that the species Sowerby had in hand was the species later described as *hiatula* Swainson, 1840. Iredale (1940) introduced the generic names *Foralepas* (*tasmaniae* Sowerby) and *Dolichoschisma* (*producta* Adams). These names have been ranked as subgenera in the *Treatise of Invertebrate Paleontology* but there is much variation among the species and they are treated here as subjective synonyms of *Macroschisma*.

Hedley (1907) figured a growth series of *M. madreporia* showing how the hole is central in juveniles but becomes posterior as a result of differential growth of the shell.

Macroschisma bakiei Adams in Sowerby, 1862

Pl. 6; fig. 7 a-b   Elongate-ovate; hole oval; sculptured with strong, rough radial ribs anteriorly and rough concentric cords posteriorly. Cream, pink or light brown, sometimes radially rayed or concentrically banded.

2.8 cm. Southern WA. Under stones; intertidal and shallow sublittoral. *M. tasmaniae* Sowerby, 1866, from southern NSW to SA, closely resembles *M. bakeiei* but with finer sculpture. The two may be variants of the same thing.

Macroschisma hiatula Swainson, 1840

Pl. 6; fig. 3   Elongate, rectangular; hole long, often half the shell length; anterior slope convex, not ridged; sculpture of granulose radial ribs, more pronounced than in *M. producta*, sometimes with 6-10 larger radial folds in the anterior half.

1.8 cm. Lord Howe I. There is a form in northern Australia (North West Cape, WA to Caloundra, Qld) which fits this description better than that of the southern *M. producta* or the western *M. munita*, but more study is needed to determine relationships.

Macroschisma madreporaria Hedley, 1907

Thin-shelled, oblong-ovate, slightly narrowing anteriorly, anterior margin slightly indented and truncate; hole long, narrow, triangular, with straight sides very nearly coming to a point anteriorly, sculpture of fine radiated threads, faintly beaded at the intersections with growth striae. Cream, often with rose rays.

9.5 mm. Described from Mast Head Reef, Qld in 30-37 m. Habitat and distribution unknown.

Macroschisma munita Iredale, 1940

Pl. 6; fig. 1 a-b   Like *M. producta* but with a central ridge on the anterior slope which develops into a spout at the anterior margin. White, variegated with pink or light brown flames or speckles.

2.2 cm. Albany to Port Hedland, WA. This may be a synonym of *M. producta* which exhibits a range of variation in shell form. Both forms occur in southern WA and two species are provisionally recognised there pending further study.

Macroschisma producta Adams, 1850

Pl. 6; fig. 2   Oblong-ovate; hole elongate, triangular, widest posteriorly with a thick raised posterior end; sculptured with fine granulose radiating striae. White or pale rose with darker rose or brown radial rays or flecking.

2.2 cm. NSW to central WA. Usually buried in sand, under stones or hiding in crevices; intertidal and shallow sublittoral. The geographic range of this species is uncertain as there is a similar species in northern Australia with which it may be confused. See remarks on *M. hiatula* and *M. munita*.

SUBFAMILY **EMARGINULINAE**

Perforation represented by a marginal slit or notch.

▲ GENUS **EMARGINULA** Lamarck, 1801
E. conica Lamarck

Conical, high to depressed, ovate to oblong; apex backward-pointing and recurved, varying in position from central to posterior; anterior slope convex, posterior slope concave below the apex; fissure a long to moderately short slit in the anterior margin, with a sunken or raised selenizone between slit and apex; cancellate sculpture of radial ribs and concentric lamellae or striae.

Cosmopolitan. Cotton (1957) introduced *Notomella* (*candida* Adams) replacing his earlier name *Entomella*, to accommodate southern Australian species. However, the range of variation is great, especially when the tropical species are considered, and *Notomella* is treated here provisionally as a synonym of *Emarginula*. There are many living Australian species, only a selection of which are illustrated here. There are also several Tertiary fossil species in south-eastern Australia. The group badly needs study and revision.

Emarginula candida (Adams, 1851)

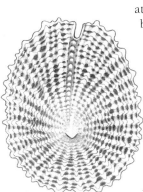

Pl. 6; fig. 11   Depressed to moderately elevated; apex low, a little behind the centre; slit relatively short, selenizone forms a humped raised ridge; with about 26 strong radial ribs and one or two secondaries in between, concentric crossbars render the radials nodulose.

15 cm. Vic. and Tas. to Fremantle, WA. Common in beach drift. NSW shells were named *hedleyi* by Thiele (1915) but they are indistinguishable from southern specimens.

Emarginula convexa Hedley, 1907

Strongly arched with a large apex spirally downturned and overhanging the posterior end, posterior slope conspicuously concave; margin finely frilled; with about 50 strong radial ribs, alternately large and small, crossed by fine concentric threads, fine nodules at the intersections; slit long, selenizone sunken between thin raised ribs, crossed by widely spaced bars.

3 mm. Southern Qld. *Notomella flindersi* (Cotton, 1930) described from 130 fathoms off Cape Jaffa, SA may be the same species. It has fewer, more distant and more spreading radial ribs.

Emarginula dilecta (Adams, 1851)

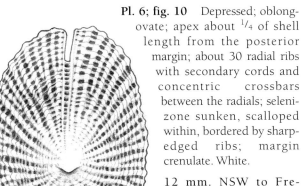

Pl. 6; fig. 10   Depressed; oblong-ovate; apex about $^1/_4$ of shell length from the posterior margin; about 30 radial ribs with secondary cords and concentric crossbars between the radials; selenizone sunken, scalloped within, bordered by sharp-edged ribs; margin crenulate. White.

12 mm. NSW to Fremantle, WA. Under stones; intertidal zone to 200 m. Hedley (1907) recorded the species from Mast Head Reef, Qld. *Emarginula amitana* Iredale, 1925, described from 135 m off Ulladulla, NSW, appears to be identical.

Emarginula cf. eximia Adams, 1843

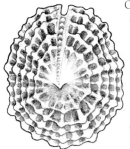

Ovate, depressed; apex almost central; slit short; selenizone raised, filled, nodulose; with simple but strong cancellate sculpture of 18-20 strong rounded radials and sub-equal well-spaced crossbars in the interspaces. White.

10 mm. Recorded from North West Cape, WA.

Emarginula gabensis Gabriel, 1961

Pl. 6; fig. 12   High; apex coiled under, at about the $^1/_3$ posterior position; margin crenulate, slightly outward-flaring; with 32 primary radial ribs and intermediate secondaries, primaries broad and low posteriorly; concentric lamellae cross the interspaces and the radial ribs making a strongly latticed sculpture; selenizone sunken, with high thin side walls and prominent transverse lamellae between. Cream.

Dimensions of paratype: length 25 mm, width 18.4 mm, height 11.5 mm. Vic. This is the largest species of the genus in Australia. Much higher, thicker-shelled, more strongly sculptured and with less outward-flaring margins than *N. patula*.

Emarginula incisura (Adams, 1853)

Pl. 6; fig. 8 a-b   Subquadrate, narrowing anteriorly; much depressed and arched in profile; apex small, almost central; margin finely serrate; slit long, bordered internally by a V-shaped callus; selenizone depressed, almost smooth, bordered by sharp-edged ridges; sculpture of many fine radial ribs crossed by crowded, imbricating, concave rugae giving a rough surface texture. Fawn to pale yellow; central interior yellow, white near the margin. The dark brown mantle partly covers the shell.

2.5 cm. Indo-West Pacific; Port Hedland, WA to Maroubra, NSW. Some specimens are heavier with smoother ribs and two rosy flames before the apex, and may not be the same species. Hedley (1913) named a small beach-worn specimen like this *E. bajula*; it may be merely a juvenile of *E. incisura*.

Emarginula cf. micans (Adams, 1852)

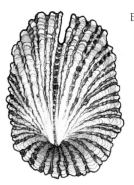

Elongate-ovate, high in profile, apex strongly recurved and located posteriorly; slit long, selenizone a shallow trough with widely spaced crossbars and flanked on each side by thin radial ridges; sculpture of strong radial ribs and widely spaced concentric bars which cross the ribs as rounded rugae. Grey.

8 mm. North Qld.

Emarginula nigromaculata (Thiele, 1930)

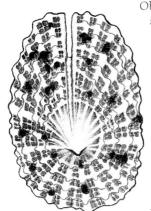

Oblong-ovate, depressed, strongly arched; apex at about the $^1/_3$ posterior position; slit long; selenizone a high, nodulose ridge; very strong cancellate sculpture of about 24 rounded radial ribs bearing round nodules, and with high, widely-spaced crossbars connecting the nodules laterally, and fine radial cords and concentric striae present in the deep, rectangular interspaces; margins crenulate. White, with large dark brown spots on a few of the nodules.

10 mm. North West Cape to Port Hedland, WA. Common in beach drift in Ningaloo Marine Park.

Emarginula patula Cotton, 1930

Pl. 6; fig. 9   Thin-shelled, depressed; apex at the posterior quarter, strongly curved and downturned; posterior margin outward flaring; radial ribs alternately large and small, crossed by slightly lamellose concentric cords producing a fine nodulose effect at the intersections; selenizone not very sunken, crossed by high concentric lamellae. Yellow-white.

Dimensions of holotype: length 19.3 mm, width 14 mm, height 5.6 mm. SA to Fremantle, WA. Dredged to 600 m.

Emarginula subtilitexta (Verco, 1908)

Thin, ovate, elevated, slightly arched; apex almost marginal; slit long; selenizone sunken between two high, sharp-edged ribs, scalloped; about 60 narrow rounded radial ribs, alternate ribs sometimes slightly smaller, with weak concentric cords in the interspaces making ribs weakly nodulose. White.

7 mm. SA to Fremantle, WA; at depths to 200 m. The posterior position of the apex, high dorsum, weaker nodules and thinner shell distinguish this shell from E. dilecta.

Emarginula superba (Hedley, 1906)

Ovate, elevated; apex at about posterior $^1/_4$, prominent; posterior slope strongly concave; margins crenulate; selenizone inconspicuous externally, filled with close-packed curved laminae between raised, sharp-edged, radial ribs, evident internally as a thick, rounded axial rib; sculpture of numerous radial ribs with crowded, curved lamellae in the narrow interspaces, forming concentrically elongated nodules where they cross the ribs. Grey.

Dimensions of the holotype: length 23.9 mm; height 9.4 mm, width 17.2 mm. NSW, eastern Vic. and Tas. The type was dredged in 457 m off Sydney.

▲ SUBGENUS SUBZEIDORA Iredale, 1924
E. connectens Thiele

Apex posterior, incurved, hidden by the curvature within; slit very long.

South Pacific and southern Australian regions.

Emarginula devota Thiele, 1915

Elongate-ovate, very thin-shelled; selenizone raised between thin ribs, crossed by well-spaced nodulose, crescentric lamellae; sculpture of about 36 radial ribs, alternate ribs sometimes smaller, interspaces crossed by thick cords with deep, rounded pits between. White.

5 mm. NSW to Fremantle, WA. Common in beach drift. E. connectans (Thiele, 1924) is a similar species from the Kermadec Is.

▲ GENUS HEMITOMA Swainson, 1840
Patella tricostata Sowerby

Low to elevated; apex behind centre, recurved and backward pointing; with a short anterior marginal notch; internally the selenizone groove is distinct from the notch to the apex.

Cosmopolitan in tropical and warm temperate seas. Three subgenera are recognised, two of them represented in the Australian fauna. Hemitoma s.s is from the Carribbean. In addition to the species illustrated here, Hedley (1910 and 1918) recorded from Qld H. cumingii Sowerby and H. aspersa Gould. The nomenclatural status of these names is uncertain.

▲ SUBGENUS MONTFORTIA Récluz, 1843
Emarginula australis Quoy & Gaimard (=subemarginata Blainville)

Posterior slope concave, anterior slope convex; selenizone forming a stout ridge on the anterior slope, often flanked on each side by a broad rib; otherwise with symmetrically arranged heavy radial ribs crossed by concentric cords.

Australia.

Hemitoma subemarginata (Blainville, 1819)

Pl. 6; fig. 34 a-c   Ovate, narrower anteriorly, anterior end sometimes truncate with sharply pinched-in sides antero-laterally; with three prominent radial ribs anteriorally, the central one usually most prominent, and about 12 additional primary radials and fine radial cords between; interspaces crossed by concentric struts giving a cancellate appearance;

radial ribs angular and finely granulose or thick, broad and heavily rugose. White or yellow.

3.2 cm Vic. and Tas. to Fremantle, WA. The extreme variation in the sculpture of this species is confusing; more than one species could be involved. Synonyms: *australis* Quoy and Gaimard, 1834; *emarginata* Blainville, 1825.

▲ SUBGENUS **MONTFORTISTA** Iredale, 1929
M. excentrica Iredale

With strong concentric cords producing lattice-like sculpture; outline subquadrate.

Northern Australia.

Hemitoma excentrica (Iredale, 1929)

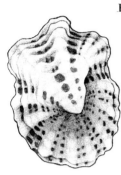

Pl. 6; fig. 33 a-c Elevated, elongate-ovate; with strong radial and concentric ribs forming a cancellate sculpture. Exterior greenish white, interior green.

1.8 cm. Central Indo-West Pacific; Exmouth Gulf, WA to southern Qld. Compare with *H. panhi*.

Hemitoma panhi (Quoy & Gaimard, 1834)

Subquadrate, high in profile; apex strongly recurved and almost overhanging the posterior margin; selenizone prominent, nodulose, flanked on each side by prominent radial folds made up of three ribs; with five major ribs anteriorly and several intermediate cords; crossbars prominent. Off-white.

1.5 cm. Western Pacific; north Qld. Walter Cernohorsky has pointed out that the name *octo-radiata* Gmelin, 1791 was based on the Born reference so that it should be used for this Pacific species, not for the American species with which it is currently associated. If this is so it would cause some nomenclatural confusion. Other synonyms: *tricarinata* Born, 1778 (not Linnaeus,1767); *clathrata* Adams & Reeve, 1850; *panihensis* Adams, 1852.

▲ GENUS **MONTFORTULA** Iredale, 1915
Emarginula rugosa Quoy & Gaimard

Ovate, outline conical in elevation; apex subcentral; slit very short, notch-like; with a distinct internal groove from lip notch to apex; sculpture of axial ribs made cancellate by crossing concentric striae; margin crenulate.

Australia. *Montfortula* is sometimes treated as a subgenus of *Clypidina* Gray, 1847 but the apex of *M. rugosa* is recurved, unlike that of *Clypidina* s.s. It seems more appropriate to use Iredale's name at full generic rank.

Montfortula pulchra (Adams, 1851)

Pl. 6; fig. 23 Apex subcentral; slit rectangular, notch-like; selenizone rounded or flat-topped, weakly nodulose, filling the space between two strong, rounded ribs; with about 34 heavily nodulose, rounded ribs, four or five posterior ribs usually larger than the others. Fawn or greenish.

1.3 cm. North West Cape, WA to southern Qld. Intertidal, often in the oyster zone.

Montfortula rugosa (Quoy & Gaimard, 1834)

Pl. 6; fig. 21 a-b Apex about central; slit rounded; ribs numerous, made finely rugose by crossing striae; upper surface white or greyish brown; spatula prominently outlined with dark green, otherwise interior white or green.

2 cm. Southern Qld to Geraldton, WA. Lives in the middle intertidal zone among mussels and *Galeolaria*. East coast shells tend to be more strongly and more rugosely ribbed. Synonym: *conoidea* Reeve, 1842.

Montfortula variegata (Adams, 1852)

Pl. 6; fig. 22 a-b Apex at about the $^{1}/_{3}$ posterior position; slit rectangular; selenizone narrow, raised, filled and nearly flat-topped except for low transverse nodules; with numerous, crowded, nodulose radial ribs, sculpture lacking in the interspaces. Brown or rose, sometimes broadly rayed.

1.2 cm. Central Indo-West Pacific; Exmouth Gulf, WA to southern Qld. The generic position of this species is problematical. The short notch-like slit seems to disqualify it from *Emarginula* but the selenizone is not deeply grooved. Placement of the species in *Montfortula* is provisional. WA specimens differ from Qld shells in having three prominent ribs anteriorly (including the raised central selenizone), four posteriorly. They may warrant taxonomic separation.

▲ GENUS **PUNCTURELLA** Lowe, 1827
Patella noachina Linnaeus

Conical, ovate; apex about central, slightly recurved and pointing backwards; anterior slope convex, posterior slope concave; perforation on the anterior slope or near the apex, entering an internal conduit or curved shelly plate; internal groove and slit scar weak or lacking.

Cosmopolitan. Synonym: *Vacerra* Iredale, 1924 (*demissa* Hedley) from New Zealand.

▲ SUBGENUS **PUNCTURELLA** s.s.

Hole near the apex; apical whorls persist in the adult.

Cosmopolitan. In addition to the species described here, *P. harrisoni* Beddome, 1882 and *P. menda* Iredale, 1924 are described from south-eastern Australia.

Puncturella galerita Hedley, 1903

Solid; height equalling or exceeding the length; apex overhanging the posterior margin, posterior slope concave; about 20 low radiating ribs render the margin crenulate, ribs evident on the interior surface, with weak growth striae crossing the ribs and intertices, whole surface finely granulate; slit narrow, at shell summit; interior septum occupying almost half the shell length. White.

Length 2.2 mm; height 2 mm. North Qld. Dredged in shallow water.

Puncturella kesteveni Hedley, 1900

Tiny, thin, opaque; summit behind the centre, excavated by a slit, with a backward-pointing projection which almost overhangs the posterior margin; internal septum occupying rear of the shell; margin expanded, externally marked by a concentric furrow; outer surface uniformly shagreened. Pale brown.

Length 2.3 mm, height 1.3 mm. The type was found in beach sand at La Perouse, NSW. For illustrations see original description (*Proceedings Linnean Society NSW*, 1900, p. 499, Pl. 25, figs. 15, 16, 17).

▲ SUBGENUS CRANOPSIS Adams, 1860
C. pelex Adams

Perforation at the middle of the anterior slope; interior groove visible.

Atlantic, Indo-West Pacific. Synonym: *Rimulanax* Iredale, 1924 (*corolla* Verco).

Puncturella corolla Verco, 1908

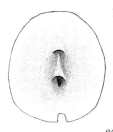

Fragile, thin, roundly ovate, depressed in elevation, margins flaring, especially posteriorly where the shell is almost horizontal; apex situated behind centre, protoconch projecting on the right side; hole elongate with sharp, elevated sides; above the hole the sunken selenizone is flanked by sharp-edged ridges and crossed by several raised laminae, below it there is a raised rib superficially fissured down its mid-line; internally a curved and raised roof covers the dorsal part of the hole; sculpture of crowded, angular radial ribs crossed by fine spiral cords. Whitish.

Dimensions of holotype: length 18 mm; width 15.4 mm; height 4.7 mm. Dredged at 130-300 fathoms off Cape Jaffa, SA. According to Cotton (1959) the fissure is a marginal slit in juveniles, becoming closed and keeping its position relative to the apex in adults. For illustration see Cotton, 1959, p. 67, fig. 30.

▲ SUBGENUS **FISSURISEPTA** Seguenza, 1863
F. papillosa Seguenza

Apex absorbed by the apical hole; internal septum strongly developed, internal groove lacking; surface sculpture weak.

Cosmopolitan. The type species is a Miocene fossil from Sicily. The single Australian species is typical of the group.

Puncturella fumarium Hedley, 1911

Tiny, translucent, oblong in plan, tall and almost symmetrical in elevation; anterior and posterior slopes slightly convex, apical hole notched; internally a deep septal plate partly covers the posterior end of the hole; sculpture of radial rows of minute tubercles.

Dimensions of syntype: length 2.15 mm; width 1.35 mm; height 1.85 mm. Known only from the type locality, i.e. 100 fathoms, 40 miles south of Cape Wiles, SA. Illustrated by Cotton, 1959, p. 69, fig. 31.

▲ GENUS **RIMULA** Defrance, 1827
R. blainvillei Defrance

Elevated, anterior slope convex, posterior slope concave; elongate-ovate in plan view; apical whorls present, downward coiled and inclined to the right; fissure a long, narrow slit on the anterior slope; internal septum lacking.

A cosmopolitan genus with a long fossil history.

Rimula exquisitor Adams, 1853

Ovate; apex at posterior ¹/₄; slit half way down the anterior slope; selenizone above the slit depressed, crossed by widely spaced crescentic lamellae; about 30 primary radial ribs alternating with subequal secondaries, ribs nodulose; also with nodulose concentric crossbars in the interspaces. Cream.

11 mm. Central Indo-West Pacific; North West Cape, WA to Mast Head Reef, southern Qld. *R. mariei* Crosse, 1866, is similar but more elongate-ovate and arched, and the apex recurves almost over the posterior margin. It is a western Pacific species recorded from Lord Howe I.

▲ GENUS RIXA Iredale, 1924
Glyphis watsoni Henn & Brazier

With the characters of the type species.

South-eastern Australia. Monotypic.

Rixa watsoni (Henn & Brazier, 1894)

Narrow, oblong, depressed, interior slope a little more than half as long as the posterior slope; apex truncate, smooth, inclined forward, with a small groove at the centre; fissure square, rounded behind, thickened by callus around its margin, partly closed by an internal septum; sculptured with numerous radial riblets; margin minutely crenulate. White, radial riblets with interrupted rays of pink.

6 mm. NSW and eastern Vic. The original specimens were dredged off Port Macquarie in 12 fathoms. For an illustration see the original description (*Proceedings Linnean Society NSW*, **19**: 177, Pl. 14, fig. 15).

▲ GENUS SCUTUS Montfort, 1810
S. antipodes Montfort

Shell very depressed, shield-like, ovate to oblong; apex small, backward pointing, located behind centre; anal notch weak; slit scar lacking; muscle scar near the margin; sculpture of only concentric striae.

Cosmopolitan. The body is large and fleshy. Synonym: *Aviscutum* Iredale, 1940 (*olonguis* Iredale).

▲ SUBGENUS SCUTUS s.s.

Shell large, smooth except for weak growth striae, truncate anteriorly.

European, Indo-West Pacific and southern Australian regions.

Scutus antipodes Montfort, 1810

Pl. 6; fig. 29 Apex at about $^1/_4$ position; anal notch very weak or lacking; concentric striae may be eroded away in adults. Shell white or off-white; animal black.

11 cm. NSW to Geraldton, WA. This remarkable creature is a feature of the southern Australian coast, where it is commonly found under stones and in crevices in shallow water. Synonyms: *anatinus* Donovan, 1820; *veitchi* Cotton, 1953.

Scutus unguis (Linnaeus, 1758)

Pl. 6; fig. 30 Apex at about $^1/_3$ position; anal notch shallow but wide and V-shaped; concentric striae characteristically pustulose. White or fawn; body brown.

4.5 cm. Shark Bay, WA to southern Qld. This species has been referred to in Australian literature as *S. granulatus* (Blainville, 1819) which is a synonym. *Scutus olonguis* Iredale, 1940, described from NSW, is almost identical to *S. unguis* except that the outline is more ovate. It may be the same species.

▲ SUBGENUS NANNOSCUTUM Iredale, 1937
N. forsythi Iredale

Small, stout; with strong concentric sculpture.

Scutus forsythi (Iredale, 1937)

Almost flat; oblong, posterior end rounded, anterior end truncate with a central sinuation representing the anal notch; apex backward pointing, close to the posterior end; slit scar lacking; dorsum and apex smooth except for minute punctations; edges thickened and with pronounced concentric growth striae. Shell white; animal black.

8 mm. Lord Howe I.

▲ GENUS TUGALI Gray, 1843
Emarginula parmophoidea Quoy & Gaimard

Much depressed; elongate-ovate to oblong; apex backward pointing, slightly recurved, far behind the centre; anal notch weak, slit scar represented by a weak groove to the apex internally and a weak ridge externally; margin crenulate within; exterior with cancellate sculpture.

Indo-West Pacific, New Zealand and southern Australian regions. There is also a monospecific subgenus, *Parmophoridea* Wenz, 1938, in Antarctica. The shells resemble those of *Scutus* in form but are easily distinguished by the presence of a slit and prominent sculpture. The animals are orange and the mantle almost covers the shell.

Tugali cicatricosa Adams, 1851

Pl. 6; fig. 32 a-b Very like *T. parmophoidea;* more depressed and oblong with slightly stronger concentric sculpture.

3.3 cm. Vic. to Carnarvon, WA. The apex in this species is said to be "cicatrised" (scarred) but there is little difference to the apex of *T. parmophoidea*. The two species are provisionally retained here as distinct but further study is needed on their relationship, especially in south-eastern Australia where the two forms occur together.

Tugali parmophoidea (Quoy & Gaimard, 1834)

Pl. 6; fig. 31 a-b Elongate-ovate, slightly narrower anteriorly, usually arched; radial sculpture dominant.

2.5 cm. NSW to SA. The apex is said to be obtuse in this species but in some specimens in the series examined it is sharp and slightly downturned.

▲ GENUS **ZEIDORA** Adams, 1860

Z. calceolina Adams

Depressed, oblong-ovate; apex posterior, recurved; internally there is a broad septum at the posterior end; anal slit marginal, short; selenizone a concave groove down the anterior slope.

Cosmopolitan. There are two or three Australian species.

Zeidora legrandi Tate, 1894

Elongate, with rounded ends; apex almost level with the posterior margin in adults, overhangs in juveniles; selenizone prominent, crossed by crescentic laminae; with delicate sculpture of fine radial riblets crossed by concentric, scalloped cords forming tiny nodules at the intersections. White.

Dimensions of holotype: length 9.5 mm; width 6 mm; height 2 mm. SA. The illustration is drawn from Tate's holotype. This may be a synonym of *Z. tasmanica* Beddome, 1883, which was described from Tasmania.

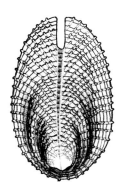 

Zeidora lodderae Tate & May, 1900

Resembles *Z. legrandi* but the radial sculpture is finer, crescentic lamellae are lacking and the apex overhangs the posterior end in adults.

1 cm. North Qld to Tas. The specimen figured by Hedley (1900, *Proceedings Linnean Society NSW*, **25**: 93) as *Z. tasmanica* is actually this species, not the one described earlier by Beddome under that name.

TROCHOIDEA

THIS is one of the largest superfamilies in the class. It has a fossil record from at least as far back as the Permian. It includes the top and turban shells and the skeneids - archaeogastropods which have no slit or sinus in the edge of the outer lip and no hole in the shell for the escape of excurrent water. The shell is typically conical but other forms occur. The aperture is roughly circular and has an entire margin. The operculum is either corneous or calcified.

Anatomically the group preserves many archaic features. The heart retains two auricles although the right gill (ctenidium) is lost. The remaining left gill is bipectinate or sometimes monopectinate. Two kidneys are also retained. The gonad discharges through the duct of the right kidney. Head tentacles are long and there are sensory flaps on the sides of the head and sensory epipodial tentacles along the sides of the foot. The radula is rhipidoglossate with a lateromarginal plate separating and interconnecting with the marginal and central teeth, and there is a semitubular food-collecting groove beneath the cusps of the many marginal teeth.

Classification of the trochoideans remains problematical. This account follows the recent work of Hickman & McLean (1990, Systematic revision and suprageneric classification of trochacean gastropods, *Los Angeles County Museum, Science Series* **35**: 1-169). Several of the groups treated as subfamilies had been classified as full families or even superfamilies by previous authors.

FAMILY **TROCHIDAE**

Top Shells

The typical form of this family is the genus *Trochus* in which the shell is conical and shaped like a top, but many trochids are turbinate or even auriform. The aperture is entire (that is, posterior and anterior canals are lacking). The operculum is thin, round or ovate, multispiral, with a central or subcentral nucleus and a very short growing edge. It is made of brown, uncalcified, horny material. The radula differs from that of the Turbinidae in that the rachidian teeth lack secondary flaps or pegs providing additional attachment to the radula ribbon. Otherwise the two families are very similar.

The animals are herbivores, carnivores or detritus feeders. Most trochids live on hard substrates but some crawl on sand or mud substrates. Their foot is rather small. Males and females lack reproductive organs for copulation and most release their sperm and eggs into the water where fertilisation occurs. In a few cases the females lay jelly egg masses in which early development takes place. There is a short planktotrophic larval stage.

The largest species in the family is *Trochus niloticus* which is fished commercially for its meat and "mother of pearl" shell. At one time the nacreous pearly shells were used to make buttons, but these days they are mostly used for shell jewellery in Australia. Torres St. Islanders and Bardi Aborigines at One Arm Point north of Broome maintain traditional trochus fisheries, selling most of their catch to the Asian market.

The taxonomy of the Trochidae is very unsatisfactory and the generic arrangement remains provisional. Hickman & McLean (1990) recognised 13 subfamilies and 11 tribes, many of which are represented in the Australian fauna.

SUBFAMILY **CALLIOSTOMATINAE**

Generally with conical shells which may be umbilicate. The aperture is typically oblique. The columella is basally truncated, meeting the outer lip at a distinct angle; it is never denticulated but may be thickened and reflected over the umbilical area. Characteristically the protoconch is sculptured with a honeycomb pattern of shallow pits bordered by a reticulate network of minute ridges. The foot is large, rounded anteriorly and tapering to a point posteriorly. These animals are carnivorous, feeding mainly on coelenterate polyps. Associated with this, the mouth is situated on a prominent pseudo-proboscis and is lined internally with cuticle. The radula is distinctive within the family, being long, with delicate, finely serrate rachidian tooth cusps separated by a sharp bend from broad, thin, flat bases; the marginal teeth are differentiated into many types with complex patterns, the innnermost marginals are greatly enlarged and laterally flattened, with a series of hooked cusps.

▲ GENUS **CALLIOSTOMA** Swainson, 1840
Trochus conulus Linnaeus

Conical, base more or less flattened, periphery angulate or subangulate; outer lip prosocline; umbilicus lacking; columella smooth, arched, meets the parietal wall at a steep angle, sometimes truncate at the base or with a small basal nodule.

Cosmopolitan. Many subgenera are recognised in this large and cosmopolitan group. Several of them are represented in Australian waters, although *Calliostoma* s.s. seems to be absent. The lack of an open umbilicus is the sole shell character distinguishing *Calliostoma* from *Astele* and it seems little and arbitrary. Anatomical study may eventually show a very different set of relationships among the species assigned to those two groups, and in the arrangement of the subgenera, but the conventional Australian system is followed here, although several species usually treated as *Calliostoma* are here moved to *Astele*.

▲ SUBGENUS **FAUTOR** Iredale, 1924
Ziziphinus comptus Adams

Spire high, acute; base more or less convex, periphery carinate to subcarinate; columella meets the outer lip at an angled corner; columellar margin thickened, usually with a small basal nodule; sculpture of spiral riblets, usually granulose, supramarginal riblets stronger.

South-eastern Australia.

Calliostoma allporti (Tenison Woods, 1876)

Pl. 10; fig. 6   Like *C. hedleyi* but taller, the periphery rounded, not angulate, the base convex not flat, the suture channelled, and the spiral ribs on the sides of the whorls granulose. Cream or buff, dotted with chestnut.

12 mm high; 10 mm wide. Vic., Tas. and SA. Dredged to 400 m.

Calliostoma columnarium Hedley & May, 1908

Height and width about equal; protoconch large, with about $1\frac{1}{2}$ smooth whorls; early teleoconch whorls spirally ribbed, ribs rounded, not nodulose, becoming wide and flat on later whorls with sculpture only of incised spiral striae, very weak spiral striae present on base; with convex sides; periphery subangulate, becoming rounded near the lip; umbilical area weakly indented; columella curved, rounded, filling behind with callus, basal tooth lacking; aperture interior smooth.

8 mm high. Tas. to SA. Type dredged at 182 m off Cape Pillar, Tas.

Calliostoma comptum (Adams, 1854)

Pl. 10; fig. 9   Tall, conical, apical angle about 46°, with flat sides, base inclined but flat, periphery subangulate; columella slightly arcuate, basally thickened; sculpture of five evenly beaded spiral ribs on the sides, plus a smooth rib close below the periphery forming a double keel, about 12 low smooth cords on the base. Apex blue, teleoconch yellow to fawn, subsutural and peripheral beads sometimes mauve, with a row of brown patches around the periphery.

Holotype: 10 mm high; 7.2 mm wide. NSW to Geographe Bay, WA. Lives on cup-shaped sponges in the sublittoral zone. Synonyms: *purpureocinctum* Hedley, 1894; *adamsi* Brazier, 1895.

Calliostoma hedleyi Pritchard & Gatliff, 1902

Pl. 10; fig. 7   With eight slightly convex whorls and slightly carinate periphery; first two teleoconch whorls latticed, remainder with flat-topped, close-set, faintly nodulose spiral ribs of variable size, some double, the ribs below the suture and at the periphery may be heavier and more nodulose than the others, about 12 ribs on the body whorl above the periphery, 12-15 on the base; outer lip finely crenulate. Yellow-brown with red-brown spots.

15 mm high. Vic. and Tas. to Cape Naturaliste, WA.

Calliostoma legrandi (Tenison Woods, 1876)

Pl. 10; fig. 8   Thin, height usually a little greater than width, sides flat, periphery angulate; sculpture of regular spiral ribs, 9-11 on the sides of the whorls and 14 on the base, often with spiral threads in the interspaces. Flesh-coloured or yellow.

13 mm high. NSW and Tas. to south coast of WA. The NSW form was given subspecies status as *C. legrandi tentabundas* by Iredale, 1925.

Calliostoma retiarius Hedley & May, 1908

Tall and flat-sided; suture channelled; base descending, almost straight, with sharply angled periphery; aperture trapezoidal, sharply cornered at the periphery; umbilical area indented but filled, columellar basal tooth weak to lacking; strong sculpture of five spiral ribs between the periphery and suture, made nodulose by broad oblique-axial folds, about 10 spiral cords on the base, plus 2-3 spiral rows of nodules in the umbilical area.

7 mm high. Tas. to SA. Dredged 170-180 m.

Calliostoma zietzi Verco, 1905

Tall, conical, with seven rounded teleoconch whorls, the first two rounded and mamillate; suture moderately deep, slightly overhung by the peripheral lirae; periphery subcarinate to rounded; sculpture of close-set spiral cords and weak oblique threads in the interspaces, with six cords above the periphery plus a strong peripheral rib forming a keel, cords narrower than the interspaces, base with eight flat ribs which are wider than the interspaces; basal columellar nodule obsolete. Horn-coloured, with a faint peripheral white band.

8 mm high; 5 mm wide. Westernport, Vic. to Hopetoun, WA. Depth range to 400 m. The taller shell and non-granulose sculpture distinguish this species from *C. comptum* and *C. allporti*. The illustration is drawn from the holotype.

▲ SUBGENUS SALSIPOTENS Iredale, 1924
Calliostoma armillatus Wood

Pyramidal, sides and base flat; periphery angulate, with a double, beaded ridge forming a distinct rim; outer lip sharp, usually lirate or crenulate within; columellar lip a little thickened and with a small basal or terminal nodule.

Australia. The prominent keeled peripheral rim is the main character distinguishing this group from *Fautor*.

Calliostoma armillatum (Wood, 1828)

Pl. 10; fig. 21 a-b   Height only a trifle greater than width; sculpture of granulose spiral cords, seven on the penultimate whorl and upper part of the body whorl, 12-14 on the base, the central ones larger, divided peripheral ridge moderately developed and weakly granulose, interstitial spiral threads sometimes present. Apex blue, teleoconch flesh-coloured, with light fawn maculations, peripheral rim alternately white and fawn-spotted.

3.4 cm high; 3.3 cm wide. NSW and Tas. to southern WA. Synonym: *meyeri* Philippi, 1848.

Calliostoma rubiginosum (Valenciennes, 1846)

Pl. 10; fig. 15   Height greater than width; spire outline flat-sided but individual whorls slightly concave; peripheral keel only weakly divided but forming a prominent rim bearing heavy transverse nodules; sides with five widely separate spiral ribs below the suture bearing even-sized rounded nodules, base with about nine nodulose ribs; umbilical area impressed so that the base is concave; columella nearly straight, ending at a terminal nodule, meeting the outer lip at a distinct corner. Base red-brown, sides cream with broad axial rays of brown; at least the first, third and fifth ribs on the sides are mauve.

3 cm high. Western Vic. and northern Tas. to Fremantle, WA. This species is infrequently collected. Synonyms: *australis* Broderip, 1835 (not Lamarck, 1822); *nobilis* Philippi, 1849; *broderipi* Philippi, 1855; *splendidus* "Philippi" Reeve, 1863.

Calliostoma similarae (Reeve, 1863)

Pl. 10; fig. 16 a-b   Like *C. rubiginosa* but with the peripheral rim less well developed and there are 7-8 nodulose spiral ribs on the sides and 8-13 on the base. Also with conspicuous oblique striae in the interspaces on the base. Fawn, with large, slightly darker blotches on the sides, regular mauve or rose spots on the periphery and a central spiral line of alternating mauve or rose and white dashes on the base.

3.5 cm high. Central Indo-West Pacific; Abrolhos, WA to north Qld. Ponder (1978) ranked this as a subspecies of the southern *C. rubiginosum* with *C. excellens* Thiele, 1930 as a synonym.

▲ SUBGENUS LAETIFAUTOR Iredale, 1929
Calliostoma trepidum Hedley

Thin-shelled and glossy externally; whorls and periphery rounded or angulate; sculpture of tuberculate, nodulose or spinose spiral ribs; columella broadening slightly at the base and bearing a low basal nodular swelling; outer lip sharp, serrate, aperture internally furrowed by the external sculpture.

Australia. Synonym: *Spicator* Cotton & Godfrey, 1935 (*spinulosum* Tate).

Calliostoma cf. rubropunctatum (Adams, 1851)

Turbinate, apical angle about 58°; teleoconch with six convex whorls, suture impressed; periphery rounded; columella slightly inclined, meeting the outer lip with a continuous curve, i.e. without any angle or terminal corner; outer lip rounded; sculpture of strongly tuberculate spiral ribs, six on the penultimate whorl and 15 on the body whorl and base, the three ribs around whorl centres are the strongest and bear almost spinose tubercles, basal ribs nodulose; axial cords making cancellate sculpture on the early whorls. Fawn, tubercles pale, with brown spots in the interspaces.

8 mm high; 7.3 mm wide. Onslow, WA to north Qld. Lives in the sublittoral zone.

Calliostoma spinulosum Tate, 1893

Conical, peripherally keeled, base flat; apical angle about 50°; suture impressed; spire whorls with three spiral ribs, the peripheral one prominent and spinose, the others nodulose; the body whorl has, in addition, thin spiral cords in the interspaces and a fourth nodulose rib just below the periphery forming a double keel; oblique ridges in the interspaces connect the spines and render the sculpture cancellate; base with weakly nodulose spiral cords. Fawn, with small red spots in the interspaces.

7 mm high; 5.6 mm wide. SA to Fremantle, WA. Dredged to 250 m. This is the type species of *Spicator* Cotton and Godfrey, 1935 but the diagnostic characters are the same as those of *Laetifautor*.

Calliostoma trepidum Hedley, 1907

Turbinate, whorls and periphery rounded, apical angle about 68°, suture channelled; protoconch of 1½ smooth whorls, teleoconch of 3½ whorls; columella slightly arcuate, meeting the outer lip at a slight angle; body whorl with six spiral rows of pointed tubercles which align also in oblique rows, eight beaded spiral ribs on the base. Pale purple apex, teleoconch pearl-grey to fawn with orange spots.

Holotype: 4.5 mm high; 3.75 mm wide. North West Shelf, WA to Mast Head I., Qld. Drawn from the holotype. *C. rubropunctatum* is much taller and has an impressed not channelled suture; *C. spinulosum* is more spinose and has a keeled periphery. Synonym: *C. deceptum* Smith, 1899.

▲ SUBGENUS BENTHASTELENA Iredale, 1936
B. katherina Iredale

With the characters of the type species.

Eastern Australia. Monotypic.

Calliostoma katherina (Iredale, 1936)

Height and width about equal; protoconch of one smooth whorl; with seven teleoconch whorls, sculptured with two primary keels at the periphery, the upper one bearing triangular spinose nodules and the lower one plain but with a finely crenulated secondary keel between, plus additional crenulate and plain spiral ribs on the upper slopes of the whorls, base with flattened spiral cords, the four surrounding the umbilical area crenulate; columella curved, forming an angular tip anteriorly where it meets the outer lip; inner lip reflected over and closing the umbilicus. Upper surface deep brownish fawn, base paler.

1.2 cm high. Dredged off Sydney, NSW in 206 m. The illustration is drawn from the holotype.

▲ SUBGENUS SINUTOR Cotton & Godfrey, 1935
Ziziphinus incertus Reeve

Periphery not carinate; sinistral.

Southern Australia. Monotypic. Sometimes treated as a full genus.

Calliostoma incertum (Reeve, 1863)

Pl. 10; fig. 11   Pyramidal; height and width about equal; suture inclined; whorls convex, periphery rounded; columella smooth, thickened, curved, meets the outer lip in a continuous curve without any nodule or corner; sculpture of numerous smooth spiral cords separated by deep narrow grooves. Uniformly flesh-coloured.

2 cm high. Vic., Tas. to southern WA. This sinistral shell cannot be mistaken for any other. That normally dextral genera may give rise to a sinistral species like this one is one of the oddities of the evolutionary process.

▲ GENUS ASTELE Swainson, 1855
Trochus subcarinatus Swainson

Like *Calliostoma* in form but usually umbilicate.

Cosmopolitan. Several subgenera are recognised.

▲ SUBGENUS ASTELE s.s.

Base flattened; umbilicus bounded by a beaded cord; sculptured with beaded spiral ribs.

Cosmopolitan.

Astele bularra Garrard, 1961

Pl. 10; fig. 17   Periphery angulate, base slightly convex; protoconch very small, 1 whorl only, inclined; nine teleoconch whorls; with cancellate sculpture on the early whorls, spiral sculpture only on the later whorls consisting of four primary rows of beaded cords plus a more prominent double row at the periphery, weak secondary rows between them, and fine threads and axial striae in the interspaces; base with nine primary beaded cords and secondaries between; columella slightly curved; umbilicus small, half covered by a reflected nacreous callus, bordered by a well-defined spiral rib. Early whorls cream, later whorls bronze; beaded ribs white, peripheral rib with regular brown spots.

2.8 cm high; 2.8 cm wide. Central Qld to southern NSW in 80-200 m. Also occurs in Japan.

Astele ciliare (Menke, 1843)

Pl. 10; fig. 18 a-b   Thin, width equal to or slightly greater than height; periphery sharply angulate; smooth, shiny, sculptured with punctate spiral striae, early whorls weakly cancellate, later whorls with wide, low and flat spiral lirae separated by incised lines, short and weak axial folds above and below the suture, and a sharp marginal keel which forms a spiral rib above the suture; aperture weakly lirate. Pink-fawn with patches of darker red-brown around the basal margin and suture and spiral lines of alternating brown and white dashes on the base and sometimes on the sides.

3.5 cm wide. Esperance to Fremantle, WA. Uncommon. Lives on sand among seagrasses. Synonym: *calliope* Cotton & Godfrey, 1938.

Astele monile (Reeve, 1863)

Pl. 10; fig. 14   Tall, whorls slightly concave; periphery angulate, base convex; early spire whorls finely granulose, later whorls glossy-smooth but with a few fine spiral threads on the sides and striae on the base; peripheral rim thick but smooth, forming a prominent spiral rib around the spire whorls. Cream with mauve spots on the suture rib and around the base.

2.5 cm high. Moderately common. Monte Bello Is, WA to eastern Qld.

Astele speciosum (Adams, 1855)

Pl. 10; fig. 19   Height and width about equal; sometimes shouldered at the suture and with the upper parts of the whorls slightly concave; periphery rounded to subangulate; with a strong, nodulose subsutural spiral rib, followed by five widely spaced, weakly nodulose cords below, and a complex of four or five weak peripheral cords; about eight weakly nodulose basal ribs on the base; columella curved, with a small terminal swelling; umbilical region impressed, bounded by tan callus. Fawn, with small reddish or tan spots and blotches.

2.5 cm high. Central Qld to northern NSW. The tan umbilical callus is characteristic. Lives among seagrasses.

Astele subcarinatum (Swainson, 1854)

Pl. 10; fig. 20 a-b   Spire slightly concave in outline, periphery angulate; glossy; about eight smooth spiral cords encircling the upper whorls and sides of the body whorl, plus about 12 on the base, central cords on the base relatively wide and slightly transversely rugose; umbilicus bordered by a weakly nodulose funicle. Fawn, with brown blotches and stronger brown patches on the peripheral rim.

4 cm wide. NSW to Rottnest I., WA. Dredged to depths of 500 m.

▲ SUBGENUS ASTELENA Iredale, 1924
Ziziphinus scitulus Adams

Small, whorls rounded, base convex, umbilicus deep, narrow, funnel-shaped, bordered by an angular rim; columella arcuate, with a slightly projecting corner at the base where it meets the outer lip.

Southern Australian Region. The two named species live in flask-shaped sponges. Whether they actually feed on the sponge tissue is unknown, although this seems likely.

Astele multigranum (Dunker, 1871)

Pl. 10; fig. 13 a-b   Like *A. scitulum* but with flatter sides and a prominently keeled and angular periphery. The sculpture consists of nearly even-sized granulose spiral cords throughout. Fawn or pink.

1.4 cm high; 1.3 cm wide. SA to Geraldton, WA. The illustrated specimens were found alive in the cavity of a brown, flask-shaped sponge at 20 m off Dunsborough, Geographe Bay, WA.

Astele scitulum (Adams, 1854)

Pl. 10; fig. 12   Spire tall, narrow, with concave outlines; whorls convex, body whorl inflated and subangulate at the periphery; early teleoconch whorls densely spirally granulate, later whorls with numerous weakly granulose spiral cords and weak oblique striae in the interspaces. Reddish brown, spiral cords with red and white dashes alternating, prominently so on the peripheral cords and those closest to the umbilicus.

1.2 cm high; 1.1 cm wide. NSW to SA. The illustrated specimen was found alive in a brown cup-shaped sponge at 20 m off North Head, Sydney.

▲ SUBGENUS **CALLISTELE** Cotton & Godfrey, 1935
Astele calliston Verco

With the characters of the type species.

Southern Australia. Monotypic. Although the original description of the type species was detailed, the diagnostic characters which purport to distinguish the subgenus are unclear.

Astele calliston Verco, 1905

Conical, high-spired, with seven flat-sided teleoconch whorls, flattened base and acutely angular periphery; umbilicus narrow, its weak marginal keel not nodulose; aperture subquadrate; columella straight, nearly vertical, with obsolete basal corner, meeting the parietal wall at a high angle; outer lip thin, simple; sculpture of fine spiral striae and a pronounced, slightly nodulose peripheral keel. Sides greenish cream with broad, expanding red axial bands, crossed by interrupted white spiral lines, keel with red spots, base rose with darker red-white spiral lines and a greenish zone around the umbilicus.

Holotype: 11.75 mm high; 9.75 mm wide. SA. This most beautiful little species is rare in collections and the extent of its distribution is unknown. Drawn from the holotype which Verco dredged in 20 fathoms in Spencer Gulf.

▲ SUBGENUS **CORALASTELE** (Iredale, 1930)
C. allanae Iredale

With the characters of the type species.

Eastern Australia. Monotypic.

Astele allanae (Iredale, 1930)

Pl. 10; fig. 10 a-b Turbinate, base flat but inclined, whorls shouldered, suture impressed; columella near vertical, slightly thickened and reflected at its abrupt end; umbilicus very small, bordered by a narrow, smooth rib; spire whorls sculptured with three spiral ribs bearing pointed nodules, body whorl with an additional peripheral rib, the second, third and fourth rib on the body whorl are about equal and form three keels, about eight spiral ribs on the base; with secondary spiral cords and fine axial lamellae in the interspaces. Cream, with red brown spots on the ribs. A thin tan periostracum covers the shell in life.

1.3 cm high; 1.25 cm wide. Southern Qld. The type specimens came from the Capricorn Group but there is little information on the species distribution or habitat.

SUBFAMILY **EUCYCLINAE**

Three tribes are recognised in this subfamily: two living and one extinct. Only the living tribe Chilodontini is recorded from the coastal marine fauna of Australia. The nominate tribe, the Eucyclini, flourished in the Mesozoic but became extinct early in the Tertiary. The subfamily is characterised by a peculiar radula which has hooded rachidian teeth and laterals with shafts that expand basally into complex interlocking pegs and flanges; the lateromarginal plates are rectangular and the outermost marginal teeth are expanded into broad mitten-shaped elements.

TRIBE **CHILODONTINI**

Shell relatively thick, columella and parietal wall usually with denticles, outer lip usually with denticles and inner lirae. The tribe originated in the Middle Jurassic and is one of the oldest groups in the Trochidae. Most of the living species live in the shallows of tropical and warm temperate seas.

▲ GENUS **DANILIA** Brusina, 1865
Monodonta tinei Calcara

Small, thin-shelled, turbinate, spire tall, whorls convex, suture impressed; with strong cancellate sculpture; columella with a prominent fold, deeply notched below; outer lip ridged within and varixed behind.

Comopolitan. The type species is from the Mediterranean. There is one described Australian species. Synonyms: *Olivia* Cantraine, 1835 (non Bertholoni, 1810); *Craspedotus* Philippi, 1847 (non Schonherr, 1844).

Danilia telebathia (Hedley, 1911)

Protoconch whorls convex, axially lamellate; teleoconch whorls with delicate oblique-axial lamellae and spiral ribs, the ribs bearing short open-sided spines where the lamellae cross although the spines are obsolete on the body whorl, three ribs around the centre of the whorls plus one at the periphery and three on the base; umbilicus lacking; outer lip thin, crenulate; parietal wall thinly calloused, the callus becoming erect near the base and continuous with the outer lip. Cream, with oblique-axial light brown flames below the suture and on the base.

Holotype: 11.3 mm high; 8.4 mm wide. Tas. and SA. Dredged to 183 m. The type was dredged in 80 fathoms east of Schouten I., Tas. but Cotton (1959, p.181) recorded the species off Cape Wiles, SA in similar depths. For an illustration see Cotton (1959, p. 180, fig. 109).

▲ GENUS **EUCHELUS** Philippi, 1847

Trochus quadricarinatus Holten (= *asper* Gmelin)

Turbinate, whorls rounded, sculpture of strong spiral ribs and axial lamellae; aperture round, smooth or lirate within, outer lip usually dentate; columella simple; operculum few-whorled.

Cosmopolitan. See remarks on *Herpetopoma*.

▲ SUBGENUS **EUCHELUS** s.s.

Juveniles with a narrow umbilicus, adults without.

Cosmopolitan. There is only one Australian species.

Euchelus foveolatus (Adams, 1853)

Juveniles narrowly umbilicate, adults erforate; sculpture strongly cancellate, ody whorl with four prominent spiral ribs, the second and third sometimes very pronounced forming keels, plus three ribs on the base, thick lamellae crossing the interspaces; suture channelled; outer lip often dentate along its inner edge. White.

8 mm high; 7 mm wide. North Qld. The illustrated specimen is one of a series collected by Hedley in the Torres St. in 1907.

▲ SUBGENUS **VACEUCHELUS** Iredale, 1929

Euchelus angulatus Pease

Umbilicus narrowly open and bordered by a spiral rib; columella simple, lacking a basal tooth; aperture smooth within.

Western Pacific and Southern Australian Region. Lack of a basal tooth on the columella is said to distinguish this group from *Herpetopoma*, and the open umbilicus from *Euchelus* s.s.

Euchelus ampullus Tate, 1893

Two smooth nuclear whorls and three rounded teleoconch whorls; sculpture of prominent, weakly nodulose primary spiral ribs, smaller intermediary secondary spirals, and strong axial lamellae in the interspaces forming a fenestrate surface; three primary ribs on the spire whorls, four on the body whorl, three on the base plus a smooth spiral cingulus bordering the umbilicus. Cream with a few prominent rose spots on the primary ribs.

11 mm high; 11 mm wide. Caloundra Qld to Fremantle, WA. The type locality given by Tate was Cambridge Gulf, WA but that seems likely to have been an error. This species resembles *H. aspersa* at first glance but the deep umbilicus, lack of a basal tooth and distinctive sculpture easily distinguish it. Some authors have placed it in *Tallorbis* but it is not thin-shelled like the type of that genus.

Euchelus profundior May, 1915

With two weakly nodulose spiral ribs on the spire whorls, four on the body whorl, the lower two most prominent, three on the base plus a smooth spiral cingulus bordering the umbilicus; strong axial crossbars in the interspaces; secondary ribs lacking. Yellow-white, brown spots on the ribs.

4 mm high; 3.2 mm wide. Bass St. to Esperance, WA. Recorded from depths between 150 and 250 m. The sculpture and taller spire distinguish this species from *E. ampullus*.

▲ GENUS **GRANATA** Cotton, 1957

Stomatella imbricata Lamarck

With the characters of the type species.

Southern Australia. Monotypic. There has been confusion about the use of this name vis à vis *Stomatella* Lamarck. Cotton (1957) concluded that Anton (1839) designated *S. auricula* Lamarck as the type species of *Stomatella*, not *S. imbricata* Lamarck, and introduced *Granata* as a new generic name for the latter.

Granata imbricata (Lamarck, 1816)

Pl. 7; fig. 9 a-b   Ear-shaped, spire small and depressed, with four rounded whorls, body whorl much enlarged, with a widely open aperture; imperforate; columella narrow, curved, simple; outer lip thin, margin crenulate; surface sculpture of numerous, crowded, scaly radial cords; operculum multi-spiral with a central nucleus, too small to cover the entire aperture. Exterior grey or cream, interior nacreous.

18 mm high; 34 mm wide. NSW to Geraldton, WA. Under stones; intertidal and shallow sublittoral. A very conspicuous species in the southern Australian fauna.

▲ GENUS **HERPETOPOMA** Pilsbry, 1890

Euchelus scabriuscula Adams & Angas

Umbilicus open or closed; columella toothed at the base; aperture lirate within; operculum many-whorled.

Australia. The presence of a basal columellar tooth is the diagnostic character distinguishing the genus from *Euchelus*.

Herpetopoma annectans (Tate, 1893)

With four rounded teleoconch whorls, suture canaliculate; sculpture of five beaded spiral ribs on the penultimate whorl, 10 on the body whorl, the last being simple; interspaces with thick axial lamellae giving a fenestrate appearance; umbilicus lacking. White with rose spots in obliquely axial rows.

5 mm high, 5 mm wide. SA to Cape Naturaliste, WA. Rocky shores. Depth range unknown.

Herpetopoma aspersa (Philippi, 1846)

Pl. 10; fig. 4   Height greater than width; whorls with approximately equal spiral rows of elevated nodules (about nine rows on the penultimate whorl) and axial cords in the interspaces; umbilicus closed; columella narrow, gently curved, with a small but conspicuous basal tooth. Cream with a few red-brown spots.

1.5 cm high. NSW to Shark Bay, WA. This is a very common species under stones in the intertidal and shallow sublittoral zones.

Herpetopoma atrata (Gmelin, 1791)

Pl. 10; fig. 3   Sculpture of many close-set, transversely beaded spiral ribs rendered scabrous by fine axial cords; narrowly umbilicate. Fawn, violet-brown to almost black, sometimes with pale spots on the ribs.

15 mm high; 13 mm wide. Indo-West Pacific; Shark Bay, WA to southern Qld. Common under stones in the intertidal and shallow sublittoral zones. The shell is often overgrown with sponges. It is the largest species of the genus.

Herpetopoma fenestrata (Tate, 1893)

With four teleoconch whorls; suture canaliculate; sculpture of three spiral ribs on the spire whorls, 13 on the body whorl and base, the two peripheral ribs being most prominent; strong axial lamellae in the interspaces forming fenestrate sculpture, and rounded nodules at the intersections.

Holotype: 4 mm high; 3.3 mm wide. SA to Fremantle, WA. Rocky shores to at least 300 m.

Herpetopoma instricta (Gould, 1849)

Thick-shelled and high-spired, with five rather flat-sided teleoconch whorls which are subangulate above the suture; narrowly umbilicate; 5-6 strong spiral ribs on the spire whorls, 10-12 on the body whorl and base, the fifth rib prominent and forms the keel; interspaces with axial lamellae which cross the ribs to produce narrow transverse ridges; outer lip wide and strongly dentate; columellar basal tooth prominent. Cream with wavy brown axial stripes and brown spots at the suture.

10 mm high. Indo-West Pacific; recorded from Carnarvon and Scott Reef, WA. Intertidal coral reefs to 140 m.

Herpetopoma pumilio (Tate, 1893)

With three rounded, teleoconch whorls and a canaliculate suture; sculpture of two thick, heavily nodulose spiral ribs on the spire whorls, five on the body whorl and base, with thick axial lamellae in the interspaces ascending onto the ribs and forming rounded nodules at the intersections; basal tooth small, sharp; umbilicus closed. White with ragged oblique tan streaks. Resembles *H. fenestrata* but with a lower spire and fewer basal ribs.

3 mm high; 3.5 mm width. Bass St. to Fremantle, WA. Rocky shores. Depth range unknown. The very heavy ribs and nodules easily distinguish this species.

Herpetopoma rubra (Adams, 1853)

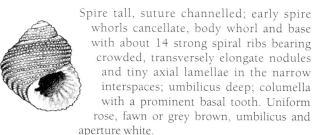

Spire tall, suture channelled; early spire whorls cancellate, body whorl and base with about 14 strong spiral ribs bearing crowded, transversely elongate nodules and tiny axial lamellae in the narrow interspaces; umbilicus deep; columella with a prominent basal tooth. Uniform rose, fawn or grey brown, umbilicus and aperture white.

9 mm high; 7 mm wide. North West Cape, WA to southern Qld. A very common shallow water tropical species. The colourful shells are unusual for the genus and render them quite conspicuous in beach drift lines.

Herpetopoma scabriusculus (Angas, 1867)

Minutely umbilicate; with four teleoconch whorls and a canaliculate suture; four unequal spiral ribs on the spire whorls, the third one carinate, about seven on the body whorl and base, plus a closely nodulose rib bordering the umbilicus; with axial cords in the interspaces and nodules at the intersections; columella deeply notched and basally toothed; inner margin of the outer lip strongly dentate. Reddish.

5.5 mm high, 5.5 mm wide. Sydney, NSW to SA. Rocky shores to the outer edge of the continental shelf. Resembles *H. vixumbilicata* which lives in the same area but distinguished by finer spiral sculpture and more conspicuous axial cords in the interspaces. It also has less convex whorls. Synonym: *tasmanicus* Tenison Woods, 1876.

Herpetopoma vixumbilicata (Tate, 1893)

With three rounded teleoconch whorls and canaliculate suture; sculpture of four subequal, finely beaded spiral ribs on the spire whorls, 13 on the body whorl and base plus one weakly nodulose rib bordering an open umbilicus; axial sculpture subdued; basal tooth thick and nodulose. White with oblique rows of rose spots.

Holotype 6 mm high; 5.8 mm wide; Shark Bay specimen 9 mm high, 8 mm wide. Western Vic. to Shark Bay, WA. To depths of at least 600 m. Found on rocky shores; in Shark Bay this species is common among dead shells on sand flats at extreme low tide. The paler colour, heavier spiral sculpture and more subdued axial sculpture serve to distinguish it from *H. scabriuscula*.

▲ GENUS **HYBOCHELUS** Pilsbry, 1890
Stomatella cancellata Krauss

Depressed, thin-shelled, whorls rounded, body whorl inflated, spire low; aperture large, round, outer lip thin, finely serrate; narrowly umbilicate; columella curved, narrow, simple, without a basal tooth or notch; sculpture cancellate.

Indo-West Pacific. Apparently close to *Euchelus* and *Tallorbis* but characterised by the very thin shell and depressed form. The shells of *H. cancellatus* and *T. roseolus* are very much alike except for the form of the latter. Anatomical studies are needed to confirm their generic distinction. *Tallorbis* is the oldest name and would have priority if the two types prove to be congeneric.

Hybochelus cancellatus (Krauss, 1848)

Pl. 10; fig. 1   With an elevated spiral rib bordering a narrow umbilicus; sculpture strongly cancellate, especially on the early spire whorls, with three strong nodulose spiral ribs on the penultimate whorl, seven on the body whorl and base, secondary ribs between, and fine cords between those, plus strong crossbars. White or fawn, with rectangular reddish brown spots on the ribs.

16 mm high; 20 mm wide. Indo-West Pacific; North West Cape, WA to Qld. Shell specimens are occasionally found in beach drift but information on habitat is lacking. The animals probably live under stones.

▲ GENUS **SYNAPTOCOCHLEA** Pilsbry, 1890
Stomatella montrouzieri (= *stellata* Souverbie)

Moderately depressed, ear-shaped, elongate, aperture longer than wide; protoconch minute, spire slightly elevated, subterminal, with convex whorls; porcelaineous; operculate.

Indo-West Pacific and Caribbean. The shells resemble *Stomatella* but the spire is more elevated, the shell less

depressed, and there is a small horny operculum. Because of the operculum, Hickman & McLean (1990) removed this genus from the Stomatellinae where most previous authors placed it. There are two Indo-West Pacific species present in northern Australia.

Synaptocochlea concinna (Gould, 1845)

Obliquely oval, last whorl slightly flaring; shiny, sculptured with finely granulose spiral threads and minute radial striae; the outer lip forms a small calloused projection where it meets the parietal wall. Colour variable, red, pink or black, usually mottled or banded.

8 mm long; 5.2 mm wide. North West Cape, WA to southern Qld. Common in the intertidal zone of coral reefs. Said to be indistinguishable from the Western Atlantic species *S. picta* Orbigny, 1842. *S. stellata* Souverbie, 1863, also recorded from north Qld, is distinguished by spirally striate sculpture.

▲ GENUS **TALLORBIS** G. & H. Nevill, 1869
T. roseolus G. & H. Nevill

Thin-shelled, turbinate and globose with very convex whorls and impressed suture, aperture round and large; columella arched, smooth except for two wide, low, spiralling nodules near the base; umbilicus closed, umbilical area bordered by a thick, spiral rib; outer lip margin sharp, obsoletely crenulate; sculpture of strong spiral ribs and fine axial lamellae.

Indo-West Pacific. Although this genus is sometimes placed in the synonymy of *Euchelus* the shell of the type species is thin and the outer lip not dentate. This author prefers to retain *Tallorbis* as a valid generic name until there is more information on the anatomy. See also remarks on *Hybochelus*.

Tallorbis roseolus G. & H. Nevill, 1869

Pl. 10; fig. 2   With the characters of the genus; early spire whorls cancellate, with strong axial lamellae between the three weakly nodulose spiral ribs; body whorl and base with about nine strong spiral ribs, plus secondary spiral cords and fine axial striae in the interspaces. Cream with prominent rose spots on the primary ribs.

13 mm high; 13.4 mm wide. Indo-West Pacific; Qld. The figure is of a Hedley specimen from Mast Head Reef, Qld which he listed (1907, *Proceedings Linnean Society NSW* **32** (3): 479) as *Euchelus lamberti* Sowerby, 1875.

▲ GENUS **TURCICA** Adams, 1855
Turcica monilifera Adams

Conical, high-spired, sides flattened, periphery angulate; suture channelled; sculpture nodulose or cancellate; columellar fold near vertical, bearing one or two nodular teeth; umbilicus closed.

Cosmopolitan.

Turcica maculata (Brazier, 1877)

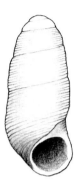

Whorls with four nodulose spiral ribs on the sides, the first prominent, the second and third smaller, the fourth very prominent forming a conspicuous peripheral keel, base with eight spiral ribs of which the one below the peripheral keel is the largest; ribs joined by oblique lamellae making the sculpture cancellate; columellar teeth obsolete. Fawn, with brown spots.

1 cm high; 9 mm wide. Bathurst I., NT to Qld. Dredged to depths of at least 100 m. *Turcica monilifera* Adams, 1855 is also recorded from Qld but specimens were not available for illustration.70

SUBFAMILY **HALISTYLINAE**

Shell small but thick, pupiform to globose, lacking interior nacre; umbilicate or non-umbilicate; aperture circular, peristome complete, its inner margin fused to the parietal wall; outer lip thickened within, tapering to a sharp edge; smooth or spirally sculptured. Gill monopectinate, with enlarged bursicles, lacking a free tip; eyes on the dorsal surface of the eye stalks.

There are only three genera in this small group, two of them endemic to Australia. The two Australian genera are represented in south-eastern fossil beds as early as the Upper Eocene.

▲ GENUS **BOTELLOIDES** Strand, 1928
Onoba bassiana Hedley

Minute, pupiform shells, lacking an inner nacreous layer; aperture circular; operculum multispiral, circular.

Circum-Australian. This is a replacement name for *Botellus* Iredale, 1924, not Moniez, 1887. The genus had been placed in the Rissoidae until Ponder (1985, *Records Australian Museum, Suppl.* 4: 99) showed its affinities with trochid genus *Halistylus* Dall. It is an endemic Australian group with five named living species. The

three species in addition to those illustrated here are: *sulcatus* Cotton, 1944 - Albany, WA to the Arafura Sea; *chrysalidus* Chapman & Gabriel, 1914 - Great Australian Bight (and with a Miocene fossil population in Vic.); *ludbrookae recens* Ponder, 1985 - SA to Rottnest I., WA.

Botelloides bassianus (Hedley, 1911)

Glossy, columnar, blunt at both ends; with 5-6 whorls, the first three short, turbinate, the last two or three broad, elongate and account for two-thirds of the shell length, wound obliquely; early whorls smooth, the last two or three with numerous fine spiral threads. Early whorls brown, last two whorls white.

4 mm. Central NSW to Fremantle, WA. The illustration is drawn from the holotype of *B. borda* Cotton, 1944, which is a synonym. Cotton's species was based on the fact that SA shells have an extra whorl but series show intergrades.

Botelloides glomerosus (Hedley, 1907)

Similar to *B. bassianus* but smaller, shorter and with fewer and stronger spiral striae.

2.9 mm. Western and northern Australia from Albany, WA to Noosa, Qld. The type locality is the Capricorn Group, Qld, where Hedley found it to be common. Hedley noted that his "novelty" is closely related to *mercurialis* Watson, 1886.

▲ GENUS **CHARISMA** Hedley, 1915
C. compacta Hedley

Small, turbinate, spire moderately high, with 3-4 rounded, spirally sculptured whorls and a channelled suture; umbilicus deep with an internal rib; aperture round, outer lip simple, lacking a varix; operculum horny, concave, multispiral with a frilled spiral lamella.

Eastern and southern Australia. This genus has been variously classified in the Liotiidae and Turbinidae but is placed here by Hickman & McLean (1990).

▲ SUBGENUS **CHARISMA** s.s

Umbilicus with an angular rim.

Eastern and southern Australia. In addition to the type species the following appear referable to this group: *C. arenacea* (Pritchard & Gatliff, 1901); *C. carinata* (Verco, 1907); *C. josephi* (Tenison Woods, 1877); *C. latebrosa* (Hedley, 1907).

Charisma compacta Hedley, 1915

Globose-turbinate, with four rounded, loosely coiled whorls; sculptured with about 20 regular spaced, sharp, spiral grooves; umbilical margin smooth, rounded, aperture side with an obscure funicle.

2 mm wide. NSW. The illustration is copied from the original description (*Proceedings Linnean Society NSW* **39**: Pl. 81; fig. 49).

▲ SUBGENUS **CAVOSTELLA** Laseron, 1954
C. radians Laseron

With characters of the type species.

NSW. Monotypic.

Charisma radians (Laseron, 1954)

Depressed-turbinate, thick translucent; protoconch minute, teleoconch with three whorls, body whorl large, rounded, sculptured with numerous, crowded spiral ridges persisting to the edge of the umbilicus; aperture round, peristome complete, inner margin reflected over the umbilicus; base with radial furrows indenting the umbilical margin.

2 mm wide. NSW; intertidal. For illustrations see original description (*Australian Zoologist* **12**: 14, figs. 30, 30 a).

▲ SUBGENUS **CAVOTERA** Laseron, 1954
C. simplex Laseron

With characters of the type species.

NSW. Monotypic. The relationship with *Cavostella* needs to be questioned.

Charisma simplex (Laseron, 1954)

Description as for *C. radians* except that the shell is half the size, the umbilicus is narrower and lacks the marginal indentations caused by radial furrows on the base, and the peripheral ridges are larger than the others, giving slight angularity to the body whorl.

1 mm. NSW; intertidal. For illustrations see the original description (*Australian Zoologist*, **12**: figs. 31, 31 a).

SUBFAMILY **SOLARIELLINAE**

Shells usually small to minute, depressed, with rounded whorls and impressed suture; cephalic lappets lacking, cephalic tentacles unusually long; oral surface of snout laterally expanded and covered with prominent papillae; anterior end of foot with prominent lateral horns; radula short, with about 10 marginal teeth per half row.

A cosmopolitan group, most common in cold to cool waters rather than the tropics. The animals live in soft substrates and feed on detrital particles on the sea floor.

▲ GENUS **SPECTAMEN** Iredale, 1924
Trochus philippensis Watson

Shoulders and periphery subcarinate, with a horizontal zone below the suture; beaded rib bordering the umbilicus very weak or lacking.

South-eastern Australian region and New Zealand. The four named Australian species of this group are very much alike. *Spectamen* is sometimes treated as a subgenus of the cosmopolitan genus *Solariella* Wood, 1842 which is distinguished by a beaded spiral rib bordering the umbilicus. *Solariella* s.s. is not represented in the Australian fauna. Synonym: *Zeminolia* Finlay, 1927.

Spectamen bellulus (Angas, 1869)

Height slightly less than width; flat subsutural zone pronounced, smooth; whorls slightly concave in the upper half, shiny and smooth except for several widely spaced, weakly incised spiral lines; periphery subangulate. Cream with wide, wavy, rose axial bands on the sides of the whorls extending onto the subsutural shelf but not the base.

1 cm wide. Central NSW. Taller than *S. philippensis*, and the sinuous profile of the whorls and more angulate periphery are distinctive.

Spectamen epithecus Iredale, 1929

Width greater than height; flat subsutural zone smooth except for six minute spiral striae; sides of the whorls with about 15 spiral striae; umbilicus rather narrow. Cream, with irregular axial red-brown flames on the sides and extending onto the base.

1 cm wide. NSW. Distinguished from *S. philippensis* by the sculpture and relatively narrow umbilicus. The illustration is drawn from the holotype.

Spectamen marsus Cotton & Godfrey, 1938

Sculpture of regular spiral striations; umbilicus with 2-3 beaded spiral cords within, distant to the weakly nodulose marginal rim. Yellow to white with strong, oblique rosy streaks.

5.2 mm high; 7 mm wide. SA and south coast of WA. Habitat unknown; depth range to at least 90 m.

Spectamen philippensis Watson, 1881

Width greater than height; flat subsutural zone with radial folds; body whorl with numerous, crowded minute spiral striae, periphery subangulate to rounded. Cream to fawn, with widely spaced, narrow, irregular rose axial stripes which extend onto the base though interrupted at the periphery.

1 cm high; 8 mm wide. NSW to Bass St. The anatomy has been described by Hickman & McLean (1990: 113).

▲ GENUS **MINOLIA** Adams, 1860

M. punctata Adams

Small, depressed-turbinate; aperture round, outer lip thin, simple; umbilicus wide, open up to the apex, surrounded by a toothed or angular funicle; columella simple, curved; spirally ribbed; periostracum marked with spots and stripes.

Southern Australian Region. The group is very close to *Spectamen*. Synonym: *Minolops* Iredale, 1929 (*pulcherrima emendata* Iredale).

Minolia arata Hedley, 1903

Spire erect, turreted, whorls with a wide, flat, subsutural zone; funicle triangular in section, weakly dentate, transversely nodulose spiral ribs present deep within the umbilicus; subsutural zone with numerous axial threads and about three thin spiral threads, sides of whorl with three or four subequal spiral cords and many spiral threads; about five spiral cords on the base increasing in size toward the funicle. Fawn-grey, with oblique red flames on the sides and base.

7.8 mm high; 9.5 mm wide. NSW. The illustration is drawn from the holotype.

Minolia cincta (Cotton & Godfrey, 1938)

Depressed; spire stepped, with a flat zone crossed by axial bars below the suture; sides of whorls spirally ribbed, base spirally corded. Cream or pale green, with red, interrupted, axial lines.

Holotype: 4.8 mm high; 6.3 mm wide. SA; to 420 m. The illustration is drawn from the holotype.

Minolia gertruda Iredale, 1936

Very like *M. pulcherrima* but spire more erect, spiral ribs less prominent and spiral threads more numerous, funicle more angular and prominent. Fawn or cream, oblique rays of red which become larger and more regular on the base.

Holotype: 7.7 mm high; 9.3 mm wide. NSW. The colouring, sculpture and form of this little shell resemble *M. arata* and it may be only a variant of that species. The description here is based on the type.

Minolia pulcherrima Angas, 1869

Whorls with prominent spiral ribs at the shoulders and centrally, plus another forming a peripheral keel on the body whorl; the upper rib forms a shoulder keel with a flat zone between it and the suture; concave above and below the central rib; weakly striate and corded in the interspaces; walls of the umbilicus finely ribbed and striate, two ribs denticulate. White or rose, ribs with prominent rose spots.

a

b

6 mm high; 9 mm wide. NSW. *Minolia pulcherrima emendata* Iredale, 1924 (fig. (a) drawn from the holotype) seems to be merely a variation in which the five spiral ribs around the body whorl are subequal.

SUBFAMILY **STOMATELLINAE**

Stomatellas

Shells of stomatellas are usually ear-shaped or slipper-shaped, without an umbilicus. The shell's aperture is very large, wide and oblique; coils of the upper whorls are visible. The lining of the aperture is nacreous. An operculum is lacking. The foot has a highly extensible posterior metapodium that cannot be retracted below the shell and has three epipodial tentacles. The dorsal surface of the metapodium is folded and papillate and has a deep anterior shell pouch. Left and right neck lobes have simple margins and may be rolled to form incurrent and excurrent troughs. The radula is similar to that of the Trochinae.

A feature of stomatellas is the practice of autotomy, whereby the rear end of the foot is broken off and left wriggling to catch the attention of an attacker while the remainder of the animal escapes. The animal later regrows the lost part. *Stomatella auricula* is one of the most expert exponents of this self-protection device.

It has been claimed that stomatellas are herbivorous molluscs. Most of the species live in shallow water, usually under stones.

Although this group is usually classified as a distinct family, Hickman & McLean (1990) argued that the radula is similar to that of the Trochinae and that the flattened shell form is convergent, so that subfamily rank within the Trochidae is more appropriate. Flattened shell form also occurs in the Eucylinae (*Granata, Hybochelus*) but those trochids are distinguished from the stomatellas by the presence of an operculum.

There has been no revision of the group for many years and, given the extreme variability of shell form, colour and pattern of most species, it remains an enigma for modern collectors and ecologists. The generic arrangement used here is provisional.

▲ GENUS **MICROTIS** Adams & Adams, 1850
M. tuberculata Adams

Thin, depressed and ovately ear-shaped; spire elevated, shouldered, body whorl wide, inflated; aperture wide, oval; columellar and outer lips thin, continuous, columellar margin visible to the apex; umbilicus lacking; sculptured with spiral ribs including two prominent, tuberculate ridges.

Indo-West Pacific. Sometimes treated as a subgenus of *Stomatia*. Species like *M. rubra* seem intermediate between *Microtis* and *Pseudostomatella*.

Microtis rubra (Deshayes, 1843)

Pl. 11; fig. 5   Spire relatively high, protoconch of $1^{1}/_{2}$ whorls, teleoconch of $2^{1}/_{2}$ whorls; subsutural area sloping, base rounded; shoulder and peripheral keels weak, with spiral sculpture of fine cords of varying size. Cream, with broad, irregular, red radial ray tending to merge on the base.

1.5 cm high; 1.8 cm wide. Indo-West Pacific; North West Cape, WA to north Qld. The taller spire, sloping subsutural ramp and weaker sculpture distinguish this from *M. tuberculata*. Placement in *Microtis* rather than *Pseudostomatella* is tentative.

Microtis tuberculata (Adams, 1850)

Pl. 11; fig. 12 a-b   Apex of about $1^{1}/_{2}$ whorls, small; teleoconch of $2^{1}/_{2}$ whorls, with a wide horizontal to slightly inclined area below the suture, a prominent angular and tuberculate shoulder, and a second but weakly tuberculate ridge just above the periphery; base rounded; sculptured with slightly prickly spiral cords and oblique-axial folds, the latter strongest on the first whorl. Fawn, sometimes cream or rose with expanding radial rays of brown, usually becoming green on the base.

1.4 cm. high; 2 cm. wide. Indo-West Pacific; Geographe Bay, WA to southern Qld. *M. heckeliana* (Crosse, 1871) is similar but much more depressed and angulate at the periphery.

▲ GENUS **PSEUDOSTOMATELLA** Thiele, 1921
Stomatella papyracea Gmelin

Thin-shelled, globose, spire conical, whorls rounded with a regular coil; body whorl inflated; aperture wide, ovate, weakly lirate within; columella narrow, curved, calloused, simple; outer lip thin, sharp-edged, weakly crenulate; with finely beaded spiral sculpture.

Indo-West Pacific. Synonym: *Stomatolina* Iredale, 1937 (*rufescens* Gray).

Pseudostomatella decolorata (Gould, 1848)

Pl. 11; fig. 10 a-b   Protoconch minute, elevated, consisting of about $2^{1}/_{2}$ glossy whorls; teleoconch of $2^{1}/_{2}$ whorls, the first sometimes angulate, sculptured with pronounced spiral cords and minute spiral striae. Green or cream with irregular wide, darker, greenish brown oblique bands, ribs sometimes spotted.

1 cm high; 1.8 cm wide. Indo-West Pacific; North West Cape, WA to southern Qld. Distinguished from *P. papyracea* by the more depressed form and more prominent spiral sculpture. *S. mariei* Crosse, 1871 is probably a synonym.

Pseudostomatella maculata (Quoy & Gaimard, 1834)

Pl. 11; fig. 9 a-b   Depressed and almost ear-shaped, with an elongate aperture; protoconch tiny, consisting of about $1^{1}/_{2}$

whorls; teleoconch of three whorls; sculptured with numerous fine spiral cords and crowded axial riblets. Cream or pale fawn, maculated with red-brown or green, sometimes with tent-shaped markings.

1.5 cm high; 2.5 cm wide. North West Cape, WA to north Qld. The form of this shell approaches that of the southern trochid *Granata* and it has an operculum so that its classification here is questionable.

Pseudostomatella papyracea (Gmelin, 1791)

Pl. 11; fig. 8 a-b Protoconch tiny, consisting of about two glassy whorls; teleoconch of four whorls, the first and second sharply corded, penultimate and body whorls with fine, crowded spiral cords, crossed by minute axial striae and weak oblique cords. Fawn, orange, or greenish, often with yellow crossbars or brown spots on some ribs, and yellowish blotches, sometimes with pronounced spiral bands.

2 cm high; 2.5 cm wide. Indo-West Pacific; Rottnest I., WA to southern Qld. The largest and tallest species of the group.

Pseudostomatella sp.

Pl. 11; fig. 11 a-b Very like *P. papyracea* but not so tall and with a larger protoconch. Cream, fawn or orange, sometimes weakly banded, ribs usually spotted.

1.5 cm high; 2 cm wide. North West Cape to Kimberley, WA. Further study may prove these specimens to be only a variant of *P. papyracea*.

▲ GENUS **STOMATIA** Helbling, 1779
S. phymotis Helbling

Elongate, obliquely ear-shaped and twisted; spire small, subterminal, body whorl enlarged, whorl plicate beneath the suture, with several spiral ribs; aperture ovate-triangular; columella narrow, curved.

Indo-West Pacific. It is not clear whether there are several species or only one very variable species in this genus.

Stomatia phymotis Helbling, 1779

Pl. 11; fig. 6 a-c Body whorl with one subsutural spiral rib connected to the penultimate whorl by prominent oblique plicae, and two nodulose ribs centrally connected to the subsutural rib by rough folds; with a thick nodulose spiral rib basally beside the columella; rough spiral threads throughout; impression of the external sculpture evident on the inner surface. Greyish white, sometimes with reddish brown patches.

3 cm. Indo-West Pacific; Fremantle, WA to southern Qld. The shells are quite variable. Cotton (1959) figured a strongly tuberculate form from WA under the name *S. australis* Adams, 1850.

▲ GENUS **STOMATELLA** Lamarck, 1816
S. auricula Lamarck (= *impertusa* Burrows)

Thin, depressed, elongate, ear-shaped; body whorl and aperture wide, excessively oblique; surface smooth or finely spirally striate; interior nacreous.

Indo-West Pacific and Southern Australian Regions. There has been debate about the type species for this generic name (see remarks on *Granata*) but Anton (1839) designated *auricula* Lamarck. The body of these odd creatures are large and not entirely covered by the shell. They are gregarious, living under stones, and are very quick-moving, frequently exhibiting autotomy when handled. Synonyms: *Gena* Gray, 1850 (*nigra* Quoy & Gaimard = *auricula* Lamarck); *Plocamotis* Fischer, 1885 (*laevis* Pease).

Stomatella impertusa (Burrow, 1815)

Pl. 11; fig. 7 a-c Spire subterminal, surface glossy smooth except for fine growth striae. Colour and pattern extremely variable ranging from uniform rose, red, yellow or black to variegated or striped patterns with combinations of these colours.

2.5 cm long. Indo-West Pacific; circum-Australian.

There is little doubt that the south-eastern *S. impertusa* Burrow, 1815 (= *strigosa* Adams, 1850) is the same as the south-western *S. auricula* (Lamarck, 1816). A dark colour form named *S. nigra* Quoy & Gaimard, 1834 may also be placed in the synonymy. Including tropical specimens as well seems more doubtful but I can find no distinguishing characters in the material before me and so the name *S. auricula* is used here for all specimens of this form and sculpture. Hedley recorded *G. striatula*, *G. lentricula* and *G. varia*, all of Adams, 1850 from Qld. All of these names are probably part of this species' very large synonymy resulting from the extreme variability of colour pattern.

Stomatella terminalis (Verco, 1905)

Smaller and less elongate than *S. auricula* and the spire is terminal.

6 mm. Spencer and St. Vincent Gulfs, SA. Dredged dead from 45 m. More material may show this to be merely another variant of *S. impertusa*. For illustrations see Cotton, 1959, p. 200, fig. 125.

Stomatella ungula (Hedley, 1907)

Sculpture of fine spiral grooves crossed by growth striae; spire minute, flat, at right angles to the shell's length, terminal, of two whorls; columella slightly thickened. White with irregular spirally arranged crimson splashes.

3.4 mm long; 1.65 mm wide. Described from Mast Head Reef, Qld. Distribution not known. Hedley noted that "the small size, narrowness and conspicuous painting seem to separate this from others of the genus".

SUBFAMILY **TROCHINAE**

Shells of this subfamily are usually thick, straight-sided, more or less flat-based and conical, with an oblique aperture, interrupted peristome, unthickened outer lip, and lirate or denticulate columella. They may be umbilicate or non-umbilicate. Characteristically the gill is bipectinate with a long dorsal membrane connecting it to the mantle skirt. The radular rachidian teeth have M-shaped tops; the first marginal teeth have heavy basal plates and the food-collecting groove beneath the marginals lacks serrations.

This is the largest trochoid group in numbers of genera and species and the shells are quite diverse in form. Hickman & McLean recognise three tribes.

TRIBE **GIBBULINI**

Shell usually of low conical or globose form, with a complete peristome and relatively low spire, umbilicate or non-umbilicate. Pigmentation subdued. Left neck lobe prominently digitate, right neck lobe with a smooth margin. There may be either three or four epipodial tentacles.

▲ GENUS **AUSTROCOCHLEA** Fischer, 1855
Monodonta constricta Lamarck

Turbinate, thick-shelled, spiral ribs widely spaced or absent; columellar fold wide but basal tooth weakly developed, non-nacreous apertural ridge weakly developed; adults usually without an umbilicus but a narrow umbilicus may be present in juveniles..

Southern Australian Region. Sometimes classified as a subgenus of *Monodonta*. The New Zealand *Fractarmilla* Finlay, 1926 (*corrosa* Adams) may be a synonym.

Austrocochlea adelaidae (Philippi, 1849)

Pl. 7; fig. 7 Depressed-turbinate, height and width about equal; sometimes narrowly umbilicate; periphery slightly angulate in sub-adults, rounded in adults; aperture lirate; columella slightly arched, basal tooth represented by a small nodule; whorls sculptured with narrow spiral grooves between flattened cords, five on the penultimate whorl, six on sides of the body whorl, five on the base weakening towards the columella. Exterior brown or grey with white or olive-green spiral and axial lines making a tesselate pattern; aperture iridescent green; columella iridescent, bordered with bright green.

2 cm high. Vic., Tas. and SA. Common on rocky shores.

Austrocochlea concamerata (Wood, 1828)

Pl. 7; fig. 4 a-b Depressed-turbinate, height equal to or less than width; whorls encircled by thin spiral cords, about 12 on the body whorl and base; basal columellar tooth represented by one or more small nodules; aperture weakly lirate. Black or grey, with small or medium-sized yellow or white spots on the ribs; aperture and columella white.

1.5 cm high. NSW to Cape Naturaliste, WA. Inhabits exposed rocky shores.

Austrocochlea constricta (Lamarck, 1822)

Pl. 7; fig. 3 a-b Turbinate, height equal to or greater than width; body whorl encircled by five or six spiral ribs, otherwise smooth (though spire often eroded); aperture strongly lirate; columella with a weak, nodular basal tooth. Exterior colour variable, usually grey or purplish black, uniformly coloured or with a few wide, wavy, oblique stripes or many thin oblique lines; columella and aperture white.

2.5 cm high. NSW to Abrolhos, WA. Abundant on rocky shores and muddy flats in sheltered bays and estuaries. This very common species is polytypic with several quite distinctive colour forms, each of which seems to prefer particular habitats. The striped form was once known as *A. zebra* Menke, 1829. *A. porcata* Adams, 1853 is a variant found in salt-marsh and estuarine habitats. *A. obtusa* Dillwyn, 1817 is another synonym.

Austrocochlea crinita (Philippi, 1849)

Pl. 7; fig. 8 a-b Turbinate, height and width about equal; periphery rounded; umbilicus wide and deep; columella with a small basal nodule; aperture wall lirate; spire whorls with four smooth spiral cords, five on the body whorl, six on the base. Fawn with radial brown rays which may be interrupted, giving a spotted appearance; base fawn with brown spots on the ribs, umbilical area green.

1.5 cm high. South coast of WA. Lives in weedy crevices and pools high in the intertidal zone on rocky shores. The identification of this distinctive species needs confirmation. It seems most closely related to the eastern *A. adelaidae* but immediately distinguished by its wide umbilicus.

Austrocochlea odontis (Wood, 1828)

Pl. 7; fig. 6 Depressed-turbinate, height usually less than width, periphery subangulate except in old individuals; juveniles sometimes narrowly umbilicate; whorls finely striate; aperture weakly lirate; basal columellar tooth lacking or represented by a small nodule. Exterior bluish-black to green, with spiral rows of green or yellowish diamond-shaped spots; aperture iridescent green; columella white, bordered on the umbilical side with bright green.

2 cm high. Vic., Tas. and SA. Rocky shores and shallow sublittoral. The finer spiral sculpture and different colour pattern distinguish this species from *A. adelaidae*.

Austrocochlea rudis (Gray, 1826)

Pl. 7; fig. 2   Turbinate; height greater or less than width; whorls smooth; aperture strongly lirate; columella smooth or with a weak basal nodule. Purple-grey to black; inner edge of lip black or dark green, columella and aperture white.

3.5 cm high. SA to Kalbarri, WA. Abundant high on rocky shores, usually exposed to strong wave action. Synonyms: *melanoloma* Menke, 1843; *torri* Cotton and Godfrey, 1934.

Austrocochlea zeus Fischer, 1874

Pl. 7; fig. 5   Globose, height greater than width; whorls finely spirally striate or sometimes weakly ribbed; columellar tooth represented by a weak basal nodule; aperture weakly lirate. Exterior cream or yellow with thin, continuous or interrupted orange-brown oblique lines, sometimes white but bordered by bright green along its umbilical border.

2 cm high. Cockburn Sound to Dampier Archipelago, WA. This species has been placed in *Chrysostoma* by some previous authors, but it lacks the reflected columellar callus characteristic of that genus and it seems better located in *Austrocochlea*. Common among rubble in seagrass beds.

▲ GENUS **CANTHARIDELLA** Pilsbry, 1889
Gibbula picturata Adams & Angas

Small, conical to turbinate; spire moderately high, whorls rounded, sometimes shouldered, periphery subangulate and keeled; sometimes narrowly umbilicate; columella simple, arched; outer lip simple, sharp, strongly prosocline; aperture ovate or round; sculpture of spiral cords.

Southern Australian Region. This group is sometimes treated as a subgenus of *Gibbula* but any relationship with *Gibbula* s.s. seems remote and full generic rank for *Cantharidella* may be preferable. The rounded, spirally corded whorls distinguish *Cantharidella* from *Notogibbula*, and the simple, edentulous outer lip distinguish it from *Eurytrochus*.

Cantharidella beachportensis Cotton & Godfrey, 1934

High-spired, apical angle about 55°, suture moderately channelled; whorls almost flat-sided, sculptured with five strong spiral cords on the sides, plus a thick rib forming a peripheral keel, and six spiral cords on the base; umbilicus closed. Cream or yellow with wide oblique red rays on the sides, periphery conspicuously spotted with yellow between red dashes, basal ribs alternately with red or yellow dashes.

6 mm high; 5 mm wide. Beachport, SA to Fremantle, WA. The flat-sided whorls distinguish this species from *C. tiberiana*. Illustrated by Cotton, 1959, p. 176, fig. 106.

Cantharidella ocellina (Hedley, 1911)

High-spired; whorls slightly flattened at the shoulders, giving the spire a turreted form; sculpture of spiral cords, the largest being at the shoulders and periphery, with three slightly smaller ribs between; about five weak cords on the base; umbilicus closed. Fawn, lightly mottled with rose, darker rose rings on the peripheral rib.

4 mm high. SA. The type was dredged at 183 m off Cape Wiles. The species name is derived from the eye-like rings on the peripheral cord. For an illustration see Cotton, 1959, p. 177, fig. 107.

Cantharidella picturata (Adams & Angas, 1864)

Apical angle about 70°, suture incised, whorls slightly convex; narrowly umbilicate; subsutural zone smooth, followed by three wide, low spiral ribs and a heavy, smooth rib at the periphery, and about six wide, smooth spiral ribs on the base. Colour pattern extremely variable, may be uniformly pink, rose, green or grey, but usually with a radial pattern of reddish lines or rays on a fawn or greenish background, and with cream and reddish blotches alternating around the peripheral rib and the subsutural zone.

7 mm high; 8 mm wide. NSW.

Cantharidella rottnestensis Verco, 1911

High-spired; umbilicus closed; characterised by two wide, angulate spiral ribs around the whorls, the lower one forming a peripheral keel; with six thin spiral cords on the base. White, with oblong red spots on the ribs and cords.

2 mm high. Described from Rottnest I., WA.

Cantharidella tiberiana (Crosse, 1863)

High-spired, apical angle about 64°, whorls convex, made a little turreted by a slightly flattened area below the suture; umbilicus just closed in adults; sculptured with six or seven smooth spiral cords plus a thicker rib at the periphery which becomes obsolete towards the lip, and seven spiral cords on the base; aperture weakly lirate within. Cream, olive or pink, usually with radial rays of red; peripheral rib and subsutural zone usually conspicuously marked with alternating red and cream blotches.

7 mm high; 6 mm wide. NSW to central Vic. The higher spire and closed umbilicus readily distinguish this species from *G. picturata*. *C. tiberiana* may be a synonym of *C. balteata* (Philippi, 1849) which Cotton and Godfrey (1959) list from SA.

▲ GENUS **CHRYSOSTOMA** Swainson, 1840

Turbo nicobaricus Chemnitz = *Helix paradoxa* Born

Globose, low-spired; whorls rounded; inner lip with callus partially concealing a false umbilicus; smooth or finely spirally striate.

Indo-West Pacific and New Zealand Regions.

Chrysostoma paradoxum (Born, 1778)

Pl. 10; fig. 5 a-b   Solid, globose, turbinate, glossy smooth; outer lip sharp, smooth within; columella curved, smooth; a thick parietal callus forms a plug over the umbilical area. Cream or fawn with pale olive-green or greyish maculations overlain by a reticulate pattern of fawn or grey-brown; aperture and inner edge of the parietal wall and columella bright orange-yellow, outer edge of parietal callus and columella bright rose.

2 cm high. Indo-West Pacific; eastern Qld. Geographically very widespread and moderately common in the intertidal zone of coral reefs. The shiny and very colourful shell make this handsome species easily recognisable.

▲ GENUS **EURYTROCHUS** Fischer, 1880

Clanculus danieli Crosse

Small, conical or depressed-turbinate, spire low to moderately high, suture impressed or channelled, periphery angulate to rounded; umbilicus narrow, bounded by a spiral ridge; sculpture of spiral ribs; columella simple, curved, with a basal thickening; outer lip dentate in adults, strongly prosocline.

Indo-West Pacific. Like *Notogibbula*, this group is sometimes treated as a subgenus of *Gibbula* but the shells are distinguished by a dentate lip. In addition to the species illustrated here, *E. townsendi* (Sowerby, 1895) is reported from Qld.

Eurytrochus concinnus (Dunker, 1889)

Conical with a relatively high, pointed, flat-sided spire and subangulate periphery; suture impressed; body whorl with three faintly nodulose spiral ribs and a thick peripheral rib, about seven cords on the base; umbilicus bordered by a spiral row of chunky teeth; outer lip teeth weak to obsolete. Fawn to cream, whorl sides with irregular, coalescing red-brown lines; base and peripheral ribs conspicuously spotted with red.

6 mm high; 5 mm wide. Western Pacific; Qld. The illustration is drawn from Hedley material from Funafuti Atoll.

Eurytrochus danieli (Crosse, 1862)

Conical, whorls convex, suture impressed, periphery rounded; body whorl with six strong, smooth spiral ribs, base with about six weaker spirals and narrower interspaces; spiral ridge bordering the umbilicus smooth, outer lip margin finely dentate. White, spiral ribs on the sides brown, basal ribs irregularly brown, those closest to the umbilicus white.

6 mm high; 5 mm wide. North Qld.

Eurytrochus macculochi (Hedley, 1907)

Body whorl with 5-6 spiral ribs above the periphery, eight basal; weak oblique cords in the interspaces, sometimes making the ribs faintly nodulose; umbilicus deep, with a broad, nodulose marginal rim; outer lip nodulose within; columella basally thickened. White, with irregular oblique chocolate-brown streaks and spots.

5 mm high; 6.5 mm wide. Qld south to Mast Head I.

Eurytrochus strangei (Adams, 1853)

Spire moderately high; body whorl with three smooth or weakly nodulose spiral ribs plus a slightly larger one forming a peripheral keel and sometimes thin cords in the interspaces; about eight spiral cords on the base, those closest to the umbilicus being widest. With alternating radiating brown and cream bands forming prominent blotches below the sutures, and thin red oblique lines, ribs spotted red or brown.

8 mm high; 8 mm wide. Southern Qld and NSW. Synonym: *docastana* Preston, 1909.

▲ GENUS **FOSSARINA** Adams & Angas, 1864

F. patula Adams & Angas

Small to minute, thin, glossy, depressed-turbinate to auriform; whorls rounded, body whorl large; narrowly umbilicate; aperture large; columella smooth, erect; operculum multispiral, completely closes the aperture.

▲ SUBGENUS **FOSSARINA** s.s.

Turbinate to auriform; aperture entire, oval.

Southern Australia and New Zealand. These little animals are usually abundant in their high intertidal habitat in sheltered bays. Synonym: *Minos* Hutton, 1884.

Fossarina patula (Adam & Angas, 1863)

Juvenile shells depressed-turbinate, becoming semi-auriform when mature; whorls rounded, basal periphery angulate in mature shells, sculptured with spiral cords; aperture wide, oblique, posterior end of the peristome reflected as a triangular wing over the parietal wall. Colour and pattern variable, usually cream with brown or reddish radial flames.

5 mm. Central NSW to eastern Vic. Synonym: *brazieri* Angas, 1871.

Fossarina petterdi Crosse, 1870

Depressed-turbinate, whorls convex, smooth; peristome not quite complete but just interrupted at the parietal wall where a reflection of the inner lip partly occludes the narrow umbilicus. Colour and pattern variable, usually black to dark reddish brown, undulately maculated with white.

4 mm Vic., Tas. and SA. Lives among weeds and mussels high in the intertidal zone of rocky shores. Synonym: *simsoni* Tenison Woods, 1875.

▲ SUBGENUS **MINOPA** Iredale, 1924
Fossarina legrandi Petterd

Turbinate; aperture more or less round, inner lip interrupted by the last whorl.

Southern Australia. The shape of the aperture is said to distinguish this group from *Fossarina* s.s. but note the intermediate condition of this character in *F. petterdi*.

Fossarina legrandi Petterd, 1879

Turbinate, periphery broadly rounded; whorls smooth, convex, suture much impressed. White, with spiral rows of large chestnut-brown patches which may coalesce axially to form radial rays.

5 mm high; 6 mm wide. NSW to southern WA. Lives among algae in the shallow sublittoral zone. The colour and pattern and smooth, polished surface distinguish this species from *F. reedi*. Illustrated by Cotton (1959, p. 173, fig. 103).

Fossarina reedi (Verco, 1907)

Depressed-turbinate, periphery subangulate; a broad spiral rib present below the suture; 10-12 spiral striae on the base, two weak spiral cords border the umbilicus. Chestnut-brown with dark brown spiral lines and tiny white spots, triangular

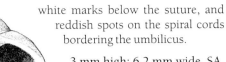

white marks below the suture, and reddish spots on the spiral cords bordering the umbilicus.

3 mm high; 6.2 mm wide. SA. A shallow water species. The depressed form, subangulate periphery and fine striations are said to distinguish this species from *F. legrandi*.

▲ GENUS **MONODONTA** Lamarck, 1799
Trochus labio Linnaeus

Globosely turbinate, thick-shelled, whorls rounded, spirally sculptured; outer lip strongly prosocline and toothed, with a strong non-nacreous apertural ridge, columellar fold wide, basal tooth strong; umbilicus lacking.

Indo-West Pacific.

Monodonta labio (Linnaeus, 1758)

Pl. 7; fig. 1 a-b   Turbinate, spire moderately high; with six granulated primary spiral ribs on the side of the body whorl, about seven on the base becoming smaller and smooth towards the columella, usually with granulated secondary ribs in the interspaces; aperture ridged and lirate below the outer lip; columella arched, with a strong basal tooth which is deeply notched below. Grey, green, brown or red with darker spots on the ribs.

4 cm high. Indo-West Pacific; Shark Bay, WA to southern Qld. Abundant on rocky shores high in the intertidal zone.

▲ GENUS **NANULA** Thiele, 1921
Gibbula tasmanica Petterd

Small, thin-shelled, globose, turbinate, whorls rounded, body whorl inflated; aperture asymmetrical, oblique, outer lip prosocline; umbilicate; columella curved, smooth; spirally ribbed.

Southern Australian Region.

Nanula flindersi Cotton & Godfrey, 1935

Turbinate; columella evenly concave, almost meets the outer lip dorsally, parietal callus weak; umbilicus narrow; whorls finely corded. Yellow, with brown blotches in a row below the suture and around the periphery, and sometimes weak oblique stripes.

4 mm high; 4 mm wide. Vic. to Bunbury, WA. Dredged to 320 m. This illustration was drawn from the holotype.

Nanula galbina (Hedley & May, 1908)

Depressed-turbinate; columella continuous with callus on the parietal wall, slightly reflected; umbilicus wide; whorls with a smooth, flattened shoulder, elsewhere finely, spirally striate. Yellow, brown or buff.

4.6 mm high; 5.6 mm wide. NSW, Tas., SA.

Nanula tasmanica (Petterd, 1879)

Globose-turbinate, whorls convex with angulate basal periphery and convex base; umbilicus narrow but deep, not bounded by an angle; columella thin, joined to the parietal callus; sculpture of fine spiral cords. Without lustre, whitish, tinged with yellow or green, unicoloured or radially maculated with brown.

5.5 mm. Sydney, NSW to southern Tas. Not illustrated.

▲ GENUS **NOTOGIBBULA** Iredale, 1924
Gibbula coxi Angas (= *bicarinata* Adams)

Turbinate, whorls rounded; body whorl with two spiral ribs, one at the shoulder and one at the periphery; umbilicus wide, spirally channelled, bounded by a spiral ridge; columella smooth, curved, continuous with the outer lip; outer lip simple, edentulous, strongly prosocline.

Southern Australian Region. This southern Australian group is sometimes ranked as a subgenus of *Gibbula* (*magus* Linnaeus, 1758) which is a European genus. The twin spiral keels, wide umbilicus, and edentulous outer lip are characteristic.

Notogibbula bicarinata (Adams, 1854)

Shoulder and peripheral keels prominent, equal, sides of whorls concave between the suture and the upper keel and between the two keels; additional sculpture of fine oblique striae, base finely lirate; umbilicus moderate. White, marbled with pink, with a few reddish axial flames and red spots on the base.

8 mm high; 9 mm wide. Moreton Bay, Qld to SA. Synonym: *coxi* Angas, 1867. Figured by Angas (1867, *Proceedings Zoological Society London*, Pl. 13, fig. 26).

Notogibbula lehmanni (Menke, 1843)

Pl. 9; fig. 8 a-b Peripheral keel prominent, shoulder keel smaller; sculptured with spiral cords, about 5 above the shoulder, 6 between the keels, 12 on the convex base. Shoulders obliquely rayed with black, white and orange in repeated series, rays

ending on the shoulder keel as conspicuous blotches but with the orange becoming claret-red; side cords alternately black, white and claret-red; peripheral keel black, white and orange; base with bright claret-red dashes on the spiral cords and sometimes a red or black band below the periphery.

9 mm high; 9 mm wide. Vic. to southern WA. This remarkably coloured species is one of Australia's most spectacular shells but little noticed because of its small size. It is common in seagrass and algal beds below low tide.

Notogibbula preissiana (Philippi, 1848)

Pl. 9; fig. 7 Shoulder and peripheral keels prominent, equal, concave between the suture and the shoulder keel and between the two keels; sculptured with fine, incised spiral lines which are strongest on the base. White, keels prominently spotted with red rectangles, elsewhere patterned with thin red oblique lines.

6 mm high; 7 mm wide. Vic. to Fremantle, WA. Very common under stones in the intertidal and shallow sublittoral zones.

TRIBE **CANTHARIDINI**

Shell with a complete peristome, relatively high spire, blunt columellar denticle, and no umbilicus. With three pairs of epipodial tentacles; foot capable of complete enrollment.

▲ GENUS **ALCYNA** Adams, 1860
Alcyna ocellata Adams

Minute, high-spired, whorls convex; aperture large, oval, smooth within; outer lip simple; umbilicus closed; columella thickly calloused, bearing a nodular plait on its apertural side; usually spirally sculptured.

Western Pacific and southern Australia. Although this group is sometimes treated as a subgenus of *Thalotia*, such a relationship seems very improbable and it is given full generic rank here.

Alcyna acia Cotton, 1948

Sculpture variable, with smooth or weakly nodular spiral ribs, five on the whorl sides, and about eight on the base, the third and fifth ribs usually prominent, making the whorls bicarinate; weak axial crossbars in the interspaces on the sides but not on the base; columellar nodule weak. Fawn or rose with brown spots and blotches.

3 mm high; 1.8 mm wide. SA to Rottnest I., WA. May be collected from weed washings in tide pools. *A. australis* Hedley, 1907 is an even smaller species from southern Qld, which has rounded, weakly sculptured whorls.

▲ GENUS **CANTHARIDUS** Montfort, 1810
Trochus iris Gmelin

Conical to pyramidal with tapering spire and steeply inclined base; whorls slightly convex, straight-sided, or concave; exterior smooth or spirally sculptured; interior brilliantly iridescent; columella smooth, curved, meeting the outer lip in a continuous arc, basal tooth lacking; imperforate or with a narrow umbilicus.

Southern Australia and New Zealand. The type species from New Zealand is imperforate. Suter (1913) has given diagnostic anatomical and radular characters for the New Zealand members of the genus. The following three southern Australian species fit the diagnostic shell characters reasonably well.

Cantharidus lehmanni (Menke, 1843)

Pl. 9; fig. 12 a-b　Conical, spire high, whorls slightly concave, periphery rounded; sculpture of fine spiral cords which become stronger and slightly nodulose on the base; umbilicus closed. Pale green or rose with broad axial bands of green-grey, often flecked with spiral rows of red dashes especially on the base; apex, lip edge and columella usually rose.

1.5 cm high. SA to Kalbarri, WA. Common in seagrass.

Cantharidus pulcherrimus (Wood, 1828)

Pl. 9; fig. 11　Spire whorls slightly convex; with wide, flat-topped or slightly angular spiral ribs, three on spire whorls though sometimes they may be divided to form four, five or six ribs, peripheral rib usually divided (double), five divided ribs on the base. Yellow, green or fawn with rows of dark rose blotches on ribs which may coalesce axially to form wide oblique stripes; columella and lip margin crimson.

2 cm high. Vic. to Jurien Bay, WA. Abundant on brown algae, intertidal and shallow sublittoral.

Cantharidus ramburi (Crosse, 1864)

Pl. 9; fig. 10　Spire whorls slightly concave, broadly carinate just above the suture; sculpture of evenly spaced, fine spiral lirae, 11-12 on the sides of the whorls, 8-10 on the base. Fawn to reddish, with dark red spots on the lirae, tending to coalesce axially in oblique rows; base usually darker, columella and lip edge rose.

2 cm high. Vic. and SA. Abundant on brown algae; intertidal and shallow sublittoral. Synonym: *lesueuri* Fischer, 1880.

▲ GENUS **JUJUBINUS** Monterosato, 1884
Trochus matoni Payraudeau = *exasperatus* Pennant

Conical, with high spire, flat sides, sharply angular periphery and inclined base; columella smooth, almost straight and vertical, terminating rather abruptly with a small thickening and meeting the outer lip at a distinct corner; umbilicus closed; aperture smooth or weakly lirate within, outer lip sharp and simple; sculpture of weak spiral lirae or incised lines.

This is an Indo-West Pacific and Mediterranean group with several tropical Australian species and one species in the temperate waters of south-western Australia. They have been assigned by some authors to *Prothalotia* (= *Calthalotia*), *Cantharidus* or *Calliostoma* but are not typical of any of these groups. The abruptly terminating columella differs from *Cantharidus*. *Calthalotia* species have a terminal thickening at the base of the columella although it is sometimes weak and close to the condition of this group. *Mirulinus* Monterosato, 1917 and *Strigosella* Sacco, 1896 are cited as subgenera by Vaught (1989, p. 9) but none of the Australian species have been assigned to them. Clearly the taxonomy of this group needs further attention. *Jujubinus* has a long fossil history with records from as early as the Cretaceous. Synonyms: *Clelandella* Winckworth, 1932; *Gravijubinus* Nordsieck, 1975; *Pictijubinus* Nordsieck, 1975.

Jujubinus crenelliferus (Adams, 1853)

Whorls nine; aperture subquadrate, a bilobed callus ridge within forming a "doorstep"; columella near vertical, with a central marginal thickening, abruptly truncate at its base; sculpture of spiral cords, nine on the penultimate whorl and sides of the body whorl, crossed by oblique threads which make the cords finely nodulose and the interspaces latticed; six widely spaced spiral cords and fine radial threads on the base. Rose-red, variegated by pale brown oblique flames.

10 mm high; 7 mm wide. North Queensland. For an illustration see Hedley, 1908, *Proceedings Linnean Society NSW*, **33**: 481, Pl. 7, fig. 5.

Jujubinus gilberti (Montrouzier, 1878)

Pl. 9; fig. 14 a-b　Tall, apex minute, apical angle about 40°; sculpture of flat-topped, smooth spiral lirae with only narrow incised lines separating them, about 12 on sides of the whorls, plus a prominent, double, smooth, peripheral keel; about 12 smooth spiral lirae on the base. Green, sides with a few flexuous white oblique-axial bands which do not always align from one whorl to the next; the three or four spiral lirae closest to the umbilicus are minutely spotted with red; peripheral keel with white blotches.

18 mm high. Northern Australia to New Caledonia. This may be synonymous with *J. crenelliferus*.

Jujubinus lepidus (Philippi, 1846)

Pl. 9; fig. 16 a-b Apical angle variable, from 50°-60°; protoconch variable, minute to medium, with 2-3 smooth whorls; sculpture of flat-topped spiral ribs separated by incised grooves, 4-8 medium ribs at centre of the whorls but with larger subsutural and peripheral ribs, the latter double and forming a rim, about nine ribs on the base, ribs minutely oblique-axially striate, especially on the base where striations are conspicuous in the interspaces; columella with a weak basal nodule; aperture smooth within. Grey, rose or greenish, ribs blotched with yellow or white, blotches usually bordered by short transverse red or brown lines, blotches tend to be axially aligned; a purple line may border the columella.

1.6 cm high. Esperance to Jurien Bay, WA. Abundant on seagrass. Given the variation in the shape and size of the protoconch, it is possible that more than one species may be involved in this complex. Synonym: *punctulosus* Adams, 1853.

Jujubinus polychromus (Adams, 1853)

Pl. 9; fig. 13 a-b Tall and slender, apical angle 37°-42°; spiral sculpture of smooth or finely granular cords, about six on the sides of the body whorl and 10 on the base, and a thick peripheral rib. Green or red, usually with irregular axial white blotches not always aligned from one whorl to the next, three or more of the lateral ribs are red-spotted.

1.6 cm high. Central Indo-West Pacific; Geographe Bay, WA (rare south of Geraldton) to the Capricorn Group, Qld, and Lord Howe I. Synonyms: *picturatus* Adams,1853; *tristis* Thiele, 1930.

Jujubinus suturalis (Adams, 1853)

Whorls six, flat-sided, last slightly descending at the aperture; sculptured with broad, flat-topped spiral ribs equal to the intervening spaces, 12 on the body whorl of which six are basal; aperture quadrate; columella with a blunt basal turbercle.

8 mm high; 8 mm wide. North Queensland.

▲ GENUS **CALTHALOTIA** Iredale, 1929
Trochus arruensis Watson (= *marginata* Tenison Woods)

Conical, high-spired; base steeply descending; periphery subangulate; whorls flat-sided or concave, narrowly umbilicate or imperforate; columella smooth, arcuate, terminating with a weak basal thickening and a shallow notch below it, meeting the outer lip at an angle.

Northern Australia. The group consists of species which live on rocks and rubble rather high in the intertidal zone. The species are characterised by the columellar features which distinguish them from *Cantharidus* and *Calliostoma*. Thiele (1930) introduced *Prothalotia* (for *flindersi*) as a subgenus of *Cantharidus* while other authors have given it full generic rank or placed it as a subgenus of *Thalotia*. However, the type of this species seems certain to be congeneric with *marginata*, so that *Prothalotia* becomes a subjective synonym of *Calthalotia*.

Calthalotia arruensis (Watson, 1880)

Pl. 9; fig. 20 a-b Periphery subcarinate, spire slightly inversely turreted; imperforate or with a very small umbilical chink; whorls with about six spiral cords bearing close round nodules and sometimes thin threads in the interspaces, the subsutural and two peripheral cords are larger than the central three; base with seven or eight subequal, roundly nodulose spiral cords. Sides with subequal, alternating wavy axial lines of reddish brown and green, which become arched radiating lines on the base.

1.8 cm high. Eastern Qld. The relationship of this to the western *C. strigata* needs investigation. *C. indistincta* (Wood, 1828) (=*marginata* Tenison Woods, 1880) is a similar species from southern Qld and NSW. Synonym: *torresi* Smith, 1884.

Calthalotia baudini (Fischer, 1878)

Pl. 9; fig. 17 Height to width ratio about 3:2; periphery subangulate, whorls concave; imperforate; with eight spiral cords on the base, two or three immediately above the periphery, cords usually obsolete on upper part of the whorls. Typically patterned with brown, reddish or greenish oblique-axial stripes alternating with white or pale fawn stripes, and heavily spotted with rose flecks, especially on the periphery.

1.5 cm high. Shark Bay, WA. The type locality was given as Bass St. but that appears to have been an error. There are no records of specimens clearly referable to this species outside Shark Bay although a central west coast WA distribution seems likely. In Shark Bay it is abundant in seagrass beds along with *C. mundula* from which it may be distinguished by weaker sculpture, darker colouring and consistently closed umbilicus. However, further study may show that they are variants of the same species.

Calthalotia comtessi (Iredale, 1931)

Pl. 9; fig. 19 Tall-spired, conical, spire flat-sided, periphery angulate, base steeply inclined and slightly convex; columella smooth, very weakly notched at the base; umbilicus closed; sides of whorls with five beaded spiral ribs, the peripheral one largest, double, and prominent, seven beaded ribs on the base. Light brown, with darker reddish brown spots on the nodules, especially conspicuous on the peripheral rib.

Holotype: 2 cm high; 1.35 cm wide. NSW. Resembles the western *J. lepidus* in form but strongly sculptured. It also resembles *Thalotia conica* in form. Iredale originally placed this common species in *Thalotia* but it lacks the columellar teeth which characterise that genus. The condition of the columella and sculpture suggest a relationship to *C. arruensis* of Qld.

Calthalotia mundula (Adams & Angas, 1864)

Pl. 9; fig. 18 a-d   Whorls flat-sided, sometimes concave, periphery sharply keeled in juveniles becoming subangulate in adults; umbilicus closed (southern part of the range) or narrowly open (Ningaloo Marine Park); with five weakly beaded spiral cords on the sides of each whorl, plus a double rib at the periphery and about six spiral cords and intermediary threads on the base. Pale fawn with tan, reddish or greenish oblique-axial rays, and flecked with tiny red or tan spots.

2 cm high. Fremantle to west Kimberley, WA. Abundant on rocky shores and shelly or stony substrates in seagrass beds. In Roebuck Bay it may be found on shelly rubble on the mud flats. The type specimen is from Shark Bay. *Trochus findersi* Fischer, 1878 is a synonym, described from the same locality. Most Shark Bay shells are imperforate although some (including the type) have a tiny open umbilicus. Further north the majority are narrowly umbilicate and the shells tend to be more depressed and peripherally rounded.

Calthalotia strigata (Adams, 1853)

Pl. 9; fig. 21 a-c   Whorls swollen just above the suture making them concave and the spire inversely turreted; periphery subcarinate; with many spiral cords of varying size, the largest being at the periphery, some cords may be weakly nodulose; narrowly umbilicate. Fawn or green, with oblique-axial red lines or bands.

2.3 cm high. Exmouth Gulf to the north Kimberley, WA. Common on shoreline rocks in muddy areas. The three paratypes from Swan Point (Kimberley, WA) are tall-spired with a few wide oblique-axial bands, while a specimen figured here from nearby One Arm Point has numerous, narrow, interrupted lines. A series examined from Exmouth Gulf are banded like the types but not so high-spired.

▲ GENUS **PHASIANOTROCHUS** Fischer, 1885
Trochus badius Wood (= *eximius* Perry)

Spire tall and tapering; whorls with straight or slightly convex sides, smooth or weakly sculptured; columellar fold well developed, basal columellar tooth present; umbilicus lacking; aperture brightly iridescent.

Southern Australian Region. Sometimes treated as a subgenus of *Cantharidus*. The species live among algae or seagrass in shallow water.

Phasianotrochus apicinus (Menke, 1843)

Pl. 9; fig. 5   Thin, translucent, shining, smooth but for fine spiral striae; with a curiously depressed body whorl and wide flaring aperture; columella truncate anteriorly forming a projecting tooth. Grey, olive-green or yellow, patterned with narrow white spiral bands interrupted by red spots, an iridescent green sutural band also interrupted by red spots, and thin, wavy axial lines on the base.

2 cm high. Vic. to Fremantle, WA. Not very common.

Phasianotrochus bellulus (Dunker, 1845)

Pl. 9; fig. 4 a-b   Stout and solid; whorls rounded, smooth; columella with a strong basal tooth, aperture strongly lirate within. Grey or brown with red or orange spiral lines, broad reddish axial bands outlined by wavy white lines, and a zone of short axial lines at the sutures. Details of pattern and colouring are variable.

1.8 cm. Vic. to Dongara, WA.

Phasianotrochus eximius (Perry, 1811)

Pl. 9; fig. 3 a-d   Thin, elongately turbinate, height greater than width; surface glossy but encircled by moderately strong incised lines, between which are very faint striae; columella smooth, nearly vertical, basal tooth weak; aperture strongly lirate, lirae sometimes end in smooth pointed nodules near the lip. Colour variable, usually fawn, olive-green, brown or rose, sometimes patterned with pale lines or stripes.

4 cm. high. NSW to Fremantle, WA. NSW specimens tend to be more solid than those from southern and western Australia.

Phasianotrochus irisodontes (Quoy & Gaimard, 1834)

Pl. 9; fig. 6 a-b   Resembles *P. bellulus* though more elongate and lacking the strong apertural lirae. Usually bright green, with thin axial red lines which may be arranged in pairs, sometimes with white-spotted spiral lines.

10 mm high. Vic. and Tas. to Geraldton, WA. Abundant on seagrasses in the sublittoral zone.

Phasianotrochus rutilus (Adams, 1851)

Stout and solid; whorls slightly convex and smooth; columellar basal tooth strong; aperture smooth within. Exterior green with crowded irregular reddish brown oblique-axial lines which may be interrupted by white flecks; interior vivid green.

1.5 cm. Vic. and Tas. to SA. The stouter form and different colour pattern distinguish this species from *P. irisodontes*. *"Cantharidus sericinus"*

Thiele, 1930 was described from a single specimen from Cockburn Sound, WA and may be only a mislocalised specimen of this species according to Ponder (1978). *Phasianotrochus ocellatus* Gould, 1861 is a similar NSW species distinguished by its eye-like pattern.

▲ GENUS **THALOTIA** Gray, 1847
Trochus pictus Wood (= *conica* Gray)

Conical, high-spired; periphery rounded or weakly angulate, whorls usually straight-sided, sometimes slightly convex or concave; columella straight or arcuate, smooth or dentate, usually ending abruptly or with a terminal nodule or thickening; umbilicus narrow or lacking; sculpture granular or spirally corded.

Central Indo-West Pacific and Southern Australian Regions.

▲ SUBGENUS **THALOTIA** s.s.

Periphery rounded; aperture small, columella toothed, ending abruptly with a terminal thickening and notch below; outer lip dentate along its inner edge.

Southern and eastern Australia.

Thalotia attenuata (Jonas, 1844)

Spire tall and acuminate; whorls convex, periphery subangulate to rounded, base sloping and slightly convex; roughly sculptured with strong, rugose axial folds crossed by sharp spiral cords; umbilical chink present. Cream to rose, with broad rose to green flames below the suture, spiral rows of tiny red or brown spots, and larger strawberry spots on the base.

2.3 cm high. Indo-West Pacific; Qld south to the Capricorn Group. A coral reef species. Synonym: *elongatus* Wood, 1828 *non* Sowerby, 1818.

Thalotia conica (Gray, 1827)

Pl. 9; fig. 2   Sculptured by spiral rows of low nodules; columella with a projecting basal tooth and several strong nodules; imperforate. Reddish brown with interrupted dark brown spiral lines, apex pink.

2.3 cm. high. Vic. to Geraldton, WA. Abundant on seagrasses. Synonym: *pictus* Wood, 1828.

▲ SUBGENUS **ODONTOTROCHUS** Fischer, 1880
Trochus chlorostomus Menke

Periphery keeled, whorls flat-sided; columella curved, sharply truncate, weakly toothed along its edge; umbilicus narrow, may be closed in adults.

Southern Australian Region.

Thalotia chlorostoma (Menke, 1843)

Pl. 9; fig. 1 a-e   Thin-shelled; sides of whorls with incised spiral lines, peripheral keel formed by a thickened rib which remains evident as a spiral rib around the suture; outer lip weakly lirate within. Colour variable, usually red or green with spiral rows of tiny arched or triangular yellow marks, and brown patches separated by yellow flames on the basal keel.

3 cm. high. SA to Geraldton, WA. Common on seagrasses. Extremely variable in form and colour pattern. Shells of a population at the Abrolhos are characteristically small, high-spired and more coarsely sculptured than others. They may represent an extreme end-of-range form.

TRIBE **TROCHINI**

Peristome incomplete because there is no parietal lip; columellar lip extending into a false umbilicus.

▲ GENUS **CLANCULUS** Montfort, 1810
Trochus pharaonius Linnaeus

Conical, more or less depressed; whorls usually rounded but periphery sometimes angulate with beaded spiral sculpture; umbilicus open, bordered by a ridge which is usually nodulose or dentate; columella toothed.

Cosmopolitan in tropical and temperate seas. Cotton and Godfrey (1934) introduced a number of subgeneric names for the southern Australian species and a key to their identification. Unfortunately the key, and the diagnostic characters for the subgenera, break down when the tropical species are considered as well. In this account the species are listed alphabetically. For reference purposes the subgenera and their type species are listed here but no attempt is made to allocate the species to them.

EUCLANCULUS Cotton & Godfrey, 1934
(*leucomphalus* Verco)

EURICLANCULUS Cotton & Godfrey, 1934
(*flagellatus* Philippi)

ISOCLANCULUS Cotton & Godfrey, 1934
(*yatesi* Crosse = *philippi* Koch).

MACROCLANCULUS Cotton & Godfrey, 1934
(*undatus* Lamarck)

MESOCLANCULUS Iredale, 1924
(*plebejus* Philippi)

MICROCLANCULUS Cotton & Godfrey, 1934
(*euchelioides* Tate)

Clanculus albinus Adams, 1853

Pl. 8; fig. 7 Globose-conical, thick-shelled; sculpture of spiral rows of fine, even-sized beads, six rows on the penultimate whorl, 18 on the body whorl and base; columella oblique, reflected, emerging from within the umbilicus, basal tooth triplicate, margin above with a fold and a denticle; aperture strongly lirate, denticle near the suture massive and tricuspid in one of the two syntypes; parietal wall calloused and bears strong ridges. Pale buff, dotted with crimson and brown on the early whorls.

18 mm high; 20 mm wide. Indo-West Pacific; north Qld. The two syntypes (fig. 7) in the British Museum (Natural History) are labelled "Sumatra". Brazier recorded the species from Fitzroy I., northern Qld in 1871, but no other Australian specimen is known to this author.

Clanculus aloysii Tenison Woods, 1876

Whorls flat-sided, sometimes somewhat shouldered, suture deeply impressed, periphery sharply angulate; whorl sides with prominent, nodulose spiral ribs below the suture and at the periphery and about six smaller ones between; base with six or seven nodulose spirals; umbilicus rim edentulous; columella slightly reflected, smooth, basal nodule oblique, single or with a second smaller one below, notch moderate, usually with a small nodule close to the parietal wall to which the columella is attached; parietal wall thinly calloused, smooth; outer lip weakly lirate within, subsutural denticle lacking. Fawn, speckled with red-brown, sometimes irregularly rayed; peripheral keel usually with prominent red-brown patches.

1 cm high; 1 cm wide. Southern NSW to Vic.

Clanculus atropurpureus (Gould, 1849)

Pl. 8; fig. 23 a-b Whorls and periphery rounded, outer lip edge slightly downturned; sculpture of strongly and evenly beaded spiral ribs, seven on the penultimate whorl, about 16 on the body whorl and base, finely striate across the interspaces; aperture regularly lirate; umbilicus rimmed by large pointed denticles, the first of which is particularly large and projects into the umbilicus; columella inclined, with several low nodules and a weak basal notch. Brown and black.

8 mm high; 11 mm wide. Indo-West Pacific; Fremantle, WA to southern Qld. In this very common species the columella differs significantly from most others in that it bears numerous nodules and emerges from deep within the umbilicus.

Clanculus bicarinatus Angas, 1880

Pl. 8; fig. 5 a-b Conical; periphery and shoulders keeled; base steeply inclined but slightly convex; upper part of whorls with four spiral rows of axially elongate nodules, the fourth at the shoulder, then four rows of rounded smaller nodules on the sides and a row of large nodules forming the peripheral keel; about nine spiral rows of nodules on the base, sometimes secondary rows in the interspaces; umbilicus deep, bordered by strong, inward-pointing denticles, the first being quite large; a small tooth on the parietal wall also projects into the umbilicus; columella slightly inclined, nodulose on its umbilical side, ending with an elongate and inclined nodule with small notch below. Dark grey.

13 mm high; 17 mm wide. Indo-West Pacific; Onslow, WA to southern Qld.

Clanculus brunneus Adams, 1853

Pl. 8; fig. 16 a-c Depressed, conical, flat-based, periphery sharply angulate; whorls encircled by six finely nodulose spiral ribs plus a sharp-edged keel at the periphery, base with about 11 narrow spiral cords; columella oblique, reflected, emerging from within the umbilicus, basal nodule bifid; outer lip lirae weak, subsutural nodule obsolete. Brown with wide radial rays and conspicuous darker blotches around the peripheral keel, sometimes uniformly fawn or rose and lacking rays.

10 mm high; 15 mm wide. NSW. Common under stones, intertidal to at least 20 m. This species is very similar to *C. limbatus* and some authors have treated it as a synonym. However, the differences in shell morphology are instantly recognised, *C. brunneus* being much more peripherally angulate and with finer sculpture. Information is needed on whether the two forms intergrade or overlap in eastern Vic. and southern NSW. They may represent isolated and divergent subspecies.

Clanculus clanguloides (Wood, 1828)

Pl. 8; fig. 1 a-b High-spired, with convex whorls and impressed suture, periphery subangulate, base inclined; whorl sides with eight rows of small rounded nodules, the fourth and sixth rows being smaller than the others, plus 12 rows on the base; umbilicus deep, narrow, bordered by thick rugae; parietal wall heavily calloused and roughly lirate, with a strong lira spiralling into the aperture; columella reflected, with thick transverse plaits and a massive tricuspid basal tooth, deeply notched below; subsutural nodule below the outer lip massive and weakly bifid, outer lip margin finely dentate, with four strong lirae within the aperture. Fawn, with oblique whitish rays, groups of nodules on some ribs are rose, especially on the base, giving a spotted appearance.

15 mm high; 14 mm wide. Indo-West Pacific; north to central Qld. There is some difficulty in interpreting this species. To begin with, the type locality is uncertain. One of the two "cotype" (i.e. syntype) lots is labelled "South America" but that is probably an error and a central Indo-West Pacific locality is more likely. Yet the Qld shells appear to be conspecific with the syntypes and Wood's name seems appropriate. The description here is based on the figured Qld specimen (fig. 1a) which differs from the syntypes (fig. 1b) in that it is much paler and higher spired. See notes on *C. stigmatarius* and *C. comarilis*.

Clanculus clangulus (Wood, 1828)

Pl. 8; fig. 15 a-b   High-spired, sides of the spire with concave outline though the sides of individual whorls are convex; periphery subangulate, base inclined; with six spiral ribs on the spire whorls, eight on body whorl, the three at the periphery being the most prominent; columella oblique, emerging from deep within the umbilicus, weakly reflexed on the umbilical side, with an angulate upper tooth, weak central nodule and prominent, bifid basal nodule, deeply notched below; outer lip with a prominent subsutural nodule. Green with red-brown maculations, apex red, periphery and base with red spots.

15 mm high; 15 mm wide. NSW. Common under stones from the intertidal zone to at least 20 m. Like *C. leucomphalus* but the shell is a little heavier. Distinctive green colour.

Clanculus comarilis Hedley, 1912

Pl. 8; fig. 4   High-spired, outline of spire slightly convex, periphery subangulate, base steeply inclined but flat; sculpture of five rows of close, prominent, rounded nodules on the sides of the whorls, the subsutural row slightly larger, six rows on the base, with fine oblique striae in the inter-spaces; umbilicus very narrow, bordered by thick, deep rugae; columella emerging from within the umbilicus, strongly reflexed, rugose, terminating with a large, double, turned down basal tooth; outer lip strongly lirate within, inner margin denticulate, bearing a massive but simple subsutural nodule; parietal wall heavily lirate. Pale fawn, with some nodules orange and some dark brown giving characteristic spotted appearance.

17 mm high; 14 mm wide. Shark Bay, WA to central Qld. Common intertidally under stones in the Dampier Archipelago, WA. This species resembles *C. clanguloides* but may be distinguished by the flatter sides and fewer rows of nodules, the bifid not tricuspid basal tooth, and by the characteristic brown spots. See also remarks on *C. unedo*.

Clanculus consobrinus Tate, 1893

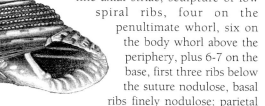

Pl. 8; fig. 21 a-b   Depressed, periphery rounded, suture subcanaliculate, with fine axial striae; sculpture of low spiral ribs, four on the penultimate whorl, six on the body whorl above the periphery, plus 6-7 on the base, first three ribs below the suture nodulose, basal ribs finely nodulose; parietal wall thinly calloused, lirae lacking; columella narrow, smooth except for a small nodule near the base and a prominent basal nodule, shallowly notched below; umbilicus with a rounded margin bearing a series of strong transverse lirae; outer lip lirate within, subsutural denticle lacking. Fawn or light brown with black dashes in the interspaces.

10 mm high, 15 mm wide. SA to Geraldton, WA. Common under stones; intertidal and shallow sublittoral.

Clanculus denticulatus (Gray, 1827)

Pl. 8; fig. 20 a-b   Depressed, periphery rounded; sculpture of six weakly granulose spiral ribs on the penultimate whorl, seven on the side of the body whorl, nine smooth ribs on the base; apertural lirae weak, subsutural nodule lacking; umbilicus narrow, rimmed by denticles, the first large, forming a prominent triangular tooth jutting into the umbilicus; parietal wall weakly calloused but lacking lirae; columella narrow, arising from the parietal wall at the side of the umbilicus, smooth except for a small basal nodule, basal notch weak. Fawn, spiral ribs brown giving a distinctive striped appearance.

8 mm high; 10 mm wide. NSW to Abrolhos, WA. There is some doubt whether this name is applicable to the species usually associated with it by Australian workers. Menke (1843) introduced the name *Monodonta lupinus* which might refer to the same thing. The types need to be checked to resolve the matter.

Clanculus dunkeri (Koch, 1843)

Pl. 8; fig. 19 a-b   High-spired; with three rows of nodules on the sides of the whorls and a fourth at the periphery, plus beaded cords in the interspaces towards the lip which may be almost as prominent as the primary rows, four rows on the base plus a weakly nodulose rib bordering the narrow umbilicus; interstitial axial striae conspicuous; parietal wall thinly calloused, smooth; columella narrow, attached to the parietal wall, smooth except for a strong, elongate basal nodule and a smaller rounded nodule on each side of it; outer lip lirate within, subsutural nodule lacking. Upper whorls yellow-green or tan, becoming axially blotched with brown on the later whorls, nodules dark brown.

11 mm high; 10 mm wide. Vic. and Tas. to King George Sound, WA. Very like *C. weedingi* but high-spired, not depressed.

Clanculus euchelioides Tate, 1893

Turbinate, whorls and periphery rounded, suture canaliculate; sculptured with 3-4 close-set, smooth spiral ribs on the spire whorls and seven weakly granular ribs on the body whorl, sometimes with an additional 2-3 spiral threads between; base with about seven very close-set spiral cords, transverse striae in the narrow interspaces; umbilicus deep, without a bordering rib but the rim is dentate; columella emerging from within the umbilicus, narrow, smooth, with a single basal nodule, deeply notched below; outer lip dentate within, subsutural denticle lacking. Reddish brown, with oblong cream spots below the sutures and small white spots around the periphery.

6.5 mm high; 6.25 mm wide. SA to Fremantle, WA. Under stones; intertidal and shallow sublittoral. The WA shells tend to have more strongly granulose sculpture and a more nodulose rim around the umbilicus. The name *C. gatliffi* Tomlin, 1924 has been applied to them but it is treated here as a synonym. Fig. (a) is drawn from the holotype of *euchelioides* (type locality Moonta Bay, SA) which has weak lip dentition and obsolete nodules around the umbilicus. The apertural view (fig. b) is drawn from a WA paratype of *gatliffi* showing the usual apertural condition of the western form.

Clanculus flagellatus (Philippi, 1848)

Pl. 8; fig. 9 a-b Periphery rounded, body whorl descending away from the suture at the margin, outer lip slightly outward-flaring; sculptured with spiral rows of rounded beads, six rows on the penultimate whorl and eight on the body whorl, the peripheral row slightly more pronounced; base with nine weakly granulose spiral lirae, plus a strongly toothed rib bordering the umbilicus; parietal wall thickly lirate; columella oblique, reflected, emerging from within the umbilicus, bearing three large, rounded nodules plus a heavy bifid basal nodule, deeply notched below. Light reddish brown, sides with darker brown blotches, some beads white, peripheral and basal lirae with red-brown dashes.

10 mm high; 12 mm wide. Vic. and Tas. to Fremantle, WA. Common under stones in the intertidal and sublittoral zones.

Clanculus floridus (Philippi, 1850)

Pl. 8; fig. 18 a-b Depressed, periphery rounded; sculpture of eight spiral rows of small granules on the penultimate whorl, nine on the body whorl, and about 10 on the base; minute oblique striae crowd the interspaces; parietal wall moderately lirate; aperture and columella oblique, columella

emerging from within the umbilicus, with a prominent central nodule and a bifid basal nodule, deeply notched below; umbilicus bordered by a row of thick nodules; aperture with short, alternating, large and small inner lirae and a prominent, bifid subsutural nodule. Fawn, with darker oblique rays; early spire whorls may be spotted with rose.

8 mm high; 11.5 mm wide. NSW and eastern Vic. Resembles *C. flagellatus* but smaller and with finer sculpture.

Clanculus granosus Brazier, 1877

High-spired and conical, suture channelled; parietal wall smooth; columella emergent from the umbilicus, not attached to the parietal wall, smooth, with a prominent but simple basal tooth; rim of umbilicus angulate, with 4-5 prominent marginal denticles; outer lip with about 12 marginal denticles which do not align with eight internal lirae, subsutural denticle lacking; whorls with four strongly beaded spiral ribs, base with eight spiral ribs, conspicuous oblique striae in the interspaces. Cream, sometimes with a greenish tinge, maculated or axially banded with red-brown.

6 mm high; 7 mm wide. Darwin, NT to Capricorn Group, Qld. A common little species found in the NT and along the mainland coast of Qld.

Clanculus granti Hedley, 1907

Pl. 8; fig. 6 Solid, conical, spire pointed; seven whorls with markedly convex sides, the body whorl slightly angulate below the shoulder, periphery rounded; umbilicus narrow, ascending to the apex; outer lip sharp, bearing a deep-seated projecting nodule below the suture; parietal callus coarsely wrinkled; columella wrinkled, with a massive basal bifid tooth and a small tubercle higher up; early whorls spirally grooved, body whorl smooth, polished; on the base four deep spiral grooves surround the umbilicus. Cream with zigzag red axial stripes.

9 mm high; 9.5 mm wide. Described from a shell collected at Mast Head I., Qld. Very few specimens are known, all of the verified localities being within the Capricorn Group or from Lady Musgrave I. This unusual shell is unlike any other Australian member of the genus with its massive bifid basal tooth on the columella and prominent subsutural nodule below the outer lip.

Clanculus johnstoni Hedley, 1917

High-spired and conical, suture impressed; parietal wall smooth; columella emergent from the side of the parietal wall within the umbilicus, smooth, with a prominent but simple basal tooth, notched below; rim of umbilicus angulate, weakly denticulate on the margin; outer lip with 8-12 weak marginal denticles, subsutural denticle lacking; whorls with four smooth or weakly beaded spiral ribs, base with eight spiral ribs, conspicuous oblique striae in the interspaces. Cream, fawn or rose, maculated or axially banded with red-brown.

4 mm high; 4.5 mm wide. Darwin, NT to Caloundra, Qld. Like *C. granosus* and found in the same areas but distinguished by finer sculpture and different umbilical and sutural structure.

Clanculus leucomphalus Verco, 1905

High-spired, sides of spire slightly concave, periphery subangulate, rather thin-shelled; with eight finely granulose spiral cords on the spire whorls, about 10 on the body whorl, cords subequal, the largest being immediately below the suture and a pair at the periphery, base with about eight even finer granulose cords; columella emerging from within the umbilicus, undercut on its inner side, smooth but with a prominent bifid basal nodule; lip denticles subequal; umbilicus wide, spirally channelled. Mottled ash-grey and fawn, with white granules on the ribs.

12 mm high; 13.5 mm wide. SA to south coast of WA. Dredged to 150 m.

Clanculus limbatus (Quoy & Gaimard, 1834)

Pl. 8; fig. 11 a-b Relatively high-spired, whorls convex, base flat, periphery angulate; with seven spiral rows of narrow nodules on the sides of the whorls plus a smooth or weakly nodulose peripheral rib on the body whorl, base with about 11 rows of less prominent nodules plus rather weak folds bordering the umbilicus, interspaces with oblique striae; parietal wall irregularly lirate, columella oblique, with 4-5 irregular transverse nodules and a strong, bifid basal nodule, the upper part of the basal nodule longer than the lower; aperture lirate within, with a small subsutural nodule. Fawn with brown patches below the suture, irregular brown bands on the sides, a row of brown dashes on the peripheral rib, and irregular, interrupted oblique rays on the base.

15 mm high; 21 mm wide. Vic. and Tas. to Rottnest I., WA. This species is very common under stones in the intertidal and sublittoral zones along the southern Australian coast. *C. brunneus* is a closely related species from NSW.

Clanculus maugeri (Wood, 1828)

Pl. 8; fig. 8 Solid, conical, rather straight-sided, basal margin angulate; sculptured by crowded spiral rows of small nodules of almost equal size except on the base where they are smaller; umbilicus narrow, shallow, bordered by strong denticles; parietal wall strongly but irregularly lirate; inner surface of outer lip lirate, subsutural denticle lacking; columella thick, reflected, emergent from the umbilicus, nodulose, terminating anteriorly in a prominent, multicuspid tooth. Reddish brown with fine white spots which are more numerous on the base.

2 cm high; 2 cm wide. NSW, Tas., Vic. This is one of the largest Australian *Clanculus*. The drawing of the aperture illustrates the complex dentition.

Clanculus maxillatus (Menke, 1843)

Pl. 8; fig. 13 a-b Moderately high-spired, whorls convex, periphery rounded; lip slightly outward-flaring but less so than in *C. flagellatus*; penultimate whorl with five spiral rows of small rounded nodules, body whorl with six rows, nodules even-sized, interspaces only minutely axially striate; base with seven spiral lirae of which only the two closest to the umbilicus are nodulose, plus a spiral row of heavy folds bordering the umbilicus; parietal wall with two thick and several lesser lirae, the larger two spiralling into the aperture; columella emerging from deep within the umbilicus, with a prominent rounded central nodule and a prominent basal nodule, deeply notched below; aperture with five long inner lirae and a prominent subsutural nodule. Pale brown or grey, nodules dark brown or white.

11 mm high; 11 mm wide. Western SA and WA as far north as Kalbarri. Very common under stones.

Clanculus ochroleucus (Philippi, 1853)

Pl. 8; fig. 17 Sculpture of close spiral ribs with rounded nodules and fine oblique striae in the interspaces, 11 ribs on the penultimate whorl, about 30 on the base and side of the body whorl, basal ribs very fine and becoming smooth near the umbilicus. Yellowish fawn.

10 mm high; 12 mm wide. Vic. and Tas. to Cape Leeuwin, WA. Not abundant.

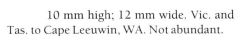

Clanculus personatus (Philippi, 1846)

Pl. 8; fig. 12 a-b Moderately high-spired and conical, periphery angulate, base flat; lip slightly outward-flaring, penultimate whorl with 5-6 spiral rows of prominent beads, six rows on the body whorl and six on the base, interspaces finely axially striate; umbilicus bordered by a spiral row of heavy folds, parietal wall lirate; columella oblique, emerging from within the umbilicus, nodulose and with a thick bifid basal nodule, deeply notched below. Cream with broad reddish radiating rays which may be broken to form peripheral and subsutural blotches.

15 mm high; 18 mm wide. Vic. to Fremantle, WA. Not common.

Clanculus philippi (Koch, 1843)

Pl. 8; fig. 22 a-b Base flat, periphery angulate and nodulose; with six spiral rows of nodules on the body whorl, the sixth peripheral, plus six on the base; umbilicus deep but narrow; parietal wall smooth; columella narrow, attached to the parietal wall, with a prominent basal nodule; outer lip lirate within, subsutural denticle lacking. White, with reddish or brown spots tending to merge together.

10 mm high; 13 mm wide. Vic. and Tas. to Rottnest I., WA. The flat base, angulate periphery and finer sculpture distinguish this species from *C. weedingi* and *C. dunkeri*.

Clanculus plebejus (Philippi, 1851)

Moderately depressed, prominently sculptured with thick spiral ribs, three on the penultimate whorl, four on the body whorl, peripheral rib double and forming a keel, two ribs closest to the suture strongly nodulose, base with 6-9 smooth spiral cords, interstices axially striate; columella attached to the parietal wall, smooth except for 2-4 elongate nodules at the base where it merges with the outer lip; outer lip with several internal lirae, subsutural nodule lacking. Pink, grey, yellow or fawn; with prominent black and white dots on the ribs which may coalesce to form oblique-radial rays.

7 mm high; 10 mm wide. Sydney, NSW to Cape Naturaliste, WA. Under stones; intertidal to 50 m. Uncommon north of Cape Leeuwin in WA but abundant along the south coast.

Clanculus ringens (Menke, 1843)

Pl. 8; fig. 10 a-b High-spired and conical, spire with pronounced peripheral keeling, base convex; sculpture of four nodulose ribs on the sides of the whorls, the upper and lower ribs much more prominent than the central two and bearing heavy, axially elongate nodules; subsutural rib ledged above; with eight close and closely beaded spiral ribs on the base, interstices obliquely striate; columella strongly reflected,

transversely lirate, with a very prominent, bifid basal tooth and deep notch below; umbilicus wide and deep, with a narrow angular rim; outer lip weakly denticulate, subsutural denticle small. Fawn, with strong brown axial blotches, basal ribs brown-spotted.

10 mm high; 10.5 mm wide. Western SA to Fremantle, WA. Very common under stones in shallow water in WA. The drawing illustrates the unusual columellar details with its very prominent, bifid, basal tooth.

Clanculus septenarius (Melvill & Standen, 1899)

Conical, spire sides flat, periphery angulate, height and width equal; aperture quadrate, columella slightly concave and inclined, simple; umbilicus deep with a concave side wall on the last whorl, and bordered by a smooth, sharply angulate rim; whorl sides sculptured with a nodulose subsutural rib, five equal rows of small, rounded nodules, evenly spaced, and a double spiral row of slightly larger nodules at the periphery; base with about 11 spiral rows of tiny close nodules. Rose or fawn, some nodules white; columella and umbilicus white.

1 cm high. North Qld. This unusually high-spired shell was originally described as an *Astele* and Iredale (1929) introduced the generic name *Pulchrastele* for it. *Clanculus* seems more appropriate even though the columella is only weakly toothed.

Clanculus stigmatarius Adams, 1853

Pl. 8; fig. 2 High-spired, whorls convex, periphery rounded, base inclined; suture slightly channelled; sculptured with subequal, beaded spiral ribs, nine on the sides, one on the base, minute oblique striae in the interspaces; parietal wall thinly calloused; umbilicus narrow, deep, bordered by thick transverse nodules, deeply excavate behind the columella; columella oblique, originating within the umbilicus, transversely rugose, basal nodule prominent, with three transverse parts, deeply notched below; outer lip lirate within, with a prominent simple or bicuspid subsutural nodule. Fawn, with conspicuous strawberry spots and dashes on the ribs.

1.3 cm high; 1.4 cm wide. Indo-West Pacific; Port Douglas to the Capricorn Group, Qld. The syntypes (fig. 2), now lodged in the British Museum (Natural History), are labelled "I. of Corigidor, Bay of Manila". The shells resemble *C. clanguloides* but have more beaded ribs per whorl, the interstitial striae are less conspicuous, and the deep excavation behind the columella is characteristic. The conspicuous strawberry spots also should easily identify this species. See also notes on *C. unedo*.

Clanculus undatoides Tenison Woods, 1879

Conical, suture impressed, whorls with a flattened subsutural shoulder so that the spire has a stepped outline; umbilicus narrow, its margin dentate only at the base; columella emerges from the side of the umbilicus, reflected but smooth except for a weak marginal nodule and a small basal nodule, basal notch shallow; outer lip lirae weak, the subsutural one is the largest; parietal wall smooth, thinly calloused; first subsutural spiral rib weak, the second and third large bearing rounded nodules, followed by another three small ribs, and finally a pair of prominent finely nodulose ribs at the subangular periphery; the nodules on the lower peripheral rib are transverse and extend obliquely onto the base; base with six subequal, finely nodulose spiral ribs. Cream with rose axial bands, strongest on the subsutural ribs, red-brown oblique patches regularly arranged around the periphery, and rose spots on many nodules on the base.

7 mm high; 9 mm wide. NSW.

Clanculus undatus (Lamarck, 1816)

Pl. 8; fig. 14 a-b   Solid, conical, depressed, with low spire, rounded whorls and subangulate basal margin; roughly sculptured with beaded spiral ribs; inner margin of the outer lip irregularly toothed, aperture spirally lirate within, subsutural denticle lacking; columella thick, reflected, emergent from the umbilicus, with strong tooth-like folds, basal tooth prominent, multicuspid; umbilicus wide but shallow. Brown with many dark red spots.

4 cm wide. NSW to Cape Naturaliste, WA. This is the largest of the Australian *Clanculus*. The western form is higher spired and usually has a green apex. Cotton & Godfrey, 1934 named it *C. occiduus* and it may warrant taxonomic separation at subspecies level.

Clanculus unedo Adams, 1853

Pl. 8; fig. 3 a-b   Solid, high-spired, base inclined, whorls convex, suture impressed; umbilicus narrow, rimmed by thick rugae; columella heavily rugose, with a massive bifid basal tooth; outer lip weakly dentate on the margin, with a massive but simple subsutural nodule, and five strong lirae deep within the aperture; parietal wall calloused, irregularly lirate; whorls with five spiral rows of rounded nodules, the subsutural row tending to be axially elongate, base with seven rows of smaller nodules. Rose-red, nodules white-topped except on the second, fourth and fifth rows on which every fourth or fifth nodule is dark red-brown.

16 mm high; 13 mm wide. Indo-West Pacific; north Qld. Taller than *C. clanguloides* and lacks the tricuspid basal tooth characteristic of that species. In shape and sculpture the shells resemble *C. comarilis* but the colouration is very distinctive. The strawberry spotted *C. stigmatarius* is also easily separated by its tricuspid columellar tooth. Synonym: *margaritarius* Philippi, 1846.

Clanculus weedingi Cotton, 1938

Somewhat depressed, periphery subangulate; sculpture of four spiral rows of nodules on the upper part of the body whorl, the fourth at the periphery, and four rows on the base plus a nodulose rib or funicle bordering the deep umbilicus; parietal wall smooth; columella attached to the parietal wall, narrow, smooth except for a prominent basal nodule; outer lip lirate within, subsutural denticle lacking. Uniformly purple-brown except for white umbilicus and columella.

9 mm high; 11.5 mm wide. SA to Fremantle, WA.

▲ GENUS **TECTUS** Montfort, 1810
Trochus mauritianus Gmelin

Conical, top-shaped, periphery angulate, base flat, umbilicus lacking; columella smooth, curved, with a strong concave spiral fold; aperture smooth within except for a thick spiral ridge and several short marginal lirae which may be present at the base near the columella.

Tropical Indo-West Pacific region. Sometimes treated as a subgenus of *Trochus*. The lack of an umbilicus and the concave spiral fold on the columella distinguishes the group. It has a long fossil record with its origins in the Upper Cretaceous.

Tectus fenestratus (Gmelin, 1791)

Pl. 7; fig. 13 a-b   Suture deeply impressed, sides sculptured with thick, rounded, closely spaced, oblique ribs, base with spiral incised lines and a calloused, polished area near the columella; outer lip made crenulate by the ribs. Cream or grey, sometimes reddish brown between the ribs.

4.5 cm high; 4.0 cm wide. Central Indo-West Pacific; Shark Bay, WA to central Qld. Inhabits rocky shores, usually in muddy areas.

Tectus pyramis (Born, 1778)

Pl. 7; fig. 17 a-b   Height and width about equal, suture impressed; sides of whorls flat, usually smooth except for growth lines crossed by fine oblique striae and weak nodules at the suture but occasional specimens are weakly spirally ribbed; base finely spirally striate except on a polished area at the centre. Sides grey, cream, pink or green; base white, green or blue-green.

8 cm high. Indo-West Pacific; Rottnest I., WA to southern Qld. Abundant in coral reef and rocky shore habitats. Synonym: *obeliscus* Gmelin, 1791.

Tectus triserialis (Lamarck, 1822)

Pl. 7; fig. 12　Tall (apical angle about 52°), spire flat-sided, suture impressed; whorls slightly concave above, convex below; sculptured with spiral rows of weak obliquely elongate ridges below the suture and at the periphery and two spiral rows of small, widely spaced nodules between, and with faint irregular oblique cords throughout; base finely corded; a spiral ridge at the base of the aperture is not developed. Sides of whorls green but with a yellow line beneath the suture; the white of the base extends onto the periphery so that there is a white line above the suture on the spire whorls ascending spirally to the apex.

3.5 cm high; 2.8 cm wide. Indo-West Pacific; north Qld. A coral reef species.

▲ GENUS **TROCHUS** Linnaeus, 1758
Trochus maculatus Linneaus

Conical, top-shaped, periphery rounded or angulate; base flat; umbilicate, the umbilicus funnel-like and lirate, calloused; columella smooth or multi-dentate, separated by a gap from the parietal wall which it meets at a high angle; corded and nodulose or granulose.

Indo-West Pacific Region. *Trochus* s.s. first appears in the fossil record in the Miocene, much later than *Tectus*.

Trochus hanleyanus Reeve, 1842

Pl. 7; fig. 14 a-c　Periphery angulate, sometimes sharply so, base sometimes concave; columella smooth or weakly toothed; sides sculptured with finely granulose spiral cords and often elongate oblique nodules widely spaced around the lower part of the whorls; base finely spirally corded. Cream or green with red oblique lines, sometimes uniformly red on the sides; base with crowded, thin, wavy, radiating lines.

5 cm high. Indo-West Pacific; Abrolhos, WA to southern Qld. Very common in coral reef and rocky shore habitats. Some specimens are difficult to distinguish from *T. maculatus*. Synonym: *lineatus* Lamarck, 1822 (not da Costa, 1778).

Trochus histrio Reeve, 1842

Pl. 7; fig. 11 a-b　Periphery angulate, whorls slightly concave, sculptured with 3-4 spiral rows of nodules and a peripheral rib with semi-spinose nodules; base with 6-7 rather widely spaced, finely nodulose spiral ribs; side wall of aperture spirally ribbed, basal part smooth except for a few small denticles near the columella.

4 cm high. Central Indo-West Pacific; North West Cape, WA to Qld. This is the smallest Australian species of the genus. It lives among living corals in the outer intertidal and sublittoral zones. Synonym: *calcaratus* Souverbie, 1875.

Trochus maculatus Linnaeus, 1758

Pl. 7; fig. 15 a-b　Periphery rounded or subangulate; columella thick, with about five strong spiral folds which end in broad rounded tubercles at the margin; outer lip smooth, aperture finely lirate, with strong spiral ridges near the base; sides sculptured with finely beaded spiral ribs; base with more regular spiral nodulose ribs. Cream to white with red-brown or green axial rays, sometimes uniformly brown.

6 cm high. Indo-West Pacific; Abrolhos, WA to southern Qld. There are many synonyms.

Trochus niloticus Linnaeus, 1767

Pl. 7; fig. 16　Adult shells smooth, very heavy with a thickened, spreading peripheral rim; body whorl with concave sides; juveniles and subadults weakly sculptured with nodulose spiral cords; columella long, curved, smooth, ending abruptly at a basal notch; outer lip and aperture smooth. Off-white with oblique reddish stripes.

15 cm high. Indo-West Pacific; Kimberley, WA to Qld. This is the commercial trochus fished for its meat and mother-of-pearl shell. Common in intertidal and shallow sublittoral zones.

Trochus cf. stellatus (Gmelin, 1791)

Pl. 7; fig. 10　Sides slightly convex, periphery subangulate; base slightly convex, aperture quadrate; whorls with three or four beaded spiral cords, the presutural row may be prominently nodulose but on the body whorl it becomes divided into several small beaded cords; base with about eight close and finely beaded spiral lirae, umbilicus finely lirate within; columella weakly toothed, aperture wall lirate. Fawn or greenish with radial reddish bands; basal cords red-spotted.

3 cm high. Central Indo-West Pacific; Dampier Archipelago, WA to central Qld. This identification is provisional; the shells are close to *T. maculatus* and may be one of the many variants of that species.

SUBFAMILY **UMBONIINAE**

Shell form is very variable and ranges from low-spired and lenticular to tall-spired and elongate. The shells are coarsely sculptured or glossy smooth, and are often highly coloured. The anatomy is distinctive: gill monopectinate; foot bifid anteriorly; snout small, tapering anteriorly and covered with prominent setiform papillae; with four pairs of epipodial tentacles along the sides of the foot; epipodial neck lobes highly modified, left neck lobe forming a tentacle-like screen which filters the water entering the mantle cavity; eyes large, terminal, on upturned stalks. Radula small, simplified with the loss of teeth in the central field, marginal teeth well developed with prominent serrate cusps.

The members of this subfamily are suspensory feeders and live on soft substrates in the sublittoral zone. The epipodial neck lobes are modified to form tentacular particle exclusion screens and may be rolled to form incurrent or excurrent siphons. The foot is highly active and produces jumping, rolling and swimming behaviour in response to disturbance. Hickman & McLean (1990) have shown that several groups previously regarded as taxonomically distant belong together here, in spite of very different shell form. They recognise three tribes.

TRIBE **BANKIVIINI**

Shells glossy, tall-spired and elongate, with or without an umbilicus. Eye stalks reduced, subequal, eyes subterminal or lacking; left neck lobe either unrolled or rolled and semitubular, right neck lobe rolled into a tubular excurrent siphon.

Leiopyrga lineolaris
(Redrawn from Hickman & McLean, 1990, p. 132, fig. 87b).

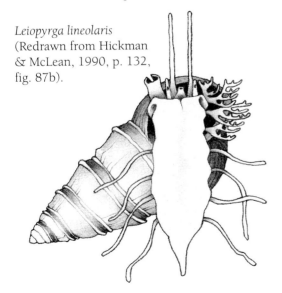

The tribe is endemic to the southern coast of Australia. Although these animals have been dubbed "kelp shells", Hickman & McLean (1990) have shown that they live in sand in the sublittoral zone and feed on suspended particles. In spite of their elongate shells their anatomy shows that they are related to the umboniums.

▲ GENUS **BANKIVIA** Krauss, 1848
Phasianella fasciata Menke

Thin-shelled, very slender with a high, turreted spire, whorls convex, glossy smooth and shining; columella arcuate, smooth, ending abruptly (truncated), umbilicus lacking.

Western Pacific and Southern Australian Regions. Monotypic.

Bankivia fasciata (Menke, 1830)

Pl. 9; fig. 9 a-j   Colour extremely variable, may be unicoloured white, pink, grey, purple, or brown, usually vari-coloured with bands or reticulate lines.

20 mm high; 8 mm wide. NSW to SA. One of Australia's most common and colourful shells, which exhibits a remarkable degree of pattern polymorphism. Dead shells are abundant in beach drift in south-eastern Australia.

▲ GENUS **LEIOPYRGA** Adams & Adams, 1863
Cantharidus lineolaris Gould

Whorls carinate, sculptured with beaded spiral riblets; narrowly umbilicate; columella curving to meet the outer lip without an abruptly truncated end.

South West Pacific and Southern Australian Regions.

Leiopyrga cingulata (Adams, 1863)

Spire whorls sharply carinate just above the sutures, slightly concave above the periphery and angled below; with a prominent spiral rib (cingulus) around the periphery and two or three weaker ribs around the upper parts of the whorls; umbilicus relatively wide and bordered by a sunken area on the base. Rose or fawn, with wavy oblique darker lines and sometimes a white spiral band below the peripheral keel and another on the base; darker rose spots often adorn the bands.

8 mm high; 4 mm wide. North Qld to central NSW.

Leiopyrga lineolaris (Gould, 1861)

Whorls almost flat-sided but rather sharply carinate at the anterior periphery so that the spire is distinctly reversely turreted, body whorl broad; glossy smooth except for a few spiral striae on the base. Cream, rose or tan, with zigzag brown lines and a row of spots below the suture and sometimes around the periphery. See also diagram on p. 91.

9 mm high; 5 mm wide. NSW to eastern Vic. Synonym: *picturata* H. & A. Adams, 1863.

Leiopyrga octona (Tate, 1891)

Whorls slightly convex, only very weakly peripherally carinate; sculpture of eight equidistant, smooth spiral ribs on the sides of the whorls, seven or eight on the base. White with oblique pink or brown lines which may coalesce on the lower parts of the whorls into a reticulate pattern.

12 mm high; 5 mm wide. Southern NSW to Cape Leeuwin, WA. The subspecies *B. octona problematica* (Iredale, 1924) described from Twofold Bay, NSW has a brown spiral band beside the umbilicus. Further study is needed to determine whether the NSW form is consistently different from the southern population.

TRIBE **MONILEINI**

Shells conical to lenticular, umbilicate. The left neck lobe is modified into a particle exclusion filter by subdivision of the margin into a series of muscular, tentacular processes.

▲ GENUS **ARCHIMINOLIA** Iredale, 1929
Monilea oleacea Hedley & Petterd

With characters of the type species.

Eastern Australia. Monotypic.

Archiminolia oleacea (Hedley & Petterd, 1907)

Turbinate, thin-shelled, with moderately high spire and rounded whorls, though the base is rather flattened and there is a narrow flat step below the suture; periphery weakly carinate; columella simple, funicle narrow and weakly nodulose; sculpture of incised spiral lines which become deep and prominent at the centre of the whorls, shoulders and base smoother; umbilicus deeply perspective, its walls spirally striate, funicle a thin, weakly nodular marginal rib. Fawn with slightly darker oblique-axial lines on the sides.

1.6 cm high; 1.2 cm wide. NSW to SA. Dredged in deep water; the type came from 456 m off Sydney.

▲ GENUS **CONOTALOPIA** Iredale, 1929
Monilea henniana Melville

Turbinate, spirally ribbed, carinate; umbilicus wide and funnel-shaped, bordered by a nodulose funicle; columella arched, simple, meeting the parietal wall at a steep angle.

Eastern Australia. Iredale introduced this generic name with *henniana* as the type species but included *tropicalis* provisionally. Both species seem close to *Minolia*.

Conotalopia henniana (Melvill, 1891)

Small, whorls prominently and sharply bi-carinate, with a wide, nearly horizontal zone above the shoulders and nearly vertical, concave sides; base convex; umbilicus with a wide funnel-shaped entrance, bordered by a weakly nodulose spiral rib, cancellate within; external surface corded. Flat upper zone fawn, sides and base fawn, with wide axial rays and small brown spots on the ribs.

2 mm high; 4 mm wide. Qld.

Conotalopia tropicalis (Hedley, 1907)

Small, rather thin, whorls angulate at the shoulder, centre and base; three strong spiral ribs encircle the centre of the whorls, with wide spaces between; the flat subsutural shelf is crossed by radial plicae; four small spiral ribs on the base plus a strong nodulose funicle bordering the wide umbilicus; the whole surface covered with fine axial threads. Grey, pinkish on the body whorl, with scattered crimson dots on the larger ribs.

3.15 mm high; 3.75 mm wide. Described from Mast Head I., Capricorn Group, Qld.

▲ GENUS **ETHALIA** Adams & Adams, 1854
Rotella guamensis Quoy & Gaimard

Suture distinct; callus plug reflected over but not entirely closing the umbilicus.

Indo-West Pacific. Synonym: *Liotrochus* Fischer, 1879.

Ethalia guamensis (Quoy & Gaimard, 1834)

Pl. 11; fig. 13 a-c Whorls convex, smooth periphery subangulate to rounded; callus plug coils around the end of the columella where it meets the parietal wall, turning into but not covering the umbilicus. Cream or white with spiral rows of tiny rose arrowhead markings, brown blotches below the suture, and broad rose flames at the periphery; an orange band surrounding the umbilicus; callus plug rose.

1.4 cm high; 1.9 cm wide. Indo-West Pacific; eastern Qld. Hickman & McLean (1990, p.128, fig. 83b) illustrated the external anatomy of this species.

Ethalia pulchella (Adams, 1855)

Pl. 11; fig. 15 a-b Conical, spire pointed and flat-sided, periphery subangulate; callus plug as in *guamensis*. Cream with wide light brown or reddish brown blotches or radial flames below the suture which sometimes extend to the periphery, usually with thin oblique or zigzag lines on the lower part of the body whorl, and a row of brown and white spots around the periphery; base plain fawn or rose, with radial lines, or mottled; umbilical plug brown.

7 mm high; 10 mm wide. Indo-West Pacific; Fremantle, WA to Qld.

▲ GENUS **ETHMINOLIA** Iredale, 1924
E. probablis Iredale

Turbinate; widely umbilicate, funicle lacking, edges of umbilicus neither crenulate nor dentate; periphery more or less angulate; columella simple.

A small Australian genus.

Ethminolia probabilis Iredale, 1924

Protoconch of about three smooth whorls, merging into spirally striate teleoconch whorls; early teleoconch whorls with flat, sharp-edged shoulders becoming less sharp-edged on the body whorl; edge of umbilicus rounded. Fawn with wide rosy-brown radial bands and zigzag lines.

6 mm high; 8mm wide. NSW. Taken in shallow water dredgings on sandy substrates. *E. elveri* Cotton & Godfrey, 1938, is said to be a larger, higher and less shouldered species from SA.

Ethminolia vitiliginea (Menke, 1843)

Pl. 11; fig. 16 a-b Thin-shelled, whorls with a flat zone at the shoulder and subcarinate periphery; sculpture of very fine spiral striae; umbilicus wide and perspective; columella thin-edged, slightly reflected where it meets the parietal wall. Olive-green, fawn, pink or yellow, usually with darker narrow spiral lines which may contain white spots.

8 mm high; 11 mm wide. Bass St. to Fremantle, WA. Synonym: *tasmanica* Tenison Woods, 1877.

▲ GENUS **ISANDA** Adams & Adams, 1854
I. coronata Adams

Small, glossy, globose; shoulders coronate; umbilicus wide, bordered by a beaded funicle; collumella arched, meeting the parietal wall.

Indo-West Pacific and New Zealand. The genus has a long fossil history and a wide Recent distribution. In the *Treatise on Invertebrate Paleontology* several generic names introduced by Iredale (1929) (*Conotalopia, Archiminolia, Parminolia, Vanitrochus*) are treated as subgenera of *Isanda* but they are given generic rank here until relationships among them are determined.

Isanda coronata Adams, 1854

Globose; glossy and smooth, sculpture lacking except for a row of pointed nodules below the suture and a double row of rectangular nodules forming a thick funicle around the umbilicus, the inner row larger than the outer; columella nearly vertical, meeting the outer lip at an abrupt angle. Colour and pattern extremely variable, usually shades of olive-green, fawn or grey with irregular axial lines.

7 mm high; 8 mm wide. Central Indo-West Pacific; Exmouth Gulf, WA to central Qld. May be extremely abundant on muddy sand flats.

▲ GENUS **MONILEA** Swainson, 1840
Trochus callifera Lamarck

Turbinate, whorls rounded or with angulate shoulder and periphery, sculptured with fine spiral cords; umbilicus narrow, bordered by a finely nodulose funicle; columella arched, terminating abruptly, its upper part thickened and reflected over the umbilicus.

Indo-West Pacific and tropical eastern Pacific. These trochids live on sandy or muddy substrates. Synonym: *Talopia* Gray, 1842.

Monilea belcheri Philippi, 1849

Pl. 11; figs. 20 a-b   Depressed, spire whorls with weakly carinate shoulders, base convex; spire whorls sculptured with 3-4 finely beaded spiral cords and oblique striae on the upper parts of the whorls, the base with numerous thin beaded spiral cords; umbilicus rimmed by a thick funicle; columellar callus small. Creamy-white, rose or fawn, cords spotted, with rose, green and brown axial streaks on the sides.

11 mm high; 17 mm wide. Western Pacific; north Qld.

Monilea callifera (Lamarck, 1822)

Pl. 11; figs. 19 a-c   Spire whorls with four weakly beaded primary spiral ribs and smaller secondaries between; body whorl with seven primary spiral ribs which have secondaries close-appressed so that they seem double, primary and secondary ribs sub-equal at the periphery, seven spiral ribs on the base, the inner three smaller, with conspicuous crossbars in the interspaces. Cream or pale grey with irregular, diffuse, oblique dark grey flames.

2 cm wide. Cockburn Sound, WA to central Qld. Lives on slightly muddy sand substrates, usually in the sublittoral zone. The sculpture is quite variable. Hickman & McLean (1990, p. 128, fig. 83c) illustrated the external anatomy. *M. lentiginosa* Adams, 1851 is a similar species from Qld; further study is needed to determine the relationship of the two.

Monilea morti Iredale, 1929

Pl. 11; fig. 18   Large and solid, with a high, turreted spire; whorls shouldered and peripherally subangulate, sculptured with many slightly nodulose spiral ribs which are crowded and of varying size; penultimate and body whorls rather flat-sided; umbilicus wide and funnel-shaped; columellar callus small. Fawn, ribs paler than the interspaces, with axial brown flames on the lower parts of the whorls.

Holotype: 2.2 cm high; 2.2 cm wide. Central Qld to NSW. This high-spired form may be only a variant of *M. callifera* or *M. lentiginosa*.

▲ GENUS **PARMINOLIA** Iredale, 1929
Monilea apicina Gould

With characters of the type species.

Eastern Australia. Monotypic.

Parminolia apicina (Gould, 1861)

Depressed-turbinate, width greater than height, spire elevated, whorls subangulate at the shoulder and with a slightly sloping subsutural zone and subangulate periphery; shoulder zone smooth, sides and base finely, spirally threaded; outer lip sharp-edged, bevelled on its inner side, aperture round; columella simple, curved, thickened behind by the umbilical fold; umbilicus deep but narrow, its wall forming a thick fold which is separated from the base by a groove. White, often with a few broad but irregular rose flames on the sides.

6 mm wide. Qld. Habitat and distribution remain uncertain. Synonym: *agapeta* Melvill & Standen, 1869.

▲ GENUS **ROSSITERIA** Brazier, 1895
Trochus nucleus Philippi

With characters of the type species.

Central Indo-West Pacific. Sometimes treated as a subgenus of *Monilea*.

Rossiteria nucleus (Philippi, 1849)

Globose, solid, whorls convex, slightly flattened below the suture; sculpture of spiral cords, crossed by axial striae; columella with 3-4 small plications; a callus plug partly covers the narrow umbilicus. Fawn, cream or rose, body whorl spotted and with a pale spiral band around the periphery.

1 cm high. Indo-West Pacific; Qld. For an illustration see Hickman & McLean, 1990, p. 127, fig. 82 d.

▲ GENUS **TALOPENA** Iredale, 1918
Monilea incerta Iredale

Spirally corded; umbilicus narrow, with a smooth, rounded margin terminating at a thickening at the base of the columella; with a second rounded shoulder deep within the umbilicus terminating at a reflected notch high on the columella just below the parietal wall.

South Pacific and eastern Australia. Sometimes treated as a subgenus of *Monilea*. The Qld shell *M. glaphyrella* (Melvill & Standen, 1895) might belong in this subgenus.

Talopena gloriola Iredale, 1929

Whorls slightly flattened below the suture, with subangulate shoulders and periphery; sculpture of numerous finely granulose spiral cords on the sides, becoming spiral striae on the base. Fawn with rows of rose blotches on the shoulders and on the periphery and interrupted spiral lines of white and brown dashes.

1 cm high. South-eastern Australia; NSW to SA.

Talopena vernicosa (Gould, 1861)

Pl. 11; figs. 17 a-b   Thin-shelled, whorls convex; finely spirally striate; umbilicus small, almost covered by the reflected columellar callus. Buff, fawn or rose, with irregular axial blotches and spiral lines of alternating white and brown dashes; columellar callus green, tan or rose.

1.2 cm high. Indo-West Pacific; Abrolhos, WA to Qld. Synonym: *lifuanua* Fischer, 1878.

▲ GENUS **VANITROCHUS** Iredale, 1929
Solariella tragema Melvill & Standen

With characters of the type species.

Central Indo-West Pacific and eastern Australia. Iredale indicated that *Trochus semiustus* Fischer, 1879 may also belong here.

Vanitrochus tragema (Melvill & Standen, 1896)

Minute, turbinate, with high spire, markedly convex whorls and conspicuous flat subsutural area; sculpture cancellate with strong spiral cords crossed by weak axials; aperture round, outer and columellar lips thin, simple and continuous; umbilicus wide and deep, spirally corded within. White, with a spiral row of reddish brown spots on the subsutural area and several spiral rows of dashes at the periphery.

4 mm high; 3 mm wide. North Qld.

TRIBE **UMBONIINI**

Shells lenticular, smooth to glossy, umbilicus completely filled by a thick callus plug. Head tentacles asymmetrical, the right tentacle longer and arising from the centre of the head, left tentacle held erect, enveloped by the rolled left neck lobe; left neck lobe rolled to form a siphon with a terminal tentacular screen, right neck lobe also rolled to form a tubular excurrent siphon.

▲ GENUS **UMBONIUM** Link, 1807
Trochus vestiarius Linnaeus

Solid, glossy smooth, lenticular, suture indistinct; umbilicus with a circular callus plug; columella simple, curved; outer lip simple, aperture smooth within.

Indo-West Pacific. Synonyms: *Globulus* Schumacher, 1817; *Rotella* Lamarck, 1822.

Umbonium vestiarium (Linnaeus, 1758)

Pl. 11; figs. 14 a-f   Small, disc-shaped, flat-based; periphery rounded; whorls glossy smooth and lacking any sculpture. Colour extremely variable, usually shades of olive-green or grey but may be rose, yellow or black, with various radial or spiral patterns.

6 mm high; 12 mm wide. Indo-West Pacific; possibly north Qld. May be extremely abundant in shallow-water sandy substrates. However, the Australian records of this species are doubtful and its presence in our northern fauna needs confirmation. No two shells seem to be patterned the same.

FAMILY **TURBINIDAE**

Turban Shells and their Relatives

The family is characterised by thick, few-whorled, turbinate or conical shells. Typically the shell interior is nacreous although nacre is lost in some of the advanced groups. The operculum is variable in form and composition, ranging from multispiral to paucispiral and from corneous to heavily calcified. The operculum has a long growing edge. Characteristically the radula has broad rachidian teeth which lack shafts and have secondary flaps or cusps providing additional attachment to the radular membrane; there are five pairs of laterals of similar size and form along each side with distinctive bends in their shafts and expanded, interacting bases.

In many respects turbinids resemble trochids but the radula is distinctive. Turbinids are herbivores. Little is known of their biology. Most are spawners but some species produce gelatinous egg-masses.

Turbinids live mostly in shallow tropical and warm-temperate seas. Some species of *Turbo* are fished commercially for their meat or mother-of-pearl shell.

There are nine subfamilies, of which seven are represented in the Australian fauna. One, the Angariinae, has an uncalcified operculum. In the Liotiinae the outer surface of the operculum is merely granulated or beaded with calcium carbonate. But in all other respects the anatomy of both groups is turbinid. The classification used here follows Hickman & McLean (1990) Systematic revision and suprageneric classification of trochacean gastropods. *Natural History Museum of Los Angeles County, Science Series*, 35: 169pp.

SUBFAMILY **ANGARIINAE**

Dolphin Shells

Dolphin shells are relatively large, turbinate, with an entirely corneous, multispiral operculum which is not enveloped by the foot and has a long growing edge. Three pairs of epipodial tentacles lie alongside the operculum; neck lobes are narrow with unsubdivided margins. The radula is markedly asymmetric with large inner marginal tooth cusps and reduced central tooth field.

This is a very small group. In the past it has been regarded as a separate family or as a subfamily of the Trochidae. There is only one living genus. In spite of the local abundance of some species in their shallow water reefy habitats, there is nothing recorded of their natural history. They are found associated with algae and it can be assumed that they are herbivorous.

▲ GENUS **ANGARIA** Röding, 1798
Turbo delphinus Linnaeus

Medium-sized to large, periphery carinate, body whorl inflated, anterior end produced, spire low-conical to almost flat; umbilicus wide and deep, may be bordered by a thick funicle; sculpture of spinose spiral ribs, aperture round, nacreous; operculum thin, round, concave, multi-spiral and horny.

Indo-West Pacific. Due to shell variability and the lack of any thorough revision it is unclear how many species there are, but at least two are common in Australian waters. Synonym: *Delphinula* Lamarck, 1804.

Angaria delphinus (Linnaeus, 1758)

Pl. 7a; fig. 1 a-b, Pl. 11; fig. 22 Spire flat, early whorls, sharply angulate at the shoulders, each with six angulate corners; body whorl rounded, sometimes only loosely attached at the suture, or even quite separate toward the lip; aperture round, peristome more or less entire, often produced anteriorly, silvery smooth within; sculptured with roughly squamose spiral cords and about six spiral rows of hollow spines, the shoulder row often long and curved. Grey, cream or rose-tinted base; the crowded cords and spines are purple-black or purple-rose so that colour may dominate.

6 cm high; 7 cm wide. Indo-West Pacific; Exmouth Gulf, WA to southern Qld. The species may be very abundant on intertidal reefs in northern Australia. There is extraordinary variation in the development of the spines and some synonyms are currently used as "form" names. For example, heavy specimens (see Pl. 11, fig. 22) having a pronounced funicle and wide, flat fronds instead of long curving spines are sometimes referred to as "forma *aculeata* Reeve, 1842". There is doubt about the technical status of this name but it may be useful. Synonyms: *laciniata* Lamarck, 1819; *atrata* Reeve, 1843.

Angaria tyria (Reeve, 1843)

Pl. 11; fig. 21 a-b Like *A. delphinus* but less heavily spinose and lighter coloured; body whorl cream but with a broad purple band on the subsutural area and another surrounding the umbilicus.

5 cm high; 6 cm wide. Fremantle to North West Cape, WA. The shells of this species are almost as variable as those of its northern cousin. Most Cockburn Sound specimens lack shoulder spines and are sculptured with crowded spiral rows of close, short, concave scales and there is no thick

funicle bordering the umbilicus. In a few specimens some of the scales are produced into hollow spines, those on the shoulder being long and curved over the spire; there is a population exhibiting this condition in Shark Bay. In fact there seems to be a trend towards greater spinosity at the northern end of the geographic range. A comparison of the biology, morphology and genetics of *A. tyria* and *A. delphinus* would make a fascinating study.

SUBFAMILY **COLLONIINAE**

The shells are small (usually less than 10 mm), thick and sturdy, with a solid, calcified operculum which is partially enveloped by the metapodium. There is a tendency for asymmetric development of the epipodial tentacles with up to six on the right side and as few as three on the left. Head lappets are absent and neck lobes are either poorly developed or absent. The radula is usually of the primitive, symmetrical turbinid type.

This group was originally introduced for fossil forms but Hickman & McLean (1990) redefined it and included several living genera previously placed in other groups. They synonymised the group previously known as the Homalopomatinae.

▲ GENUS **ARGALISTA** Iredale, 1915
Cyclostrema fluctuatum Hutton

Minute but solid, turbinate with rounded, spiral, striate whorls; aperture approximately round; umbilicus narrow but deep; columella vertical, curved, united by a callus with the parietal wall; operculum with 4-6 whorls, outside calcareous, glassy-smooth, translucent, slightly concave in the centre.

Southern Australia and New Zealand. The type species is from New Zealand. May be treated as a subgenus of genus *Homalopoma* Carpenter, 1864.

Argalista corallina (Cotton & Godfrey, 1935)

Whorls four, closely spirally ridged. White, dotted with rose spots and with broad radial rose flames formed by alignment of short spiral lines.

2 mm wide. St. Vincent Gulf, SA to Rottnest I., WA. Other southern Australian species are *roseopunctata* (Angas, 1880), *rosea* (Tenison Woods, 1876) and *A. fugitiva* (Hedley, 1911). Affinity of this species with *Collonista* needs to be reviewed.

▲ GENUS **COLLONISTA** Iredale, 1918
L. costata Pease (= *verruca* Gould)

Small, solid, turbinate, whorls rounded, spirally ribbed; aperture round, columella arcuate and wide; umbilicus narrow or lacking; operculum thick, multispiral.

Indo-West Pacific. Iredale introduced *Collonista* as a replacement name for *Leptothyra* Pease 1869 (not Cooper, 1867). In addition to the type species the following species have been reported from Australian waters: *laeta* Montrouzier, 1863; *nanina* Sowerby, 1864; *picta* Pease, 1869.

Collonista cf. **verruca** (Gould, 1845)

Spiral ribs weak, of varying size, with striae in the interspaces; whorls five, suture indistinct; outer lip thick and smooth. White, with spiral bands of red and brown patches, sometimes entirely brown or red.

5 mm wide; 5 mm high. Indo-West Pacific; Margaret River, WA northwards. Synonym: *costata* Pease, 1869.

SUBFAMILY **GABRIELONINAE**

Shells small to minute and smooth with a distinct apertural ridge and an internal palatal sulcus. The operculum is calcified, exteriorly concave and finely sculptured, with a prominent bordering ridge.

There are only two genera in this subfamily. Their anatomy is unknown. *Gabrielona* has been allied to *Tricolia*. The naticoid operculum led some authors to the opinion that *Gabrielona* belongs to the Naticidae. Most recently Hickman & McLean (1990) erected a new subfamily within the Turbinidae for it and the Caribbean genus *Eugabrielona*.

▲ GENUS **GABRIELONA** Iredale, 1917
Phasianella nepeanensis Gatliff & Gabriel

Small, globose, apex insert or barely exsert; teleoconch whorls rounded, suture impressed; aperture semi-circular, with a spiral suclus below the suture, outer lip sharp-edged; columella with 0-3 low denticles; umbilicate; operculum with a pronounced marginal ridge and concave centre.

Indo-West Pacific and Southern Australian Regions. There are four described species, one in south-eastern Australia. Protoconch details suggest planktotrophic development. The genus was reviewed by Robert Robertson (1973, *Indo-Pacific Mollusca* 3 (14): 41-61).

Gabrielona nepeanensis (Gatliff & Gabriel, 1908)

Height slightly greater than width; two small denticles present on the columella; whorls slightly flattened below the suture; sculpture lacking. Pink with variable pattern of wavy stripes or spots, a spiral white band present at the shoulder.

2 mm high. Vic. and SA. Although common in beach litter, so far this species has been taken alive only rarely and little is recorded about its habitat or its range. There are two other species in the Indo-West Pacific: *G. pisinna* Robertson, 1973 and *G. raunana* Ladd, 1966. Both could occur in northern Australia. The former is a much smaller shell than *G. nepeanensis*, with a lower spine and finer colour pattern. It is known from New Caledonia, as is *G. raunana*, distinguished by prominent axial sculpture on the early whorls.

SUBFAMILY **LIOTIINAE**

Liotines and Cyclostremes

Shells are minute to small, usually depressed-turbinate or lenticular, with prominent axial and spiral sculpture, thickened outer lip, descending suture on the body whorl, wide umbilicus, and nacreous interior. The periostracum is calcified. Most characteristic is the round, multispiral, horny operculum impregnated with calcareous granules or beads arranged in a radiating pattern on the outer surface. Most of the gill is bipectinate and the anterior tip is free. There are three to four pairs of epipodial tentacles. The radula is symmetrical with equally developed central and marginal tooth fields but the rachidians are varied.

The unique operculum distinguishes the group from the Colloniinae in which the operculum has a solid calcareous outer layer. However, opercular details are not known for many species. Nor is there much information on the natural history of these tiny molluscs.

Many small Australian species are assigned to this subfamily but their taxonomy is very confused. Hickman & McLean (1990) have shown that the cyclostremes belong with the liotines, that is, Cyclostrematidae is a synonym of Liotiinae, although many genera commonly attributed to the former group actually belong to the family Skeneidae.

Early Australian workers named quite a number of species as members of the cosmopolitan genera *Liotia* Gray, 1847 (*cancellata* Gray) and *Cyclostrema* Marryat, 1818 (*cancellatum* Marryat) but most if not all of them belong to other genera and other families. (See also the family Skeneidae.) Some representative genera and species are included here.

References

Laseron, C. (1954) Revision of the Liotiidae of New South Wales. *Australian Zoologist* **12** (1): 1-25.

Laseron, C. (1958) Liotiidae and allied molluscs from the Damperian zoogeographical province. *Records Australian Museum* **24**: 165-182.

Jenkins, B.W. (1984) Southern Australian Liotiidae. *Australian Shell News* **47**: 3-5.

Jenkins, B.W. (1984) Northern Australian Liotiidae. *loc. cit.* **48**: 8-9.

Tate, R. (1899) Revision of Australian Cyclostrematiidae and Liotiidae. *Transactions Royal Society SA* **22** (2): 223-230.

▲ GENUS **AUSTROLIOTIA** Cotton, 1948
Liotia botanica Hedley

Shell depressed, moderately heavily sculptured, aperture trumpet-shaped, not strongly varixed; umbilicus wide, perspective; operculum multispiral, with traces of calcareous granules.

Southern Australia and New Zealand. The above diagnosis is taken directly from Cotton (1959). See notes on *Liotina*.

Austroliotia australis (Kiener, 1839)

Subdiscoidal, depressed, spire very low, whorls and periphery rounded, body whorl conspicuously descending near the lip; outer lip thick with a double, tuberculose, reflected varix; peristome complete; whorls finely spirally ribbed, radially lamellose between the ribs; umbilicus wide, with a sharply dentate funicle. Yellow.

1 cm high; 1.4 cm wide. Sydney, NSW to Fremantle, WA. Under stones in shallow water.

Austroliotia botanica (Hedley, 1915)

Subdiscoidal; whorls 4-5, rounded, body whorl suddenly descending at the lip; sculpture cancellate, with seven spiral cords, two on the shoulder, three on the periphery, one at the margin of the umbilicus and one within it, and about 20 radials which project over the suture and form polished knots

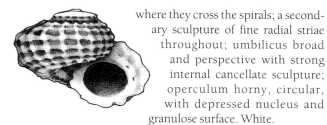

where they cross the spirals; a secondary sculpture of fine radial striae throughout; umbilicus broad and perspective with strong internal cancellate sculpture; operculum horny, circular, with depressed nucleus and granulose surface. White.

7 mm wide; 5 mm high. Frazer I., Qld to northern Tas. and SA. Common under rocks in the intertidal and shallow sublittoral zones. Brian Jenkins (1984) pointed out that the Philippine species named *clathrata* Reeve, 1843 is very similar. Laseron (1958) named a small, high-spired subspecies from the NT *darwinensis*.

Austroliotia densilineata (Tate, 1899)

Depressed-turbinate; whorls rounded, body whorl descending at the lip; spiral sculpture obsolete, almost lacking on the base, radial striae prominent, crowded, forming a corona at the suture; umbilicus wide, sharply dentate around the rim; aperture round, lip slightly thickened but not varicose; columellar margin detached from the umbilical rim. White.

4.5 mm high; 7 mm wide. Lakes Entrance, Vic. to Albany, WA. For an illustration see Cotton, 1959, p. 213, fig. 131.

Austroliotia pulcherrima (Reeve, 1843)

Subdiscoidal, upper surface convex, lower surface concave, periphery rounded, body whorl suddenly descending from the suture near the lip; sculptured with seven rounded spiral ribs, crossed by radial ribs of almost equal size, about 30 on the body whorl, the radials narrow, sharp and lamellose, forming angulate nodules where they rise and cross the spirals and continue over the base onto the sides of the umbilicus, secondary sculpture of dense, radial, lamellose striae throughout; umbilicus wide and funnel-like, bordered by a thick spiral rib made slightly nodulose by the crossing radial ribs; aperture round, peristome complete, strongly varixed and with a secondary, crenulate inner lip. Yellow to cream.

7 mm wide. Sydney, NSW to Abrolhos, WA. Although this beautiful little species is quite common, at least in southern WA, it has been confused with *A. botanica* in most collections. It differs from that species in its sculptural details.

Austroliotia saxa (Laseron, 1954)

Depressed-turbinate, suture deep, with four whorls, angular at the periphery and shoulder; sculptured with four spiral ribs, those at the periphery and shoulder forming slight keels, crossed by broad, rounded, radial costae, fading on the base,

about 12 on the body whorl, rising into protuberances where they cross the spiral ribs, secondary sculpture of fine, sharp, closely packed radial lamellae extending across the base and into the umbilicus; aperture oblique, peristome complete, with a moderate varix; umbilicus wide and deep. Yellowish.

5 mm high; 7 mm wide. NSW. Dredged. For an illustration see the original description (1954, p. 2, figs. 2, 2 a).

Austroliotia scalaris (Hedley, 1903)

Turbinate, whorls rounded but made slightly angulate at the shoulder and basal periphery by moderate keels, body whorl descending at the lip; spiral sculpture of three ribs plus lesser cords, crossed by strong radial lirae forming pointed nodules at the intersections, radial lirae continue across the base with crowded, lamellose, radial striae throughout including into the umbilicus; umbilicus bordered by two spiral ribs, the upper one with prominent rounded nodules, the lower one less nodulose; peristome complete, outer lip moderately varixed. White.

6.5 mm high; 8 mm wide. Frazer I., Qld to Port Kembla, NSW. Dredged on the outer shelf. Hedley originally described this as a variety of *Liotia tasmanica* but later authors have shown that the resemblance is superficial. Cotton (1959) placed it in *Munditia* while Iredale & McMichael (1962) put it in *Austroliotia*.

▲ GENUS **CIRCUMSTELLA** Laseron, 1958
Liotia devexa Hedley

Shell small, flat above with a low, turreted spire; sculpture primarily of strong spiral keels, radial sculpture also present; aperture detached, oblique, twisting downwards, with a heavy double varix; umbilicus deep, rendered stellate by sharp, angular ribs, separated by deep pits which descend into the cavity.

Northern Australia.

Circumstella devexa (Hedley, 1901)

Turbinate, solid, nearly square in transverse section, body whorl descending rapidly near the lip, base flattened; suture deeply impressed, its edge crenulate; periphery flattened between upper and lower keels, the upper crenulate; umbilicus wide, bordered by thick, angular nodules; sculpture between the keels of fine radial threads; aperture semilunate, very oblique, with two massive lips, one within the other, the outer massive and forming a varix. Creamy white.

3.3 mm high; 4.5 mm wide. Darwin, NT to Lindeman I., Qld. Synonym: *walkeri* Sowerby, 1908. For a figure see Hedley's original description (*Proceedings Linnean Society NSW, 1901* (1): Pl. 2, figs. 4-6).

▲ GENUS **DENTARENE** Iredale, 1929
D. sarcina Iredale

Turbinate, suture channelled, umbilicus wide with an oblique strut connecting the outer wall to the lower end of the columella; periphery squared, with prominent nodules.

Indo-West Pacific. In addition to the type species, Iredale included *Dentarene munitus* Iredale, 1929 in his genus.

Dentarene sarcina Iredale, 1929

With three whorls, subsutural slopes smooth, convex, base convex, periphery squared, bearing stellate nodules which are oddly hollow on the apical sides; edge of suture crenulate, with fine, crowded, lamellose striae in the interspaces; umbilicus funnel-shaped, with angulate nodules at the rim and a thick strut emerging from deep within to connect with the lower end of the columella; aperture circular, heavily varixed behind, the varix surmounted by 5-6 spiral cords making its crest nodulose. Cream.

9 mm high; 12 mm wide. Indo-West Pacific; Shark Bay, WA to Capricorn Group, Qld. This was a replacement name for *crenata* Kiener, 1839, not Sowerby, 1833.

▲ GENUS **MUNDITIA** Finlay, 1926
Liotina tryphenensis Powell

Depressed-turbinate to flat and discoidal; umbilicus wide; protoconch of one whorl; outer lip usually varixed; body whorl keeled and nodulose; operculum multispiral, horny.

Southern Australia and New Zealand. The type species is from New Zealand. There are four named Australian species assigned to this group. They resemble the genus *Liotina* but their opercula are said to be simple and horny. However, this detail needs verification in some cases and the distribution of species between the two genera needs further study.

Munditia hedleyi (Pritchard & Gatliff, 1899)

Discoidal, spire almost flat, whorls angulated at the shoulder and basal periphery by strong keels; with one additional subsutural and one basal rib, the latter forming the umbilical funicle; radial ribs strong, about 20 per whorl, forming a ladder-like pattern between the peripheral keels and continuing onto the base and into the umbilicus, with rounded nodules at the intersections; secondary sculpture of

fine radial striae; umbilicus very wide and perspective, sections concave between the inner keels; aperture round, peristome complete, moderately varixed. Cream, tinted with brown.

2 mm high; 4 mm wide. Bass St. to SA.

Munditia mayana (Tate, 1899)

Depressed-turbinate, four whorls, rounded at the basal periphery, slightly angulate at the shoulder; suture not excavated; columellar margin detached from the umbilical rim, umbilicus wide, funicular rim dentate; aperture round, outer lip slightly varixed; sculptured with nodulose spiral ribs, one subsutural, four on the periphery, and one on the base, plus weak radial costae and numerous fine radial striae. White.

5 mm high; 7 mm wide. Lakes Entrance, Vic., Tas. and west to the south coast of WA. Resembles *M. subquadrata* but the suture is not excavated and the columellar margin and umbilical rim are not attached.

Munditia subquadrata (Tenison Woods, 1878)

Depressed-turbinate; suture channelled, crossed by radial lamellae; with four whorls, angular shoulders, a subsutural rib, two peripheral keels and two spiral ribs on the base, crossed by thick radials; finely striate throughout; umbilicus wide, with a thickly dentate edge, the rim connected to the lower end of the columella by an oblique strut; aperture round, outer lip heavily varixed, the varix reflected on the adapical side. Grey-white.

6 mm high; 9 mm wide. Vic. and Tas. to south coast of WA. Dredged to at least 300 m. See note on *M. mayana*.

Munditia tasmanica (Tenison Woods, 1875)

Discoidal, spire depressed, suture impressed; body whorl with two keels forming upper and lower peripheral angulations, sculptured with a nodulose subsutural spiral cord, two weak spiral cords between the keels, and thick radial lirae which form nodules where they cross the keels; peristome entire, body whorl descending near the lip; umbilicus very wide, spirally dentate, bordered by a heavy angulation. White.

3 mm high; 8 mm wide. NSW to SA. Dredged at depths to at least 170 m. Synonym: *incerta* Tenison Woods, 1876.

▲ GENUS **LIOTINA** Munier-Chalmas in Fischer, 1885
Delphinula gervillei De France

Moderately large; with a well-developed outer lip varix; peristome continuous; spirally and radially ribbed, making cancellate sculpture as in *Liotia*; umbilicus spirally ridged within, bordered by a thick, rugose rib, pitted outside; operculum with a calcareous outer layer of granules, tessellated, edges bristly.

Cosmopolitan. The spirally ribbed umbilicus and operculum details distinguish the group from *Liotia*. *Globarene* Iredale, 1929 (*cidaris* Reeve) is sometimes used as a subgenus. Shells of *Munditia* are similar but generally are more planorbid and their operculum is horny. The distinction between this group and *Austroliotia* also seems difficult to sustain.

Liotina crassibassis Smith, 1880

Turbinate, very thick and solid, with flat, sloping subsutural area, squared periphery and inclined base; radial sculpture dominant on the upper parts of the whorls, 10 thick radial costae cross the subsutural slope and periphery, weaken on the base in juveniles, absent there in adults; with two strong spiral keels at the shoulder and basal periphery, keels sharply angulate on the early whorls, thick and rounded on the body whorl of adults, nodulose where the radials cross, plus numerous spiral cords and fine, crowded, radial lamellae in the interspaces; base with small bead-like nodules arranged in spiral and radial rows; umbilicus narrow, funicle sharply dentate bordering the umbilicus, massive and curves to join the massively varixed outer lip to form an almost continuous circle around the aperture. Cream.

1.2 cm high; 1.5 cm wide. Shark Bay to Troughton I., WA. Common in the shallow sublittoral. The massive funicle-outer lip and beaded base distinguish this large species from *L. peronii* which sometimes occurs in the same habitats.

Liotina hermanni (Dunker, 1869)

Turbinate, whorls convex, suture impressed; radial sculpture dominates, consisting of strong angulate folds from the suture to the shoulder, then diminishing as they cross the periphery and the base to the thick funicular rib, spiral sculpture consisting of moderate cords crossing the spiral folds, secondary sculpture of minute, crowded axial striae throughout, prominent pits in the interspaces on the periphery and behind the funicular rib; aperture round, peristome complete, varix moderate; umbilicus narrow but deep, smooth internally. Cream.

9 mm high; 9 mm wide. Qld. Although this name has been used for the species it appears to be a *nomen nudum*. The umbilicus is not spirally ridged, making affinity with *Liotina* doubtful.

Liotina peronii (Kiener, 1839)

Subdiscoidal, spire low, base deep and inclined; with four whorls, made subangulate at the periphery by two double spiral ribs forming a double keel; base with two spiral ribs and a thick, nodulose funicle forming the margin of the umbilicus and continuing to meet the lip varix at the base of the columella; radial sculpture very prominent, consisting of strong costae across the subsutural slope, 10 on the body whorl, forming pointed protuberances where they cross the peripheral keels, and continuing across the base to join the funicle; secondary sculpture of subequal spiral and radial cords producing a cancellate pattern; peristome complete, outer lip with a double varix; operculum concave, with a tessellated edge and concentric rows of bead-like granules on the outer surface. Cream.

1.3 cm high; 1.9 cm wide. Indo-West Pacific; Fremantle, WA to northern NSW. This is the largest Australian member of the subfamily. It is common in the shallow sublittoral, especially among seagrasses. Hickman & McLean (1990: p. 40) described the unusual radula of this species.

SUBFAMILY PHASIANELLINAE

Pheasant Shells

Shells large, thin, glossy, with a variegated pattern and fine capillary lines; parietal wall with a weak ridge supporting the oval, calcareous operculum. Radula lacking rachidian teeth, laterals with a broad rectangular area of secondary attachment. Head lappets present and the neck lobes are broad and capable of rolling to form semitubular siphons. There are three pairs of epipodial tentacles and the front end of the foot is very active and may be rolled to form a slender proboscis-like structure.

Pheasant shells take their name from their gaudy colours and patterns. They are also called "painted ladies" for the same reason. The spiral capillary lines in the surface of the shell are unique to the group.

The subfamily is widespread in the Indo-West Pacific and southern Australian regions but has few species and only one genus. At one time *Gabrielona* and *Tricolia* were placed here also but they are now considered to represent separate subfamilies. All three groups have been sometimes classified as distinct families.

The animals are very active and fast moving. Little is known of their biology but they are herbivorous and generally associated with seagrasses or algal beds. Spawning and development details are not known.

▲ GENUS **PHASIANELLA** Lamarck, 1804
Buccinum australe Gmelin

Conical, elongate to sub-globose; whorls convex, glossy, brightly coloured and complexly patterned; periostracum lacking; non-umbilicate; aperture oblique, tear-drop shaped, angular above, rounded below; outer lip sharp-edged; inner lip narrow, formed by porcelaineous callus on the parietal wall, meets the outer lip in a continuous curve; operculum oval, pointed posteriorly, paucispiral, made of thick calcareous material on a thin horny inner layer.

Indo-West Pacific and Southern Australian Regions. Synonyms: *Orthomesus* Pilsbry, 1888 (*variegata* Lamarck); *Mimelenchus* Iredale, 1924 (*ventricosa* Swainson).

Phasianella australis (Gmelin, 1791)

Pl. 11; fig. 1 a-b   Tall-spired; protoconch of two tiny whorls; teleoconch of six whorls with deep sutures; height about twice the width. There are several distinct colour patterns; base colour usually fawn, rose or yellow with white spiral bands bearing axial rose arrow marks or irregular blotches, sometimes there are irregular axial bands.

10 cm. Bass St. to about Geraldton, WA. This is the largest and most spectacular of the species, living in seagrass beds, usually in sheltered bays.

Phasianella solida (Born, 1778)

Pl. 11; fig. 3 a-b   Tall-spired; protoconch minute, of about one whorl; with $4^{1}/_{2}$ teleoconch whorls; height almost twice the width. Colour and pattern extremely variable, usually with a base colour of pink or rose, and irregular axial bands of yellow which may be continuous or broken up into spots and flecks, characterised by many red spiral lines which may be continuous or broken up into rows of dots or dashes.

2 cm. Indo-West Pacific; south coast of WA to NT. Abundant on intertidal reefs and algal beds in shallow bays and lagoons. A small shell from central Vic. and SA may be regarded as a subspecies with the name *P. s. angasi* Crosse, 1864. See remarks on *P. variegata*.

Phasianella variegata Lamarck, 1822

Pl. 11; fig. 4 a-e   Ventricose, height to width ratio variable but usually about 1.5:1; protoconch relatively large, of about two whorls. Colour pattern extremely variable like that of *P. solida* ; early teleoconch whorls usually purple.

2 cm. Cheyne Beach to Dampier, WA. Very common among algae in the intertidal and shallow sublittoral zones. There is difficulty in distinguishing this small species from *P. solida*. The ranges of size and colour pattern are about the same. *P. variegata* tends to be more ventricose. The diagnostic character is the size of the protoconch, which is minute and of barely one whorl in *solida* but larger and of two whorls in

variegata. However, some gastropod species are known to be polymorphic in this character and further study is needed to determine whether there are indeed two species. If they were shown to be the same then the Born name would take precedence. Synonym: *montebelloensis* Preston, 1914.

Phasianella ventricosa Swainson, 1822

Pl. 11; fig. 2 a-b   Spire moderately high, with $4^{1}/_{2}$ teleoconch whorls, body whorl inflated; height about $1^{1}/_{2}$ times the width. Several colour forms occur, usually with a base colour of rose, formed by a dense pattern of thin axial lines, and wide irregular axial bands of pink, which are often interrupted, and spiral lines of yellow flecks, and dashes; uniformly red-brown or rose-pink shells occur.

4 cm. NSW to about Geraldton, WA. Resembles *P. australis* but stouter and with fewer whorls. Synonym: *perdix* Wood, 1828.

SUBAMILY **TRICOLIINAE**

Tricolias

The shells of tricolias are small to minute (rarely more than 1 cm), smooth, sturdy, bulimoid to naticoid in form, and often with a weak umbilical chink. The peristome is incomplete and the parietal wall lacks a lamella and palatal sinus. The inner surface of the shell is not nacreous. The operculum is oval, smooth or with radial ridges, with a convex external calcified surface. There are two or three slender epipodial tentacles, the neck lobes are usually digitate and there are no head lappets. The radula has membranous rachidian teeth in which cusps are lacking or reduced, and usually two to five varied pairs of laterals. In some of the Australian species the central tooth is reduced to a longitudinal rod, while in others fusion of the innermost lateral teeth with the central tooth forms a "pseudocentral".

These small marine snails closely resemble phasianellids but show many important anatomical differences. The shell pigments (porphyrins) are different to those of phasianellids and the shell surface lacks the spiral capillary lines which characterise that subfamily.

Tricolias are either spawners with small eggs and short-lived planktonic (lecithotrophic) larvae, or brooders. *T. speciosa* of the Mediterranean lays a gelatinous egg tube but whether there is a planktonic larval stage after hatching is unknown. The temperate Australian species *T. gabiniana* and *T. rosea* have large protoconchs suggesting that they may brood their larvae.

Most of the modern species live among seagrasses or algae in shallow water. It may be assumed that they are herbivores although it has been suggested that the Indian species *T. indica* may be a carnivore.

Sexual dimorphism is reported in shell form and radular dentition of some species.

There is only one genus in the subfamily which has been thoroughly reviewed by Robert Robertson (1985, *Monograph of Marine Mollusca,* **3**: 1-103).

▲ GENUS **TRICOLIA** Risso, 1826
Turbo pullus Linnaeus

Small, glossy, globose to elongate-conic shells lacking a nacreous interior; usually brightly coloured with porphyrin pigments; umbilicate; operculum calcareous, paucispiral.

World-wide in temperate and tropical seas; five species in Australian waters. The Australian species have often incorrectly been placed in *Pellax* Finlay, 1926 which belongs to the mesogastropod family Eatoniellidae.

Tricolia fordiana (Pilsbry, 1888)

Juvenile shells globose, adults elongate, glossy, smooth, lacking sculpture except near the apex; protoconch barely exsert; teleoconch whorls inflated; columellar callus thin to thick; umbilicus wide or narrow. Colours variable; patterns spotted, striped or mottled.

4.5 mm. Central Indo-West Pacific; SA, WA, NT and Qld south to Moreton Bay. A very common species of rocky-shore habitats. Synonyms: *minima* Melvill, 1896; *elachista* Melvill, 1901.

Tricolia gabiniana (Cotton & Godfrey, 1938)

Elongate-ovate to subglobose; protoconch fairly exsert, not demarcated; teleoconch whorls inflated, rounded, but may be flattened at the periphery, generally smooth though 5-6 incised spiral lines may be present on the body whorl; first teleoconch whorl unusually large; columellar callus thin to thick; umbilicus narrow or absent; axial rib present within the aperture. Periostracum consisting of short tubular hairs arranged in spiral rows along incised spiral lines. Shades of pink, orange-pink, brown, cream or yellow; pattern variable, usually with prominent spiral lines or bands and subsutural flames.

7.4 mm. Recherche Archipelago to Cape Naturaliste, WA. This is the largest of the Australian species.

Tricolia rosea (Angas, 1867)

Elongate-ovate to attenuate; protoconch fairly exsert, prominently sculptured with spiral threads, and abruptly demarcated from the teleoconch; teleoconch whorls sometimes flattened at the periphery and shouldered, smooth; umbilical chink present; a thin spiral thread usually present within the aperture. Bright red or pink; pattern variable, usually with irregular axial bands or stripes.

4.6 mm. Central NSW south to Tas. and west to Shark Bay, WA. The bright red or pink colour is usually sufficient to distinguish this from the other species.

Tricolia tomlini (Gatliff & Gabriel, 1921)

Triangularly ovate, apex pointed; protoconch moderately exsert, slightly bulbous; teleoconch whorls inflated but with slightly flattened shoulders; sutures deeply impressed; sculpture of punctate spiral sulci, 4-15 on the body whorl; periostracum hirsute; columellar callus thin to thick; umbilicus chink-like and usually narrow. Colours and patterns variable, usually spirally banded with complex pattern of V-shaped or curved axial lines.

6.5 mm. Cape Yorke, SA to Fremantle, WA. Figured Robertson (1985, *loc. cit.*, Pls. 38-42). Synonym: *irritans* Thiele, 1930.

Tricolia variabilis (Pease, 1861)

Shell form extremely variable, elongate-ovate to globose; protoconch slightly or prominently exsert, smooth or spirally striate, demarcated by prominent growth lines; teleoconch whorls rounded or angulate at the shoulders, suture deeply impressed; sculptured with spiral threads or cords; umbilical chink narrow to wide; outer lips of male shells may be strongly everted; an axial ridge present within the aperture; periostracum smooth.

5 mm. Indo-West Pacific; Fremantle WA across northern Australia and south to Tas. There is doubt whether the Tasmanian records represent permanent populations. Colour and pattern extremely variable. Synonyms: There are many synonyms - see Robertson (1985, *loc. cit.* pp. 98-99). Two used in the Australian context are: *virgo* Angas, 1867; *gregaria* Laseron, 1955.

SUBFAMILY **TURBININAE**

Turban and Star Shells

Turban and star shells are of medium to large size and thick. Juvenile shells are usually bicarinate. The interior surface is nacreous. There is a circular to oval operculum with a thickly calcified exterior surface which may be smooth or elaborately sculptured. A prominent pseudoproboscis is strongly developed. Epipodial tentacles are short and stubby, neck lobes broad with simple, undivided margins. Head lappets are usually present. The radula has a reduced central tooth field, pronounced asymmetry, and enlarged inner marginal tooth cusps.

Although it has been customary in recent years to separate the star shells as a separate subfamily Beu & Ponder (1979, *Records Australian Museum* **32**: 1-68) have shown that the distinction is arbitrary. This view has been followed by Hickman & McLean (1990 *loc. cit.*).

▲ GENUS **TURBO** Linnaeus, 1758
Turbo petholatus Linnaeus

Turbiniform, solid, body whorl inflated, smooth or sculptured, inner lip wide and calloused; with or without umbilicus; operculum more or less circular, thick, convex.

Cosmopolitan, in tropical and temperate waters. The subgenera listed here could perhaps be regarded as genera but there is, as yet, little comparative anatomical information about them and a conservative approach to their classification seems preferable.

▲ SUBGENUS **TURBO** s.s

With the characters of the type species.

Indo-West Pacific. There is only one species.

Turbo petholatus Linnaeus, 1758

Pl. 12; fig. 10   Height equal to or slightly greater than width; whorls convex, with impressed suture; surface smooth and highly polished; umbilicus closed; operculum smooth. Colour and pattern variable, usually brown, red or green with darker wavy axial zones and broad dark spiral lines which may contain pale axial bars or arrowhead marks; columella yellow, orange or green; outer surface of operculum with a green centre and brown margin.

6 cm high. Indo-West Pacific; Shark Bay, WA to southern Qld. The operculum is known as the "cat's eye" and is extensively used in shell jewellery. Fairly common in shallow coral reef and rocky shore habitats.

▲ SUBGENUS **DINASSOVICA** Iredale, 1937
D. verconis Iredale (= *jourdani* Kiener)

Shell large, globose, solid; whorls convex, smooth or with spiral rows of nodules; umbilicus closed; funicle not developed; operculum massive, outer surface smooth or minutely granulose, nucleus close to the margin.

Southern Australia. There are two species in the group.

Turbo imperialis Gmelin, 1791

Pl. 12; fig. 18   Body whorl may bear a wide but low spiral rib at the shoulder and another at the periphery; the ribs may bear short, open spines or tubercles; iridescent columellar callus may spread over the parietal wall; operculum usually finely granulose. Green, usually with reddish spiral stripes; columella and outer lip edge green; operculum white, suffused with green.

10 cm. Southern Qld and NSW. Synonym: *militaris* Reeve, 1848.

Turbo jourdani Kiener, 1839

Pl. 12; fig. 17   Body whorl inflated, smooth; height slightly greater than width; operculum smooth. Early whorls cream with brown patches and interrupted brown spiral lines, later whorls rich uniform red-brown; operculum white, very thick, smooth on the outer surface, covered with a thin, brown, horny layer on the inner surface.

20 cm high. Western SA to Geraldton WA. Iredale (1937) named the SA shells *verconis* but there is no obvious basis for this distinction. The species lives mainly in the sublittoral zone among leafy brown algae but is found occasionally in tidepools. The large fleshy animal is red-brown.

▲ SUBGENUS **LUNATICA** Röding, 1798
T. olearius Linnaeus (= *marmoratus* Linnaeus)

With the characters of the type species.

Indo-West Pacific. Monotypic.

Turbo marmoratus Linnaeus, 1758

Very large and heavy; spire moderately high, shoulders and basal periphery heavily ribbed, subsutural ramp slightly concave; umbilicus open, bordered by a massive spiral funicle; outer lip thickened and flaring outward where it meets the columellar base to form a broad, flat, spout-like anterior projection; operculum massive, smooth or finely granular. Uniformly dark green or mottled dark green and brown with cream blotches; operculum white.

20 cm high. Indo-West Pacific; possibly north Qld. This species is common and extensively used for shell jewellery in Asia. No authentic Australian specimen was available for illustration.

▲ SUBGENUS **NINELLA** Gray, 1850
Turbo torquatus Gmelin

With characters of the type species.

Southern Australia. There is only one living species.

Turbo torquatus Gmelin, 1791

Pl. 12; fig. 16 a-b Width greater than height; suture sub-canaliculate; whorls rounded or carinate; columella with a callus surrounding a widely open umbilicus, callus excavated, sculptured with low spiral ribs and close-set, high, oblique lamellae; operculum oval, nucleus eccentric, outer surface prickly, dished at the centre and with two thick spiral ribs. Sand-coloured, sometimes with green or orange mottling in young specimens; operculum and columella white.

11 cm. NSW to Port Gregory, WA. For many years this species was known as *T. stamineus* Martyn but the names introduced by that author are not available. Shells from WA tend to be strongly carinate (fig. 16 b) while those from south-eastern Australia lack shoulders (fig. 16 a). The name *N. whitleyi* Iredale, 1949 has been used for the carinate WA form but there are intermediate forms on the south coast and subspecies status may be all that is warranted. Very common.

▲ SUBGENUS **SUBNINELLA** Thiele, 1929
T. undulata Lightfoot

With characters of the type species.

Southern Australia.

Turbo undulatus Lightfoot, 1786

Pl. 12; fig. 14 Width greater than height, thick-shelled, globose, spire low and domed, whorls round and lacking shoulders; deeply umbilicate; columella wide and calloused, the callus spreading around the umbilicus; sculpture obsolete, consisting of low, wide spiral ribs, usually more evident on the early whorls; operculum oval, low, smooth with a central hump encircled by a shallow channel. Pale green with dark green wavy axial lines; operculum white.

6 cm wide. NSW to Bremer Bay, WA. A common intertidal species. The distinctive colouring and more simply constructed operculum easily distinguish this species from *T. torquatus,* which is often found in the same places. Synonym: *anguis* Gmelin, 1791.

▲ SUBGENUS **EUNINELLA** Cotton, 1939
T. gruneri Philippi

Columella calloused, wide, bounded by a spiral funicle; umbilicus lacking; sculptured with granulose spiral ribs of varying thickness; operculum elliptical, outer surface smooth but with low wrinkles, paucispiral, with depressed marginal ledge.

Southern Australia.

Turbo gruneri Philippi, 1846

Pl. 12; fig. 15 Height and width about equal; spire moderately elevated; whorls rounded with granulose spiral ribs of varying thickness; umbilicus closed; columella wide, calloused, bordered by a spiral funicle; operculum elliptical, with a depressed marginal rim, outer surface smooth except for low wrinkles. Pink or fawn, mottled with brown and white; columella and operculum white.

3.5 cm wide. Vic. to Abrolhos, WA. Rarely found intertidally but commonly taken on lobster pots set among algae-covered rocks in deeper water.

▲ SUBGENUS **MARMAROSTOMA** Swainson, 1829
T. chrysostomus Linnaeus

Spire relatively high, conical, whorls rounded; aperture round, columella narrow, curved, simple; sometimes narrowly umbilicate; sculpture of spiral ribs and axial striae; operculum more or less circular with subcentral nucleus, outer surface smooth to granulose.

Indo-West Pacific. Synonym: *Senectus* Swainson, 1840.

Turbo argyrostomus Linnaeus, 1758

Pl. 12; figs. 1, 3 Height and width about equal; umbilicus usually open, bordered by a thick, double funicle; whorls convex, shoulder keel weak or lacking; sculptured with axial threads and strong spiral cords which may bear small prickles; operculum granulose. Exterior brown or fawn, often green between the ribs; operculum with a green centre and brown outer margin.

7 cm. Indo-West Pacific; Browse I., WA to north Qld. See comments on *T. perspeciosus.*

Turbo brunneus Röding, 1791

Pl. 12; fig. 5 a-b Height slightly greater than width; suture deeply impressed to channelled; whorls convex, shoulders lacking; spiral ribs rather strong, four or five on the spire whorls, about 10 primary ribs plus intermediate threads on the side and base of the body whorl; umbilicus open, bordered by a moderately strong funicle; operculum with a central perforation and a spiral ridge near the outer margin, the whole upper surface granulose. Fawn or pale green with dark brown axial bands and spots on the ribs; operculum with a purple-brown centre, fawn on the ridge, pale otherwise.

5 cm high. Central Indo-West Pacific; Exmouth Gulf to Kimberley, WA; probably also in Qld. The operculum is very similar to that of *T. squamosus* and although the sculpture is different a close affinity seems likely. Compare *T. haynesi.*

Turbo chrysostomus Linnaeus, 1758

Pl. 12; fig. 7 a-b   Height greater than width; anterior end outward flaring; whorls with angular shoulders bearing short spines, weak spiral cords which may be weakly spinose, and fine scaly axial threads; umbilicus closed or minutely chinked, bordered by a thick funicle; operculum smooth except for oblique striae on the outer margin. Exterior light brown with axial flames of brown and green, sometimes brown spots on the ribs; interior white or gold; operculum brown, yellow or dark green, golden brown at the outer margin.

6 cm high. Indo-West Pacific; Rowley Shoals, WA to southern Qld. Inhabits the intertidal and shallow sublittoral zones of coral reefs.

Turbo crassus Wood, 1828

Pl. 12; fig. 4   Height greater than width; anterior end produced and slightly reflected; whorls subcarinate with a prominent rib forming a shoulder keel and a concave subsutural ramp; sculptured with flat spiral ribs and fine axial striae; operculum finely granulate, granules finer toward the margin. White or yellow, with irregular brown or green flames on the ramp and rectangular brown or green blotches on the ribs; interior pearly white; columella with a green border; operculum dark brown at the centre, yellow to green marginally.

8 cm. Indo-West Pacific; north Qld. The shell illustrated here was collected by Hedley in 1907 at Murray I., Torres St. *Turbo setosus* Gmelin, 1791 is a very similar species, though with a shorter spire, which is also reported from far north Qld.

Turbo gemmatus Reeve, 1848

Pl. 13; fig. 2   Height slightly greater than width; whorls rounded, shoulders lacking; umbilicus closed; sculptured with spiral rows of close rounded nodules, four on the spire whorls and nine on the sides and base of the body whorl, plus a thick double funicle and sometimes spiral threads in the interspaces; fine axial striae cross the interspaces; operculum not seen. Pink or fawn with darker axial bands.

1.5 cm high. Central Indo-West Pacific; Qld south to the Capricorn Group. There are very few Australian records of this beautiful little species. It appears to live in deeper water. Its placement in *Marmarostoma* is provisional.

Turbo haynesi Preston, 1914

Pl. 12; fig. 6 a-c   Height slightly greater than width; suture moderately impressed, whorls convex, shoulders lacking; spiral ribs rounded, irregular and suppressed, sometimes weakly nodulose, axial striae weak; umbilicus closed or open only by a tiny chink, bordered by a weak funicle; operculum subcircular, upper surface convex, weakly and evenly granulose. Pale green or fawn, with wide irregular brown axial rays and tan spots on some ribs; operculum white or sometimes flushed with pale green or orange.

3 cm. Shark Bay, WA to about Hervey Bay, Qld. Common among algae and stones in the intertidal and shallow sublittoral zones. *Turbo menkei* Thiele, 1930 is a synonym described from Shark Bay. *T. nivosa* Reeve, 1848 is probably also the same species but that name was preoccupied. Iredale, 1929 proposed the alternative name *necnivosa* for Qld shells but if *nivosa* and *haynesi* are the same, then Preston's alternative name has priority. It is often difficult to distinguish this species from the *Turbo* identified here as *brunneus*, but the imperforate operculum and weaker sculpture will usually suffice.

Turbo lajonkairii Deshayes, 1839

Pl. 12; fig. 11   Height slightly greater than width; base drawn out; umbilicate; surface with weak irregular spiral cords and two rows of thick, prominent, hollow spines, one row at the shoulder and another at the periphery, the latter usually double with double spines; funicle thick and also spinose; the spines form open channels in the outer lip; operculum weakly granulose. Brown or green, ribs usually green; operculum white, brown around the outer margin, greenish centrally.

9 cm high. Eastern Indian Ocean. This species is not known from the Australian mainland but is quite common at Christmas I. It is one of the most handsome of all turban shells.

Turbo perspeciosus (Iredale, 1929)

Pl. 12; fig. 2   Height greater than width; umbilicus closed or open by a mere chink, bordered by a rib forming a weak funicle; whorls convex, shoulder keel lacking or represented by a rib larger than the others; spiral ribs moderately strong, usually prickly or spinose, and with axial scales giving a scabrous appearance, operculum granulose, with oblique striae at the outer margin. Brown or fawn, some ribs green, often with brown spots and brown axial flames; operculum white, brown at the outer margin.

7 cm. Northern Australia from Point Cloates, WA to southern Qld. A common intertidal species on coral reefs. Reeve (1848) named this species *T. speciosus* from WA material but the name had been used previously and Iredale provided an alternative. There is some doubt about the status of this species as it may be only a variant *T. argyrostomus*.

Turbo pulcher Reeve, 1842

Pl. 12; fig. 13   Height greater than width; suture impressed, whorls convex without shoulders; sculptured with low spineless spiral ribs and secondary cords and close-packed, low, wavy lamellae in the interspaces; umbilicus closed or open by only a tiny chink, bordered by a weak, double funicle; operculum circular, convex, covered with numerous blunt tubercles. Fawn with wavy green-brown axial stripes, ribs may be fawn or pink with brown or green spots; operculum white.

8 cm high. Esperance to Point Quobba, WA. This species has an unusually wide habitat from tide pools of the rocky shores of the temperate south coast to the coral reefs of the subtropical Abrolhos. Synonym: *intercostalis* Menke, 1843.

Turbo radiatus Gmelin, 1791

Pl. 12; fig. 12 a-b Tall-spired, teardrop shaped, base steeply descending; umbilicus closed; sculptured with fine spiral ribs, those at the shoulder and basal periphery bearing short compressed spines, those in the umbilical area relatively thick; finely axially striate; operculum finely granulate. Fawn or greenish, with brown patches and brown spots on some ribs; operculum green, sometimes with an orange patch centrally.

4 cm high. Central Indo-West Pacific; Ashmore Reef, NT to north Qld. Lives on intertidal coral reefs.

Turbo squamosus Gray, 1847

Pl. 12; fig. 8 High-spired, with deeply channelled sutures; aperture round, inner lip entire, separated from a deep umbilicus; sculpture of distant spiral ribs, those at the centre of the body whorl with erect tubercles formed by tight folds of raised axial lamellae which dominate the sculpture; operculum granulose, subcircular, exterior perforate over the nucleus, with a spiral rib setting apart a thin peripheral zone. Fawn with dark brown blotches. Operculum fawn with greenish centre and a lateral brown patch.

5 cm high. Exmouth Gulf, WA to Qld and Papua New Guinea. This distinctive species lives among rocks in shallow muddy water. Synonyms: *foliaceus* Philippi, 1847; *lamellosus* Philippi, 1846; *laminiferus* Reeve, 1848.

▲ SUBGENUS **LUNELLA** Röding, 1798
T. versicolor Gmelin (= *cinereus* Born)

Thick-shelled, globose with rounded whorls and low, domed spire; umbilicus narrow but deep; aperture round, outer lip sharp-edged, thickened and flaring strongly outward where it meets the columellar base forming a flat spout-like anterior projection; columella narrow, rounded, simple; sculpture of finely nodulose weak spiral riblets and faint axial-oblique striae; operculum subcircular, high and domed, obsoletely granulose, multispiral.

Circum-tropical.

Turbo cinereus Born, 1778

Pl. 12; fig. 9 a-b Thick-shelled, globose, spire depressed and domed, whorls rounded and lacking shoulders; aperture round, outer lip sharp-edged; columella wide and flat, outward-flaring where it meets the outer lip forming a flat spout-like anterior projection; umbilicus open with a narrow spiral cord on one side; sculptured with finely nodulose spiral cords and faint axial striae; operculum nearly circular, domed, minutely granulose. Cream, grey or fawn, with green or brown spots on some ribs, and green and brown blotches; operculum white, with a crescent-shaped band of dark green at the outer margin.

3.5 cm Indo-West Pacific; Exmouth Gulf, WA to southern Qld. Lives among rocks rather high in the intertidal zone. Synonyms: *porphyrites*, *mespilus* and *versicolor*, all introduced by Gmelin, 1791; *picta* Röding, 1798; *lugubris* and *porcatus* Reeve, 1848.

▲ SUBGENUS **CARSWELLENA** Iredale, 1931
Turbo exquisita Angas

Characters of the type species.

Eastern Australia. Iredale introduced *Carswellena* because of the open umbilicus and operculum details.

Turbo exquisitus Angas, 1877

Pl. 13; fig. 1 Thin-shelled; height greater than width; whorls with angular shoulders and slightly convex subsutural ramp; aperture round, columella slightly reflected over the narrowly open umbilicus which is bordered by a small funicle; sculptured with weakly nodulose spiral ribs and axial lamellae in the interspaces which may ascend and cross the ribs; the rib at the shoulder is the most prominent, there are six small ribs on the ramp, five strong ribs on the side of the body whorl, and nine weak ribs on the base. Fawn or pink, with orange and cream oblique flames on the ramps, sometimes extending onto the sides, ribs sometimes spotted.

1.2 cm high. Southern Qld and NSW. This truly exquisite shell is rather rare, at least in collections.

▲ GENUS **BOLMA** Risso, 1826
Turbo rugosus Linnaeus

Turbinate; with one or two angulations on the body whorl, the upper or peripheral one usually bearing nodules or spines; columellar callus simple, usually spreading over the base; operculum oval or circular, simple, markedly convex, smooth or granular, sometimes with a central depression.

Cosmopolitan. The earliest fossil *Bolma* is from the New Zealand Palaeocene. The genus was reviewed by Beu & Ponder (1979, *Records Australian Museum* **32**: 1-68). There are only a few Australian records and further dredging in tropical areas is sure to extend the geographic ranges of the species.

Bolma aureola (Hedley, 1907)

Pl. 13; fig. 12 Width greater than height; suture channelled; umbilicus closed, umbilical area encircled by a smooth funicular rib; columella narrow, rounded, arched; periphery with broad, downward-pointing hollow spines; whorls obliquely plicate below the suture, otherwise sculptured with scaly spiral cords, about nine more regular cords on the base;

operculum smooth, oblong, nucleus subterminal, hollow in the centre below a thick arched rib. Fawn with brown subsutural and peripheral bands; funicle yellow-orange, operculum pink, suffused with mauve.

10 cm high; 9 cm wide. Southern Qld to northern NSW. The type was dredged in 37 m off Mast Head I. but the species has been dredged more recently as far south as Tweed Heads.

Bolma guttata millegranosa (Kuroda and Habe, 1958)

Pl. 13; fig. 14   Height and width about equal; whorls weakly convex, suture channelled; periphery of body whorl with 22-28 hollow, rather short spines; basal angle distinct; sides of body whorl between periphery and suture with 6-8 gemmate spiral cords, one weakly gemmate cord between the basal angle and the periphery, six weakly gemmate spiral threads on the base; basal callus very thin, white to transparent; operculum oval with steep concentrically striated sides and flattened finely pustulose centre. Pink or fawn with purple-brown blotches; operculum white.

4.6 cm high. The nominate form is a deepwater shell from Japan; the subspecies *millegranosa* may be widespread in the Western Pacific and has been recorded off southern Qld at 201 m. Beu & Ponder (1979) figured a juvenile of another form which appears to belong to this species complex from the North West Shelf of WA in which the peripheral spines are longer, the basal angulation has a strongly gemmate cord, and the operculum is nearly circular with irregular concentric grooves at its margin and a coarsely granular surface.

Bolma henica (Watson, 1885)

Pl. 13; fig. 13 a-b   Height greater than width; spire flat-sided, suture deeply channelled; periphery angulate, bearing long curved hollow spines; basal angle sharp, base nodulose with up to 10 finely beaded spiral cords, sides above the periphery with up to seven rows of small beaded spiral ribs; basal callus narrow, white, not distinct from the columella; operculum oval, steep-sided, flat-topped, weakly granulose at the centre. Yellow or pink; operculum white.

2.6 cm high. Japan and central Indo-West Pacific Region; eastern Australia from north Qld to northern NSW. Dredged 130 to 600 m. Synonyms: *abyssorum* Schepman, 1908; *gloriosum* Kuroda in Kira, 1955.

Bolma tamikoana (Shikama, 1973)

Pl. 13; fig. 15   Whorls slightly convex, suture indistinct; periphery only weakly angulate, with or without small spines; basal angle sharp; sides of whorls with granular spiral cords; basal callus thin, glazed in mature shells; operculum almost circular with a high, concave concentrically sculptured edge, upper surface convex and pustulose. Pinkish fawn with purplish blotches, basal callus orange-yellow.

3.5 cm high. Japan and Central Indo-West Pacific Region including the Arafura Sea. There is a single Australian record from 125 m off Arnhem Land.

▲ GENUS **ASTRALIUM** Link, 1807

A. deplanatum Link (= *calcar* Linnaeus)

Low to high-spired; periphery angulate; sculpture of radial folds or ribs which are sometimes produced into short spines; umbilicus closed; basal callus usually narrow and separated from the columella; operculum smooth, flat.

Australia. In the *Treatise on Invertebrate Paleontology* this group is treated as a subgenus of *Astraea* Röding, 1798. Beu & Ponder (1979, *Records Australian Museum* **32**: 1-68) showed that it is a distinct group worthy of generic rank. Only fossil species of *Astraea* occur in Australia and the only living species of the genus (*imperialis*) occurs in New Zealand. All the living Australian species are referable to *Astralium*.

Calcar Montfort, 1810 (*calcar* Linnaeus) is a synonym. *Distellifer* Iredale, 1937 (*wallisi* Iredale = *rhodostomus* Lamarck), *Pagocalcar* Iredale, 1937 (*pileolum* Reeve), *Rugastella* Iredale, 1937 (*rotularius* Lamarck), *Bellastraea* Iredale, 1924 (*kesteveni* Iredale), and *Micrastraea* Cotton, 1939 (*aureus* Jonas) are genus-level names which may prove to be worthy of generic or subgeneric status. But in the absence of a review they are simply noted here and treated as synonyms of *Astralium*.

Astralium aureum (Jonas, 1844)

Pl. 13; fig. 3   Depressed conical, width greater than height; sides slightly convex, periphery angulate; whorls with oblique-axial rugae below the suture which sometimes divide and extend to the periphery, and beaded spiral cords, basal ribs numbering about six are more strongly beaded; parietal wall calloused; columella arched, narrow, with a shallow channel behind; operculum oval, smooth. Yellow or golden; operculum white or gold-tinted.

1.8 cm wide. Vic. and Tas. to Cape Naturaliste, WA. Common in the intertidal and shallow sublittoral zones. The SA *rutidoloma* Tate, 1893 may be only a variant of this species.

Astralium pileolum (Reeve, 1842)

Pl. 13; fig. 5 a-b   Depressed, width much greater than height, whorls with concave sides, smooth except for oblique rugae and granulose spiral threads; periphery sharply angulate, with an expanded, thin, fluted rim; base convex, finely sculptured with scaly spiral cords; columella rather narrow, its iridescent callus not spreading onto the parietal wall; operculum oval, depressed over the nucleus, but with an expanding spiral ridge which becomes a blunt knob at the end. Fawn or grey; columella white; operculum dark purple-brown, greenish towards the outer margin.

6.5 cm. Exmouth Gulf, WA to north Qld. Common among rocks intertidally. Synonyms: *cucculatus* and *limbiferus*, both of Kiener, 1850.

Astralium rhodostomum (Lamarck, 1822)

Pl. 13; fig. 11 a-b High-spired and conical, whorls flat-sided, with oblique rugae below the suture and two spiral rows of prominent spines, the lower row being at the angulate periphery; 5-6 spiral rows of sharp nodules on the base becoming finer towards the columella; columella arched, with a narrow callus zone behind; operculum ovate, thick, with a small depression over the nucleus and a spiral central ridge which becomes thicker towards the end. White, green or fawn; parietal wall; and callus zone behind the columella blue, mauve or pink; operculum blue or mauve.

3 cm high. Indo-West Pacific; Rowley Shoals and Kimberley coast, WA to Lady Elliot I., Qld. Common among *Acropora* corals in the lower intertidal and shallow sublittoral zones. Synonym: *wallisi* Iredale, 1937.

Astralium rotularia (Lamarck, 1822)

Pl. 13; fig. 6 Depressed; width nearly double the height; whorls with concave sides, smooth or with weak oblique rugae and very fine spiral threads, periphery angulate but with a row of axially compressed, blunt-ending, foliose spines; base with 7-8 strongly imbricate spirals, becoming smaller towards the columella; columella wide, its iridescent callus spreading onto the parietal wall; umbilical area swollen and finely threaded; operculum oval, smooth with a wide spiral central rib and narrow marginal channel. Exterior fawn, grey or white; columella silver; operculum dark green, dark purplish or black.

4.5 cm wide. Exmouth Gulf, WA to north Qld. Rather common among rocks in the intertidal zone.

Astralium squamiferum (Koch, 1844)

Pl. 13; fig. 4 a-b Depressed-conical with acutely keeled periphery; sides of whorls flat, obliquely plicate, fluted above the flanged peripheral keel and with finely granulose spiral cords; base with crowded scaly spiral cords; columella arched, smooth, iridescent; umbilical area depressed, iridescent, bordered by an arched, low funicular rib which is porcelaineous; operculum oblong, concave centrally, with a spiral rib near the margin. Yellow or grey; operculum white.

3 cm wide. Southern NSW to about Geraldton, WA. Very common in shallow water, especially in seagrass beds. *A. fimbriatum* Lamarck is an older name once commonly used but it is preoccupied. Synonym: *kesteveni* Iredale.

Astralium stellare (Gmelin, 1791)

Pl. 13; fig. 8 a-b Solid, high-spired; whorls flat-sided, with obscure, oblique rugae and wide, flaring spines which may be

concave on the underside around the sharply angulate periphery; base with many fine, scaly, spiral threads; parietal wall and area behind the columella covered with callus; operculum oval, smooth, depressed over the nucleus, with a narrow and shallow channel near the margin. Fawn pale green or white exteriorly, aperture and columellar margin iridescent silver; parietal wall, umbilical area and operculum turquoise, violet or blue.

4 cm high. Indo-West Pacific; Exmouth Gulf, WA to north Qld. Moderately common on rocky substrates in the intertidal and shallow sublittoral zones.

Astralium tentoriformis (Jonas, 1845)

Pl.13; fig. 9 Tall-spired, periphery sharply angulate and fluted by the ends of oblique folds on the sides of the body whorl, base deeply concave; sculptured with finely squamose cords; operculum with a strong spiral rib, sometimes ending in a callosity. Cream sides and base, parietal wall and columella pearly; operculum white centrally, blue to purple along the side beyond the spiral rib.

4.5 cm. Central Qld to southern NSW. The deeply concave base and coloured operculum distinguish this from *A. squamiferum*. Synonym: *sirius* Gould, 1851.

Astralium tentorium (Thiele, 1930)

Pl. 13; fig. 7 High-spired; whorls flat-sided, with finely scaled spiral threads and a sharply angulate and slightly expanded peripheral rim; base with numerous scaly spiral cords and a rather wide callus over the umbilical area; operculum oval, smooth. Fawn, umbilical area, parietal wall and columella blue.

5 cm. Hopetoun to Shark Bay, WA. Common in the sublittoral zone.

▲ GENUS **GUILDFORDIA** Gray, 1850
Australium triumphans Philippi

Subdiscoidal, spire low, with peripheral spines or nodules; upper surface sculpture of fine, low, closely spaced gemmae ending below the peripheral spines or nodules; base smooth, with a callus pad, the pad with a central depression; operculum flat and rather thin.

Central Indo-West Pacific. A deepwater group which lives on soft substrates.

Guildfordia yoka Jousseaume, 1888

Pl. 13; fig. 10 Spines long and curved, 9-10 around the periphery of the body whorl; upper surface spiral lirae irregular, with 4-5 spiral rows of tiny nodules below the suture. Upper surface light reddish brown to pink, base white, nacreous around the umbilicus.

4 cm diam. (excluding spines). Central Indo-West Pacific. Recently discovered along the edge of the North West Shelf.

FAMILY **SKENEIDAE**

Shells minute (less than 5 mm), thin to sturdy, unpigmented and lacking internal nacre. Peristome complete or incomplete, aperture oblique, outer lip thin or thickened; umbilicate or without an umbilicus. Operculum corneous, multispiral, not enveloped by the foot. Epipodial tentacles long; head lappets lacking, development of neck lobes variable; gill monopectinate, lacking bursicles and free tip. Radula variable.

The taxonomy of this group has been terribly confused in the Australian literature with genera originally placed in a number of families. A history of the matter may be found in Hickman & McLean (1990, Systematic revision and suprageneric classification of trochean gastropods, *Natural History Museum of Los Angeles County*, No. 35: 169pp.) who provide a provisional but incomplete classification of the family. There is no anatomical information available for most of the genera and it is likely that the group as represented here is polyphyletic.

In his papers on NSW and Dampierian "Liotidae" (1954, *Australian Zoologist* **12**: 1-25; 1958, *Records Australian Museum* **24**: 165-182) Charles Laseron introduced a number of new generic and species names, some of which seem attributable to the Skeneidae.

Genus-level names which may be attributable to the Skeneidae are: *Brookula* Iredale, 1912 (*stibarochila* Iredale); *Liotella* Iredale, 1915 (*polypleura* Hedley); *Liocarinia* Laseron, 1954 (*disjuncta* Hedley); *Cirsonella* Angas, 1877 (*australis* Angas = *weldii* Tenison Woods); *Crossea* Adams, 1865 (*miranda* Adams); *Crosseola* Iredale, 1924 (*concinna* Angas); *Dolicrossea* Iredale, 1924 (*labiata* Tenison Woods); *Daronia* Adams, 1861 (*spirula* Adams); *Eudaronia* Cotton, 1945 (*jaffaensis* Verco, 1909); *Elachorbis* Iredale, 1915 (*tatei* Angas); *Lissotesta* Iredale, 1915 (*micra* Tenison Woods); *Lodderena* Iredale, 1924 (*minima* Tenison Woods); *Lodderia* Tate, 1899 (*lodderae* Petterd); *Partubolia* Iredale, 1936 (*blancha* Iredale); *Pseudoliotia* Tate,1898 (*micans* Adams); *Putilla* Adams, 1867 (*lucida* Adams); *Zalipais* Iredale, 1915 (*lissum* Suter).

A single species, *Brookula densilaminata* (Verco, 1907), is illustrated here to represent the family.

SUPERFAMILY
SEGUENZIOIDEA

FAMILY **SEGUENZIIDAE**

This small family exhibits features relating it to the archaeogastropods but setting it apart in its own suborder.

The shells are trochoid, nacreous, and the outer lip has a sinus. The radula has a reduced number of marginal teeth. Ctenidium monopectinate; oesophageal pouches lacking. The family is represented in the Australian fauna by two genera.

▲ GENUS **SEGUENZIA** Jeffreys, 1876
S. formosa Jeffreys

Globular or conical; columella wide, truncate and terminally toothed, notched below; outer lip with a deep, narrow, subsutural sinus and another at the basal corner; umbilicus present or absent; whorls carinate.

Cosmopolitan. Minute species inhabiting deep water.

Sequenzia polita Verco, 1906

Turbinate; spire stepped, whorls with a strong shoulder keel; body whorl tricarinate, concave between the keels; base spirally corded; imperforate. White.

Holotype: 3.5 mm high. The holotype came from 550 m off Cape Jaffa, SA.

▲ GENUS **BASILISSA** Watson, 1879
B. lampra Watson

Thin-shelled, conical or turbinate, carinate, suture channelled, base spirally ribbed; usually umbilicate; outer lip with a wide, shallow sinus below the suture and another at the basal corner; columella curved, straight or oblique, not markedly toothed at the end.

Cosmopolitan. A genus of tiny deepwater snails. There is a Miocene species in the Victorian Muddy Creek fossil beds named *B. radialis* Tate, 1890.

Basilissa bilix Hedley, 1905

Spire flat-sided, conical, periphery subangulate, base flat; suture canaliculate; sculpture of spiral cords and a double spiral rib at the periphery, rendered cancellate by undulating oblique-axial cords; base finely spirally and obliquely corded; umbilicus open, bordered by elongate nodules. Fawn.

4 mm high. NSW to SA; dredged to 550 m. Synonym: *B. bombax* Cotton & Godfrey, 1938.

Basilissa nicetorum Hedley & May, 1908

Turbinate, periphery rounded, base descending; suture deeply and broadly canaliculate; whorls with elevated, narrow spiral keels, deeply canaliculate between, three keels on the spire whorls, four on the body whorl, plus several spiral ribs on the base; umbilicus lacking; outer lip sinuses poorly developed.

4.3 mm high. Tas. Dredged to 220 m. The strong carinate sculpture, closed umbilicus and weak lip sinuses make it unlikely that this is correctly placed in *Basilissa*.

Basilissa superba Watson, 1879

Conical, whorls slightly convex, periphery finely nodulose and sharply angulate, base flat; sculpture of fine spiral threads crossed by axial threads; outer lip sharply cornered at the periphery, sinuses shallow; columella sharply edged, curved; umbilicus wide and deep. Cream.

2 cm high. North Qld. Described from Cape York; also recorded from the Coral Sea. Dredged in very deep water. No specimen was available for illustration.

SUPERORDER **CAENOGASTROPODA**
ORDER **NEOTAENIOGLOSSA**

SUPERFAMILY

CAMPANILOIDEA

*I*N THIS superfamily there is a single family and only one living genus. Until recently it was usually classified as a cerith but Houbrick (1989, *American Malacological Bulletin* **7**: 1-6), separated the family from the Cerithioidea and elevated it to superfamily rank.

Females of the living species deposit egg masses consisting of large gelatinous tubes attached to the substratum, containing spirally arranged capsules each of which contains several eggs. It is not yet known whether development is direct or there is a late pelagic larval stage.

Details of the anatomy of *C. symbolicum,* and a discussion of its taxonomic position, may be found in Houbrick, 1981, *Malacologia* **21**: 263-289.

FAMILY **CAMPANILIDAE**

Lighthouse Shells

The Campanilidae originated in the Late Cretaceous or Early Tertiary and its species were prominent in the tropical Sea of Tethys throughout most of the Tertiary. Although there are many fossil species, belonging to several genera, only one species survives, i.e. *Campanile symbolicum* of southern Western Australia. In popular jargon, this is a true "living fossil".

Some of the fossil species are among the largest of the gastropods. Shells of some of the fossil species were as much as a metre long. Their shells are multiwhorled and attenuate, with a chalky periostracum. The aperture has a central anterior canal and sets at an angle of about 45° to the shell axis. The general aspects of the anatomy of the living *C. symbolicum* are like those of the Cerithioidea. However, it has a short, thick snout, tiny eyes, and a deep, ciliated pedal gland around the entire margin of the sole of the foot. There are small papillae around the entire mantle edge. The short oval osphradium is bipectinate. These and other anatomical characters indicate that the group warrants separation from the Cerithioidea.

▲ GENUS **CAMPANILE** Bayle in Fischer,1884
Cerithium giganteum Lamarck

Shells large, elongate, turreted, with straight-sided whorls, sinuous growth lines, and a chalky periostracum; body whorl truncate; aperture narrow, aligned at 45° to the shell axis, anterior canal moderately long, twisted to the left; outer lip thin, smooth, sinuous, with an anal notch; early whorls sculptured with spiral cords and rows of nodules, later whorls smooth; operculum ovate, with few whorls and eccentric nucleus.

Fossil distribution cosmopolitan; the sole living species confined to southern WA. The type species of the genus is an Eocene fossil from Europe.

Campanile symbolicum Iredale, 1917

Pl. 15; fig. 9   Very large, acutely turreted, with slightly concave sides; whorls smooth except for weak growth lines and a low spiral ridge just behind the suture; anterior canal nearly horizontal; columella smooth and sinuous. Chalky white; interior glossy white.

20 cm. Esperance to Geraldton, WA. Abundant in the lower intertidal and shallow sublittoral zones. Lives on sand among algae, usually close to rocks. Synonyms: *leve* Quoy & Gaimard, 1834; *truncatum* Gray, 1834.

CERITHIOIDEA

ECENT reviews of the classification of this large and important group have clarified the relationships of the families and genera (see Houbrick, 1988, *Malacological Review, Supp.* **4**: 88-128).

FAMILY **BATILLARIIDAE**

Southern Mud Creepers

In most previous accounts this group has been treated as a subfamily of the Potamididae but Houbrick (1988, *loc. cit.*) has argued that it should be given full family rank. The animals have similar habits and intertidal and shallow-water estuarine habitats to the potamidids but are found in temperate waters and have a different radular structure. Batillariids have cusps on the lower plate of the central radular tooth, while the potamidids lack this feature. The shells are like those of the Cerithiidae but the anterior canal is very short and the operculum is multispiral.

▲ GENUS **BATILLARIELLA** Thiele, 1931
Bittium estuarinum Tate

With the characters of the type species.

Southern Australia. Sometimes treated as a subgenus of *Batillaria* Benson, 1842. The shell resembles those of ceriths such as *Bittium* but the round, multispiral operculum is diagnostic. *Eubittium* Cotton, 1937 (*lawleyanum* Crosse) is a synonym.

Batillariella estuarina (Tate, 1893)

Tall, elongate, with convex whorls and deeply impressed suture; sculptured with axial ribs and several spiral cords; aperture ovate, outer lip simple, not reflexed, smooth within; posterior canal a deep notch bordered by a flat callus on the parietal wall; anterior canal deep, very short, turning to the left; columella smooth, concave; operculum thin, round, multispiral, with a roughly central nucleus. Red-brown or grey.

3 mm. Bass St. to the Abrolhos, WA. Abundant in southern estuaries. Synonyms: *lawleyanum* Crosse, 1863 (type species of *Eubittium*); *insculptum* Reeve, 1865.

▲ GENUS **PYRAZUS** Monfort, 1810
P. baudini Montfort = *ebeninus* Bruguière

With characters of the type species.

Eastern Australia. Monotypic.

Pyrazus ebeninus (Bruguière, 1792)

Pl. 15; fig. 20 Elongate with finely, spirally ribbed whorls; early whorls bearing axially elongate nodules, later spire whorls and body whorl angulate and usually bearing a single spiral row of high nodules or tubercles; outer lip widely flaring and weakly lirate internally; anterior canal very short, posterior canal present; columella with a broad, smooth, reflected shield. Exterior grey or brown, columella white, deep interior purple or brown.

11 cm. Port Curtis, Qld to eastern Vic. Abundant on estuarine mudflats. Once commonly known under the species name "*herculea*" Martyn, 1784; but Martyn's names are not available for technical reasons, so the name given by Bruguière applies. This is the "Hercules Club Whelk" of authors. It was one of the Australian shells taken back to Europe by the Cook voyages, from which source Martyn obtained his specimens.

▲ GENUS **VELACUMANTUS** Iredale, 1936
Cerithium australis Quoy & Gaimard

With the characters of the type species.

Southern Australia. There is only one species.

Velacumantus australis (Quoy & Gaimard, 1834)

Pl. 15; fig. 11   Elongate, with convex whorls, deeply impressed suture, prominent beaded spiral cords and oblique folds, just below the suture the beads may become nodules; anterior and posterior canals not pronounced; parietal-columella callus narrow, columella smooth, interior weakly lirate. Uniformly dark brown or grey, or with a central white spiral line; interior brown.

4.5 cm. Southern Qld to SA; also found in the Swan Estuary and Cockburn Sound, WA where it may have been introduced. Abundant in sandy substrates among green algae and seagrasses.

▲ GENUS **ZEACUMANTUS** Finlay, 1927
Cerithidea subcarinata Sowerby

Elongate, with convex whorls and impressed suture; sculpture of weak spiral cords; outer lip thin, not flaring, posterior canal not developed, anterior canal very short; parietal wall thinly calloused, columella short, without an erect shield.

Southern Australia and New Zealand.

Zeacumantus diemenesis (Quoy & Gaimard, 1834)

Pl. 15; fig. 13 a-b   Sharply tapering, whorls with four spiral rows of nodules which align axially, base with four additional nodulose spiral ribs. Grey or brown with paler nodules.

2.5 cm. Vic., Tas., SA and southern WA as far west as Augusta. Abundant on estuarine mud flats. *Z. subcarinatus* (Sowerby, 1855) is a similar species from New Zealand, apparently introduced to NSW early this century.

FAMILY **CERITHIIDAE**

Creepers

The true ceriths, commonly known as creeper shells, generally have long, tapered, many-whorled shells, with a distinct anterior canal which may be drawn out ("produced") and upturned. A posterior (anal) canal is usually present as a notch where the lip edge joins the body whorl. The operculum is horny, subcircular and has an off-centre nucleus. The body lacks epipodial tentacles. Cerith shells may be intricately sculptured and some are colourful.

Most ceriths are gregarious detrital feeders and live in sandy or muddy substrates. They may be extremely abundant in patches. The radula is typically taenioglossate, with the two long marginal teeth on each side folding over the laterals when not in use. They scrape fine particles of detritus, micro-algae and bacteria from the surface of the substrate, along with a quantity of indigestible particles. The stomach is large, complexly folded and ciliated, and there is a short crystalline style. Each species seems to specialise in a particular size fraction of food particles, which may help explain why they are found in particular habitats characterised by sediment grain size.

Female ceriths lay their eggs in rods or twisted strings of jelly and the young hatch as free-swimming veliger larvae. The males have no penis and produce their sperm in packets (spermatophores).

A feature of the family is the exteme variability of sculpture and colour pattern in the shells of many species. While this makes the ceriths an interesting group, it means that the shells in hand may be very difficult to identify. The variation may be both geographic (i.e. between different populations) and local. Some species, e.g. *Clypeomorus bifasciata*, are poltypic with several distinct colour forms within a single population. There is also evidence that some species may hybridise.

The family is most strongly represented in shallow tropical waters, but there are many species (usually small ones) which live in cooler waters of the temperate zones and the outer edge of continental shelves.

Although several subfamily groupings have been recognised the relationships within the Cerithiidae are not yet fully resolved, and no attempt is made here to arrange the genera in subfamilies. Dr Houbrick of the Smithsonian Institution has reviewed many of the genera and references to his works are given in the appropriate places in this book.

▲ GENUS **ARGYROPEZA** Melvill & Standen, 1901
A. divina Melvill & Standen

Shells thin and glassy, turreted; suture deeply impressed; sculptured with axial riblets and two sharply nodulose spiral cords; protoconch multispiral, sculptured with spiral lirae and pustules; aperture ovate, anterior canal distinct, short and broad, posterior canal weak; with a prominent varix on the body whorl opposite the outer lip and former varixes on the spire whorls.

Indo-West Pacific. There are five living species, three of which have been recorded from northern Australia. They live in mud or fine sand in deep water. The genus has been reviewed by Houbrick, who described the anatomy in detail (1980, *Smithsonian Contributions to Zoology*, No. 321: 1-30).

Argyropeza izekiana Kuroda, 1949

Apical angle 35°; whorls flat-sided, with two weakly beaded spiral cords, the lower one sometimes weak; 12 to 14 axial ribs per whorl. Shining white.

5 mm. Central Indo-West Pacific; recorded from several localities in northern WA and Qld at depths from 100 to 360 m. *A. verecunda* (Melville & Standen, 1903); *A. schepmaniana* Melvill (1912) have also been recorded from northern Australia but from fewer localities.

▲ GENUS **BITTIUM** Leach in Gray, 1847
reticulatum da Costa

Small, turreted, whorls varicose, body whorl small; aperture ovate, anterior canal shallow, wide, not extended; posterior canal lacking or indistinct; sculpture of axial and spiral ribs, usually beaded; columella simple, outer lip thin and not expanded.

Cosmopolitan. Although some southern Australian species have been assigned to the subgenus *Cacozeliana* there has been no review and other species remain in the literature simply as *Bittium*.

▲ SUBGENUS **BITTIUM** s.s.

Operculum with many whorls.

Cosmopolitan. In addition to the illustrated species the following names have been used for Qld material: *diplax* Watson, 1886; *porcellanum* Watson, 1886; *pupiforme* Watson, 1880; *xanthum* Watson, 1886.

Bittium elegantissimum (Hedley, 1899)

Tall and slender, whorls flattened, the last long and no broader than the penultimate whorl; suture deeply excavated; sculpture of weak axial ribs and spiral cords, gemmate at the intersections, prominently varixed; aperture shelved within, columella arched and calloused, anterior canal wide, very short, slightly recurved. Brown, ribs picked out in white.

5 mm. Western Pacific; Described from Funafuti Atoll with records from north Qld. Synonym: *torresiense* Melvill & Standen, 1899.

Bittium zebrum (Kiener, 1842)

Relatively stout, sides of spire convex in outline; anterior canal oblique; sculpture of spiral rows of small, equi-sized beads which may become axially aligned to form nodulose axial ribs. White or brown, usually with two narrow darker brown spiral lines.

6 mm. Indo-West Pacific; Qld. Very common under coral stones or among coral rubble; intertidal.

▲ SUBGENUS **CACOZELIANA** Strand, 1928
Bittium granarium Kiener

With many granular whorls; operculum with few whorls.

Cosmopolitan in temperate seas. *Cacozeliana* is a replacement name for *Cacozelia* Iredale 1924 (*lacertina* Gould = *granarium* Kiener). In addition to the species illustrated here, the following NSW species are assigned to this subgenus: *furva* Watson, 1886; *fuscocapitulum* Hedley & Petterd, 1906; *variegatum* Henn & Brazier, 1894.

Bittium granarium Kiener, 1842

With about 11 whorls, each with four spiral rows of rounded beads giving the shell a granular appearance. Brown, fawn or grey.

3 cm. NSW to Fremantle, WA. Very common among weeds and rocks in estuaries and sheltered bays. Hedley (1916) considered that the NSW form is less tall and more closely granulose than the southern form and recommended that the name S. *lacertinum* Gould, 1861 should be used for it.

Bittium icarus (Bayle, 1880)

With about nine whorls bearing axial folds crossed by spiral cords, making the surface roughly nodulose.

1.2 cm. NSW to the south coast of WA.

▲ GENUS **CERITHIUM** Bruguière, 1789

C. adansonii Bruguière (= *nodulosum* Bruguière)

Turreted, with a short to moderately short anterior canal which is usually oblique and slightly upturned; posterior canal usually distinct; columella simple or slightly elevated, central spiral plait lacking but there may be a spiral nodule on the parietal wall beside the posterior canal; usually strongly sculptured.

Cosmopolitan in tropical and warm temperate waters. The genus has been comprehensively monographed by Houbrick (1992, *Smithsonian Contributions to Zoology* **510**: 211pp.). The monograph includes reviews of the radulae, reproductive biology and ecology of cerithiid species. Full synonymies are also given in that publication and they are not repeated here except where they have special relevance in the Australian context.

Most species live in the shallows. There are very many species with quite a lot of variation in shell form among them. Synonym: *Contumax* Hedley, 1899 (*decollatus* Hedley = *nodulosum* Bruguière).

Cerithium atromarginatum Dautzenberg & Bouge, 1933

Stoutly ovate, anterior canal deep but very short; protoconch of five whorls, the later whorls sculptured with two spiral cords and subsutural plications; teleoconch whorls without sculpture or sculptured with 11-13 smooth or finely beaded spiral cords and widely spaced, axially elongate nodules below the sutures; body whorl with a weak oblique varix, slightly constricted at the anterior canal; columella concave, with a slight callus; posterior canal defined by a prominent columellar plait. White, cords may bear thin spiral brown dashes, often with red-brown blotches, characterised by large dark brown patches at the lip below the suture and at the anterior end.

1.5 cm. Indo-West Pacific; North West Cape, WA to Caloundra, Qld. Lives in the intertidal zone in areas exposed to high wave action, often associated with algal turfs. The stout form, less granulose spiral cords and posterior spot distinguish this small species from *C. egenum* which often occurs in the same places.

Cerithium balteatum Philippi, 1848

Pl. 17; fig. 9 a-b   Elongately fusiform, pagoda-like; protoconch comprising $2\frac{1}{2}$ smooth whorls; anterior canal long, tubular, straight or slightly oblique; posterior canal small but well defined by a parietal tooth; base deeply excavated, demarcated by a prominent spiral rib around the periphery; whorls slightly convex, sculptured with axial ridges (about 10

per whorl) which may be roughly nodulose, crossed by 0-4 thin spiral cords; columella concave, with a narrow callus. White to yellow, usually with a brown subsutural band and often another round the periphery of the body whorl, nodes and ribs usually white.

3 cm. Point Cloates, WA to Townsville, Qld (but with a record also from Twofold Bay, NSW). Lives in shallow, sandy habitats, often associated with coral reefs. This is a common and very variable species. The two shells illustrated here indicate the range of variability, the first (*a*) from Dampier, WA. and the second (*b*) from Orpheus I., Qld Compare with *C. torresi*.

Cerithium citrinum Sowerby, 1855

Pl. 17; fig. 1 a-b   Elongate, turreted, whorls convex; outer lip greatly inflated, thick, crenulate; anterior canal long, tubular, horizontal, reflexed to the left; posterior canal well defined by a distinct parietal tooth; columella concave, with a thick callus forming an elevated columellar lip; sculptured with broad, inflated axial ribs, 8-11 on the penultimate whorl, crossed by incised lines and about 13 finely granulose spiral cords. Lemon-yellow, sometimes with diffuse brown blotches, anterior canal often dark brown or bluish.

4.5 cm. Indo-West Pacific; North West Cape, WA to Bunker Group, Qld. Sandy habitats in the sublittoral zone, usually in backreef or lagoonal situations. It does not live in dense populations like most species of the genus.

Cerithium columna Sowerby, 1834

Pl. 17; figs. 3, 4   Broad or slender, fusiform, turreted, base excavated; whorls angulate to straight-sided, sculptured with strong axial ribs, about 18 on the penultimate whorl, and granulose spiral cords which are strongest at the periphery of the body whorl; outer lip foliated; large, randomly placed varixes usually present; anterior canal long, moderately tubular, slightly upturned and reflected to the left; posterior canal distinct, bordered by a strong parietal plait; columella concave, heavily calloused, with a thick lip. White or cream, finely flecked with brown or purple.

4 cm. Indo-West Pacific; Abrolhos, WA to Gladstone, Qld and Lord Howe and Norfolk Is. Lives in intertidal and sublittoral sandy and rubble habitats, usually in reefal areas. The shells are extremely variable and there are many synonyms. Fig. 4 represents an extreme form which is probably a variant of this species.

Cerithium coralium Kiener, 1841

Pl. 17; fig. 12 a-b   Elongate, turreted, suture deeply impressed, varixes weak; protoconch of four whorls with a strong sinusigeral notch; sculptured with granulose spiral cords, three on the adult whorls, plus fine threads in the interspaces, crossed by 11-21 axial ribs; multiple varixes placed at random intervals; aperture hemispherical, outer lip thickened, weakly crenulate; anterior canal short, wide, slightly recurved; posterior canal deep, bordered by a strong parietal denticle and callus; columella slightly concave, with a thick callus and lip. Black to purple-brown or grey, granulae darker; interior white or light purple-brown.

4 cm. Indo-West Pacific; Shark Bay, WA to Long Reef, NSW. Abundant in mangroves, often associated with potamidids in these high intertidal, muddy habitats.

Cerithium dialeucum Philippi, 1849

Pl. 17; fig. 2   Stoutly acuminate, light, base excavated, whorls convex to angulate, with impressed suture; later whorls sculptured with nodulose axial folds, those on the body whorl with elevated pointed nodules, crossed by 5-6 beaded spiral cords, and a subsutural band formed by two or three beaded spiral threads; varix strong, opposite the lip; aperture ovate, wide, lip margin crenulate; anterior canal short, reflected, constricted; posterior canal well defined, bordered by a strong parietal tooth; columella concave, with a thick callus and lip. Cream, tan or fawn, usually striped with darker brown or purple-black; aperture white but with the inner side of the outer lip striped.

3 cm. Indo-West Pacific; Broome, WA to Keppel Bay, Qld. Lives in intertidal and shallow sublittoral, silty habitats of bays and lagoons, usually associated with algae.

Cerithium echinatum Lamarck, 1822

Pl. 16; figs. 1, 2 a-b   Stout, robust; protoconch of $2^{1}/_{2}$ whorls, the second sculptured with spiral cords, a deep sinusigeral notch present; teleoconch whorls inflated, usually with a central spiral cord bearing prominent spinose nodules and 3-4 secondary nodulose, spiral cords, interspaces between the spiral cords with deeply incised spiral lines; aperture ovate, outer lip crenulate, spinose along the edge, its anterior end partly reflected over the canal; anterior canal short, tubular, reflected dorsally and to the left; posterior canal wide, formed between a constriction of the outer lip and a large parietal tooth; columella slightly concave and with a thick callus and well-formed lip. Cream with purplish brown spots, aperture white.

8 cm. Indo-West Pacific; Rottnest I., WA to the Capricorn Group, Qld. Very common; lives in rocky areas in the intertidal and shallow sublittoral zones. The central spinose nodules are sometimes lacking.

Cerithium egenum Gould, 1849

Small, narrow, turreted, body whorl concave below the suture; protoconch of three whorls, the last elaborately sculptured and with a prominent sinusigeral notch; adult whorls convex to angulate, sculptured with incised spiral lines and 5-7 beaded spiral cords of unequal size; aperture ovate, outer lip smooth; anterior canal very short, reflected to the left; posterior canal moderate, bordered by a small parietal tooth; columella concave, slightly calloused. White with axial

chestnut markings on the base, sometimes spirally banded or with brown blotches.

8 mm. Indo-West Pacific; Geraldton, WA to Woolgoola, NSW. Lives in the intertidal zone, usually associated with algae on rubble substrates. Resembles *Cerithium atromarginatum* and may occur in the same places. It may be distinguished by its more slender shape, granulose spiral cords, the chestnut flames on the base, and the lack of the brown anterior spot on the outer lip which characterises the latter species.

Cerithium interstriatum Sowerby, 1855

Elongate, whorls convex; protoconch of $3^{1}/_{2}$ corded whorls, with a sinusigeral notch; adult whorls sculptured with 3-4 beaded spiral cords and thin spiral threads; base weakly excavated; aperture small, round, outer lip thin, weakly crenulate; anterior canal short, tubular, sharply reflected upwards and to the left; posterior canal small, bordered by a tiny parietal tooth; columella strongly concave, with a prominent callus.

1.6 cm. Indo-West Pacific; the only Australian record so far is from Middleton Reef, Coral Sea. This is a widespread but uncommon species which lives in silty sand of coral reef lagoons. See Houbrick, 1992, *loc. cit.*, figs. 64-66.

Cerithium munitum Sowerby, 1855

Pl. 17; fig. 5   Elongate, turreted, rather thin-shelled, whorls flat-sided to centrally angulate; outer lip flaring, anterior canal broad but short, slightly oblique; adult whorls with a spiral row of subsutural plicae, numerous fine spiral striae, axial folds becoming high nodulose ridges on the body whorl, and two semispinose, keel-like ribs around the periphery of the body whorl; strongly varixed; aperture ovate, outer lip flared, crenulate on the margin, lirate within; anterior canal of moderate length, reflected dorsally and to the left; posterior canal deep, bordered by a strong parietal tooth; columella concave, with a strongly calloused lip. White, cream or tan, sometimes brown, with thin spiral lines or lightly flecked with brown on the nodules.

3.9 cm. Indo-West Pacific; eastern Qld at least as far south as Townsville. A coral reef species which lives in sandy substrates in the sublittoral zone. Not very common. The double, nodulose peripheral keel is characteristic.

Cerithium nesioticum Pilsbry & Vanatta, 1906

Elongately fusiform; protoconch of $3^{1}/_{2}$ whorls, with a deep sinusigeral notch; adult whorls with convex sides, sculptured with four smooth or finely beaded spiral cords and weak axial ridges below the sutures; varix wide and low, opposite the aperture; aperture small, lenticular, outer lip smooth or weakly crenulate; anterior canal broad, very short; posterior canal distinct, bordered by a weak parietal tooth; columella concave, with a distinct callus and lip. White, sometimes with spiral rows of brown spots, at least at the suture.

2 cm. Indo-West Pacific; Cape Naturaliste, WA to Caloundra, Qld. This species is widely distributed but rarely abundant. It lives in sand associated with rocky substrates in the intertidal and shallow sublittoral zones. This is the species Hedley, 1899 named *C. spiculum* (from Funafuti Atoll) and Iredale, 1929 named *C. collacteum*.

Cerithium nodulosum Bruguière, 1789

Pl. 16; fig. 17 Elongate, robust and heavily sculptured; protoconch of two whorls, with a prominent sinusigeral notch; adult whorls angulate, each bearing a single spiral row of prominent tubercles and a number of weakly nodulose spiral ribs, strong nodulose spiral ribs also present at the anterior end of the body whorl; base strongly excavated, anterior canal short, constricted, reflexed; aperture large, outer lip thick, flaring, crenulate, spirally channelled within, with a claw-like anterior extension crossing over the anterior canal and a strong parietal ridge beside a deep posterior canal; columella concave, with a thick callus and distinct though narrow lip. Exterior white with brown blotches but usually thickly covered with calcareous growths, interior and columella white.

1.5 cm. Indo-West Pacific; Rowley Shoals and Scott Reef, WA to the Bunker Group, Qld. This is a very widespread, common and easily recognised species which lives in sand and rubble on intertidal reef flats, usually near the reef edge. It is the largest living species of the genus. Hedley, 1899 named a topless specimen *C. decollata*.

Cerithium novaehollandiae Adams in Sowerby, 1855

Pl. 16; fig. 4 a-b Slender and tapering, turreted; early whorls with three spiral threads, later whorls with spiral cords and axial ribs, 10-16 on the penultimate whorl, the ribs often nodulose; aperture oval, outer lip thick and crenulate; anterior canal short, reflected to the left; posterior canal distinct but small, bordered by a small parietal plait; columella concave, with a thick callus and distinct lip. White, with a broad, brown, subsutural band.

5 cm. Abrolhos, WA to Moreton Bay, Qld. An intertidal and shallow sublittoral species which seems to prefer silty sand habitats. Although common in northern Australia there are a few records from elsewhere, namely from New Caledonia and Papua New Guinea.

Cerithium pacificum Houbrick, 1992

Elongately pupate, shiny, sides of spire convex, body whorl constricted subsuturally, rounded at the base; adult whorls weakly sculptured with axial ribs and beaded spiral cords, sculpture becoming obsolete on the body whorl; aperture narrowly ovate, outer lip smooth, thick, weakly flaring; anterior canal moderately long, constricted, reflexed to the left; posterior canal well formed, bordered by a parietal tooth. Cream to tan, sometimes with axial flames of darker tan.

2.7 cm. Western Pacific; recorded in Australia so far only from Log Reef in far north Qld. A sublittoral coral reef lagoon species. See Houbrick, 1992; *loc. cit.*, fig. 103.

Cerithium phoxum Watson, 1880

Elongate, tapering, sharply pointed; adult sculpture finely cancellate, consisting of weakly beaded spiral cords, 13-18 axial folds, and numerous spiral striae; base only moderately excavated; aperture narrowly ovate, outer lip thick, slightly crenulate, finely striate within; anterior canal of moderate length and moderately constricted, sharply recurved to the left; posterior canal deep, defined by a strong parietal tooth; columella concave, thinly calloused, lip distinct.

3 cm. Western Pacific; recorded in Australia from Torres St. south to Mackay, Qld. Mainly a sublittoral species, living on coarse sand and rubble substrates in reefal areas. See Houbrick, 1992, *loc. cit.*, fig. 105.

Cerithium punctatum Bruguière, 1792

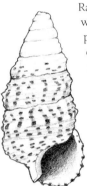

Rather stout, conic-ovate; protoconch of three whorls; adult whorls angulate centrally, penultimate whorl with 2-5 dominant spiral cords, the central one strong and weakly nodulose, numerous striae present in the interspaces; body whorl with an irregular, oblique varix; aperture ovate, outer lip thick, smooth; anterior canal deep but very short, oblique; posterior canal shallow, bordered by a weak parietal plait; columella concave, with a slight callus and weak lip. White, with regular dark brown spots on the ribs tending to align in axial rows; columella purple to violet.

2 cm. Indo-West Pacific; Abrolhos, WA to Lady Elliot I., Qld. Lives associated with algae on intertidal rock flats. It is quite easy to confuse this little species with *Clypeomorus nympha* Houbrick, 1985 but this shell is not pupate or beaded.

Cerithium rostratum Sowerby, 1855

Elongately fusiform, whorls inflated and with strongly convex to angulate sides; protoconch large, of three elaborately sculptured whorls, with a deep sinusigeral notch; adult whorls sculptured with wide to narrow axial ribs which become obsolete on the body whorl, crossed by crowded, irregular, spiral cords, sometimes forming tiny beads at the intersections; varix low, opposite the lip; aperture ovate, outer lip strongly convex, crenulate, adpressed anteriorly to form the anterior canal tube; anterior canal short to moderate, tubular, oblique; posterior canal weak, bordered by a small parietal plait; columella concave, with a thin and narrow callus. White to cream, fawn or mauve, with thin brown spiral lines interrupted by the axial ribs, anterior canal mauve or brown.

2 cm. Indo-West Pacific; Ningaloo Marine Park, WA to Caloundra, Qld. Lives in lagoonal habitats, usually associated with seagrass beds and abundant in its preferred habitats. This is the type species of *Ischnocerithium* Thiele, 1929, distinguished on the basis of the long siphonal canal, but Houbrick (1992, *loc. cit.*, p. 161) relegated the name to the synonymy of *Cerithium*.

Cerithium salebrosum Sowerby, 1855

Pl. 17; fig. 7 a-c   Unusually slender, fusiform; adult whorls shouldered and flat-sided, sculptured with 10-14 axial ribs rendered sharply nodulose where they are crossed by 4-5 spiral cords making a very prickly surface; with large, randomly placed varixes; base sharply excavated, demarcated by a spiral rib around the body whorl periphery; aperture narrowly ovate, outer lip crenulate, with incised spiral lines internally; anterior canal short, upturned, narrowly constricted; posterior canal deep, well defined by a parietal tooth; columella concave, with a thick callus and distinct lip. White, sometimes with fawn bands or blotches.

2.5 cm. Indo-West Pacific; Scott Reef, WA to eastern Qld at least as far south as Townsville. Lives sublittorally in clean sandy substrates, usually associated with coral reefs. Dead shells are commonly found drilled by naticids.

Cerithium tenellum Sowerby, 1855

Pl. 17; fig. 8 a-d   Thick, fusiform, turreted; whorls slightly convex, sometimes angulate centrally; sculpture of adult whorls very variable, usually consisting of fine spiral cords and 4-5 spiral rows of elevated nodules, the first and second rows below the suture often more prominent but all rows may be sub-equal; varix well formed, placed oposite the lip; aperture fusiform-ovate, wide, outer lip thick, crenulate, slightly concave below the suture; anterior canal short, oblique; posterior canal well developed, bordered by a strong parietal plait; columella concave, with a thickly calloused lip. White, sometimes pink, usually with brown pepper spots in spiral rows in the interspaces, spots sometimes coalesce to form spiral lines.

3 cm. Indo-West Pacific; Abrolhos, WA to Bunker Group, Qld. An inhabitant of shallow, sandy lagoons, usually associated with coral reefs. Strongly angulate and nodulose forms from Qld were named *C. tomlini* Hedley, 1914 (later replaced with *C. mysterium* Hedley, 1917) but, according to Houbrick, they fall within the range of variation of this species.

Cerithium torresi Smith, 1884

Pl. 17; figs. 10, 11 a-b   Rather slender, turreted, with a slightly channelled suture; protoconch of $2^1/_2$ whorls, with a deep sinusigeral notch; sculpture cancellate, spire whorls with three spiral rows of rounded nodules aligning axially and sometimes forming axial ridges, connected by spiral cords and with striae in the interspaces, body whorl with an additional two rows of smaller nodules around the periphery, plus some spiral cords; varixes well formed, randomly placed; aperture wide, ovate, outer lip thick, flaring, crenulate, dentate within; anterior canal moderate, oblique, constricted; columella concave, with a strong callus and moderate lip. Off-white, overlain with brown or fawn, with or without brown spots or blotches in the interspaces.

2.5 cm. Indo-West Pacific; Abrolhos, WA to Stradbroke I., Qld. Very common on muddy intertidal sand flats and the shallow sublittoral zone.

Cerithium traillii Sowerby, 1855

Pl. 16; fig. 3 a-b   Broad at the base, turreted, whorls slightly convex; base of body whorl rounded; varixes well formed on all whorls, usually at about every half turn; sculptured with numerous spiral rows of small rounded nodules on narrow spiral ribs, five primary rows on the body whorl, plus several smaller secondaries, nodules often aligning axially and forming 24-45 axial ribs; aperture wide, ovate, outer lip flaring, varixed, weakly crenulate; anterior canal short, tubular, oblique; posterior canal large, bordered by a strong parietal plait; columella concave, with a thick lip. White to fawn, varixes paler, sometimes banded, nodules brown giving the shell a spotted appearance.

4 cm. Central Indo-West Pacific; Exmouth Gulf, WA to Whitsunday Group, Qld. This is a very distinctive and moderately common cerith. It lives in moderately muddy situations near mainland type islands.

Cerithium zonatum (Wood, 1828)

Pl. 17; figs. 6 a-b, 13 a-c   Broad to slender, turreted, whorls flat-sided to inflated, sometimes angulate, base rounded not excavated; protoconch of two weakly sculptured whorls and a sinusigeral notch; adult whorls sculptured with many weak axial cords crossing 3-4 major spiral cords bearing rounded or pointed nodules, fine threads in the interspaces; varix on the body whorl opposite the lip, moderately strong; aperture ovate, outer lip slightly thickened marginally, sharply crenulate, spirally lirate within; aperture oval, outer lip lirate and thickened; anterior canal short, tubular, recurved; posterior canal well developed, bordered with a parietal plait; columella concave, with a thick callus and moderate lip. White, with alternating brown and white bands and spiral rows of brown spots; columella brown, aperture white.

3 cm. Indo-West Pacific; Shark Bay, WA to Capricorn Group, Qld. An extremely variable intertidal species living in sand and rubble substrates, often among rocks. *C. probleema* Iredale, 1929, introduced as a new name for *C. lemniscatum* Quoy & Gaimard, 1834, is one of many synonyms.

▲ GENUS **CLYPEOMORUS** Jousseaume, 1888
C. clypeomorus Jousseaume (= *bifasciata* Sowerby)

Shells not exceeding 3 cm, short, squat, with 8-10 inflated whorls; columella deeply concave; anterior siphon canal very short, deeply incised and directed leftward at about 45° to the shell axis; posterior canal distinct; shell sculpture of two or three nodulose spiral cords, axial sculpture weak.

Indo-West Pacific. The genus has 12 living species, some of which are extremely variable. Nine species are recorded from northern Australia. They are all confined to the intertidal zone, usually on coarse, rocky substrates, and they are able to withstand long periods of exposure to sun and air at low tide. The animals are gregarious and populations are sometimes extremely dense. The Latin generic name refers to the mulberry-like dark, beaded appearance. Synonyms are not given here but may be found in the recent comprehensive review published by Houbrick, 1985, *Smithsonian Contributions to Zoology*, **403**: 1-131.

Clypeomorus admirabilis Houbrick, 1985

Pl. 17; fig. 15 Elongate, spire whorls with two thick nodose spiral cords, nodes spinose or blunt, aligned to form about nine axial ribs per whorl; body whorl with 5-6 nodose spiral cords, axial ribs prominent; entire shell covered with fine spiral threads; outer lip varixed, inner side bevelled and with four pairs of spiral plicae, varix extending slightly onto the penultimate whorl; former varix present opposite the lip; anterior canal deflected at 45° to the shell axis.

2.6 cm. Endemic to north-western Australia between Carnarvon, WA and Coburg Peninsula, NT. Sand and rubble on intertidal flats near mangroves.

Clypeomorus batillariaeformis Habe & Kosuge, 1966

Pl. 17; figs 17 a-d Body whorl rounded in cross-section, not dorso-ventrally flattened; spire whorls rather straight-sided; beaded sculpture usually not axially aligned; former varix of the body whorl situated high on the left side not opposite the lips; lower part of the outer lip seems to close over the anterior canal when viewed anteriorly. Dark brown to buff, often greyish with black or brown beads which may be drawn out to form transverse lines.

1.5 cm. Indo-West Pacific; Shark Bay, WA to Moreton Bay, Qld. Lives in dense clusters in the upper part of the midtidal zone, in both marine and estuarine situations. The high position of the body whorl varix, narrower lip, and rounded ventral side distinguish this species from *C. bifasciata*. For data on the biology of this species (identified as *C. moniliferum* Kiener) see Cannon, 1975, *Pacific Science*, **29**: 353-359; and 1979, *Australian Journal Marine and Freshwater Research*, **30**: 365-374.

Clypeomorus bifasciata (Sowerby, 1855)

Pl. 17; fig. 22 a-b Elongate to obese; body whorl flattened dorso-ventrally; spire whorls with three major, equally spaced, heavily beaded spiral cords and several finer spiral threads; body whorl with six major heavily beaded cords, plus two on the siphonal constriction and many fine spiral threads; beads aligned in axial rows; outer lip crenulate, bevelled, with numerous internal denticle-like lirae; when viewed anteriorly the outer lip lower than the canal and not closing over it; former varix of the body whorl directly opposite the outer lip. Colour extremely variable, usually grey, with brown or black blotches on paler background; beads usually dark; whorls sometimes spirally banded.

3 cm. Indo-West Pacific on continental coasts or high islands; Geraldton, WA to Moreton Bay, Qld. Lives high in the intertidal zone; often extremely abundant. The name *Cerithium morus* Lamarck, 1822 (not Bruguière, 1792) is commonly misused for this species.

Clypeomorus brevis (Quoy & Gaimard, 1834)

Pl. 17; fig. 16 a-b Squat, body whorl inflated, width a little more than half shell length; spire whorls with two major spiral cords, six on the body whorl; spiral cords heavily beaded where they cross strong axial ribs creating a cancellate appearance; outer lip thick, strongly varixed on the outer side, crenulate, with thick interior denticles; former varix on left side of the body whorl directly opposite the lip; varixes rarely present on the spire whorls. White with brown or grey spiral lines, spots or blotches, sometimes banded.

2.4 cm. Oceanic atolls and low islands of the western Pacific; recorded from Ashmore I. but not yet from mainland Australia. Lives on hard substrates such as beachrock; often extremely abundant.

Clypeomorus irrorata (Gould, 1849)

Pl. 17; fig. 14 Thick-shelled, body whorl inflated, about half length of shell, with many spiral cords and threads and axial ribs on upper third; spire whorls with 9-15 broad axial ribs crossed by spiral cords and threads; outer lip smooth, thick, internally lirate, and with a thick outer varix; former varixes lacking. Colour variable; white, brown, grey or black, with darker spiral dotted lines.

2.4 cm. Indo-West Pacific; north Qld. Lives in sand among rocks between mid and high tide zones. Few Australian records but it may be widespread across the north.

Clypeomorus pellucida (Hombron & Jacquinot, 1852)

Pl. 17; fig. 21 a-b Elongate, body whorl large, inflated, sculptured with eight or nine strong beaded spiral cords and a number of weaker ones, spire whorls with three nodose spiral cords crossing strong axial ribs; former varixes present on the spire whorls, that on the body whorl dorsal, large and angular; outer lip thick, flaring, crenulate or toothed on the inner surface. Brown, spiral cords usually white.

3 cm. Western Pacific; Qld south to Stradbroke I. An estuarine species which lives on aerial roots of mangroves at about high tide level.

Clypeomorus petrosa (Wood, 1828)

Pl. 17; fig. 20 Adult shell obese, body and penultimate whorls inflated; body whorl with eight or nine beaded or nodose spiral cords and numerous spiral threads, spire whorls with three nodose spiral cords and numerous threads, nodes sometimes spinose; strong former varixes present; outer lip thin, moderately crenulate, lower part appears to extend slightly over the anterior canal when viewed anteriorly. White, cream, yellow or brown; nodes dark brown or black giving spotted pattern; sometimes banded.

3 cm. South-western Pacific; eastern Australia from Torres St. to northern NSW. Lives in sand among rocks from low to midtidal zones.

Clypeomorus purpurastoma Houbrick, 1985

Fusiform, whorls moderately inflated, suture straight and indistinct; spire whorls with three beaded spiral cords, spiral lirae and numerous incised threads; body whorl with eight beaded spiral cords; moderate varix on right dorsal side; outer lip varixed, bottom partially crosses anterior canal. White with dark brown beads; columella and inner side of lip purple or violet.

2.7 cm. Indo-West Pacific; central Qld. Houbrick did not record this species from Australia but Mollie Bowman has sent me specimens from Tryon I. with the characteristic purple aperture, a character which readily distinguishes it from *C. batillariaeformis*. For illustrations see Houbrick, 1985 (*loc. cit.*, figs. 38-40).

Clypeomorus subbrevicula (Oostingh, 1925)

Pl. 17; fig. 18 Stout, obese; aperture wide, nearly half shell length; spire whorls with two major beaded cords, and several spiral threads which may be weakly beaded; body whorl inflated, with six or seven beaded spiral cords that are more or less axially aligned; entire shell with numerous incised lines and fine, incised axial lines giving a scaly appearance; outer lip thick, crenulate, interior bevelled and bearing numerous spiral plicae; former varixes lacking. Grey or brown with dark brown nodes.

2.5 cm. Western Pacific; recorded from north Qld. Lives on intertidal rocks, often in wave-exposed positions.

▲ GENUS COLINA H. & A. Adams, 1854
Cerithium macrostomum Hinds

Shells small, elongate, pupate, usually decollate, midwhorls gibbous, body whorl constricted; protoconch small, of one whorl, smooth; columella with a strong internal plait; aperture narrowly ovoid, with an expanded outer lip; operculum corneus, ovoid, paucispiral, nucleus subterminal.

Indo-West Pacific. The genus has been reviewed by Houbrick (*Nautilus* 104 (2): 35-52). There are three living species, at least one of them present in northern Australia. The decollate apex and constricted body whorl together distinguish this group from other cerithiid genera. The paucispiral protoconch suggests that these animals have direct development. They live in the shallow sublittoral zone, apparently associated with seagrass beds.

Colina macrostoma (Hinds, 1844)

Elongate, turreted, with about 15 inflated whorls; upper whorls sculptured with three spiral cords and numerous fine axial striae, later whorls with four major and four minor spiral cords, the central one carinate, crossed by 12-13 weak to strong axial ribs forming a cancellate pattern, intersections sometimes beaded, body whorl strongly constricted, with broad, flattened cords; aperture ovately elongate, with a flaring outer lip. Pinkish tan to light brown, with dark brown spiral lines and bands, blotched with white, especially on the beads.

1.9 cm. Indo-West Pacific; North West Cape, WA to Mackay, Qld. Synonyms: *pupiformis* A. Adams, 1853; *costatum* A. Adams in Sowerby, 1855; *pupaeforme* A. Adams in Sowerby, 1855; *costiferum* Sowerby, 1855; *gracilis* H. Adams, 1866; *pygmaea* H. Adams, 1867; *coarctatum* Sowerby, 1866; *ringens* Bayle, 1880.

▲ GENUS GLYPTOZARIA Iredale, 1924
Turritella opulenta Hedley

Shell thin, slender, turreted, suture deep, whorls with cancellate sculpture; protoconch smooth; aperture ovate with slight anterior canal, anal canal lacking; operculum ovate, moderately spiral, nucleus central. Animal with broad head, large eyes, bilobed and papillate mantle edge.

Western Pacific. This is a deepwater genus. Its taxonomic position has been problematical but resolved by Houbrick (1981, *Proceedings Biological Association Washington*, 94 (3): 838-847), who studied the anatomy and radula of the type species and showed that it is a cerithiid. There are two living species and one Miocene fossil species.

Glyptozaria opulenta (Hedley, 1907)

Apical angle about 16°; protoconch of $1^1/_2$ smooth whorls with a straight lip; teleoconch whorls about 10, angulate, with cancellate sculpture of spiral cords and weak axial riblets bearing pointed beads at the intersections, body whorl with a strong varix opposite the lip, base spirally corded; outer lip slightly crenulate at the margin. Off-white.

8 mm. Western Pacific; eastern Australia south to eastern Bass St. Dredged at depths to over 530 m. There is one confirmed record from Japan, indicating that the species is probably widespread. *G. columnaria* Cotton & Woods, 1935 is a second species of the genus from Tas. and SA.

▲ GENUS **GOURMYA** Bayle in Fischer, 1884
Cerithium gourmyi Crosse

Stout, solid, with inflated whorls and moderately impressed suture; smooth to spirally striate, outer lip recurved anteriorly over (but not attached to) the columella, so covering the anterior canal which is short and tubular; operculum oval, rather thick, with few whorls and eccentric nucleus.

South West Pacific. There is only one living species but there are several Tertiary fossil species appearing first in the Eocene. The unusual formation of the anterior canal resembles that of the potamidid genus *Terebralia* but the similarity is apparently due to convergent evolution as the anatomy of *G. gourmyi* is definitely cerithiid. For an account of the group, including details of anatomy and fossil record, see Houbrick (1981, *Nautilus* 95 (1): 2-11).

Gourmya gourmyi (Crosse, 1861)

Pl. 17; fig. 19   Smooth except for minute spiral striae; with nine teleoconch whorls; posterior canal bordered by a sharp-edged callus ridge on the parietal wall. Early whorls white or pink, later whorls white, with irregular but roughly axially aligned brown blotches and streaks; periostracum moderately thick and covers most of the shell in living specimens.

4.5 cm high; 2.5 cm wide. New Caledonia, Loyalty Is., Chesterfield Is., Brodie Cay and Marmion Reef, Qld. Found on the reef-front slopes of coral reefs at depths of 5-30 m, usually on algal-covered rocky surfaces.

▲ GENUS **PSEUDOVERTAGUS** Vignal, 1904
Cerithium aluco Linnaeus

Large, turreted shells with a sharply reflexed and elongated anterior canal; posterior canal distinct, columella concave, smooth, central plait lacking but with a heavy parietal callosity; outer lip margin approximately straight; aperture ovate.

Indo-West Pacific. There are two extinct fossil subgenera, as well as *Pseudovertagus* s.s. which originated in the Miocene and has five living species, four of which occur in Australia. They inhabit sandy or rubble substrates in the sublittoral zone. References: Houbrick, 1978, *Monographs of Marine Mollusca* 1: 99-120; Wilson, 1975, *Records W.A. Museum* 3: 327-333.

Pseudovertagus aluco (Linnaeus, 1758)

Pl. 16; fig. 13 a-b   Apical angle 35°; with 11-15 whorls; on the early whorls a steep sutural ramp is bordered by a subsutural spiral row of nodules which vary in size, later whorls and body whorl relatively smooth; anterior siphon canal wide, deep, sharply reflexed at 90°, anal canal distinct; columella smooth, heavily calloused, the anterior end of the outer lip turns and extends over the canal. White or cream, heavily blotched and finely spotted with purplish brown.

9.5 cm. Central Indo-West Pacific; Shark Bay, WA to Whitsunday Group, Qld. Lives in seagrass beds and coral rubble in the intertidal and shallow sublittoral zones. Synonym: *cumingi* Adams, 1855.

Pseudovertagus clava (Gmelin, 1791)

Pl. 16; fig. 14   Apical angle 25°; with 15-18 whorls; post-nuclear whorls with a broad sutural ramp and 2-5 prominent spiral ribs which are roughly and irregularly nodulose, and many intermediate cords crossing prominent axial folds; subsutural rib wide, bearing axial folds; body whorl with about nine spiral ribs, flat on the ventral side; aperture heavily enamelled, outer lip thick, weakly crenulate, lower portion extending slightly over the siphon canal; columella thick, heavily calloused and flanged; anterior siphon canal moderate, recurved 75°, anal canal well developed and extended posteriorly, bordered on the columellar side by a prominent plait. White with tan blotches.

14 cm. South West Pacific Ocean; central Qld. Lives in sand to 30 m. Though locality records are few, this species may be very abundant in its preferred habitat. As no Australian specimen was available, one from Tahiti was used to illustrate the species. See remarks on *P. peroni*.

Pseudovertagus peroni Wilson, 1975

Pl. 16; fig. 15   Teleoconch of about 13 whorls with almost straight sides; suture deeply incised; axial folds lacking; sutural ramp represented by a broad, raised spiral band divided by axial grooves into a series of rather flat axial ribs; later whorls with flattened spiral cords separated by narrow incised spiral lines; spiral sculpture slightly nodulose on the early whorls; with a weak former varix on the left side of the body whorl; outer lip flaring, sharp-edged anteriorly, thickened and reflected upwards posteriorly, weakly denticulate; anterior canal high, near vertical; columella moderately calloused, smooth except for a parietal plait bordering the deep and long anal canal. Shiny, ivory white with chestnut brown blotching between the spiral cords and distinct thin spiral brown lines.

13.5 cm. South-western Australia. Habitat unknown but probably lives in sand among sublittoral rocky ridges. Only a few dead specimens have been collected so far, mostly from rock-lobster pots in the vicinity of Garden I. Houbrick (1978, *loc. cit.*) treated the WA population as an end-of-range variant of *P. clava*. This may be correct although the differences in shell sculpture are considerable.

Pseudovertagus phylarchus (Iredale, 1929)

Pl. 16; fig. 16   Apical angle 30°m; with 14-18 whorls; post-nuclear whorls with spiral cords and axial ribs, later whorls smooth except for four weak spiral incised lines and a broad, thickened band behind the suture representing a sutural ramp, former varixes lacking; outer lip moderately thick, smooth, basal portion extends slightly over the anterior canal; columella moderately calloused, flanged at the base; anal canal deep, bordered by a strong columellar plait; anterior canal long, near vertical. White, with about five spiral rows of quadrangular brown spots on each whorl and brown blotches on the sutural ramp.

11 cm. Central Indo-West Pacific; central Qld. Lives in seagrass beds. The name *sowerbyi* (Kiener, 1841) has been used frequently for this species but it was used earlier for a fossil.

▲ GENUS **RHINOCLAVIS** Swainson, 1840

Murex vertagus Linnaeus

Turreted; with a sharply reflexed and elongated anterior canal and a well-defined anal canal; columella with a central oblique plait which is visible in the aperture and extends along the entire axis of the shell internally; outer lip margin approximately straight.

Indo-West Pacific and tropical eastern Pacific regions. Houbrick (1978, *Monographs of Marine Mollusca* 1: 1-97) has reviewed the genus, giving full synonymies and recognising four subgenera, two of which are represented in the Australian fauna.

▲ SUBGENUS **RHINOCLAVIS** s.s.

Anterior canal very long and reflexed; median columellar plait prominent; aperture not greater than one fifth length of shell.

Indo-West Pacific. There are eight living species all of which occur in northern Australia.

Rhinoclavis articulata (Adams & Reeve, 1850)

Pl. 16; fig. 10   Slender, light, glossy; 12-17 whorls, strongly sculptured, adult whorls with three primary beaded spiral cords separated by three smaller, smooth, spiral bands; former varixes indistinct; aperture oval, anterior canal relatively short; columella with a flange near the canal, columellar plait weak; outer lip slightly crenulate. Cream to white, sometimes irregularly blotched with tan; brown spots between the nodes on the first spiral cord.

5.5 cm. Indo-West Pacific; North West Cape, WA to Whitsunday Group, Qld. Subtidal sand and sandy rubble to 91 m; occasionally intertidal.

Rhinoclavis aspera (Linnaeus, 1758)

Pl. 16; fig. 5 a-b   Slender and tapering; strongly sculptured with angular axial ribs crossed by nodulose cords; anterior canal long and recurved to almost vertical; columella with a weak parietal ridge and one or more central spiral folds. Cream, usually spirally lined with brown.

6 cm. Indo-West Pacific; NT to Capricorn Group, Qld. Subtidal to 28 m in sandy substrates; a very common species which is usually associated with coral reefs. Synonyms: *granulatum* Linnaeus, 1758; *lineatum* Lamarck, 1822; *comptus* Sowerby, 1914.

Rhinoclavis bituberculata (Sowerby, 1865)

Pl. 16; fig. 8 a-b   Rather stout, sharp-spired, with 10-14 whorls; whorls finely spirally striate, bearing several weakly nodulose spiral cords, microscopically cancellate sculpture between the cords; anterior canal short but almost vertical; columella heavily calloused, with a strong central spiral fold and a low parietal ridge beside the posterior canal. Exterior fawn, spiral ribs and nodules white, apex blue-grey, interior and columella white.

6 cm. Cape Leeuwin, WA to southern Qld. Very common in sand substrates associated with rocky shores in the intertidal and shallow sublittoral zones. The spiral ribs are more heavily nodulose in specimens from the north of WA than in southern specimens. Synonyms: *semigranosum* Lamarck, 1822; *cordigerum* Bayle, 1880. Lamarck used the name *semigranosum* twice, the first time in 1804 for a different fossil species, so that this name is not available for the living species referred to here.

Rhinoclavis brettinghami Cernohorsky, 1974

Pl. 16; fig. 7 a-b   Resembles *R. bituberculata* but crowned with a spiral row of large nodules at the suture and sculptured elsewhere with numerous finely pustulose spiral cords.

6 cm. Barrow I., WA to Capricorn Group, Qld. Also recorded from Lord Howe I. *Rhinoclavis brettinghami* and *R. bituberculata* live together on intertidal reefs in northern WA. The name is a replacement for *pulchrum* Sowerby, 1855 (not Adams, 1852).

Rhinoclavis diadema Houbrick, 1978

Solid, with 13-15 whorls; sculptured with three prominent nodulose spiral cords, the sutural cord the largest bearing 14 or 15 spinose tubercles and forming a distinct sutural ramp; outer lip weakly crenulate; anterior canal rather short for the genus; anal canal indistinct; columellar plait moderately developed. White, spiral cords with brown spots.

2.5 cm. Western and Central Indian Ocean, Western Pacific; recorded from Cartier I., NT and Lady Elliot I., Qld. Subtidally to 7 m on sandy substrates. Synonym: *nitidum* Hombron & Jaquinot, 1854.

Rhinoclavis fasciata (Bruguière, 1792)

Pl. 16; fig. 12 a-c Solid, slender and acutely tapering with a high, nearly vertical anterior canal; early whorls faintly axially ribbed, later whorls glossy, smooth, sculptured only with fine spiral striae; columella calloused, with a broad central plait. External colour extremely variable, usually white with brown spiral lines or bands, sometimes uniformly white, interior and columella white.

9.5 cm. Indo-West Pacific; Shark Bay, WA to Capricorn Group, Qld. Sandy substrates to 18 m. Synonyms: *carminatus* Röding, 1798 (part); *bandatum* Perry, 1811; *martinianum* Pfieffer, 1840; *procerum* Kiener, 1841; *pharos* Hinds, 1844; *mountfordae* Cotton, 1964.

Rhinoclavis sinensis (Gmelin, 1791)

Pl. 16; fig. 6 a-b Solid, usually glossy, with 10-15 whorls; strongly sculptured, with four nodulose spiral cords on later whorls, fine sprial striae between the cords, and prominent nodes below the suture, nodes sometimes spinose, weak axial ribs sometimes present; body whorl with 8-19 nodulose cords; outer lip smooth; columella heavily calloused, with strong central plaits; anterior canal long. Cream with brown blotches and darker brown spots on the spiral cords between the nodes.

7 cm. Indo-West Pacific; NT to central Qld. Intertidal to 23 m; sandy or coral rubble substrates. The nodulose sculpture easily distinguishes this species. Synonyms: *obeliscus* Bruguière, 1792; *muricatus* Röding, 1798; *cedonulli* Sowerby, 1855; *chinensis* Chemnitz, 1882.

Rhinoclavis vertagus (Linnaeus, 1758)

Pl. 16; fig. 9 Rather solid and stout, with 10-13 whorls that are smooth anteriorly but more ventricose and axially plicate near the suture; early whorls rugose; former varixes present on the early whorls but weak or absent on the later whorls; aperture oval to fusiform, elongate, columellar plait prominent; siphon canal long, recurved almost 90°. Cream-white or yellow, sometimes with a broad light-tan spiral band behind the suture.

7 cm. Western Pacific and S. E. Asia; Shark Bay, WA to Moreton Bay, Qld. Sandy substrates to 13 m. May be extremely abundant. Synonyms: *varia* & *volvax* Humphrey, 1797; *carminatus* Röding, 1798 (part); *virgatum* Montfort, 1810; *despectum* Perry, 1811; *vulgaris* Schumacher, 1817.

▲ SUBGENUS **PROCLAVA** Thiele, 1929
C. pfefferi Dunker = *sordidula* Gould

Anterior canal relatively short and only moderately reflexed; shell long, tapering; anal canal weak; aperture small, one-fourth length of shell; sculpture beaded.

Indo-West Pacific. There are two living species.

Rhinoclavis kochi (Philippi, 1848)

Pl. 16; fig. 11 a-b Solid, shiny, with 14-17 whorls; sculpture of three strong, beaded spiral cords and a fourth weaker one close to suture, cords separated by deeply incised spiral lines; outer lip crenulate, columellar plait weak, anterior canal reflexed at 80° from shell axis. Brown, usually with a cream band on the first spiral cord, beads usually white; incised lines darker brown.

5 cm. Indo-West Pacific; North West Cape, WA to north Qld; subtidal to 60 m on muddy to sandy substrates. This species has established itself in the eastern Mediterranean. Synonyms: *subulatum* Lamarck, 1822 (not Lamarck, 1804); *recurvum* Sowerby, 1855; *kochi* var. *polita* Preston, 1908. The other member of *Proclava*, *R. sordidula*, was listed from Qld by Charles Hedley many years ago but I have not been able to confirm its presence there. It differs from *R. kochi* by its smaller size, narrower spire and the absence of incised spiral lines on the whorls.

▲ GENUS **ROYELLA** Iredale, 1912
Cerithium clathratum Sowerby (= *sinon* Bayle)

With characters of the type species.

Indo-West Pacific. Monotypic. Houbrick (1986, *Veliger* 28 (4): 429-435) has described the anatomy and shown that the genus belongs in the Cerithiidae.

Royella sinon (Bayle, 1880)

Turreted, suture deeply impressed, whorls angulate; early whorls cancellate, later whorls with two nodulose spiral cords and weak axial riblets; aperture circular, with a short anterior canal and poorly developed posterior canal. White.

3 cm. Indo-West Pacific; recorded in Australia from the Abrolhos, WA, Norfolk and Lord Howe I. and northern Qld. It lives on rubble substrates below low tide.

FAMILY **DIALIDAE**

Dialids

Dialids are tiny snails which are normally very abundant, especially in silty lagoons of coral reefs in the tropics. Because of this abundance they are an important element of shallow-water ecosystems.

The classification of the genus *Diala* has been problematical for many years. It has been assigned to many different cerithioidean families and has been used as a "grab bag" genus for a range of unrelated species. Ludbrook (1941) elevated it to family rank. Since then Dr Winston Ponder has described the anatomy of a southern Western Australian species and discussed the relationships of the genus with other Cerithioidea (1991, in Wells *et. al.*, *The Marine Flora and Fauna of Albany, Western Australia* **2**: 499-519, Western Australian Museum). This study showed that *Diala* is probably most closely related to the Litiopidae but warrants separation as a distinct family.

The Dialidae is distinguished from the Cerithiidae by an entire aperture, i.e. lack of an anterior canal or notch. It differs from the Litiopidae by a smooth protoconch and the lack of epipodial tentacles. Other diagnostic anatomical characters may be found in Ponder (1991).

Ponder has suggested that several genera besides *Diala* should be included in this family, viz.: *Dialopsis* Cossmann, 1889; *Laevitesta* Laseron, 1950; *Mellitesta* Laseron, 1950; *Rissoalaba* Oyama, 1954; *Paradilia* Laseron, 1956. Conversely some genera once associated with the dialids, e.g. *Alaba*, *Litiopa* and *Argyropeza*, have now been moved to other families.

An account of the NSW species of *Diala* has been given by Laseron (1950, *Records of the Australian Museum* **22**: 284-286). Only a few examples of *Diala* are included here to represent the group.

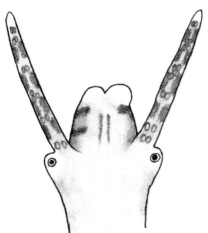

Head of *Diala* sp. (Redrawn from Ponder, 1991, fig. 4b).

▲ GENUS **DIALA** Adams, 1861
D. varia Adams

Tall-spired, conical, smooth, translucent; protoconch glossy, smooth; aperture oval, entire, columella straight or with a slight sinus near the end.

Cosmopolitan. There are many species in this genus.

Diala lirata Laseron, 1950

Elongate, conical, usually rather broad, thin; with nine slightly rounded whorls and impressed suture, sculptured with numerous, very fine, distinct, regular, spiral striae, slightly pitted; outer lip convex, columella almost straight. Shell transluscent-yellow with a few faint brown spots when alive, colourless when found dead.

5.5 mm. NSW. According to Laseron the species is "exceedingly common, living on weed in many localities, both inside the bays and on the outer coast".

Diala magna Tate, 1891

Protoconch of $2^1/_2$ purple-black, smooth, rounded whorls; teleoconch whorls flat-sided, smooth except for fine spiral grooves above the deeply excavated suture, body whorl with a rounded base, bearing 6-9 flat, encircling ridges; outer lip thin, curving to meet the columella with barely any canal development. Fawn, with brown axial blotches below the suture and spiral rows of brown dashes.

1.3 cm. South-eastern Australia.

Diala monile Adams, 1862

Whorls rather flat-sided but angulate and nodulose at the periphery, spire is pagoda-like. Shiny, white, with irregular red-brown axial streaks and a spiral row of red-brown spots at the periphery.

8 mm high. NSW to about Fremantle, WA. Laseron (1950) noted that the nodulose forms from NSW generally known under this name are merely aberrations of several different species. *D. lauta* Adams,1862 may be a synonym.

Diala varia Adams, 1861

Glossy, conic-ovate, with a subangulate periphery and inclined base, flat-sided whorls, and narrow, channelled suture; smooth except for minute growth striae on the early whorls and eight spiral threads on the base. Brown, with a white and brown spiral band just below the suture and a paler band at the periphery.

2 cm high. Indo-West Pacific; Qld.

FAMILY **DIASTOMATIDAE**

Diastomas

Like the Campanilidae, the family Diastomatidae has many fossil species in Tertiary sediments deposited during the time of the Sea of Tethys but is represented today by a single living species surviving on the southern coast of WA.

The shells are elongate and turreted, with convex whorls and a wide, shallow, anterior canal. Anatomically the living *Diastoma melanioides* is typically cerithioidean but there are differences warranting separation of the family. (See Houbrick, 1981, *Proceedings Biological Society of Washington* **94** (2): 598-621.) There is little information on the natural history of this species although it is known that it lives among seagrasses in shallow water and it may be assumed that it is herbivorous, probably feeding on micro-algae or detrital particles.

▲ GENUS **DIASTOMA** Deshayes, 1850

Melania costellata Lamarck

Turreted, elongate, with convex whorls and oval aperture; outer lip simple, anterior canal not distinct from the base of the lip; columella concave, with a slight central fold; operculum ovate, corneous, with few whorls and an eccentric nucleus.

Cosmopolitan fossil distribution. The type species is a Miocene fossil from the Paris Basin and the genus had its origins at least as early as the Paleocene. There are three fossil species in Australian Tertiary fossil beds. The single living species confined to the south coast of WA. Synonym: *Neodiastoma* Cotton, 1932 (*melanioides*).

Diastoma melanioides (Reeve, 1849)

Pl. 15; fig. 10 a-b Teleoconch with about 12 flat to slightly rounded whorls and weakly channelled suture; aperture tear-shaped, base rounded; spire whorls with axial folds that become obsolete on the body whorl, folds crossed by numerous fine, sharp cords. Cream or white, flecked with chestnut spots on the spiral ribs.

4 cm. Western SA and southern WA about as far west as Albany. Lives in sand among seagrasses in the shallow sublittoral zone. Not common.

FAMILY **LITIOPIDAE**

Litiopids

Litiopids have usually been allied with *Diala* as a subfamily of the Cerithiidae but there are significant differences. Of the externally visible features the most characteristic of the group are very long and narrow head tentacles, a very long, extensible foot, with long epipodial tentacles, and a mucous gland at the rear of the foot which produces a mucous string used to attach the animal to its algal habitat. The animals live on algae in the shallows or floating at the surface. Their shells are thin, with inflated body whorls and strongly sculptured protoconchs.

Litiopid reproductive anatomy is typical of the Cerithioidea, the male lacking a penis. The egg masses are flat, dome-shaped structures formed by tight spiral coils of jelly strings, attached to algal fronds. Development is rapid, hatching taking place after about six days, after which there is a planktotrophic larval stage.

A recent review by Houbrick (1987, *Nautilus* **101**: 9-18) acknowledges only two genera in this family of tiny snails, i.e. *Litiopa* and *Alaba*. He has tentatively assigned several other genera to the synonymy of these groups as indicated below.

There is only one *Litiopa* described from Australia so far but many *Alaba*. Much remains to be done to clarify the taxonomy of these and other small cerithioideans.

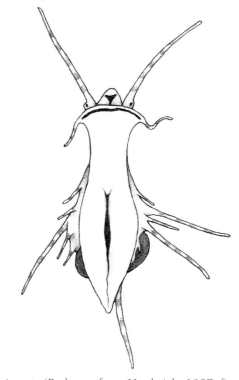

Alaba incerta (Redrawn from Houbrick, 1987, fig. 2)

▲ GENUS **ALABA** H. & A. Adams, 1853
Rissoa melaneura Adams

Shell glassy, tall and turreted, weakly sculptured; aperture quadrangular, with a very weak anterior canal; thick varixes present on several whorls; protoconch axially ribbed; operculum oval, with few whorls and an eccentric nucleus.

Cosmopolitan. Synonyms: *Gibborissoa* Cossmann in Sacco, 1895 (*costellata* Grateloupe); *Diffalaba* Iredale, 1936 (*opiniosa* Iredale); *Obstopalia* Iredale, 1936 (*lixa* Iredale); *Australaba* Laseron, 1956 (*bowenensis* Laseron); *Styliferina* Adams, 1860 (*orthochila* Adams); *Dialessa* Iredale, 1955 (*translucida* Hedley). Shells of the type species of some of these synonyms, and one other, are illustrated here for reference but there are many species in the Australian fauna.

Alaba fragilis (Thiele, 1930)

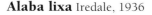

With a tall, conical spire and rounded whorls, body whorl inflated; aperture wide, columella almost straight, slightly thickened. White with light brown axial and spiral lines.

6 mm. Swan Estuary to Shark Bay, WA.

Alaba lixa Iredale, 1936

Slender, awl-shaped, with about 1¹/₂ protoconch whorls and nine slightly ventricose teleoconch whorls, suture impressed; sculptured with obscure axial ribs on the early whorls, spiral grooves on the later whorls; outer lip thin, columella a little twisted; imperforate.

1.25 cm. NSW. The original material was dredged "from 70 fathoms off Green Cape".

Alaba opiniosa (Iredale, 1936)

Slender, with 10 whorls which are thin, a little rounded, subvaricose, varixes irregular and showing as raised, rounded ribs only, with sculpture of fine grooves with flat-topped lirae, about six on the penultimate whorl and 15 on the body whorl; aperture oval, outer lip thin, columella sinuate; imperforate. White with a few brown streaks.

5.5 mm high. NSW. The original material was dredged in Sydney Harbour.

Alaba translucida (Hedley, 1905)

Glossy, very thin, usually transparent enough for the axial pillar to be seen through the shell; rather stout, spire conical, protoconch minute, with about nine teleoconch rounded whorls, body whorl sometimes subcarinate at the basal periphery; aperture oblong, sometimes effuse anteriorly, columella straight, slightly thickened. Transparent white, often with a row of opaque spots at the periphery.

6 mm. NSW to southern WA. The transparent shell distinguishes this species from most others. There is a smaller but otherwise similar species in Qld known as *A. goniochila* Adams.

▲ GENUS **LITIOPA** Rang, 1829
Litiopa melanostoma Rang

Turreted, body whorl inflated; protoconch microscopically, spirally lined between strong axial ribs; teleoconch whorls with weakly incised spiral grooves; columella with a small basal tooth.

Cosmopolitan.

Litiopa limnophysa Melvill & Standen, 1896

Whorls convex, suture impressed; protoconch of about five prominently axially ribbed whorls, teleoconch of four whorls, finely axially striate between the 6-7 spiral cords; columella furrowed, with a bilobed callus ridge in adults.

4.5 mm high. North Qld to central NSW.

FAMILY **FOSSARIDAE**

Fossarids

In the past this small family has been associated with the Hipponicidae and Vanikoridae, but Ponder & Warén (1988, *Malacological Review Supp.* **4**: 295) have argued that it has anatomical features more like those of the Planaxidae, i.e. a cephalic brood pouch, open genital ducts and no penis. For this reason the family is included with the cerithioideans in this account.

The shells of fossarids are small, turbinate, usually umbilicate, and have a prominent protoconch and a horny operculum. There are two genera in the Australian fauna.

Houbrick (1990, *Açoreana, Supp*: 59-73) has described the anatomy, reproductive biology and systematic position of the Atlantic species *Fossarus ambiguus* (Linnaeus, 1758). That species has dwarf males and is probably a protandric hermaphrodite.

▲ GENUS **FOSSARUS** Philippi, 1841
Turbo ambiguus Linnaeus = *adansoni* Philippi

Subglobose to ovate, spire moderately high, whorls spirally corded; umbilicate; aperture semicircular, outer lip curved, undulate; operculum subconcentric.

Cosmopolitan. Synonym: *Anafossarus* Iredale, 1936 (*sydneyensis* Hedley).

Fossarus brumalis Hedley, 1907

Globose-turbinate; protoconch of one smooth, dome-shaped whorl; teleoconch of four strongly keeled whorls, five keels on the body whorl, with wide, flat, deep interspaces and close, minute spiral striae; aperture large, subquadrate, outer lip sharp-pointed at the termination of each keel; columella straight, slightly reflexed over the wide umbilicus. Coffee-brown.

1.4 cm. Described from Mast Head and Hope Is., Qld. *F. cereus* Watson, 1880 is a similar species from Qld.

Fossarus sydneyensis Hedley, 1900

Rather solid, spire short and turreted; with four whorls, slightly flat or concave below the suture, inflated at the periphery, gently rounded at the base; body whorl with 10-11 sharply elevated spiral cords and wide interspaces; aperture oblique, ovate, angled above but rounded anteriorly, furrowed within; outer lip sharp, denticulate; columella arched, broad, faintly channelled anteriorly; umbilicus narrow. White.

4.5 mm. Central NSW. Common in beach sand.

▲ GENUS **LARINOPSIS** Gatliff & Gabriel, 1916
Larina turbinata Gatliff & Gabriel

Fragile, turbinate, many-whorled, with deeply impressed suture; umbilicate; peristome complete, outer lip slightly reflexed anteriorly, canals lacking; operculum with a submarginal nucleus; periostracum olive-green.

South-eastern Australia. There are two described species. May be a synonym of the cosmopolitan genus *Megalomphalus* Brusina, 1871.

Larinopsis ostensus Iredale, 1936

With five ventricose whorls; protoconch of one small, incurved, smooth whorl; later teleoconch whorls uncoiled; sculpture of fine growth striae only. Transparent, white.

1.7 cm high. NSW. Dredged on the outer continental shelf.

Larinopsis turbinatus (Gatliff & Gabriel, 1909)

With six ventricose whorls, including two tiny, minutely pitted nuclear whorls; suture moderately impressed; sculpture of dense, undulating spiral striae; umbilicus deep, partly hidden by the reflection of the columella. White beneath periostracum.

2 cm high. Bass St. Dredged in moderately shallow water. The type specimen was collected alive and found to contain about 400 minute embryos within the brood pouch.

FAMILY **MODULIDAE**

Modulids

The family consists of a single genus with about six species in shallow waters of tropical and subtropical regions. The shells differ from most other cerithioideans in being turbinate instead of elongate and turreted. However, the males lack a penis and the pallial oviducts are open - features characteristic of the Cerithioidea. The males produce packets (spermatophores) of sperma-tozoa which are shed through the exhalent siphon. These are drawn into the mantle cavity of the females by means of ciliary water currents, and the spermatozoa emerge and pass into the female reproductive tract. After fertilisation the female lays her eggs in tiny worm-like gelatinous tubes which are attached to plant fronds.

Modulids feed on diatoms, microphytic algae and detritus associated with seagrass and algal beds in the lower littoral and shallow sublittoral zones.

The anatomy and life history of the Atlantic species *Modulus modulus* has been described in detail by Houbrick (1980, *Malacologia* **20** (1): 117-142).

▲ GENUS **MODULUS** Gray, 1842
Trochus modulus Gmelin

Turbinate, spire low, solid, with a prominent tooth at the base of the columella; operculum horny, multispiral.

Tropical waters of the Indo-West Pacific, Eastern Pacific and Western Atlantic. One described species is found on the northern coast of Australia.

Modulus tectum (Gmelin, 1791)

Pl. 14; fig. 15 a-b Ovate, flat-topped; shoulder subangulate, bearing axially elongate nodules which may descend as oblique folds to the suture; with thin, slightly nodulose spiral cords; aperture almost circular, outer lip finely crenulate, columellar callus spreads onto the parietal wall, columellar tooth prominent.

3 cm wide. Indo-West Pacific; Shark Bay, WA to southern Qld. These animals are quite common in sand on algal mats on intertidal rocky reefs and in seagrass beds. Synonym: *candidus* Petit, 1853.

FAMILY **OBTORTIONIDAE**

This is a very small family whose relationships remain unsettled. Ponder & Warén (1988, *Malacological Review, Supp.* **4**: 294) have indicated that the genus *Obtortio* (= *Fenella* Adams, 1864), upon which the family name was based, deserves family status. Previous authors had placed the genus in the Rissoidae or Dialidae. The genus is represented in the Australian fauna by several species.

▲ GENUS **OBTORTIO** Hedley, 1899
Rissoa pyrrhacme Melvill & Standen

Elongate and turreted; protoconch multispiral, apex of two minute whorls; early teleoconch whorls spirally threaded, later whorls finely cancellate; lower part of the outer lip slightly flaring; varixes lacking.

Cosmopolitan.

Obtortio lutosus Hedley, 1914

Teleoconch of eight convex whorls with impressed suture; later whorls sculptured with narrow, elevated spiral cords, about 14 on the body whorl, sometimes also with thin axial lamellae. Buff, spirally banded with chestnut or hazel-brown.

3.3 mm. North Qld to central NSW. For an illustration see the original description (1914, *Proceedings Linnean Society NSW*, **39**: Pl. 81, fig. 53).

FAMILY **PLANAXIDAE**

Clusterwinks

The Planaxidae is a small family of amphibious marine snails which inhabit rocky shores of the intertidal zone. There are only six genera and about 20 species in the family, most of them tropical or subtropical. The species tend to be gregarious and widespread. The vernacular name is derived from a tendency of the animals to cluster in colonies. Superficially the shells are like those of littorinids but there is a conspicuous anterior canal. The animals feed on micro-algae growing on the substrate.

Perhaps the most interesting feature of their natural history is their reproduction. It has been reported that planaxids are parthenogenic but this seems to have been an error. Like other cerithioideans, planaxid males have no penis but produce tiny packages (spermatophores) of spermatozoa. Females brood the larvae in special brood pouches. In most species the larvae are released as planktotrophic veligers but in some development is direct.

The anatomy, reproductive biology and phylogeny of the family have been documented by Houbrick (1987, *Smithsonian Contributions to Zoology* **445**: 1-57).

▲ GENUS **ANGIOLA** Dall, 1926
Planaxis periscelida Dall

Shell small, thick, smooth, glossy; spirally banded. Periostracum thin. Rachidian tooth wider than tall, with a single serrated triangular cusp and basal plate with two cusps. Luminous organ present.

Indo-West Pacific and western Atlantic. Houbrick (1987, *loc. cit.*, p. 20) noted an obvious close relationship with *Hinea*.

Angiola fasciata (Pease, 1868)

Conical-ovate, spire tall, whorls slightly convex, base curved; surface sculptured with minute spiral striae; outer lip broad and inclined, aperture lirate within, columella concave. White or yellow, often with orange or brown spiral lines which may be closely or widely spaced; columella sometimes tinged with violet

1 cm. Indo-West Pacific; Quobba, WA to Caloundra, Qld. This little species found in the high intertidal zone has been known in Australian literature as *Planaxis lineolatus* Gould, 1851 but when moved to *Angiola*, Gould's name becomes a homonym of *A. lineolatus* Risso, 1826. For illustrations see Cernohorsky, 1972, *Marine Shells of the Pacific* **2**: pl. 12, figs 18, 18a.

▲ GENUS **FISSILABIA** Macgillivray, 1836
Fissilabia fasciata Macgillivray (=*decollata* Quoy & Gaimard)

With the characters of the type species.

Indo-West Pacific. Monotypic. Synonyms: *Quoyia* Gray, 1839 (*decollata* Quoy & Gaimard); *Leucostoma* Swainson, 1840 (not *Leucostoma* Meigen, 1803).

Fissilabia decollata (Quoy & Gaimard, 1833)

Pl. 14; fig. 18 a-b   Shell thick, decollate, spirally grooved; aperture narrow, ovate, columella slightly calloused, with a prominent parietal fold. Radula long, cutting edge of the central tooth with a single, large, wide, blunt cusp, basal plate with two medium-sized cusps and two bifurcate lateral extensions. Larger portion of the brood pouch located on the right side of the head-foot. Tan to dark brownish grey; interior purple-brown, columella white.

3 cm. Indo-West Pacific; north Qld. Lives high in the intertidal zone on exposed rocky shores where there is wave action, and under stones and rubble in more protected areas.

▲ GENUS **HINEA** Gray, 1847
Planaxis mollis Sowerby (= *brasiliana* Lamarck)

Shell ovate, smooth, high spired, whorls flat-sided; outer lip grooved internally, folded over the anal canal, thickened where it extends onto the side of the previous whorl, parietal wall thickened below the suture. The central radular tooth is long, with a large, single, serrated basal plate bearing two lateral extensions and two tiny basal cusps. In females the brood pouch is small and located on the right side of the foot.

South-west Pacific. There are two species.

Hinea brasiliana (Lamarck, 1822)

Pl. 14; fig. 16 a-b   Protoconch of three whorls, each sculptured with a spiral cord above the suture and axial plates below it; outer lip thick, smooth along the edge, denticulate within. White below a thick, velvety, yellow periostracum.

2 cm. Southern Qld to SA; also at Lord Howe, Norfolk and Kermadec Is. Gregarious in discrete populations in the upper intertidal zone among rocks and rubble along rocky shores in wave-exposed areas. It has been reported that the animal glows brightly in the dark when it is agitated (Ponder, 1988, *Journal Molluscan Studies* **54** (3): 361 and Loch, 1990, *Australian Shell News* **69**: 8). Bioluminesence is uncommon in gastropods. Synonym: *mollis* Sowerby, 1823.

▲ GENUS **PLANAXIS** Lamarck, 1822

Buccinum sulcatum Born, 1780

Conical, whorls inflated, body whorl large, sculptured with spiral grooves; outer lip with a scalloped edge, deeply grooved within, columella concave, with a large parietal tooth. Central radular tooth with a single, broad, blunt cusp, a pair of tiny, midlateral basal cusps, and two moderately elongate lateral extensions on a low basal plate. Brood pouch large, divided into many lamellar chambers on both sides of the head-foot and over the buccal mass.

Pantropical. There are several species, one in tropical Australia. Synonym: *Proplanaxis* Thiele, 1929 (*planicostatus* Sowerby).

Planaxis sulcatus (Born, 1780)

Pl. 14; fig. 17 a-b   Protoconch with about two whorls; teleoconch with about seven whorls, suture deeply incised, base moderately constricted; sculptured with incised spiral lines and grooves; anterior canal short, wide, outer lip with a smooth edge, denticulate within; columella concave, slightly calloused, parietal tooth prominent; periostracum thin, brown; operculum large, with few whorls and subterminal nucleus. White but with crowded, spiral, dark brown blotches giving an overall dark appearance; interior and columella white.

3.5 cm. Indo-West Pacific; North West Cape, WA to northern NSW. A very common shell on rocks in the intertidal zone. Houbrick (1987, *loc. cit.*, p. 5) noted that there appears to be a geographic difference in development strategy within the range of this species. Populations from the north-western Indian Ocean have been cited as having viviparous development, whereas western Pacific populations brood the early larvae but release them later as planktotrophic veliger.

This implies that the genetic variability of the species may be much more complex than is evident from the shells. Synonyms: *pyramidale* Gmelin, 1791; *undulata* Lamarck, 1822; *buccinoides* Deshayes, 1828.

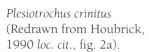

Plesiotrochus crinitus (Redrawn from Houbrick, 1990 *loc. cit.*, fig. 2a).

FAMILY **PLESIOTROCHIDAE**

There is a single genus in this family. Early authors had difficulty in classifying it, assigning it variously to the Littorinidae, Planaxidae and Cerithiidae. Recently Houbrick (1990, in Wells *et al. Marine Flora and Fauna of Albany, WA,* **1**: 237-249, WA Museum) described the anatomy of a species from southern Western Australia, showing that it is very different from that of other cerithioideans, and elevated the group to family rank accordingly.

The animals have many long papillae along the mantle edge, a fringe of epipodial tentacles surrounding the foot and a prominent mucous gland on the sole. There is a hairy periostracum. The jaws are large, toothed, sickle-shaped, with a peg-like basal arm. The radula is short and wide, with large teeth. The animals are believed to be herbivorous. At least one species (*P. crinitus*) is a simultaneous hermaphrodite. There is no penis and fertilisation is achieved by exchange of small spermatophores between mating pairs.

▲ GENUS **PLESIOTROCHUS** Fischer, 1878

P. souverbianus Fischer

Conical, trochiform, with sharply angulate whorls and turreted spire; aperture subquadrate, outer lip thin and simple; posterior canal weak, anterior canal wide, short, oblique; columella straight; periostracum hairy.

Indo-West Pacific and southern Australia. The type species was figured by Houbrick (1990, *loc. cit.*, p. 240, fig. 1 a). There are about 11 species but it is uncertain how many of them occur in Australian waters. Synonyms: *Hemicerithium* Cossmann, 1893 (*imperfectum* Deshayes); *Trochocerithium* Cossmann & Sacco, 1896 (*turritum* Bonelli); *Hypotrochus* Cotton, 1932 (*monachum* Crosse & Fischer).

Plesiotrochus crinitus Thiele, 1930

Pagoda-like; protoconch unknown; teleoconch of 6-7 broad, angulate, sloping whorls, slightly concave posteriorly, body whorl large, slightly constricted basally; early whorls with a broad peripheral spiral cord, body whorl with two peripheral cords; suture slightly overhung by anterior whorls; aperture ovate, outer lip thin, angulate, columella slightly concave, thinly calloused; anterior canal wide, moderately short; operculum ovate, thin, semitransparent, paucispiral, nucleus eccentric; periostracum hairy.

5.5 mm. Two Peoples Bay to Shark Bay, WA. Lives on brown algae in tide pools on rocky shores. The species range may extend beyond these limits. The type locality is Shark Bay but Houbrick (1990, *loc. cit.*) studied living specimens collected in the vicinity of Albany, referring to them as "*P. cf. penitricinctus*". *P. penetricinctus* (Cotton, 1932) was described from SA and is like this but with radiating axial ribs. Further study is needed to determine whether the two species overlap in southern WA. See Houbrick (1990, *loc. cit.*) for an illustration.

Plesiotrochus monachus (Crosse & Fischer, 1864)

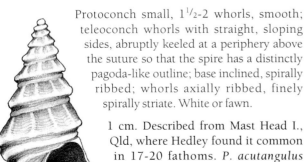

Turreted, thin, varixed; protoconch large, bulbous; spire about ²/₃ total shell height; with 6-8 carinate adult whorls; sculpture of strong spiral striae, body whorl bicarinate, the peripheral keel prominently nodulose, the one below it smaller and only weakly nodulose; anterior canal oblique; varixes present or absent; periostracum thick and hairy. Fawn or cream, with chestnut flecks and sometimes axial chestnut lines.

2.4 cm. Southern NSW to Fremantle, WA. Lives among seagrasses in the shallow sublittoral.

Plesiotrochus pagodiformis Hedley, 1907

Protoconch small, 1¹/₂-2 whorls, smooth; teleoconch whorls with straight, sloping sides, abruptly keeled at a periphery above the suture so that the spire has a distinctly pagoda-like outline; base inclined, spirally ribbed; whorls axially ribbed, finely spirally striate. White or fawn.

1 cm. Described from Mast Head I., Qld, where Hedley found it common in 17-20 fathoms. *P. acutangulus* Yokoyama is a similar species also recorded from northern Australia.

Mud Creepers

Members of this family are intertidal snails of muddy, estuarine habitats, some living an amphibious existence. Although there are not many species in the family, they may be very conspicuous animals living in vast numbers in mangroves or saltmarshes, where they graze on organic detritus or micro-algae.

In both shell and body morphology, potamidids are very similar to the ceriths. They are mainly tropical creepers. Until recently (see Houbrick, 1988, *Malacological Review, Supp.* **4**: 88-128) the mainly temperate family Batillariidae was ranked as a subfamily of the Potamididae. The two groups are distinguished by very different radular characters. Batillariidae have cusps on the lower basal plate of the rachidian tooth, while the Potamididae lack this feature.

Like all cerithioideans, the males lack a penis and sperm are transferred during mating in spermatophores via the siphons. The eggs are laid in jelly strings on leaves or sticks. In some species development is direct but in others there is a planktotrophic larval stage.

▲ GENUS **CERITHIDEA** Swainson, 1840
Melania lineolata Griffith & Pidgeon (= *obtusa* Lamarck)

Elongate, turreted, thin, sculptured with spiral ridges and axial ribs; anterior canal short, lower lip projects beyond the columella base; operculum thin, corneous, circular, multispiral, with a central nucleus.

Circumtropical. There are three subgenera. For a discussion of the biology and taxonomy of the genus, see Houbrick (1984, *American Malacological Bulletin* **2**: 1-20).

▲ SUBGENUS **CERITHIDEA** s.s.

With prominent axial ribs, with or without spiral sulci; aperture wide, outer lip smooth, anterior canal rudimentary.

Indo-West Pacific. Live in mangroves at or above high tide level. There are four species in northern Australia.

Cerithidea anticipata Iredale, 1929

Pl. 15; fig. 16   Relatively small, thin and narrow, with about eight whorls; outer lip thin, flaring; base with both spiral and axial sculpture; whorl sides with rather straight axial ribs and prominent spiral cords but with the axials dominant. Grey or light brown; aperture, light brown.

4.5 cm. Admiralty Gulf, WA to southern Qld. Synonym: *kieneri* Hombron & Jacquinot, 1853.

Cerithidea largillierti (Philippi, 1849)

Pl. 15; fig. 12   Whorls convex, with deeply impressed suture and weak curved axial ribs which become obsolete on the body whorl; spiral sculpture of very weak cords which are most numerous on the base; with a weak varix opposite the lip; outer lip thin, not thickened or flared; columella nearly vertical. Light brown.

4 cm. Indo-West Pacific; north Qld. Although it has been assigned to *Cerithidea*, recent authors have questioned whether this species belongs there (Houbrick, 1984).

Cerithidea obtusa (Lamarck, 1822)

Pl. 15; fig. 15   Thin with about seven convex whorls; outer lip thick, shelf-like; base with spiral cords but no axial sculpture, axial ribs on the sides of the whorls slightly sinuous, spiral sculpture poorly developed. Brown; aperture light brown.

5 cm. Indo-West Pacific; north Qld. Apparently not common in Australia; most literature records are based on misidentified specimens of *C. reidi* or *C. anticipata*.

Cerithidea reidi Houbrick, 1986

Pl. 15; fig. 17 a-b   Resembles *C. obtusa* but grows to a larger size, the shell is relatively thin, with a thin and flaring outer lip, both spiral and axial sculpture on the base, less sinuous ribs, spiral cords and incised lines crossing the axial ribs, and a spiral row of nodules below the suture. Tan to flesh-coloured, aperture purple.

6 cm. Exmouth Gulf to Admiralty Gulf, WA. Lives on the trunks and roots of mangroves, sometimes together with *C. anticipata*. The subsutural nodules and purple aperture serve to distinguish between the two species.

▲ SUBGENUS **CERITHIDEOPSILLA** Thiele, 1929
C. fluviatilis Potiez & Michaud = *cingulata* Gmelin

Sculptured with axial ribs divided into spiral rows of nodules by three spiral grooves; posteriorly the outer lip forming a flaring, wing-like process extending onto the parietal wall.

Indo-West Pacfic. Live on intertidal mud flats, usually associated with mangroves.

Cerithidea cingulata (Gmelin, 1791)

Pl. 15; fig. 14 a-c   Sides of spire rather flat; body whorl with a thick varix opposite the lip; the subsutural row of tubercles

larger than the two anterior rows, tubercles rounded, obsolete on the body whorl; base with six smooth spiral ribs. Colour variable, usually grey or brown, subsutural rib, and tubercles white, aperture brown.

2.5 cm. Kimberley, WA to central Qld. Abundant on mud flats near mangroves.

▲ GENUS **TELESCOPIUM** Montfort, 1810
Trochus telescopium Linnaeus

Shell large, thick, conical, nonvaricate; sculptured with numerous flat-sided whorls and spiral grooves, base with concentric cords and a deep channel around the columellar pillar; aperture narrowly ovate, tangential to the shell axis, columella twisted; siphonal canal short; operculum corneous, circular, multispiral, nucleus central.

Indo-West Pacific. There is only one living species but the genus is represented by several fossil species in the Middle and Late Tertiary. For a comprehensive account of the anatomy and biology of the genus see Houbrick, 1991, *Malacologia* 33 (1-2): 289-338.

Telescopium telescopium (Linnaeus, 1758)

Pl. 15; fig. 21   Conical, sides straight, base broad and rather flat, periphery angulate; whorls short, sculptured with several deep spiral striae, varixes lacking; anterior canal and columella very short, columella twisted, with a strong central spiral ridge and a deep channel around the pillar; outer lip thin, not flared, smooth except for weak serrations anteriorly. Brown, often with a single spiral central white line on each whorl; columella yellow, interior brown.

11 cm. Indo-West Pacific; Exmouth Gulf, WA to central Qld. Abundant intertidally in mud among mangroves. Wells (1986, *Journal Molluscan Studies* 52: 83-90) and Lasiak & Dye (1986, *Journal Molluscan Studies* 52: 174-179) have described the ecology of this species in northern WA and Qld respectively. Synonyms: *indicator* Montfort, 1810; *fuscum* Schumacher, 1817; *mauritsi* Butot, 1974.

▲ GENUS **TEREBRALIA** Swainson, 1840
Strombus palustris Linnaeus

Solid, spire tall and flat-sided, varicate, base rounded; outer lip slightly thickened and outward flaring; posterior canal a conspicuous channel, siphonal canal encircled, or nearly so, by a left turn of the anterior extremity of the outer lip; columella with paired internal plaits.

Indo-West Pacific. The fossil record of this genus dates from the Miocene. There are three living species in northern Australia. For a comprehensive account of the anatomy and biology of the genus see Houbrick, 1991, *Malacologia* 33 (1-2): 289-338.

Terebralia palustris (Linnaeus, 1767)

Pl. 15; fig. 18   Whorls with low axial folds, deeply incised suture, and three or four narrow deeply incised spiral grooves; base of shell with numerous spiral cords; outer lip reflexed, partly occluding (but not surrounding) the canal; columella with a small parietal ridge and a callus beside the anterior canal. Brown; columella fawn, interior brown.

19 cm. Indo-West Pacific; Shark Bay, WA to southern Qld. Abundant and conspicuous in the mud of mangroves. Synonyms: *trisulcatus* Forskol, 1775; *agnatus* Gmelin, 1791; *carnaticum* Perry, 1811; *crassum* Lamarck, 1822; *caledonica* Jousseaume, 1884; *palustre* var. ? *tryoni* Kobelt, 1895.

Terebralia semistriata (Mörch, 1852)

Pl. 15; fig. 19 a-b   Turreted, with 10-12 flat-sided whorls, varixes randomly spaced; suture distinct, slightly inset into each successive anterior whorl; body whorl very large with a strong varix opposite the outer lip; flaring outer lip forming a complete peristome and enclosing the siphonal canal; early whorls axially ribbed, developing spiral cords and becoming cancellate; later whorls sculptured with flattened spiral cords and 4-5 minor spiral cords, axial ribs lacking. Dark to light brown; aperture tan with shiny brown and whitish patches on the columella and outer lip.

5.7 cm. South coast of New Guinea and northern Australia from Shark Bay to Keppel Bay, Qld. Until recently the presence of this third species in northern Australia had been overlooked. It has a larger, heavier shell than *T. sulcata*, with weaker sutures and sculpture, and the adult whorls lack the axial ribs and nodules which characterise that species.

Terebralia sulcata (Born, 1778)

Pendant-shaped, suture deeply incised, with several randomly spaced varixes; early whorls cancellate, later whorls sculptured with 4-5 flattened spiral cords and deeply incised spiral lines, overlain by numerous axial ribs forming squarish nodules, also with beaded spiral cords on the body whorl; outer lip smooth, widely flaring and completely surrounding the anterior canal so that the latter appear as a round hole in the anterior end of the shell. Grey or brown, often stained green, lip and varixes white; deep interior rich brown to cream, shiny.

6 cm. Indo-West Pacific, Shark Bay, WA to southern Qld. Abundant in the mud of mangroves. In northern Australia this species may be found at the same localities as *T. semistriata* but further study is needed to determine whether the two are ecologically differentiated.

FAMILY **SCALIOLIDAE**

Another of the very small families of very small cerithioidean snails. Although usually placed in the Dialidae, Ponder and Warén (1988, *Malacological Review, Supp.* **4**: 294) have suggested that the group deserves family status.

▲ GENUS **SCALIOLA** Adams, 1860
S. bella Adams.

Conical to conic-ovate; protoconch of three or more whorls, the first smooth but the others axially ribbed or threaded, sometimes carinate; teleoconch whorls convex, with impressed suture, smooth but agglutinated with sand grains.

Indo-West Pacific. There can be no mistaking this remarkable genus. Although the shell form resembles that of *Diala*, the sand collecting habit, reminiscent of the shell and debris collecting habit of the xenophorids, is characteristic. Several species of the genus are common in coral lagoons in the region; one is recorded from NSW: *Scaliola caledonica* Crosse, 1870. No specimen was available for illustration and so a copy of a figure of *S. lapillifera* Hedley, 1899 (from Funafuti Atoll) is reproduced here to represent the genus.

Scaliola lapillifera. Redrawn from Hedley, 1899, *Australian Museum Memoir* **3**: p. 415, fig. 8.

FAMILY **TURRITELLIDAE**

Screw Shells

Screw shells are long, turreted and many-whorled, resembling terebrids except that the anterior end of the aperture is entire, lacking an anterior canal notch. There is a sinus in the outer lip of most genera which is a useful character for classification. The whorls are usually strongly sculptured with spiral ribs or keels. In most species the axis of the protoconch is offset at an angle to the teleoconch axis. There is a horny operculum. The anatomy is typically cerithioidean in most respects but there are specialisations relating to their unusual feeding habits.

These animals live in soft substrates (usually mud), most often in deeper water although a few species sometimes extend into the intertidal zone. They gather fine particles of organic detritus which are trapped in mucous sheets on the gills after being sucked into the mantle cavity by means of ciliary water currents. There are small tentacles at the entrance to the mantle cavity, and tiny bristles along the edge of the operculum, which assist in the sorting of particles. At least some species are ovoviviparous or viviparous.

There are several subfamilies, two of which are represented in the Australian fauna. Marwick (1957, *Proceedings Malacological Society London* **32**: 144-166) published a generic revision of the family. T. A. Garrard (1972, *Journal Malacological Society of Australia* **2** (3): 267-338) revised the Australian Turretellidae. Garrard later published supplementary notes in the same journal, including descriptions of additional species, viz.: 1974, **3** (1): 15-17; (1982) **5** (3-4): 195-200. Illustrations of species not figured here may be found in those publications.

SUBFAMILY **TURRITELLINAE**

▲ GENUS **ARCHIMEDIELLA** Sacco, 1895
Turritella archimedis Brongniart

Brightly coloured, medium-sized to large; protoconch with about 1½ asymmetrical whorls, clear, glassy; whorls weakly convex, sculptured with medium to strong spiral cords, base flat to convex; aperture subcircular to subquadrate, outer lip with a broad, shallow, orthocline sinus, columella arched.

Indo-West Pacific.

Archimediella dirkhartogensis Garrard, 1972

Teleoconch with deeply impressed suture, about 14 convex whorls, each broadest just above the lower suture so that the spire is reversely turreted, base almost flat with a subangulate periphery; sculpture of medium spiral cords and finely granulated threads, base spirally threaded; aperture subcircular. White with red-brown axial flames showing mainly on the spiral cords giving a spotted appearance.

3 cm. Eucla to Shark Bay, WA. Dredged on the middle part of the continental shelf.

Archimediella fastigiata (Adams & Reeve, 1848)

Pl. 14; fig. 11 a-c Teleoconch of 18-20 whorls, suture deeply impressed, sides of whorls sloping below the upper suture, convex at the centre, with a keeled basal periphery and weakly convex base; sculptured with about 12 moderate spiral cords, those at the centre and periphery the strongest, base with 18-20 irregular spiral threads; aperture subcircular, columella with a reflected inner edge.

6 cm. Indo-West Pacific; Shark Bay, WA to the NT.

Archimediella maculata (Reeve, 1849)

Pl. 14; fig. 1 a-c Teleoconch of about 18 whorls, suture slightly impressed, base weakly convex; each whorl with three strong spiral cords and many fine threads, sides concave between the cords, a fourth cord at the periphery of the body whorl forming a double basal keel, plus 2-3 on the base; aperture subquadrate. Off-white, threads light brown, cords brown spotted, with occasional irregular, light brown flames.

7 cm. Indo-West Pacific; NT and north Qld. There are few Australian records.

Archimediella occidua (Cotton & Woods, 1935)

With about 14 convex whorls, rounded base, slightly impressed suture; early whorls with three strong spiral cords and fine striae, on the later whorls the three primary cords remain dominant but the early threads become secondary cords, base spirally threaded; aperture subquadrate. Light brown, sometimes with white axial flames.

3 cm. Hopetoun to King Sound, WA. There are few records and the distribution is likely to be greater than indicated.

▲ GENUS **COLPOSPIRA** Donald, 1900
Turritella runcinata Watson

Small to medium-sized; protoconch of 1½-2 whorls, asymmetrical, rounded to globose, clear or translucent, nucleus usually slightly submerged; teleoconch whorls usually flat-sided or weakly convex, finely or coarsely spirally corded; aperture round to subquadrate, outer lip sinus orthocline, deep, narrow.

Indo-West Pacific. This is a very large group with many Australian species. The deep, orthoclinal outer lip sinus is the main definitive character of the group. The characters used as diagnostic of the subgenera seem trivial but are accepted here for convenience.

▲ SUBGENUS **ACUTOSPIRA** Kotaka, 1959
Turritella okadai Nagao

Spiral sculpture consisting of sharp-edged keels.

Indo-West Pacific. In addition to the species of this subgenus illustrated here, there are two described by Garrard (1982) from the vicinity of the Capricorn Group, Qld, that he named *reefiana* and *swainsiana*.

Colpospira accisa (Watson, 1881)

Protoconch of two smooth, globose whorls, nucleus not submerged; spire straight-sided, teleoconch of about 16 whorls, base rounded; whorls with three strong, sharp, primary keels, the one at the periphery of the body whorl usually double, plus secondary threads; columella straight, reflected. Cream, fawn or light red-brown, often with irregular reddish spots on the keels.

3 cm. Central NSW to the south coast of WA. Common on the middle part of the continental shelf. The strongly keeled shell of this species readily distinguishes it from *C. runcinata* which is superficially similar but has rounded spiral cords. The two species have similar range and habitat.

Colpospira atkinsoni (Tate & May, 1900)

Pl. 14; fig. 6   Protoconch very small, of 1¹/₂ glassy, convex whorls; teleoconch of 14 whorls that are convex and inflated but made angulate at the centre by a sharp and prominent keel, with two lesser keels below the suture and another two at the base making the periphery angulate, base concave; columella reflected. Cream to light brown.

2.2 cm. Central NSW to Bass St. Common on the middle part of the continental shelf. Synonym: *godeffroyana* Donald, 1900.

Colpospira smithiana (Donald, 1900)

Protoconch of two glassy, globose whorls, nucleus not submerged; teleoconch of 12 convex whorls, made angulate by a very strong central keel, with two lesser keels above and one below, plus a rib at the subangulate periphery, base straight, sloping, irregularly striate; columella forming a callus on the parietal wall, reflected at the base. White, translucent.

1 cm. Southern Qld to the south coast of WA. Dredged on the continental slope to 1500 m. This species has a wide geographic and depth range. Great variability in the relative strengths of the spiral keels. See remarks on *C. yarramundi*.

Colpospira yarramundi Garrard, 1972

Protoconch translucent, of 1¹/₂ rounded whorls, nucleus not submerged; with 10 teleoconch whorls, the later whorls convex, suture impressed; sculpture of strong keels, four on the body whorl, base with several fine spiral cords and threads; outer lip curving downward to form a slight anterior canal, columella reflected. White.

6 mm. Southern Qld to eastern Bass St. Dredged at depths from 75-550 m. Resembles *C. smithiana* but the three main sharp-edged keels appear one after the other and do not become of equal prominence until about the sixth whorl. In *C. smithiana* the three main keels all appear at about the same time on the second and third whorls.

▲ SUBGENUS **COLPOSPIRA** S.S.

Spiral sculpture of rounded cords and threads.

In addition to the species of this subgenus illustrated here the following species are recorded from Australian localities: *deliciosa* (Watson, 1881) from the Torres St. and Bunker Group, Qld; *indigena* Garrard, 1972 from off Cape Moreton, Qld.

Colpospira aquamarina Garrard, 1972

Protoconch small, of 1¹/₂ almost globose, translucent whorls; teleoconch with 14 rather flat-sided whorls, suture very slightly impressed, base flat, sloping, with an angulate periphery; sculpture variable, usually with strong, flat, spiral ribs above and below the suture, concave between, forming a strong double basal rib around the periphery of the body whorl, finely striate. Fawn, usually with a purplish central band, and a subsutural row of white or white and brown spots.

1.5 cm. Moreton Bay, Qld to Wooli, NSW at depths from 35 to 122 m. Although this species superficially resembles *Gazameda tasmanica* it has a typical *Colpospira* protoconch.

Colpospira bundilla Garrard, 1972

Protoconch of two whorls, flatly globose; sides of spire concave, teleoconch of 10-12 whorls, suture at the base of a V-shaped notch; sculpture of two weak but subangulate spiral cords, smooth and slightly concave between, peripheral cord becoming more prominent on later whorls, base spirally striate; columella reflected at the base. Light yellow-brown, sometimes red-brown or white.

5 mm. Southern Qld to Eucla, WA. Dredged at depths from 100-550 m. A similar species from southern Qld, *C. indigena* Garrard, 1972, is a little broader, larger and more heavily built, with bolder ridges, the lower one bifurcate, and lacks a central colour band.

Colpospira cordismei (Watson, 1881)

Prototoconch small, of 1½ convex whorls; teleoconch of about 13 whorls, spire and whorls flat-sided, suture deeply impressed; sculpture of spiral cords, 3-4 per whorl plus a thick, angulate peripheral cord and fine striae; periphery subangulate, base flat to concave; columella reflected throughout its length. Cream to light chestnut, often with alternating brown and white spots on the cords.

1.2 cm. NSW to Bass St. Southern specimens tend to be darker and with stronger sculpture.

Colpospira curialis (Hedley, 1907)

Protoconch large, of two rounded, dome-shaped whorls; teleoconch of nine slightly concave whorls, with an impressed suture within a deep V-shaped notch, made crenulate by nodules on prominent subsutural and presutural spiral ribs, often with two additional angulate spiral cords at the centre of the whorls; columella arcuate, reflected. Cream.

7 mm. Bunker Group, Qld to central NSW. This is a very deepwater species dredged on the continental slope at 457-1464 m. Only a few specimens are known. Synonyms: *crenulata* Donald, 1900; *reevei* Cossmann, 1912; *joannae* Hedley, 1923.

Colpospira decoramen (Iredale, 1936)

Protoconch small, of 1½ slightly globose whorls, translucent; teleoconch of 12-14 whorls, spire narrow at the top with concave sides, whorls flat-sided but concave between the ribs at the centre; sculpture variable, usually with a strong subsutural spiral cord and another at the periphery, the latter often double, plus fine to coarse spiral and axial striae, base spirally striate; columella narrowly reflected over the previous whorl. Early whorls white, later whorls mottled brown and white, often with light brown axial flames, red and white spots on the spiral cords.

2 cm. Southern Qld to Bass St. Very common on the outer part of the continental shelf and slope. Variable but readily recognised by the narrow apex and concave sides.

Colpospira mediolevis (Verco, 1910)

Protoconch large, of 1½ slightly convex whorls; teleoconch of about nine flat-sided whorls, suture deeply impressed and within a prominent V-shaped notch, basal periphery rounded, slightly convex below; sculpture of 2-3 weak threads above and below the suture, centre smooth, base spirally striate; columella narrowly reflected. Translucent, creamy white, fawn about the suture and on the base and columella.

5 mm. SA and south coast of WA on the outer part of the continental shelf. The lip sinus tends to become deeper and narrower in older specimens.

Colpospira moretonensis Garrard, 1972

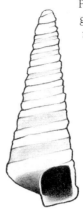

Protoconch large, of 1½ glassy, somewhat globose whorls; teleoconch of eight whorls or more, spire flat-sided, whorls shouldered, slightly concave at the centre, basal periphery subangulate; sculpture of low, rather flat, spiral ribs at the shoulder and basal periphery, sometimes with a single fine striation on the basal rib; columella reflected at the base. Translucent pale cream to amber.

6 mm. Southern Qld. Dredged at depths from 36-123 m. The blunt apex and broad-shouldered whorls characterise this shell.

Colpospira runcinata (Watson, 1881)

Pl. 14; fig. 4   Protoconch small, translucent, of two slightly globose whorls; teleoconch of about 15 whorls, spire flat sided, suture impressed, within a wide V-shaped notch, base flat and sloping; early whorls smooth, later whorls sculptured with strong spiral cords and striae, base finely striate; columella thin, rather straight and nearly vertical. Apex yellow, remainder white, speckled with yellow.

3 cm. Central NSW, southern Tas. and west to at least as far as Rottnest I. on the WA coast.

Colpospira sinuata (Reeve, 1849)

Protoconch of 1½ subglobose whorls; teleoconch of about 14 whorls, spire flat-sided but with a wide V-shaped sutural notch, basal periphery rounded; early whorls smooth, later whorls with spiral threads; columella straight, slightly reflected. Off-white with faint, fawn, axial patches and alternating red-brown and white subsutural spots.

1.8 cm. Southern Qld to Bass St. Moderately common on the middle part of the continental shelf. The subsutural spots are useful for identification of this species.

Colpospira sophiae (Brazier, 1883)

Protoconch slightly globose, of 1½ clear, glassy whorls; teleoconch of 13 whorls, base subangulate, early whorls slightly convex, smooth, later whorls almost flat-sided, shouldered, sculptured with strong spiral cords, the basal cord double, base with 7-8 flat cords, flat spaces between; columella reflected. Apex white, remainder cream to chestnut with red-brown patches on the shoulder ridge.

1.1 cm. Southern Qld to central NSW. Dredged on the middle part of the continental shelf. Iredale (1925) thought the type was a juvenile *C. sinuata* but Garrard (1982) disagreed. Synonym: *incisa* Tenison Woods, 1878 (not of Reeve, 1849).

Colpospira translucida Garrard, 1972

Protoconch of two glassy, subglobose, carinate whorls; teleoconch of 10 convex whorls with well impressed suture; later whorls with a broad, low, rounded, central rib and a lesser cord on each side of it, base spirally striate; columella reflected at the base. Either translucent-white or clear and glassy.

6 mm. Recorded from off Port Davey, Tas., Cape Wiles, SA and Eucla, WA at depths from 79-183 m, but probably widespread along the south coast. Resembles *C. mediolevis* but lacks colour and has more convex whorls, a deeper suture and stronger sculpture.

Colpospira wollumbi Garrard, 1972

Protoconch globose, 1½ glassy whorls, nucleus not submerged; teleoconch of about nine convex whorls with well impressed suture, basal periphery rounded; sculptured with angulate spiral cords, four per whorl, the centre two usually more pronounced, base spirally striate; columella slightly reflected. White or pale chestnut, usually with a darker subsutural band.

5 mm. Southern Qld to about as far west as Cape Leeuwin, WA. Dredged at depths from 77-237 m.

▲ SUBGENUS CTENOCOLPUS Iredale, 1925
Turritella australis Lamarck

Shells very small to medium; spiral cords usually gemmate.

South-eastern Australia.

Colpospira australis (Lamarck, 1822)

Protoconch small, subglobose, of 1½ whorls; teleoconch of 13 whorls, spire straight-sided, suture deeply incised, sculptured with subsutural and peripheral ribs, both gemmulate, with a distinctly concave area between, peripheral rib on the body whorl double, base subangulate. Yellow-brown, darker at the suture, with a brown band on the base.

2 cm. Central NSW to southern Tas. and SA. Common on the continental shelf. Synonyms: *granulifera* Tenison Woods, 1876; *terebellata* Tate, 1893.

Colpospira guilleaumei (Iredale, 1924)

Pl. 14; fig. 8   Protoconch small, of 1½ dome-shaped whorls; 11 teleoconch whorls, the later whorls concave, periphery angulate, suture impressed within a V-shaped notch; smooth except for spiral ribs above and below the suture, the lower rib overhanging the suture giving a pagoda-like effect; columella straight, reflected. Pinkish white, suffused with light brown, alternate brown and white spots on the ribs.

1.5 cm. NSW to Bass St. Common on the middle part of the continental shelf.

▲ SUBGENUS PLATYCOLPUS Donald, 1900
Turritella (Colpospira) quadrata Donald

Very small to medium-sized; outer lip sinus broad, shallow and medial.

Indo-West Pacific. In addition to the species described here, Garrard (1982) named *C. musgravia* from the Capricorn and Bunker Groups, Qld.

Colpospira circumligata (Verco, 1910)

Protoconch of two globose, translucent whorls, nucleus prominent; teleoconch of 11 rather flat-sided whorls, suture slightly caniculate in the early whorls, periphery rounded, base convex; sculpture of a rounded subsutural rib, followed by two weak spiral cords and a broad, rounded presutural rib forming a prominent basal rim around the periphery of the body whorl, base spirally striate; columella arcuate, reflected. Pale yellow-brown with darker brown spiral bands between the ribs and cords, and a broad brown spiral band on the base.

1.7 cm. Bass St. to SA. Dredged at depths of 58-200 m.

Colpospira congelata (Adams & Reeve, 1848)

Pl. 14; fig. 2   Shell thin; protoconch small, of two rounded whorls with prominent nucleus; teleoconch of 16 slightly convex whorls, suture impressed, basal periphery subangulate, base flat, sloping; sculpture of three moderate spiral cords and weaker threads; columella straight, slightly reflected at the base. White, cords edged with brown.

3.6 cm. Indo-West Pacific; Onslow, WA to Gladstone, Qld. Dredged at 40-91 m.

Colpospira quadrata (Donald, 1900)

Protoconch of two smooth, convex whorls, nucleus not submerged; teleoconch of 11 whorls, broad at the base, periphery subangulate, base convex, later whorls flat-sided, with deeply impressed suture within a deep, V-shaped notch; sculpture of subsutural and presutural cords and spiral threads, base with faintly gemmate spiral threads; columella strongly arcuate, reflected. Light yellow, with brown flames and brown spots on the cords.

2 cm. Central NSW to SA. Dredged at depths of 82-860 m.

▲ GENUS GAZAMEDA Iredale, 1924
Turritella gunnii Reeve

Medium-sized to large, slender; protoconch with two whorls, asymmetrical, nucleus usually raised and prominent; teleoconch whorls flat-sided to concave, spire usually straight-sided, base flat to concave; aperture quadrate; outer lip sinus prosocline, broad, moderately deep; columella arcuate, reflected at the base.

Indo-West Pacific.

Gazameda declivis (Adams & Reeve, 1848)

Pl. 14; fig. 12   Teleoconch of 15-18 whorls, suture lightly impressed, later whorls flat-sided or slightly concave centrally; sculpture of a broad, rounded, subsutural rib, two lesser cords at the centre, and a strong, double peripheral rib, base concave with 10-12 fine spiral cords. Off-white with irregular red-brown subsutural spots which tend to elongate axially on the later whorls.

6 cm. Indo-West Pacific; Rottnest I., WA around northern Australia and down the east coast to eastern Bass St. Dredged on the middle part of the continental shelf. Synonym: *captiva* Hedley, 1907.

Gazameda gunnii (Reeve, 1848)

Pl. 14; fig. 5 a-b   Teleoconch of about 18 whorls, suture deeply incised, basal periphery subangulate; sculpture very variable, with a strong basal spiral rib and 3-6 fine cords above, base flat with weak spiral striae. Off-white to light brown, sometimes with purple-brown spiral bands and wavy axial flames.

5.6 cm. Southern Qld to eastern Vic. and Tas. Very common on the inner shelf at depths of 8-140 m. In deeper water shells the peripheral rib may be very prominent. Iredale (1924, *Proceedings Linnean Society NSW*, **49**: 247) noted that this species is viviparous. Synonym: *philippensis* Watson, 1881.

Gazameda iredalei Finlay, 1927

Pl. 14; fig. 3   Teleoconch of about 18 whorls, sculptured with two erect spiral keels, one at the centre, one at the periphery, with two deeply concave areas between them, suture hidden by the lower keel; base flat, sculptured only with fine growth lines. Purplish brown, keels white with brown spots.

4 cm. Bass St. to south coast of WA. The prominent spiral keels give this remarkable shell the appearance of an awl or corkscrew. Sometimes found in beach litter and presumably lives in the shallow sublittoral zone. Synonym: *clathrata* Kiener & Fischer, 1873 (name preoccupied).

Gazameda tasmanica (Reeve, 1849)

Spire flat-sided, teleoconch of about 15 flat-sided whorls, suture moderately impressed, basal periphery angulate, base flat; with a moderate subsutural cord, a strong peripheral cord that may obscure the suture, and fine spiral striae between, base spirally striate. Early whorls translucent white, later whorls purple-brown with brown axial flames and brown spots on the ribs.

4 cm. Southern Qld to the south coast of WA. Common on the inner shelf at depths of 8-90 m. Synonyms: *subsquamosa* Dunker, 1871; *acuta* Tenison Woods, 1876; *lamellosa* Watson, 1881; *oxyacris* Tate, 1897.

▲ GENUS **HAUSTATOR** Montfort, 1810
Turritella imbricataria Lamarck

▲ SUBGENUS **KUROSIOIA** Ida, 1952
Turritella kurosia Ida

Small to medium-sized, thin shells with slightly convex whorls, early whorls strongly keeled, later whorls spirally corded or striate; base weakly convex to weakly concave; aperture subcircular to subrhomboid, outer lip sinus moderately deep and broad, slightly prosocline; protoconch of 3-5 minute, convex whorls, nucleus low, rounded.

Indo-West Pacific.

Haustator cingulifera (Sowerby, 1825)

Pl. 14; fig. 9 a-b   Protoconch mammillate, of three whorls; teleoconch of about 18 whorls, suture moderately incised, basal periphery subangulate, base convex; sculpture variable, usually with three, sometimes more, subangulate, primary spiral cords and numerous spiral threads, base striate; aperture almost circular, columella straight, thin, not reflected. Off-white to light brown with chestnut-brown bands at the suture.

2.5 cm. Indo-West Pacific; Onslow to central NSW. Synonyms: *fascialis* Menke, 1830; *tricarinata* King, 1832; *bicolor* Adams & Reeve, 1848; *fusco-tincta* Petit, 1853; *gracillima* Gould, 1861; *fragilis* Kiener, 1873; *parva* Angas, 1877.

Haustator gilletti Garrard, 1974

Protoconch of three or more opaque, narrow, convex whorls; suture incised, teleoconch of nine or more whorls, strongly unicarinate, the keel slightly below centre, concave between the keels, base concave, periphery angulate; aperture subrhomboid, columella almost straight and perpendicular. Opaque or translucent, white, sometimes with minute subsutural brown spots.

1.4 cm. Described from 63-73 m, Swain Reefs, Qld.

Haustator leeuwinensis Garrard, 1972

Protoconch of five whorls, smooth and translucent; teleoconch of 11 or more convex whorls, suture impressed, base flat; sculpture of a single spiral cord on the early whorls, vanishing on the later whorls which are weakly striate; aperture subquadrate, columella slightly arcuate, reflected at the base. Upper parts of whorls pale yellow, lower parts chestnut, with a thin brown line between.

1 cm. Cape Leeuwin, WA to the Timor Sea. Dredged on the middle part of the continental shelf.

▲ GENUS **MAORICOLPUS** Finlay, 1927
Turritella rosea Quoy & Gaimard

Small to large, with more or less straight-sided whorls; protoconch of 3-4 slightly globose whorls, diameter rapidly increasing; aperture subquadrate, base flat to concave; outer lip sinus broad, moderately deep, prosocline; sculpture of spiral threads and cords.

Eastern Australia and New Zealand. There is only one living species in Australian waters but several fossil species occur in Tertiary sediments of the south-east.

Maoricolpus roseus (Quoy & Gaimard, 1834)

Pl. 14; fig. 14 Teleoconch of about 18 whorls, slightly concave at the centres, sculpture of three spiral cords on the early whorls, diminishing on the later whorls, the lower cord remaining dominant, with numerous spiral threads and striae; periphery angulate; columella straight, reflected at the base, spreading as a thin callus on the parietal wall. Fawn to creamy-pink, threads usually reddish brown, often with red-brown axial flames.

8 cm. Southern Qld to Tas.; also New Zealand. Common.

▲ GENUS **TURRITELLA** Lamarck, 1799
Turbo terebra Linnaeus

Large and solid; whorls convex, base convex, periphery rounded, sculptured with spiral cords and threads; outer lip sinus broad, very shallow, opisthocline or arched backward, continued as a straight line across the base.

Pantropical. Most Australian and New Zealand species once placed in this genus have now been reclassified into other genera but in other parts of the world *Turritella* is still used in the broad sense.

Turritella terebra (Linnaeus, 1758)

Pl. 14; fig. 13 Protoconch invariably missing in adult shells; teleoconch of 20-25 strongly convex and spirally corded whorls, three prominent cords near the base of the later whorls, suture incised; aperture almost circular. Early whorls pale cream or fawn, later whorls brown.

17 cm. Indo-West Pacific; North West Cape, WA to central Qld. Synonyms: *cerea* Reeve, 1849; *spectrum* Reeve, 1849.

▲ GENUS **ZARIA** Gray, 1847
Turbo duplicata Linnaeus

Large and solid; whorls convex, later whorls sharply keeled; aperture subcircular, outer lip sinus prosocline, forming a wide arc.

Indo-West Pacific.

Zaria duplicata (Linnaeus, 1758)

Protoconch invariably missing in adult shells; suture deeply impressed, teleoconch of about 16 convex whorls, bulbous toward the lower suture; base convex, periphery rounded;

sculpture of three strong, angulate spiral keels on the lower part of the whorls, the keel at the periphery becoming prominent, plus secondary cords and threads; columella arcuate. Reddish cream, upper parts of the whorls slightly darker.

18 cm. Indo-West Pacific; Kimberley, WA to north Qld. An inner shelf species. Synonym: *acutangulus* Linnaeus, 1758.

▲ GENUS **ZEACOLPUS** Finlay, 1927
Turritella (Haustator) vittata Hutton

Medium-sized; protoconch of about 1½ whorls, asymmetrical, nucleus slightly submerged; teleoconch whorls convex to flat-sided, spirally corded, base weakly convex; aperture subquadrate, outer lip sinus prosocline, broad, moderately deep.

Eastern Australia and New Zealand.

Zeacolpus capricornius (Garrard, 1972)

Teleoconch of 10 convex whorls, suture impressed, base slightly convex, spirally corded; sculpture of three primary cords and several secondaries; columella arcuate, reflected at the base. Light brown to chocolate.

2 cm. Described from Hervey Bay, Qld. Apparently lives in the shallow sublittoral zone. Garrard classified this species in the subgenus *Stiracolpus* Finlay, 1929 (*symmetrica* Hutton).

SUBFAMILY **TURRITELLOPSINAE**

▲ GENUS **TURRITELLOPSIS** Sars, 1878
Turritella acicula Stimpson

Very small, frail, translucent or clear; protoconch of 3-4 minute, convex whorls; teleoconch whorls convex, base convex and smooth; sculpture of spiral keels; aperture elongate-ovate, outer lip sinus lacking.

Cosmopolitan in high latitudes. This coldwater group is best known from the northern hemisphere. Synonym: *Kimberia* Cotton & Woods, 1935 (*kimberi* Verco).

Turritellopsis kimberi (Verco, 1908)

Protoconch narrow, conical, of four tall whorls; teleoconch of eight convex whorls; sculpture of subequal spiral cords, 5-6 on the penultimate whorl; columella reflected as callus on the parietal wall. Translucent white.

8 mm. SA. Dredged at 36-170 m. Verco (1910) also described from SA *K. neptunensis*. It has more whorls, the early whorls tending to be subangulate at the centre, and has since been shown to be distributed from southern Qld to SA.

FAMILY **SILIQUARIIDAE**

Slit Worm Shells

These are perhaps the most aberrant and least known of the cerithioideans. At one time they were regarded as vermetids but are now acknowledged as a distinct family within the Cerithioidea. Their shells are tubular, loosely coiled, at least in the later stages, and have a row of tiny excurrent holes or a slit along one side. There are no internal septae. The operculum consists of a spirally wound band forming a cone, with bristles around the edges of the turns. Siliquariids live embedded in the bodies of sponges. Presumably they feed on particulate matter suspended in the water by means of mucous traps like those of vermetids.

The family is found in tropical and warm temperate seas. Recent treatments list a single genus with two subgenera but the family needs revision. The following account of Australian siliquariids is very provisional.

▲ GENUS **SILIQUARIA** Bruguière, 1792
Serpula anguina Linnaeus

With the characters of the family.

Cosmopolitan in tropical and temperate seas. *Tenagodus* Guettard, 1770 is an earlier name but unavailable because the publication did not follow binominal principles.

▲ SUBGENUS **PYXIPOMA** Mörch, 1860
Siliquaria lactea Lamarck

Protoconch of a little more than one whorl, coiled planispirally; teleoconch coiled in a loose, irregular, corkscrew spiral; longitudinal fissure a narrow slit open for most of the shell length; mouth wide, circular, trumpet-shaped; operculum a tall dome, margin simple.

Indo-West Pacific, southern Australian and New Zealand regions. Sometimes treated as a separate genus.

Siliquaria weldii (Tenison Woods, 1875)

Pl. 15; fig. 2   Slit filled for the first three whorls, open thereafter; surface of smooth appearance but microscopically striate. White, yellow or pale pink.

3 cm. Northern NSW to the midwest coast of WA; also in New Zealand. An Indo-West Pacific species known as *S. cumingi* Mörch, 1860 may be the same.

▲ SUBGENUS **SILIQUARIA** S.S.

Early whorls spirally coiled but later whorls irregularly twisted; fissure consisting of a series of holes which may be separate or narrowly connected, open from the apex to the mouth.

Cosmopolitan. **Three species are listed here but there may be more than this in the Australian fauna.**

Siliquaria anguina (Linnaeus, 1758)

Pl. 15; figs. 3, 4   Early whorls smooth and regularly coiled; later whorls rapidly uncoiling and irregular, sculptured with spinose axial cords; fissure begins as separate holes but becomes a continuous slit in the last half. Off-white, pink or mauve.

Tube diameter up to 6 mm. Indo-West Pacific; Shark Bay, WA to Qld.

Siliquaria australis (Quoy & Gaimard, 1834)

Early whorls loosely coiled but quickly becoming irregularly twisted; tube finely striate on the inner side, rugose and transversely fissured on the outer side; fissure begins as a series of holes but becomes a continuous slit. Creamy white.

Tube diameter up to 1 cm. Northern NSW to at least as far north as Geraldton in WA. The distribution of this species needs further study. A similar shell in Qld has been referred to under the same name but the identity needs confirmation.

Siliquaria ponderosa (Mörch, 1860)

Pl. 15; fig. 5   First three whorls rapidly expanding in diameter, almost planispirally coiled; the next three even-sized, tightly and evenly coiled around a narrow "umbilicus"; the tube then turning abruptly tangentially, becoming more or less straight and downward sloping; fissure closed in the first three whorls, then becoming a series of barely connected holes, then a continuous slit, and finally a closed slit terminally; inner side of the tube smooth, outer side transversely fissured, axially corded toward the end. White.

Tube diameter up to 1.3 cm. North West Cape, WA to Qld. An Indo-West Pacific distribution is likely but information is scanty. Shells of this species are often cast up on beaches. They are among the most interesting objects to be found on the sea shore.

LITTORINOIDEA

*P*ONDER & Warén (1988, *Malacological Review, Supp.* **4**: 288-326) assigned four families to this superfamily, of which two are represented in the Australian marine fauna. One of these, the Littorinidae, is a significant element on our rocky shores and mangrove forests. The other, the Skeneopsidae, is a very small and litle known family.

FAMILY **LITTORINIDAE**

Periwinkles

As the name implies littorinids live in the intertidal zone, or sometimes in the splash area above true high tide level, i.e. the supralittoral zone. They are conspicuous on rocky shores almost everywhere from the tropics to the cold coasts of high latitudes. Some species live in tidal marshes or mangroves.

Littorinids have strong, turbinate shells without an umbilicus, and a thin horny operculum. They are herbivores, grazing on algae growing on their substrate. Their radula is typically taenioglossate. The shells of many littorinids show marked sexual dimorphism, females being significantly larger, on average, than males.

Males possess a penis and fertilisation is internal. The male testis produces two types of sperm in some species, the typical type responsible for fertilising the eggs, and specialised, atypical sperm known as nurse cells which are believed to nourish the others. In many species the females release the eggs in small capsules during high tide periods and there is a planktotrophic veliger stage after hatching. Sometimes the egg capsules are embedded in a gelatinous egg mass when first released but the jelly soon dissolves and the capsules become free-floating. Other species produce gelatinous egg masses which are attached to the substrate, and the veligers hatch from there. Ovoviviparity is common, the eggs being retained between the lamellae of the gills in the mantle cavity until the larvae have reached an advanced stage of development before they are released as advanced veligers.

Bandel & Kadolsky published a reclassification of the family (1982, *Veliger* **25**: 1-42). Five subfamilies are recognised. The late Dr Joseph Rosewater reviewed the Littorininae, Tectariinae and Echinininae of the Indo-West Pacific region (1970, *Indo-Pacific Mollusca* **2** (11): 417-506; 1972, *ibid.* **2** (12): 507-533). Reid (1988, *Records Australian Museum* 40: 91-150) reviewed the Bembiciinae. The fifth subfamily, the Lacuninae, sometimes ranked as a family, is not represented in the Australian fauna. Further notes on Australian littorinids were added by Ponder & Rosewater (1979, *Proceedings Biological Society of Washington* **92** (2): 773-782). David Reid published an entire book on the anatomy, ecology and taxonomy of the littorinids of mangrove forests in the Indo-Pacific region (1986, British Museum of Natural History, Publ. No. **978**: 228 pp.).

SUBFAMILY **BEMBICIINAE**

Top-shaped inhabitants of rocky shores in marine and estuarine environments. According to Reid (1986, *loc. cit.*) this group is the most deviant in the family. The subfamily is endemic to the Southern Australian and New Zealand regions. There are two genera, *Bembicium* and the New Zealand endemic *Risellopsis*, both revised by Reid (1988, *loc. cit.*).

▲ GENUS **BEMBICIUM** Philippi, 1846
Trochus melanostomus Gmelin

Trochiform-conical, with a flat base and angulate periphery; sculpture of spiral lines or nodules and sometimes radial folds; aperture oblique, outer lip thin; small umbilicus usually present in juveniles, lacking in adults; operculum with few whorls.

Southern Australia and Norfolk and Lord Howe Is. There are also fossil records from New Zealand and the Kermadec Is. Synonym: *Risella* Gray, 1847.

The egg masses of *Bembicium* species are small (2-3 mm long), gelatinous, either ovoid or irregular, and are attached in closely packed clusters to hard substrates. The number of eggs in each mass is variable, e.g. 60-100 in *B. auratum*; 100-200 in *B. nanum*. The larvae of these species hatch as veligers after 10-12 days (Anderson, 1962, *Proceedings Linnean Society NSW* **87**: 62-68). The eggs of *B. vittatum* are larger, with only 8-30 per egg mass, and probably have lecithotrophic larval development (Anderson, 1958, *Australian Journal Marine & Freshwater Research* **9**: 546-568).

Bembicium auratum (Quoy & Gaimard, 1834)

Pl. 18; fig. 21 a-b   Equilaterally conical, sides straight to slightly convex, base flat to slightly convex, periphery sharply angulate and tuberculate or nodulose; sculptured with rugose spiral ribs. Cream to brown with grey mottling or wavy oblique stripes; base pale yellow-brown, columella white to pale brown, aperture cream interiorly, black near the margin.

2 cm. Lizard I., Qld to Abrolhos, WA. Inhabits less exposed shores than *B. nanum*; usually on mangroves and often in estuaries. North Qld shells are paler and have a lower spire and strongly keeled periphery lacking folds. Synonyms: *imbricata* Gray, 1839; *cicatricosus* "Jonas" Philippi, 1843; *kielmanseggi* Zelebor in Dunker & Zelebor, 1866; *nodulosum* Musgrave, 1929.

Bembicium flavescens (Philippi, 1851)

Pl. 18; fig. 22 a-b   Equilaterally conical or taller, spire sometimes turreted, sides slightly convex, base flat; periphery strongly keeled, often crenulated by strong radial folds. Upper half of each whorl grey, lower part cream, spaces between radial folds dark brown sometimes forming stripes; base white to cream, brown-spotted; columella and parietal callus pale orange to pink, aperture brown above the periphery, white below.

1.6 cm. Lord Howe and Norfolk Is. This is an example of the local endemism characterising these two small oceanic islands. Synonym: *plicatula* Philippi, 1851.

Bembicium melanostomum (Gmelin, 1791)

Pl. 18; fig. 24 a-b   Approximately equilaterally conical, sides slightly convex, base flat to slightly concave, periphery angulate or flanged; sculptured with fine spiral threads and strong radial folds which may crenulate the periphery of the last whorl. Upper surface cream to blue-grey, sometimes with dark brown oblique stripes, base plain cream or speckled with brown; columella orange, parietal callus cream to pale orange; interior dark brown above, cream below the periphery.

1.5 cm. Central Vic. and Tas. Lives in sheltered bays and estuaries. Synonyms: *luteus* Quoy & Gaimard, 1834; *lividum* Philippi, 1846; *squamiferus* Gould, 1852.

Bembicium nanum (Lamarck, 1822)

Pl. 18; fig. 19 a-e   Depressed, conical, base flat, whorls convex, periphery sharply angulate and sometimes flanged; sculpture weak. Upper surface light brown with oblique, irregular, black bands or stripes which tend to bifurcate toward the periphery; columella white, cream or pale orange; aperture pale within, showing the external black stripes at the edge.

1.2 cm. Central Qld to Tas. and west to Spencer Gulf, SA. Lives on moderately exposed rocky shores. Synonyms: *acuminatus* Wood, 1828; *planus* Quoy & Gaimard, 1834; *australis* Gray, 1839; *pictum* Philppi, 1846; *grisea* Philippi, 1851; *crassa* Dunker, 1861.

Bembicium vittatum Philippi, 1846

Pl. 18; fig. 23 a-b Low to equilaterally conical, sides slightly convex, base flat; periphery strongly keeled, sometimes flanged, sometimes rounded on the last whorl; juvenile sculpture of radial folds may persist to the last whorl, with fine spiral threads. Cream to lilac-grey, sometimes paler at the periphery, flecked with darker grey; base unmarked or brown-flecked; columella and parietal callus orange or cream; aperture orange-brown above the periphery, paler below.

2 cm. SA to Abrolhos, WA. On rocky shores, mangroves or marshes in sheltered inlets. Synonyms: *fimbriata* Philippi, 1851; *bruni* Crosse, 1864.

SUBFAMILY **LITTORININAE**

Shells turbinate, umbilicus lacking, operculum with few whorls. Radula of the generalised littorinid type.

▲ GENUS **LAEVILITTORINA** Pfeiffer, 1886
Littorina caliginosa Gould

Small, ovate-conical, thin; outer lip sharp-edged, continuing onto the parietal wall as a thin callus.

Circumpolar in high latitudes of the southern hemisphere. There are two species in southern Australia.

Laevilittorina johnstoni (Cotton, 1945)

High-spired, suture impressed; protoconch blunt, of two whorls; teleoconch of four smooth, slightly convex whorls, base gently rounded; columella narrow, slightly thickened and arched. Protoconch red-brown; adult whorls with a wide, white subsutural band, crossed by about nine brown axial bands which broaden anteriorly, with a second white spiral band on the base, decorated throughout with oblique rows of red-brown spots.

6 mm. Venus Bay, SA to Cape Naturaliste, WA. Synonym: *burni* Ponder, 1976.

Laevilittorina mariae (Tenison Woods, 1876)

Similar to *L. johnstoni* but less spectacularly coloured and with more ventricose whorls and blunter spire.

5mm. Southern NSW to eastern SA. Common on algae growing on sheltered reefs.

▲ GENUS **LITTORARIA** Griffith & Pidgeon, 1834
Turbo zebra Donovan

Shell sculpture of spiral ribs and grooves, nodulose sculpture lacking; colour pattern usually of short spiral dashes aligned to form axial stripes; operculum thin, with few whorls.

Pantropical. Live in the supralittoral zone, many in mangroves, some on rocky shores. There are three named subgenera, all represented in the Australian fauna.

Reid (1986, *loc. cit.*) has shown that thin-shelled, colourful, mangrove-inhabiting littorinids commonly regarded as a single, highly variable species known as *"Littorina scabra"* in fact belong to two distinct subgenera containing no less than 36 species, 20 of which live in the Indo-West Pacific region, including nine in northern Australia. The differences between these very common littorinids are often mainly anatomical and collectors and professional biologists will have great difficulty telling them apart on shell characters. Reid gives an excellent key to the species, plus anatomical details.

▲ SUBGENUS **LITTORARIA** s.s.

Shells solid, varixes lacking; microsculpture, if present, of spiral striae; colour variable but rarely polymorphic. Sperm nurse cells not flagelate, penial base bifurcate or simple; bursa copulatrix opens anteriorly or posteriorly; capsule glands present. Development oviparous.

Indo-West Pacific.

Littoraria coccinea (Gmelin, 1791)

Pl. 18; fig. 1 a-b Elongate to conic-turbinate; base slightly flattened, periphery slightly keeled; suture distinctly impressed; smooth except for minute spiral striae; columella wide, concave, pinched where it meets the outer lip anteriorly. White, lacking pattern except for faint blotches on the early whorls; interior and columella orange or brown.

2.6 cm. Western Pacific and Cocos Keeling; north Qld. Lives on rocky shores, often together with *L. undulata*. Synonyms: *obesa* Sowerby, 1832; *limax* Gray, 1839.

Littoraria undulata (Gray, 1839)

Pl. 18; fig. 2 a-d Elongate-turbinate; base not flattened, periphery rounded; suture impressed, whorls convex; sculpture of incised spiral striae and minute spiral threads; columella wide, oblique, only slightly concave, pinched at the base. Colour pattern extremely variable, usually yellowish gray to brown, may be banded or with oblique brown wavy lines or rows of spots; interior yellow-brown, columella usually greyish violet, sometimes white.

2.4 cm. Indo-West Pacific; North West Cape, WA to central NSW. This is a very common little shell, living high in the intertidal zone of rocky shores. It is readily identified by its colour pattern. Synonyms: *tenuis* Philippi, 1846; *columna* "Jonas" in Philippi, 1847; *acuminata* Gould, 1849; *scabra concolor* Weinkauff, 1878; *undulata contracta, u. sulcatula* and *conica subintermedia* Nevill, 1855.

▲ SUBGENUS **LITTORINOPSIS** Mörch, 1876

Littorina subangulata Lamarck (error for *angulifera* Lamarck)

Shell often thin, sometimes varixed, microsculpture, if present, of spiral striae in the grooves, shell colour often polymorphic. Sperm nurse cells not flagellate, penial base bifurcate, bursa copulatrix opens anteriorly, capsule glands absent. Development ovoviviparous.

With 11 species in the Indo-West Pacific and two in the Atlantic.

Littoraria cingulata (Philippi, 1846)

Pl. 18; figs. 7 a-d, 8 a-b Spire outline straight, whorls with convex sides, basal periphery made carinate by a prominent rib; suture impressed; columella of moderate width, rounded; primary grooves 5-6, with microsculpture of regular spiral striae; primary ribs rounded, prominent, 11-13 on the body whorl, grooves of equal or almost equal width, secondary riblets on body whorl only. Apex blue-grey, remainder whitish, cream or blue-grey, with a marbled pattern of orange-brown on the ribs, columella dark purple, lilac or pink, parietal callus often also purple.

2.5 cm. The nominate subspecies ranges from Exmouth Gulf to King Sound, WA. A population in Shark Bay distinguished by more numerous ribs (40-60) on the body whorl, rounded periphery and the frequent presence of axial stripes, has been given subspecies status as *L. cingulata pristissini* Reid, 1986 (Pl. 18; figs. 7 a-d). On the northern coast of WA, *L. cingulata* may be confused with the superficially similar *L. sulculosa* but the latter has axial not spiral striae in the interspaces and lacks colouring on the columella.

Littoraria filosa (Sowerby, 1832)

Pl. 18; fig. 12 a-d Shell thin, spire outline almost flat, peripheral keel conspicuous until whorl seven, basal periphery rounded; columella narrow, rounded, not excavated; with 5-7 primary grooves, spacing markedly unequal; body whorl with 9-11 strong, narrow, keel-like cords, interspaces spirally grooved. Colour polymorphic, shades of yellow and brown frequent, sometimes pink or orange, patterned with brown dashes on the ribs, base pale except for a dark band around the columella, parietal callus and columella dark purple-brown.

3.5 cm. Exmouth Gulf, WA to Botany Bay, NSW. Also found in Indonesia and southern PNG. Typically lives on leaves and trunks in the *Avicennia* fringe of mangroves.

Littoraria intermedia (Philippi, 1846)

Pl. 18; fig. 18 Spire outline gently convex, basal periphery rounded; columella wide, excavated, edge slightly convex, sharply pinched at the base; primary grooves 8-10, narrow, microsculpture indistinct; body whorl with 17-32 low ribs, posterior rib on spire whorls may be prominent. Usually grey, sometimes pale brown, cream, whitish or orange-pink, with a pattern of discrete, closely spaced, black or brown dashes on the ribs, often roughly axially aligned; columella and parietal callus dark purple or pink.

3.2 cm. Indo-West Pacific; Darwin, NT to southern Qld. As the name suggests this is a variable species and is commonly misidentified. The diagnostic characters are the grey ground colour and large number of dark dashes on the body whorl. Synonyms: *ambigua* Philippi, 1848; *newcombi* Reeve, 1857; *frazeri* Reeve, 1857.

Littoraria luteola (Quoy & Gaimard, 1833)

Pl. 18; fig. 17 a-b Shell thin, spire tall, outline convex, with 0-2 varixes, outer lip flaring; columella narrow, rounded; with 8-10 primary grooves, body whorl with 18-23 rounded ribs, the two at the periphery most prominent. Colour polymorphic, frequently brown, yellow or red, with dark brown dashes on the ribs which may align axially, columella purple or red-brown.

2.3 cm. Torres St. to Merimbula, NSW. Lives on leaves and trunks of *Avicennia*. Synonym: *filosa* var. *subcingulata* Nevill, 1885.

Littoraria pallescens (Philippi, 1846)

Pl. 18; fig. 14 a-d Spire outline gently convex; peripheral keel evident in young shells, indistinct on later whorls; columella wide, excavated, inner lip calloused; with 9-10 primary grooves, microsculpture faint; body whorl with 21-26 rounded or flattened ribs. Shades of brown, yellow or pink, may be striped or banded, usually with brown dashes or flecks on the ribs; columella purple, whitish at the edge.

2.5 cm. Indo-West Pacific; Barrow I., WA to Moreton Bay, Qld. Often misidentified as *L. filosa* which is also extremely variable in colour. *L. pallescens* is distinguishable by its wide columella and weak secondary sculpture. Typically on leaves of *Rhizophora* at the seaward edge of mangroves. Synonyms: *sieboldii* Philippi, 1846; *arboricola* Reeve, 1857.

Littoraria philippiana (Reeve, 1857)

Pl. 18; fig. 13 a-b Spire gently convex; columella wide, excavated; with 8-9 primary grooves; body whorl with 13-18 spiral cords and 1-3 riblets in the interspaces plus microsculpture of spiral striae. Colour polymorphic, dark brown shells predominate, sometimes orange, pink or yellowish, with a pattern of darker, diffuse dashes and flecks aligning at the suture to form indistinct oblique stripes.

3.5 cm. Southern PNG and eastern Australia as far south as Sydney. Records from WA and the NT need confirmation. Typically lives on trunks of *Rhizophora*.

Littoraria scabra (Linnaeus, 1758)

Pl. 18; fig. 16 Spire outline concave near the apex; peripheral keel strong; columella wide, excavated; primary grooves 9-11, spirally striate; with 36-41 low ribs on the body whorl with prominent spiral microsculpture in the grooves. Pale with more or less dense pattern of black or dark brown dashes, aligned at the suture to form oblique stripes numbering 9-14 on the body whorl; columella white.

3.5 cm. Indo-West Pacific; Point Cloates, WA to Stradbroke I., Qld. The white columella and strong spiral microsculpture are diagnostic. Lives on trunks and roots at the seaward edge of mangroves. Many other mangrove littorinids have been lumped under this name. Synonyms: *lineatum* Gmelin, 1791; *foliorum* Gmelin, 1791; *novaehiberniae* Lesson, 1831.

▲ SUBGENUS PALUSTORINA Reid, 1986
Littorina melanostoma Gray

Shell solid, microsculpture, if present, of spiral striae on the ribs and axial striae in the grooves, shell colour variable but not polymorphic. Sperm nurse cells flagellate, penial base simple, not bifurcate, bursa copulatrix opens posteriorly, capsule gland present. Development oviparous.

With seven species, all in the Indo-West Pacific.

Littoraria articulata (Philippi, 1846)

Pl. 18; fig. 6 a-b Spire relatively low, slightly convex in outline, sides of whorls and basal periphery rounded; columella wide, excavated; with 8-10 primary grooves, 20-33 flat ribs on the last whorl, grooves usually impressed lines only. Cream-yellow, grey or white with a pattern of black or dark brown dashes on the ribs which may align to form axial stripes posteriorly, columella and parietal wall dark purple, sometimes white or pink, columellar edge white.

2 cm. Central Indo-West Pacific; Exmouth Gulf, WA to Moreton Bay, Qld. Lives in the outer parts of mangrove forests, occasionally found on pilings and rocks. Synonyms: *sinensis* Pilsbry, 1895; *blanfordi* Dunker, 1871; *strigata* Lischke, 1871.

Littoraria sulculosa (Philippi, 1846)

Pl. 18; fig. 15 a-b Shell thick, spire outline straight, whorls slightly convex, sometimes turreted, basal periphery moderately keeled by the most prominent rib; columella wide, scarcely excavated, edge concave, constricted at the base producing a small knob; with 5-6 primary grooves, 9-11 prominent and rounded primary ribs on the body whorl, secondary sculpture usually lacking. Cream to pale fawn, ribs fawn, pale orange-pink or grey-brown, often unicoloured or marked with long, indistinct dashes.

2.3 cm. Exmouth Gulf to Vansittart Bay, WA. Lives mostly on leaves and trunks of *Avicennia* trees, close to the ground, in the middle and rear parts of mangrove forests.

▲ GENUS LITTORINA Ferussac, 1821
Turbo littoreus Linnaeus

The subgenus *Littorina* s.s. is a northern hemisphere and Arctic group. The genus is represented in the Australian fauna by the subgenus *Austrolittorina*.

▲ SUBGENUS AUSTROLITTORINA Rosewater, 1970
Littorina unifasciata unifasciata Gray

Turbinate-conical, smooth or spirally lirate, periphery carinate, base flattened or hollow forming a crescent-shaped area; columella flattened. Penis with basal enlargement containing a single penial gland and an accessory flagellum.

Indo-West Pacific and middle and high latitudes of the southern hemisphere. There are four species in the subgenus, three of them in Australia.

Littorina acutispira Smith, 1892

Spire tall, convex, suture distinct, whorls convex, base slightly flattened with a weakly carinate periphery; columella thick, slightly concave, pinched at the junction with the lip; spiral sculpture of widely spaced, weak striae. Light yellow-brown to dark brown, usually with a wide brown or grey spiral band, often speckled with white spots; interior and columella red-brown.

7 mm. Central Qld to western Vic. Lives high on rocky shores, often together with *L. unifasciata*. Synonym: *infans* Smith, 1892.

Littorina praetermissa May, 1909

Pl. 18; fig. 4 a-b Globose-turbinate, spire low, whorls rounded, base very slightly flattened, periphery carinate; columella moderately excavated, with a small tooth-like swelling near the base; surface uneven, sculpture of indistinct spiral striae. Greyish white, usually with brown zigzag oblique lines; interior brown, with white marks at the lip margin and a white spiral band interiorly, columella yellow to white.

1.8 cm. Vic., Tas. and eastern SA. Lives lower on rocky shores than *L. unifasciata*. Synonym: *paludinella* authors (not Reeve, 1857).

Littorina unifasciata Gray, 1826

Pl. 18; fig. 3 Elongate-turbinate, spire moderately tall to low, whorls convex, base slightly flattened, basal periphery keeled; columella with a shallow excavation; sculpture of weak spiral striae. White or grey with a broad, central, pale blue band; interior and columella brown, with a crescent-shaped brown area on the base adjacent.

2.5 cm. Southern Qld to North West Cape, WA. Lives high on rocky shores. This southern Australian form is the nominate subspecies. Other subspecies occur in New Zealand and south-eastern Pacific. Synonyms: *diemenensis* Quoy & Gaimard, 1833; *acuta* Menke, 1843; *mauritiana crassior* Philippi, 1847; *diemenensis pseudolaevis* Nevill, 1885.

▲ GENUS **MACQUARIELLA** Finlay, 1927
Paludestrina hamiltoni Smith

Very small, ovate to subglobose, with a narrow crescentic umbilicus. Radula with a narrow central tooth bearing a single cusp.

New Zealand and south-eastern Australia.

Macquariella kingensis (May, 1924)

Spire low; four teleoconch whorls, rapidly increasing, body whorl ventricose; aperture semilunar, sharply pointed above; columella narrow, and only slightly thickened beside the umbilicus. Light yellow-chestnut with a darker band below the suture, a light, broader band centrally and a patch of dark colour behind the umbilicus.

2.5 mm. Bass St.., Tas. Originally described as a *Natica*.

▲ GENUS **NODILITTORINA** von Martens, 1897
Littorina pyramidalis Quoy & Gaimard

Shell thick, turbinate-conical, heavily nodulose or at least with prominent spiral sculpture.

Indo-West Pacific and tropical western Atlantic. There are three described subgenera, two in the Indo-West Pacific including Australia.

▲ SUBGENUS **NODILITTORINA** s.s.

With 2-3 rows of nodules.

Indo-West Pacific.

Nodilittorina australis (Quoy & Gaimard, 1826)

Pl. 18; fig. 10 a-b Turbinate to ovate, spire low to moderately high, suture only moderately impressed, whorls rounded, surface rough, sculptured with fine spiral striae and usually low axial folds; columella hardly excavated but with a flattened crescent-shaped area adjacent. Yellow-grey, sometimes bluish; interior yellow or light brown, columella usually light violet to tan.

2 cm. Esperance to North Kimberley, WA. Lives high on rocky shores. Synonym: *rugosa* Menke, 1843.

Nodilittorina nodosa (Gray, 1839)

Pl. 18; fig. 11 a-b Diamond-shaped in outline, base rather flattened, periphery moderately keeled; sculpture of fine spiral striae and two bands of large nodules, the lower band around the basal periphery. Light brown, nodules red-brown to white; interior dark brown with white spots near the lip, columella brown.

1.3 cm. Geraldton to North Kimberley, WA. Thought to hybridise with *N. australis* and occupies the same zone as that species on rocky shores. The typical shells differ from *N. australis* in their shape and the two bands of large nodules.

Nodilittorina pyramidalis (Quoy & Gaimard, 1833)

Pl. 18; fig. 5 a-b Turbinate to pryramidal, spire moderately high, suture not deeply impressed, whorls convex; spiral sculpture varying from simple to moderately granulose with raised spiral cords, 18-20 on the body whorl, often axially wrinkled, and with closely spaced, fine, wavy axial threads; columella hardly excavated but with an adjacent crescent-shaped area on the base. Blue-grey with fawn nodules; interior and columella brown

1.5 cm. Indo-West Pacific; Fremantle, WA to eastern Vic. Lives high in the intertidal zone of rocky shores. Synonyms: *nodulosus* Gmelin, 1791 (not of Solander, 1766); *trochiformis* Dillwyn, 1817 (not of Brocchi, 1814); *trochoides* Gray, 1839; *vilis* "Menke" Philippi, 1846; *malaccana* Philippi, 1847; *cecillei* Philippi, 1851; *monilifera* Eydoux & Souleyet, 1852.

▲ SUBGENUS **GRANULITTORINA** Habe & Kosuge, 1966
Littorina millegrana Philippi

Globose, with multiple rows of low nodules.

Indo-West Pacific. Rocky shore dwellers. There are about seven species but only one in Australia.

Nodilittorina millegrana (Philippi, 1848)

Pl. 18; fig. 9 a-b Whorls rounded, spire low, suture well impressed, base not flattened, periphery rounded; sculpture variable, sometimes smooth but usually with spiral nodulose spiral cords; columella broad, thick, with a shallow excavation, sometimes bulging centrally. Greyish white, usually spotted or axially and spirally striped with red-brown marks; interior and columella brown.

1.4 cm. Indo-West Pacific; North West Cape to NSW. Lives high on rocky shores. Synonyms: *granularis* "Gray" of authors, not Gray,1839; *radiata* Eydoux & Souleyet, 1852; *novaezelandiae* and *granocostata* Reeve, 1857; *vidua* Gould, 1859; *melanacme* Smith, 1876; *erronea* Nevill, 1885; *granicostata* Smith, 1887; *insularis* Smith, 1889; *ventricosa strubelli* von Martens, 1897; *eudeli* Sowerby, 1915; *chaoi* Yen, 1936.

▲ GENUS **PEASIELLA** Nevill, 1884

Trochus tantilla Gould

Small, depressed-conical with a more or less flat base and carinate periphery; umbilicate; sculptured with spiral grooves and radial folds which may crenulate the periphery; operculum concentric, multispiral; sometimes with periostracal hairs.

Indo-West Pacific and eastern Pacific. Species of this genus live on rocks and mangroves high in the intertidal zone. The shell form of these little shells resembles that of *Bembicium* and species of the genus have sometimes been associated with that group. Reid (1988, *Records Australian Museum* 40: 91-150) has shown that *Peasiella* should be classified in the Littorininae. There are at least two species in the Australian fauna.

Peasiella roepstorffiana (Nevill, 1885)

Suture moderately impressed, with a subsutural sulcus forming shoulders on the whorls, sides of whorls and base flat, periphery sharply carinate; sides of whorls and base with five incised grooves. Fawn to yellowish brown.

3 mm. Indo-West Pacific; north Qld. The shell is more depressed than that of *P. tantilla* and the base is flatter.

Peasiella tantilla (Gould, 1849)

Suture impressed, sides of whorls convex, base slightly convex, periphery made carinate by a prominent rib; sides of whorls with five spiral cords separated by narrow grooves, base similarly corded. Fawn with prominent red-brown spots on the peripheral keel and below the suture and obscure radial bands; base of columella light brown.

3 mm wide. Indo-West Pacific; Qld south at least to Wellesley Group.

SUBFAMILY **TECTARIINAE**

Shells with or without an umbilicus; operculum more or less round with a central nucleus and a moderate number of whorls. Radular lateral teeth partitioned. Males with papillose surfaces on the penis and multiple penial glands. Development of the larvae takes place within a pelagic capsule.

▲ GENUS **TECTARIUS** Valenciennes, 1832

T. coronatus Valenciennes

There are two living subgenera, the monotypic *Cenchritis* of the tropical western Atlantic and *Tectarius* s.s. in the Indo-West Pacific.

▲ SUBGENUS **TECTARIUS** s.s.

Shells pyramidal to turbinate, strongly spinose, nodulose or papillose, umbilicus lacking, aperture plicate within, columella with a tooth-like swelling.

Indo-West Pacific. Synonyms: *Pagodus* Gray, 1839 (*pagodus* Lamarck = *pagodus* Linnaeus); *Pagodella* Swainson, 1840 (*pagodus* Linnaeus); *Echinella* Swainson, 1840 (*grandinatus* Gmelin); *Fectaria* Philippi, 1846 (*pagodus* Linnaeus); *Hamus* "Klein" H. & A. Adams, 1858 (*pagodus* Linnaeus); *Echinellopsis* Rovereto, 1899 (new name for *Echinella* Swainson).

Tectarius rusticus (Philippi, 1846)

Pl. 18; fig. 20   Solid, broadly conical, spire straight-sided; whorls flat, encircled by four spiral ribs bearing angulate nodules and numerous striae, the most prominent rib at the periphery making it angulate and nodulose, base steeply sloping, sculptured with one spiral rib and oblique, rugose folds. Yellow, sometimes with orange stripes; interior and columella white, sometimes tinged with pink or orange, columellar tooth sometimes stained brown.

4 cm high. Kimberley coast of WA. Lives very high on rocky shores, well above mean high tide level.

FAMILY **SKENEOPSIDAE**

There are two genera in this small family. Ponder (1988, *Malacological Review Supp.*4: 129-166) showed that these tiny snails are related to the littorinids and placed them in this superfamily. The two genera are *Skeneopsis* Iredale, 1915 (*planorbis* Fabricius) and *Starkeyna* Iredale, 1930 (*starkeyae* Hedley). *S. starkeyae*, an estuarine species from NSW, is illustrated here to represent the family.

Starkeyna starkeyae (Redrawn from Hedley, 1899, *Proceedings Linnean Society* NSW **24**: 433, fig. 4).

CINGULOPSOIDEA

THREE families of minute marine snails are included in this group. Many of the species are abundant in the shallows along Australian shores, and play important roles in food and energy cycles of coastal ecosystems, but they are outside the scope of this work. In recent years Dr Winston Ponder of The Australian Museum has published many papers on these and other minute gastropods. References to his publications may be found in the bibliographies of the papers cited below. For reference purposes the three families are:

FAMILY **CINGULOPSIDAE**

Shells brown-tinted, pupoid to depressed-trochiform, up to 5 mm in length; with or without an umbilicus; protoconch paucispiral. Operculum paucispiral, horny, transparent, with an inner peg.

Ponder & Yoo (1980) recognised four genera in the family and about 30 species and subspecies in the Australian and Indo-West Pacific regions. The genera are *Eatonina* Thiele, 1912 (*pusilla* Thiele); *Eatoniopsis* Thiele, 1912 (*paludinoides* Smith); *Tubbreva* Ponder, 1965 (*exigua* Ponder); *Pseudopisinna* Ponder & Yoo, 1980 (*gregaria* Laseron). Although their shells are like those of other families in the superfamily, these snails have distinctive radular and anatomical characters.

Reference

Ponder, W. F. & Yoo E.K. (1980) A review of the genera of the Cingulopsidae with a revision of the Australian and tropical Indo-Pacific species. *Records Australian Museum* **33** (1): 1-88.

FAMILY **EATONIELLIDAE**

Shells ovate to conical, aperture D-shaped to circular, outer lip strongly contracted. Protoconch paucispiral. Operculum oval, horny, pegged. Cephalic tentacles long, simple, tapering and lash about as the animal crawls.

Ponder & Yoo (1977) recognised two genera in Australia: *Eatoniella* Dall, 1876 (*kerguelenenis* Smith) and *Crassitoniella* Ponder, 1965 (*carinata* Ponder) and provided a key for the identification of the 20 known living Australian species.

References

Ponder, W. F. 1965. The family Eatoniellidae in New Zealand. *Records Auckland Institute & Museum* **6**: 47-99.

Ponder, W.F. & Yoo, E.K. (1977). A revision of the Eatoniellidae of Australia. *Records Australian Museum* **31** (15): 606-658.

FAMILY **RASTODENTIDAE**

Shells conical, smooth or spirally corded, whorls distinctly margined, operculum oval and pegged. There are several undescribed species in Australia but none published (pers. comm. Winston Ponder, March, 1993).

Reference

Ponder, W. F. (1966). A new family of the Rissoacea from New Zealand. *Records Dominion Museum* **5**: 177-184.

SUPERFAMILY

TRUNCATELLOIDEA

*T*HERE are fourteen families of living gastropods in this large group, including marine, estuarine and terrestrial species. For the most part they are minute snails and will not be dealt with in this work. Representative species of some families are illustrated for reference. Families of the superfamily which have Australian marine species but which are not referred to below are: Stenothyridae and Adeorbidae.

FAMILY **ASSIMINEIDAE**

Shells up to 1 cm high, brown, smooth, conical or turbinate with rounded whorls, operculum, when present, horny and with few whorls. Amphibious, air-breathing snails mostly living in marshes in brackish areas. Some species have become truly terrestrial.

There is a single, cosmopolitan genus, *Assiminea* Fleming, 1828 (*A. grayana* Fleming) in the family. Two marine species are described in the southern Australian fauna: *A. brazieri* Tenison Woods, 1876 and *A. tasmanica* Tenison Woods, 1876, both distributed from NSW to southern WA. Two subgenera are recognised. *Metassiminea* Thiele, 1927 (*philippinea* Boettger, 1893) was proposed for South-East Asian and Australian marine species. *Austroassiminea* Solem *et al.* 1982 was introduced for a terrestrial species (*A. letha*) described from shaded, moss-covered cliffs near the coast in the Naturaliste-Leeuwin National Park, WA.

Reference

Abbott R.T. (1958). *Proceedings Academy Natural Sciences Philadelphia* **110**.

FAMILY **BARLEEIDAE**

Minute marine snails with solid, broadly ovate to elongately conical shells which may be with or without an umbilicus and smooth or sculptured. The aperture has a simple peristome and may be with or without a varix. There is an inner chitinous layer. The dome-shaped protoconch has few whorls and a fine micro-sculpture of minute pits. The corneous, oval operculum has a convex outer surface, an eccentric nucleus, and may be with or without internal ridge and peg. There are two subfamilies, the Barleeinae and the Anabathrinae, the latter having its centre of distribution in Australasia with six genera represented in the region. The family is convergent with the Rissoidae and has often been lumped with it. See figure below.

Reference

Ponder, W. F. (1983). Review of the genera of the Barleeidae. *Records Australian Museum* **35**: 231-281.

Amphithalamus (Notoscrobs) *liratus* Thiele, 1930.

FAMILY **CAECIDAE**

Small shells (up to 5 mm long) which begin life with a coil but grow into a curved tube in the adult stage. The embryonic coil usually breaks off and the broken end is plugged with callus. Adult shells may be smooth or sculptured with ring-like ribs; there is a horny operculum.

These strange little shells are quite common in soft substrates but are easy to overlook. At first sight you would not take them to be gastropods. Little is known of their natural history. Several genera are described but their taxonomy is confused, and most modern authors prefer to use the original name *Caecum* Fleming, 1813 (*imperforatum* Kanmacher) in the broad sense for all the species, and other names as subgenera.

There has no been no review of the Australian caecids although several species are listed in our fauna: *amputatum* Hedley, 1894 from NSW to southern WA; *lilianum* Hedley, 1903 from NSW; *angustum, attenuatum, chinense, eburneum, microcyclus, subflavum* and *succineum,* all of de Folin, 1879 and listed by Charles Hedley from Qld.

FAMILY **EPIGRIDAE**

There is only one genus, *Epigrus* Hedley, 1903 (*cylindracea* Tenison Woods), in this marine family which was named because of the very unusual radula of the type species. There are three, possibly four species, apparently all confined to Australian waters, viz.: *cylindracea* Tenison Woods, 1878; *dissimilis* Watson, 1886; *obesa* Laseron, 1956; *columnaria* May, 1910.

The shell of the type species is elongately oval, with lightly convex whorls, shining and smooth surface, D-shaped aperture with a complete peristome, and a domed, few-whorled protoconch. The operculum is horny, thin, and has an eccentric nucleus. The radula is very broad, short, with minute, simple, very narrow, unicuspid central teeth and rectangular lateral teeth bearing several small, blunt cusps. Dead shells are commonly encountered in dredgings.

Reference

Ponder, W. F. (1985). Review of the genera of the Rissoidae. *Records Australian Museum, Supp.***4**: 101.

FAMILY **HYDROBIIDAE**

A large family with some estuarine genera and species and many in inland waters. It is currently under review by Ponder and will not be dealt with here.

FAMILY **HYDROCOCCIDAE**

Small, ovate shells with rounded whorls. Possess a metapodial tentacle like of the rissoids. The family contains a single genus inhabiting supralittoral mud flats in estuaries of the southern Australian coast.

Reference

Ponder, W. F. (1982). The anatomy and relationships of *Hydrococcus brazieri* (T. Woods). *Journal Molluscan Studies* **48**: 64-79.

FAMILY **IRAVADIIDAE**

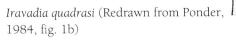

The members of this family are small and estuarine or marine. Their shells are ovately conical, usually non-umbilicate, lack an inner chitinous layer and are smooth or with spiral sculpture dominant. The operculum has an eccentric, marginal or submarginal nucleus. The animals typically have a strongly pigmented head-foot with long and slender cephalic tentacles. Most species lack pallial tentacles. Fairbankiidae is an alternative name for the family. Four of the nine genera have living representatives in the Australian fauna: *Iravadia* Blanford, 1867 (*ornata* Blanford); *Chevallieria* Cossmann, 1888 (*labrosa* Cossmann); *Nozeba* Iredale, 1915 (*emarginata* Hutton); *Rissopsis* Garrett, 1873 (*typica* Garrett).

Iravadia quadrasi (Redrawn from Ponder, 1984, fig. 1b)

Reference

Ponder, W. F. (1984). A review of the genera of the Iravadiidae (Gastropoda: Rissoacea) with an assessment of the relationships of the family. *Malacologia* **25** (1): 21-71.

FAMILY **RISSOIDAE**

The rissoids are a group of small to minute epifaunal marine snails. Species may be found from the littoral zone to the abyssal deeps, but they are most diverse in the shallows where they live on algae and hard substrates.

Rissoids themselves feed on the micro-algal film growing on their substrate. As you would expect with such a large and diverse family, the shells and anatomy of rissoids are varied. Head tentacles are long to short, often with a tuft of cilia at their tips, and with eyes at the bases. The foot typically has a metapodial tentacle. Females lay lens-shaped, oval or spherical capsules containing one or many eggs; development is direct or there is a planktotrophic larval stage. The shells are elongate-ovate to conical, smooth or sculptured. There is a horny, few-whorled, oval or subcircular operculum which often has a peg. The radula is taenioglossate. Two

subfamilies and 31 genera are recognised. Ponder (1985) provides a key to the genera.

Reference

Ponder, W. F. (1985). A review of the Genera of the Rissoidae. *Records Australian Museum, Supp.* **4**: 1-221.

FAMILY **TORNIDAE**

Several marine genera are assigned to this family including the Australian *Naricava* Hedley, 1913 (*angasi* Adams). Hedley cited the following additional Australian species to that genus: *vincentiana* Angas, 1880; *angulata* Hedley, 1905; *kimberi* Verco, 1907. Cotton & Godfrey (1932) added *flindersi*.

FAMILY **TRUNCATELLIDAE**

Marine and estuarine snails, usually in supralittoral habitats. Shells ovate to elongate-cylindrical, smooth or with strong axial sculpture, usually with the apex lost in adults; operculum horny, with few whorls. Gills reduced, with two long tentacles with eyes at the base.

The family is represented on both the tropical and temperate shores of Australia by the genus *Truncatella* Risso, 1826 (*subcylindrica* Linnaeus). Common temperate species are: *brazieri* Cox, 1868 from NSW; *scalarina* Cox, 1868 and *vincentiana* Cotton, 1942 from SA and southern WA. In his Qld list Hedley included *ferruginea* Cox, 1868; *teres* Pfeiffer, 1857 and *yorkensis* Cox, 1868.

FAMILY **VITRINELLIDAE**

Minute, umbilicate, usually low-spired and flattened shells with a glassy appearance, at least in life. The operculum is horny, multispiral and circular.

There are many genera but there has been no review. Several Australian genera are assigned here including *Pseudoliotia* Tate, 1893 (*angasi* Crosse, 1864; *micans* Adams, 1850, type species; *speciosa* Angas, 1871) and *Callomphala* Adams & Angas, 1864 (*alta* Laseron, 1984; *lucida* Adams & Angas, 1864, type species).

STROMBOIDEA

THREE families are classified together in this superfamily but only the largest of them, the Strombidae, has living representatives in the Australian fauna. The Struthiolariidae has a single living genus, *Struthiolaria*, which is endemic to New Zealand, while the Aporrhaiidae, also with a single living genus, is a northern hemisphere group.

FAMILY **STROMBIDAE**

Strombs

The strombs are a tropical and subtropical family with five living genera. The largest of these is *Strombus*, a group with colourful shells and about 25 species in northern Australia, most of which are widespread in the Indo-West Pacific region. One species of *Strombus* extends well south into the temperate waters of western and eastern Australia.

Strombs are herbivorous animals, feeding on algae or detrital material on muddy, sandy or rubble substrates. They tend to be gregarious and patchily abundant where conditions suit them. The egg masses are long jelly tubes, usually covered with sand grains and coiled, looped or twisted into knotted tangles. The eggs are tiny and numerous and the larvae hatch as planktotrophic veligers.

Shells of adult *Strombus* and *Lambis* have a flaring or thickened outer lip with a deep, U-shaped notch near the anterior end. The notch is used as a peep hole for the right eye. The remarkable eyes of strombs are large and located on long stalks that may be swivelled about. The columella of stromb shells is smooth and the siphonal canal is short to moderately long and may be upturned. There is a long, serrate, horny operulum which the animal uses as a weapon of defence and as a lever to push itself along or flip itself over when it has been rolled onto its back.

The small, related genera *Tibia* and *Rimella* are found in deeper water. They are also confined to the Indo-West Pacific region. A few specimens of *Tibia powisi* are said to have been trawled from deep water off North Reef, Qld but specimens are not available for this study. One species of *Rimella* is found on the north coast of WA. In both genera the shells are tall-spired with a long to extremely long, rather straight siphonal canal.

Terebellum is the fifth genus in the group. Some authors recognise a separate family, the Seraphidae, based on the closely related fossil genus *Seraphs* Montfort, and including the living Indo-West Pacific *Terebellum*. Although there are many available species names for the living members of the genus, Jung & Abbott (1967) admit only one widespread, extremely variable, living species. The shells are glossy, slender and bullet-shaped, without any spines or sculpture.

References

Abbott, R. Tucker (1960). The genus *Strombus* in the Indo-Pacific. *Indo-Pacific Mollusca* **1** (2): 33-144, Pls 11-117.

Abbott, R. Tucker (1961). The genus *Lambis* in the Indo-Pacific. *Indo-Pacific Mollusca* **1** (3): 147-174, Pls 118-134.

Jung, P. & Abbott, R. Tucker (1967). The genus *Terebellum* (Gastropoda: Strombidae). *Indo-Pacific Mollusca* **1** (7): 445-454, Pls 318-327.

▲ GENUS **LAMBIS** Röding, 1798
Strombus lambis Linnaeus

Large shells with flaring outer lip bearing prominent spines; siphonal canal long, recurved; posterior mantle filament much reduced or absent.

Indo-West Pacific. There are three subgenera and nine living species, five of them occurring in Australian waters. Commonly known as spider and scorpion shells. They are like strombs in most respects but distinguished by the spines and extended siphonal canal. They live among coral rubble and algae on intertidal reefs and in the sublittoral. Synonyms: *Pterocera* Lamarck, 1799 (*lambis*); *Heptadactylus* Mörch, 1852 (*lambis*).

▲ SUBGENUS **LAMBIS** s.s.

Columella smooth, inner wall of aperture lacking spiral lirae; outer lip with seven spines.

Lambis crocata (Link, 1807)

Pl. 19; fig. 4 Similar to *L. lambis* but smaller, with longer spines; aperture orange and entirely smooth, shoulders of the spire whorls with a nodulose keel; second, third and fourth spines are hooked towards the posterior.

15 cm. Indo-West Pacific; Rowley Shoals, WA to north Qld. Synonyms: *aculeatus* Perry, 1811; *aurantia* Lamarck, 1822; *aurantiacum* Sowerby, 1825.

Lambis lambis (Linnaeus, 1758)

Pl. 19; fig. 1 a-b Spire moderately high, pointed; stromboid notch deep; body whorl rough with spiral striae and heavy shoulder nodules; shoulders of the spire whorls sharply angulate; columella and ventral surface of the body whorl glazed. Exterior colour variable, white-cream with brown or bluish patches; interior rich pink, orange or purple-tan. (White morphs are common at Cocos Keeling.)

20 cm. Indo-West Pacific; Barrow I., WA to southern Qld. This species is perhaps the best known of the "spider shells". It is moderately common throughout its very wide range. Sexual dimorphism is pronounced in most populations. Males tend to be smaller and their three lower spines may be short and hooked. Females usually have longer, upwardly curved spines and large shoulder knobs. Synonyms: *lobata, cerea, hermaphrodita, laciniata, maculata* and *lamboides* all of Röding, 1798.

Lambis truncata (Humphrey, 1786)

Pl. 19; fig. 6 Larger than *L. lambis*; spire high in the Australian forms and strongly nodulose; body whorl spirally ribbed, shoulder nodules weak; posterior canal forming a shallow groove across the parietal callus to the base of the

posterior spine; outer lip much thickened, ventral side of the body whorl heavily calloused. Cream, sometimes brown speckled; deep interior white, lip and ventral callus creamy pink, sometimes nearly purple at the edges of the aperture.

35 cm. Indo-West Pacific; Rowley Shoals, Scott Reef, WA to Qld. The form found in Australian waters is the subspecies *L.t.sebae* Kiener, 1843, which has a pointed apex. The nominate subspecies is found in the western Indian Ocean (excluding the Red Sea) and in that form the spire apex is truncate, not pointed. Synonyms: *bryonia* Gmelin, 1791; *davilae* and *radix* Röding, 1798; *bengalina* Grateloup, 1840; *sowerbyi* Mörch, 1872.

▲ SUBGENUS **MILLEPES** Mörch, 1852
Lambis millepeda Linnaeus

Aperture with pronounced spiral lirae and 6-10 lip spines.

Indo-West Pacific. There are five species but only one is common in Australia and another doubtfully occurring here.

Lambis millepeda (Linnaeus, 1758)

Pl. 19; fig. 3 Outer lip with 10 short spines, including the twisted anterior canal, central and anterior spines hooked posteriorly; early whorls carinate and nodulose at the shoulders, body whorl finely spirally striate, with nodulose shoulder and two central spiral rows of smaller nodules; columella thick and lirate anteriorly. Exterior fawn or rosy cream; aperture mauve-brown with white lirae, white internally.

14 cm. Central Indo-West Pacific. So far not recorded in Australian waters but the species is common in the Lesser Sunda Is. of Indonesia and might be expected to occur on the Australian reefs of the Sahul Shelf. The illustrated specimen is from eastern Java and is included here for reference.

Lambis scorpius (Linnaeus, 1758)

Pl. 19; fig. 2 a-b Outer lip with seven spines, including the long anterior canal, spines flattened and nodulose; spire whorls weakly carinate, body whorl spirally corded and with moderately heavy shoulder nodules and two central spiral rows of smaller nodules; outer lip dentate marginally, spirally lirate deep within, columellar lip strongly lirate. Exterior cream or fawn with brown patches; aperture and columella purple with white lirae; outer lip orange, sometimes brown at the edge.

17 cm. Indo-West Pacific; north Qld. The Australian form is typical of the species and is characterised by a flat, ear-like lobe on the left side of the rear spine. This character is lacking in western Indian Ocean specimens which are given subspecific status as *L. s. indomaris* Abbott, 1961. Synonym: *nodosa* Lamarck, 1816.

▲ SUBGENUS **HARPAGO** Mörch, 1852

Strombus chiragra Linnaeus

Aperture spirally lirate; with six very strong spines, the anterior and posterior spines (i.e. the two canals) recurved to the left side.

Indo-West Pacific. Monotypic.

Lambis chiragra (Linnaeus, 1758)

Pl. 19; fig. 5 a-b   Body whorl rough with heavy nodules on the shoulders and spiral rows of smaller nodules anteriorly; posterior canal a deep, narrow, winding channel almost closed by folds of parietal and labial walls. White with brown patches and flecks, aperture rose-pink or orange, parietal wall glazed, orange-violet or cream.

25 cm. Indo-West Pacific; North West Cape, WA to north Qld. Male specimens are usually smaller and have stronger lirations on the columellar side of the aperture. The Australian form is of the nominate subspecies. Another subspecies, *L. c. arthritica* Röding, 1798, is found in the western Indian Ocean. Synonyms: *harpago* and *undulata* Röding, 1798; *rugosum* Sowerby, 1842; *kochii* Freyer, 1855.

▲ GENUS **STROMBUS** Linnaeus, 1758

Strombus pugilis Linnaeus

Outer lip thick, flaring, with a deep U-shaped notch anteriorly; spire whorls varicose; operculum horny, elongate and pointed, serrate on one side.

Tropical and subtropical waters of the Atlantic, Pacific and Indian Oceans. Several subgenera are recognised, most of them represented in the Australian fauna. The subgenus *Strombus* s.s. is an exception, being confined to the Atlantic and Eastern Pacific.

▲ SUBGENUS **CANARIUM** Schumacher, 1817

Strombus urceus Linnaeus

Rather small, fusiform to ovate; outer lip neither flaring nor winged; operculum with 5-9 well-developed serrations.

Indo-West Pacific. There are 18 living species in this group and a wide range of shell form. Synonyms: *Strombidea* Swainson, 1840 (*urceus* Linnaeus); *Canarium* Jousseaume, 1888 (*mutabilis* Swainson).

Strombus dentatus Linnaeus, 1758

Pl. 21; fig. 14 a-b   Elongate, with a tall, pointed spire and rounded, weakly nodulose shoulders; aperture short, the outer lip meeting the parietal wall well below the suture; whorls glossy smooth, weakly axially plicate; outer lip bearing 3-4 pointed teeth anteriorly which obscure the notch;

columella weakly lirate at each end. Colour variable, usually white, cream or shades of pink or rose, maculated with brown; aperture yellow or rose, outer wall purple-brown with white lirae.

6 cm. Indo-West Pacific; Scott Reef, WA to north Qld. This is a coral reef species which is never abundant. Synonyms: *tridentatus* Gmelin, 1791; *samar* Dillwyn, 1817; *samarensis* Reeve, 1851.

Strombus erythrinus Dillwyn, 1817

Pl. 21; fig. 9   Elongate with a high, shouldered and nodulose spire; body whorl axially ribbed and spirally corded; aperture and inner edge of the columella finely lirate. Exterior white or cream, mottled or banded with brown; aperture dark purple-brown, outer half of the columella white.

5 cm. Indo-West Pacific; north Qld. Synonyms: *elegans* Sowerby, 1842; *radians* Duclos, 1844; *ruppelli* Reeve, 1850; *rubicunda* Perry, 1811; *haemastoma* Sowerby, 1853.

Strombus fragilis (Röding, 1798)

Pl. 21; fig. 13 a-b   Thin-shelled, spire moderately tall, with smooth convex whorls; body whorl almost without shoulders, very weakly nodulose below the suture, finely spirally striate anteriorly; outer lip meeting the parietal wall well below the suture, notch broad and shallow; aperture finely lirate near the lip, columella smooth. Cream with pale tan bands and blotches, becoming dark at the lip and anterior end; aperture and columella dark brown.

5 cm. Western Pacific; north Qld. A coral reef inhabitant which is rather uncommon. Synonym: *bulbulus* Sowerby, 1842.

Strombus haemastoma Sowerby, 1842

Pl. 21; fig. 6   Small, stout, elongate-ovate; spire height equal to or greater than aperture length, with 12-14 former varixes; spire whorls sculptured with fine spiral threads and axial cords, subsutural thread weak or absent; shoulders slightly angular, bearing 4-5 prominent tubercles, a second row of tubercles present around the centre of the body whorl; lip varix swollen, threaded; columella narrow, thickened, slightly concave, lirate; inner side of the outer lip lirate. White with pale yellow blotches, brown to violet on the outer lip and columella, apex usually violet.

2 cm. Indo-West Pacific; recorded from Cartier I., NT. Although this little stromb is widespread it is rare, at least in collections, and the figured specimen is the only one this author has seen from an Australian locality. Its nearest relative seems to be *S. helli* Kiener, 1843 from Hawaii.

Strombus labiatus (Röding, 1798)

Pl. 21; fig. 3 a-b   Spire moderately high, shouldered and nodulose; body whorl with prominent elongate nodules on the shoulders and weak spiral lines anteriorly and near the lip;

shoulder nodules becoming axial ribs on the ventral side; inner surface of the outer lip with many fine spiral lirae; columella calloused and crossed by fine lirae throughout its length. External colour variable, usually white, green or grey-blue, with brown or orange bands or patches, inner side of outer lip violet or orange, columella yellow or orange.

5 cm. Central Indo-West Pacific; eastern Qld. *S. urceus* resembles this species but lacks lirae at the centre of the columella and axial ribs on the ventral side of the body whorl. The subspecies *S. l. olydius* Duclos, 1844 comes from the western Indian Ocean. Synonyms: *otiolum* Iredale, 1931.

Strombus microurceus (Kira, 1959)

Pl. 21; fig. 4 a-b   Small, subquadrate, spire moderately low; early spire whorls with weak axially elongate nodules, finely incised spiral lines and 3-5 former varixes, suture with a spiral cord; body whorl smooth, with 3-6 blunt tubercles at the shoulder, the last two large and located distant from the outer lip; columella slightly concave, smooth except for a few anterior and posterior lirae. Cream to white background, overlain with brown or yellow blotches and reticulate marks and sometimes spiral bands and lines, anterior end bluish brown on the back; aperture dark brown near the lips, yellow and brown striped deep within, columella yellow to orange on the outer side, dark purplish brown on the inner side.

2.8 cm. Central Indo-West Pacific and Western Pacific; northern Australia from Ashmore I., NT to Hayman I., Qld. Also recorded from Lord Howe I. The shape of this shell is like that of *S. mutabilis* but the colouring is distinctive, particularly that of the columella and the dark patch on the back at the anterior end.

Strombus mutabilis Swainson, 1821

Pl. 21; fig. 8 a-b   Solid, with a rather low spire and broad nodulose shoulders; body whorl smooth except for spiral striae anteriorly and sometimes a weak spiral cord around the centre; aperture finely lirate; columella calloused and crossed by well-marked lirae throughout its length. External colour extremely variable, usually cream with patches, spots or bands of yellow, brown, black or orange; interior usually pink.

4 cm. Indo-West Pacific; Cape Leeuwin, WA to central NSW. Usually abundant where it occurs on sandy and rubble areas of coral and rocky reefs in the intertidal and shallow sublittoral zones. This is one of the most widespread of the Indo-West Pacific molluscs, extending well into the temperate zones. Synonyms: *floridus* Lamarck, 1822; *epimellus* Duclos, 1844; *flosculosus* Mörch, 1852; *flammeus* Link, 1876; *floridus zebriolata* Adam & Leloup, 1938.

Strombus terebellatus Sowerby, 1842

Pl. 21; fig. 15   Thin-shelled, elongate; spire tall, with smooth rounded whorls which lack shoulders, weak spiral cords present anteriorly; outer lip thin, meets the parietal wall well below the suture, anterior notch hardly evident at all;

columella weakly calloused, smooth; aperture smooth. Spire tip pink; whorls cream or light tan to orange, with mid-tan blotches; aperture same colour as the exterior.

5 cm. Indo-West Pacific; north Qld. The nominate subspecies is confined to the western Pacific but a subspecies *S. t. afrobellatus* Abbot, 1960 occurs in the western Indian Ocean. In form this uncommon species resembles *Terebellum* but Abbott, argues that it is not related to that genus. It is more slender than *S. fragilis*, and lacks the dark and calloused columella typical of that species.

Strombus urceus (Linnaeus, 1758)

Pl. 21; fig. 5 a-c   Solid with a high turreted spire; body whorl finely corded anteriorly and near the lip; axial folds sometimes present on the ventral side; aperture finely lirate; columella calloused and lirate at the anterior and posterior ends. Exterior brown, white, cream or greenish with darker spiral lines, bands or blotches; deep interior white, purple or almost black, often becoming yellow towards the lip and on the columella; columellar and outer lip margins white; anterior canal tip black.

5 cm. The nominate form is restricted to the central Indo-West Pacific region. The Australian subspecies is *S. orrae* Abbott, 1960 which is common from North West Cape, WA to the Gulf of Carpentaria. It is more quadrate than *S.u. urceus*, has a shorter anterior canal, and fewer but larger shoulder nodules. Synonyms: *reticulata* Link, 1807; *ustulatum* Schumacher, 1817; *incisus* Wood, 1828; *anatellus* Duclos, 1844; *muricatus* Watson, 1885; *ustulatus laevis* Dodge, 1946.

Strombus wilsoni Abbott, 1967

Pl. 21; fig. 7   Resembles *S. mutabilis* but with strong axial riblets on the spire whorls, 5-7 axially-elongate nodules on the shoulder of the body whorl, columella smooth at the centre; aperture tan not pink.

2.5 cm. Indo-West Pacific; northern WA. The figured paratype came from the Dampier Arch., WA. The holotype came from Zanzibar. The species has been recorded also from Fiji but not yet from Qld.

▲ SUBGENUS **CONOMUREX** Fischer, 1884
Strombus luhuanus Linnaeus

Conical, spire low, aperture little flared.

Indo-West Pacific. There are two species.

Strombus luhuanus Linnaeus, 1758

Pl. 20; fig. 5 a-b   Outer lip slightly inturned instead of flaring; body whorl smooth, spire whorls with axial ribs; anterior canal a deep siphonal notch; periostracum thick, rough, green-brown; aperture smooth, rich orange or red; columella black-brown.

7 cm. Central Indo-West Pacific; Scott Reef, WA to Qld and northern NSW. Abundant among seagrass in intertidal and shallow sublittoral habitats of coral reefs. The western Indian Ocean species *S. decorus* (Röding, 1798) is closely related. Synonyms: *luguanus* Herbst, 1788; *pusillus* Anton, 1839.

▲ SUBGENUS **DOLOMENA** Iredale, 1931
Strombus plicatus pulchellus Reeve

Small to moderate size, with expanded, flaring lip and deep stromboid notch.

Indo-West Pacific. May be a synonym of *Labiostrombus* as the distinguishing characters are minor. There are six living species, each of which has geographically separated subspecies.

Strombus dilatatus Swainson, 1821

Pl. 20; fig. 6 a-c   Spire moderately high; body whorl glossy, sculptured with minute spiral striae and weak shoulder nodules; spire whorls axially plicate; aperture strongly lirate but the calloused columella bears only a few obsolete lirae at the ends; posterior canal groove deep, curves across the ventral surface of the spire. Exterior cream with pale brown blotches and bands, interior and columella white except for a purple-brown patch deep within the aperture.

6 cm. Central Indo-West Pacific; Carnarvon, WA to north Qld. Synonym: *orosminus* Duclos, 1844.

Strombus plicatus (Röding, 1798)

Pl. 21; fig. 11 a-b   Spire long, pointed, spire whorls axially ribbed and strongly shouldered; body whorl axially ribbed on the left and ventral sides; outer lip wide and flaring centrally, attaching to the penultimate whorl forming a shallow posterior canal; strongly lirate on its inner side, lirae bifurcate, columella calloused, smooth at the centre but lirate toward the ends. Cream or pale brown, faintly banded; interior purple-brown, columella white or orange.

5 cm. Indo-West Pacific; Dampier Arch., WA to north Qld. The Australian form is the subspecies *S. p. pulchellus* Reeve, 1851. The nominate form and 2 other subspecies occur in the western Indian Ocean. Synonym: *malekulensis* Abrard, 1946.

Strombus variabilis Swainson, 1820

Pl. 21; fig. 10 a-b   Rather thin, with a moderately high, shouldered, nodulose spire, and a widely flaring outer lip notched posteriorly as well as anteriorly; body whorl smooth except for spiral striae anteriorly and shoulder nodules that become elongate on the ventral side; aperture and calloused columella smooth. Exterior white with 5 or 6 spiral bands of wavy brown lines or patches, interior glossy white; often with a conspicuous brown patch on the columella.

6 cm. Common. Indo-West Pacific; NT to Qld. The absence of internal lirae distinguishes this species from *S. vittatus* and *S. campbelli* which it resembles superficially. Synonym: *lituratus* Menke, 1829.

▲ SUBGENUS **DOXANDER** Iredale, 1931
Strombus vittatus Linnaeus

Spire tall and pointed: outer lip flared.

Indo-West Pacific. There are 3 living species and several fossil species.

Strombus campbelli Griffith & Pidgeon, 1834

Pl. 20; fig. 3 a-c   Spire high, spire whorls flat-sided with a conspicuous subsutural spiral groove and flat axial ribs; body whorl smooth except for spiral grooves anteriorly and sometimes weak axially elongate shoulder nodules; aperture smooth or weakly spirally lirate on the outer wall, columella smooth but calloused. Exterior usually white with spiral and axial brown bands or zigzag lines, pink and mauve forms are common; interior white.

7 cm. Abrolhos, WA to northern NSW. Some authors treat this as a subspecies of *S. vittatus* but both forms occur in Qld.

Strombus vittatus Linnaeus, 1758

Pl. 20; fig. 2   Spire very high; body whorl smooth except for a number of spiral striae anteriorly, spire whorls bear prominent axial ribs; weak lirae present in the aperture; columella slightly calloused and smooth. Exterior pale yellow-brown, sometimes with flecked darker spiral bands or faint zigzag lines, interior glossy white.

8.5 cm. Central Indo-West Pacific; north Qld. Synonyms: *sulcatus* Holten, 1802; *australis* Schröter, 1805; *turritus* Lamarck, 1822.

▲ SUBGENUS **EUPROTOMUS** Gill, 1870
Strombus aurisdianae Linnaeus

Medium-sized, spire high, characterised by a prong-like posterior extension of the outer lip and smooth columella.

Indo-West Pacific. There are three living species. The generic name *Monodactylus* "Klein" Hermannsen has been sometimes misapplied to this genus.

Strombus aurisdianae Linnaeus, 1758

Pl. 20; fig. 7 a-b   Spire and body whorl heavily nodulose at the shoulders, with many fine spiral ribs on the body whorl, plus one or two stronger nodulose ribs; anterior canal long and curved upwards at an angle of about 75°; weak posterior lirae present in the aperture. Ventral side of the body whorl and the lower spire whorls glazed dark brown or orange; exterior mottled cream, brown and blue-grey; interior rich orange.

8 cm. Indo-West Pacific; north Qld. The typical form is a coral reef inhabitant. An elongate form with less recurved siphonal canal, smaller knobs, and darker orange-brown aperture occurs in coastal waters of Qld and has been known as *S. aratrum* (Röding, 1798) (fig. 7a). Some authors regard it as a subspecies of *aurisdianae* but it is treated here as only a variant. Synonyms: *aratrum*, *buris* and *stiva* Röding, 1798; *lamarckii* Gray, 1842; *chrysostomus* Kuroda, 1942.

Strombus bulla (Röding, 1798)

Pl. 20; fig. 9 a-b   Spire whorls finely cancellate but usually this is completely covered by a thick glaze spreading from the parietal wall and ventral side; body whorl with a single row of nodules at the shoulder, several obsolete central spiral cords and weak spiral threads anteriorly, otherwise glossy and smooth; posterior lip projection rather flat, straight, anterior canal upturned at 80-90°; aperture smooth, columella smooth and thickly calloused. Exterior fawn with fawn to mauve bands appearing toward the lip; columella and spire glaze white, sometimes mauve at the ends; interior orange.

7 cm. Western Pacific; Ashmore Reef, NT to north Qld. Not abundant; lives on sandy substrates near coral reefs, usually in deeper water. Synonyms: *laevis* Perry, 1811; *guttatus* "Martini" Kiener, 1843.

Strombus vomer Röding, 1798

Pl. 20; fig. 8 a-b   Spire high, with sharp shoulder nodules; body whorl heavily nodulose at the shoulder and elsewhere spirally ribbed, ribs nodulose; anterior canal almost vertical, posterior lip projection inward curving; aperture strongly lirate on the outer lip side, anterior end of the columella usually weakly lirate. Exterior cream or grey, mottled with brown; interior yellowish or off-white, sometimes becoming orange or dark brown near the lip edge and on the columella.

8 cm. Indo-West Pacific. The Australian form is the subspecies *S. v. iredalei* Abbott, 1960 which is found from Fremantle, WA to the NT; the nominate subspecies comes from southern Japan and New Caledonia. Moderately common in sandy substrates and often found in the intertidal zone. Distinguished from *S. aurisdianae* by the lirate aperture, shorter anterior canal, and shorter posterior lip prong. Synonyms: *acutus* Perry, 1811; *zelandiae* Gray, 1826; *pacificus* Swainson, 1821; *chemnitzii* Pfeiffer, 1840; *novae zelandiae* "Chemnitz" Reeve, 1842; *donnellyi* Iredale, 1931; *hirasei* Kuroda, 1942.

▲ SUBGENUS GIBBERULUS Jousseaume, 1888
Strombus gibberulus Linnaeus

Shell asymmetrically coiled; spire moderately high and varicate; anterior end attenuate and twisted.

Indo-West Pacific. Monotypic.

Strombus gibberulus Linnaeus, 1758

Pl. 21; fig. 12 a-b   Outer lip only slightly thickened and flared; penultimate whorl characteristically swollen and distorted; body whorl smooth and curiously flattened; spire whorls bearing fine spiral striae and white varixes; aperture lirate, columella calloused, smooth; exterior white with spiral bands of brown blotches, interior white, or tinted with violet, brown or yellow; columella often with an elongate brown patch.

6 cm. Indo-West Pacific. The Australian form, found from Exmouth Gulf to southern Qld, is *S. g. gibbosus* (Röding, 1798) which is distributed in the Western Pacific. The nominate form occurs in the Indian Ocean, apparently including Cocos Keeling. However, the WA shells are quite like their Qld cousins and seem typical of *gibbosus*. Abundant in intertidal and shallow sublittoral.

▲ SUBGENUS LABIOSTROMBUS Oostingh, 1925
Strombus epidromis Linnaeus

Outer lip rising evenly, attaching to the base of the penultimate whorl; posterior canal lacking or only weakly developed; outer lip widely flaring.

Central Indo-West Pacific. There is a single living species but several fossil species in the region. Synonym: *Gallinula* "Klein" Hermannsen, 1847 (not Brisson, 1760).

Strombus epidromis Linnaeus, 1758

Pl. 20; fig. 4   Rather light for its size, with a moderately high spire; outer lip slightly thickened and widely flaring; body whorl smooth except for weak spiral lines anteriorly and a few low, elongate nodules on the shoulders; spire whorls spirally striate and with small, distinct shoulder nodules; aperture smooth, columella calloused. Exterior white or cream, with faint flecks of yellow-brown, interior glossy white.

9 cm. Central Indo-West Pacific; North West Cape, WA to north Qld. Synonym: *expansa* "Martini" Tryon, 1885.

▲ SUBGENUS LAEVISTROMBUS Kira, 1955
Strombus canarium Linnaeus

Rotund, smooth; outer lip thick, smooth and flared; columella smooth.

Indo-West Pacific. There is only one living species but there are several fossil species in India and Indonesia.

Strombus canarium Linnaeus, 1758

Pl. 21; fig. 2 a-b   Heavy, broad, with angular shoulders and a widely flaring outer lip; spire moderately high and tapering to a sharp point; spire whorls weakly ribbed and varicate;

body whorl smooth except for a few striae anteriorly; aperture and columella smooth. Exterior white, grey or yellow-brown usually with a reticulate pattern of brown lines; interior white.

9 cm. Central Indo-West Pacific; Qld. Abundant in somewhat muddy sand in shallow water. Fished commercially for food in many parts of South East Asia. Synonyms: *turturella* Röding, 1798; *isabella* Lamarck, 1822; *vanikorensis* and *taeniatus* Quoy & Gaimard, 1834; *gibbus* Issel & Canefri, 1876.

▲ SUBGENUS LENTIGO Jousseaume, 1886
Strombus lentiginosus Linnaeus

Large, thick-shelled, somewhat quadrate, with 3-4 rows of knobs on the body whorl; operculum weakly serrate.

Indo-West Pacific, eastern Pacific, west Africa. There are five living species (two in Australia) and a fossil record as far back as the Lower Miocene.

Strombus lentiginosus Linnaeus, 1758

Pl. 20; fig. 11 a-b Solid, with large protruding nodules on the shoulders, especially on the body whorl, and spiral rows of smaller nodules elsewhere on the body whorl; outer lip only slightly flaring, anterior canal short, posterior canal represented by a deep channel along the side of the spire; aperture and columella smooth, columella projecting anteriorly. Exterior white with large chestnut spots in front of shoulder nodules, mottled with green-grey or grey-brown; interior orange, becoming cream toward the margins; ventral surface and columella glazed.

10 cm. Indo-West Pacific; Dampier Archipelago, WA to north Qld. An inhabitant of coral reefs and shallow water. Synonym: *rana* Röding, 1798.

Strombus pipus (Röding, 1798)

Pl. 21; fig. 1 a-b Outer lip thickened but only slightly flared; heavy shoulder nodules and several spiral rows of smaller nodules present on the body whorl; spire low, spire whorls nodulose; inner side of the outer lip strongly lirate, columella smooth and calloused, ventral surface of the body whorl heavily glazed. Exterior white or cream with orange-brown patches, interior dark purple; columella cream.

7 cm. Indo-West Pacific; Ashmore Reef, NT to north Qld. Synonym: *papilio* "Chemnitz" Dillwyn, 1817; *exustus* Swainson, 1822; *adustus* "Swainson" Reeve, 1851.

▲ SUBGENUS TRICORNIS Jousseaume, 1886
S. tricornis Humphrey

Large, heavy shells with prominent knobs, smooth columella and a rather thick periostracum.

Tropical Atlantic, Pacific and Indian Oceans. There are 12 living species, including three in Australian waters.

Synonyms: *Lobatus* "Swainson" Iredale, 1921 (*bituberculatus* Lamarck); *Aliger* Thiele, 1929 (*gallus* Linnaeus); *Eustrombus* Wenz, 1940 (*gigas*).

Strombus latissimus Linnaeus, 1758

Pl. 20; fig. 10 Outer lip massive and widely flaring, including posteriorly where it forms a wide flat plate covering the ventral side of the spire; stromboid notch deep; spire whorls carinate and nodulose, body whorl with one very large but low and rounded knob on the shoulder, spirally ridged toward the lip; aperture smooth, columella thickened, smooth. Exterior cream, densely mottled and axially streaked with brown; aperture white, becoming pinkish tan towards the lip; columella glazed, pinkish tan.

20 cm. Western Pacific. Scott Reef, WA. This is the largest and heaviest species of *Strombus* in the Indo-West Pacific Region. At Scott Reef it is found among sand and rubble in deeper water. Synonyms: *picta* Röding, 1798; *alata* Schumacher, 1817.

Strombus sinuatus Humphrey, 1786

Pl. 20; fig. 1 Rather flat-based and high-spired; outer lip rather thin, with flat, projecting blades at its posterior edge; spire whorls with heavy shoulder nodules, body whorl smooth except for a high central dorsal knob and two other smaller shoulder nodules; aperture smooth. Exterior cream with orange-brown irregular patches which become spiral bands near the lip; interior rich purple-brown becoming pink or orange near the margin of the aperture.

13 cm. Central Indo-West Pacific; Ashmore Reef, NT and north Qld. Reported from North West Cape, WA but this needs confirmation. Synonyms: *palmata* Fischer, 1807; *laciniatus* Dillwyn, 1817; *cristatus* lamarck, 1822.

Strombus thersites Swainson, 1823

Pl. 20; fig. 12 Massive, with a tall, pointed spire; outer lip thick, moderately flaring, with a deep stromboid notch and a shallow posterior groove turned across to the spire where it meets the penultimate whorl; body whorl with one large and two small knobs, otherwise lacking sculpture except for weak spiral cords; columella, aperture, and outer lip smooth; periostracum thin. White, with sparse, zigzag yellow-brown or rose streaks; aperture and columella white, often glazed.

14 cm. Western Pacific; recorded from north Qld but very rare. Information is needed on its geographic range and habitat in Australia. The illustrated specimen comes from the Cook Is. Synonym: *ponderosus* Philippi, 1842.

▲ GENUS RIMELLA Agassiz, 1840
Rostellaria fissurella Lamarck = *cancellata* Lamarck

Fusiform, with high spire, rounded whorls and cancellate sculpture; outer lip only slightly thickened, forming a conspicuous posterior canal where it meets the parietal wall; stromboid notch lacking; columella smooth, calloused; anterior canal short, attenuate.

Indo-West Pacific. Some authors treat this group as a subgenus of *Tibia*.

Rimella cancellata (Lamarck, 1816)

Pl. 21; fig. 18 a-b   Spire high and attenuate; anterior canal projecting forward almost horizontally; sculpture strong, consisting of axial folds with fine spiral riblets in the interspaces, and usually two varixes on each of the spire whorls; posterior canal groove long and running backwards over two or more of the spire whorls on the right side; inner edge of the outer lip lirate. Exterior mauve or yellow-brown between the white or pale yellow axial folds, with narrow yellow-brown bands; aperture mauve; columella white.

3.5 cm. Central Indo-West Pacific; Exmouth Gulf to Broome, WA. Uncommon but occasionally found in sandy habitats of the intertidal and sublittoral zones.

Rimella crispata (Sowerby, 1842)

Pl. 21; fig. 17   Very like *R. cancellata* but less elongate, with a shorter, straighter anterior canal and less prominent spiral sculpture. Posterior canal curling upwards on the penultimate whorl, not crossing onto the earlier spire whorls.

2.5 cm. Central Indo-West Pacific. Strictly speaking this species is not recorded from Australian waters but has been dredged around the eastern Indonesian island of Aru (on muddy substrates) and is part of the fauna of the Australasian continental shelf.

▲ GENUS **TEREBELLUM** Röding, 1798
Conus terebellum Linnaeus

Elongate, cylindrical, glossy, sculpture lacking; spire short, aperture long, straight, narrow, anterior canal wide; stromboid notch lacking.

Indo-West Pacific. There is only one living species. Synonyms: *Terebrina* Rafinesque, 1815 (emendation of *Terebellum*); *Lucis* Gistel, 1848 (*subulatus*); *Artopoia* Gistel, 1848 (*subulatus*).

Terebellum terebellum (Linnaeus, 1758)

Pl. 21; fig. 16 a-c   Colour and pattern extremely variable, usually with a pale yellow, brown or cream background and red-brown bands, zigzag lines, spiral lines, dots or flecks.

7 cm. Indo-West Pacific; Abrolhos, WA to north Qld. The many colour forms have been named separately but a recent study has shown that they are all one species. The species is common, burrowing in sand, usually in deeper water. WA specimens are often very narrow and small. For a full synonymy see Jung and Abbott (1967, *loc. cit.*).

SUPERFAMILY
VANIKOROIDEA

THERE are two families assigned to this superfamily in the contemporary classification (Ponder & Warén, 1988, *Malacological Review Supp.*4: 299). The Fossaridae, once included here, are removed to the Cerithioidea.

FAMILY **HIPPONICIDAE**

Horse Hoof Limpets

These are conical, limpet-like snails, with a backward-pointing apex. They live fixed to their substrate of rocks or other shells. Their vernacular name comes from their horseshoe-shaped muscle scar. They have a long, extensible proboscis with which they reach out to obtain algae, detritus or the pseudofaeces of other molluscs for food. Females retain egg capsules resembling bunches of balloons within the mantle cavity and development is direct. Some species are thought to be protandrous hermaphrodites. There are two genera in Australia.

▲ GENUS **ANTISABIA** Iredale, 1937
Hipponix foliaceus Quoy & Gaimard

With the characters of the type species.

Southern Australia. There is some doubt whether this genus should be separated from *Hipponix* although the free-living habit supports the view that it deserves to be retained as a genus. The American species *Hipponix antiquatus* Linnaeus is very similar to *A. foliaceus*.

Antisabia foliacea (Quoy & Gaimard, 1835)

Cap-shaped but irregular with a posterior, overhanging, slightly recurved, noncoiled, usually eroded apex; sculptured with prominent, foliating laminae and radiating striae; periostracum thick and hairy. White or yellow.

2 cm. NSW to Geraldton, WA. Lives attached to the underside of stones; very common in the intertidal and sublittoral zones. Synonym: *erma* Cotton, 1938.

▲ GENUS **HIPPONIX** Defrance, 1819
Hipponix cornucopiae Defrance

Cap-shaped; apex originally minutely coiled but eroded away in adults, posterior, recurved; with rough radial ribs; usually with a furry periostracum.

Cosmopolitan. Several American species are assigned to this genus. In Australia there are probably two species. Knudson (1991 in Wells *et al.* The Marine Flora and Fauna of Albany, WA. *Western Australian Museum* 2: 641-660) discussed the taxonomy, anatomy and reproduction of the Australian species. Synonyms: *Sabia* Gray, 1847 (*conica*); *Amalthea* Schumacher, 1817; *Capulonix* Iredale, 1929 (*danieli* Crosse).

Hipponix australis (Lamarck, 1819)

Pl. 22; fig. 11 a-c Axial ribs flat and broad with narrow interspaces; adult shells often wrinkled and irregular. Exterior cream or white, often with brown lines between the ribs, margin brown; interior brown at the centre, usually white toward the margin.

2.5 cm. Southern Australia. A conspicuous feature of the southern Australian intertidal and shallow sublittoral fauna where it lives in groups on the shells of other gastropods. In northern Australia there is a similar species, probably *H. conicus* (Schumacher, 1817), in which the shells are smaller and more finely sculptured.

FAMILY **VANIKORIDAE**

Vanikoros

These animals live under stones or among rubble in the intertidal or shallow sublittoral zones. Little is known of their anatomy or life history. Most of the species are found in the tropical Indo-West Pacific Region. Current classifications admit only one genus.

▲ GENUS **VANIKORO** Quoy & Gaimard, 1832
Sigaretus cancellata Lamarck

Globose, turbinate; protoconch small; aperture large, outer lip simple; columella simple, smooth; usually narrowly umbilicate; early whorls with strong axial and spiral sculpture; operculum with few whorls, thin, horny.

Pantropical. Synonym: *Merria* Gray, 1839.

In addition to the species illustrated here the following names have been used in the Australian context: *foveolata* Souverbie & Montrouzier, 1866 from NSW; *deshayesiana* Récluz, 1843, *clathrata* Récluz, 1845, *gaimardi* Adams, 1853, and *quoyiana* Adams, 1853 from Qld; *expansa* Sowerby and *orbignyana* Récluz and *helicoidea* le Guillou, 1842 from northern WA and coastal Qld.

Vanikoro cancellata (Lamarck, 1822)

Pl. 22; fig. 16 a-b   Thin; sculptured with lamellate axial ribs and spiral threads, body whorl sometimes lacking axial ribs. White, periostracum yellow-brown.

2.5 cm. Indo-West Pacific; North West Cape, WA to central Qld. Hedley noted that Hermann, 1781 may be an earlier valid use of this name.

Vanikoro gueriniana Récluz, 1845

Pl. 22; fig. 15   Thick and solid, compressed, aperture wide, umbilicus wide and deep; sculpture of very strong curved axial rugae extending to the lip, prominent spiral threads in the interspaces. White.

1 cm. Torres St., Qld to Sydney, NSW. Synonym: *distans* Récluz, 1844.

Vanikoro sigaretiformis (Potiez & Michaud, 1838)

Like *V. cancellata* but with strong, oblique rugae on the early whorls which cease abruptly on the third whorl and are replaced by delicate spiral threads.

1.1 cm. NT to NSW. Synonym: *recluziana* Adams & Angas, 1864.

SUPERFAMILY
CALYPTRAEOIDEA

CONICAL to limpet-like shells characterised by a more or less central, sometimes slightly coiled apex. There is usually a bristly periostracum. The animals are sedentary, attaching to stones or other shells. They are ciliary feeders with very large gills that function as the filtering organ. At least some of the species are protandrous hermaphrodites, that is, they undergo change of sex from male to female with an amazing resorbtion of the male organs and replacement by female organs. There are two families.

FAMILY **CALYPTRAEIDAE**

Slipper Shells

The shells have an internal shelf-like septum which helps support the visceral organs. The females brood their young. There are five genera in the Australian fauna but few species.

▲ GENUS **CALYPTRAEA** Lamarck, 1799
Patella chinensis Linnaeus

Approximately circular, conical; apex high, coiled, central or eccentric; body whorl large; septum with an oblique, more or less straight margin and a twisted columella.

Cosmopolitan. The Australian representative of the group has been placed incorrectly in the genus *Sigapatella* Lesson, 1830 (*novaezelandiae* Lesson).

Calyptraea calyptraeformis Lamarck, 1822

Pl. 22; fig. 12 a-b   Depressed-conical, outline round, whorls convex, smooth, periostracum present. White or cream, apex sometimes violet; septum white.

2.5 cm. NSW to Fremantle, WA. Lives attached to stones and the inner sides of large dead shells.

▲ GENUS **CHEILEA** Modeer, 1793
Patella equestris Linnaeus

Conical, roughly circular or oval, apex subcentral not coiled; internal septum shaped like a half funnel projecting vertically from the apex.

Indo-West Pacific and southern Australian regions. The shells are extremely variable in form, tending to take on the shape of the uneven surface to which they are attached. The American genus *Crucibulum* Schumacher, 1817 is similar but the septum forms a complete funnel-shaped structure, giving that genus the vernacular name "cup and saucer shells".

Cheilea equestris (Linnaeus, 1758)

Pl. 22; fig. 3 a-b   Approximately round but irregular in plan view, apex minute, a little behind centre; upper surface rough, sometimes crumpled close to the apex, becoming radially striate toward the margin making the margin seem finely crenulate. White.

3 cm. Indo-West Pacific; Shark Bay, WA to southern Qld. Shape and sculpture is variable. Many names have been introduced for the various forms. Reeve (1858, *Conchologia Iconica* 11, *Calyptraea*) listed 33 "species" but it is impossible to determine which of these represent true species. The name *C. porosa* Reeve, 1858, which appears in the Iredale & McMichael NSW list, refers to a form in which the entire surface is honeycombed with wrinkles. In another common northern Australian form the entire upper surface is irregularly crumpled.

Cheilea occidua Cotton, 1935

Pl. 22; fig. 5   Relatively small, with very irregular outline, rough, irregular folds, and strong radial cords; apex anterior, the anterior slope steep and almost concave; posterior slope long and convex near the apex, becoming flat near the margin; apex blunt and smooth. White.

2.5 cm. Esperance to Fremantle, WA. Smaller and more irregular than the SA *C. flindersi*.

Cheilea flindersi Cotton, 1935

Pl. 22; fig. 4   Large, with strong, irregular growth lines and very many fine, flexuous, close-set radial striae; margin minutely dentate along the inner side. White.

5.3 cm. Port Phillip, Vic to SA. The radial striae on the holotype are much finer than in any of the tropical *C. equestris* examined and this may be a "good" species.

▲ GENUS **CREPIDULA** Lamarck, 1799
Patella fornicata Linnaeus

Low, top convex, apex coiled and posterior; internal septum horizontal.

Cosmopolitan. *Bostrycapulus* Olsson & Harbison, 1953 was introduced as a subgenus for *aculeata* and this may be an appropriate arrangement.

Crepidula aculeata (Gmelin, 1791)

Pl. 22; fig. 13 a-c   Depressed, ear-shaped to irregularly ovate; apex coiled; exterior with spiral rows of small scaly spines; margin simple; septum with a central axial ridge, concave on either side and a sinuous edge. Brown with a white spiral band on the right side; interior brown or white, margin brown, septum white.

2.5 cm. Pantropical. Shark Bay, WA around the northern coast and as far down the east coast as eastern Vic. This species, assuming that it is the same, is reported in American literature from North Carolina to Brazil in the Western Atlantic, and from central California to Chile in the Western Pacific. It lives in sandy or muddy habitats of the intertidal and shallow sublittoral zones, attached to other shells, stones or mangrove trunks. The taxonomy of the species needs careful examination.

▲ GENUS **ZEACRYPTA** Finlay, 1927
Calyptraea monoxyla Lesson

Very depressed and thin-shelled, ovate, approximately symmetrical; apex terminal, not coiled; margin of the septum straight or weakly sinuous.

Southern Australia and New Zealand. There are two species. The type species is from New Zealand.

Zeacrypta immersa (Angas, 1865)

Pl. 22; fig. 14 a-b   Apex tiny, median, overhanging the posterior end; very flat, upper surface convex or concave depending on whether the animal is attached to the inside or the outside of the host shell; margin thin and sharp; exterior smooth except for minute growth lines. Brown or fawn, sometimes very pale or with faint radial stripes; interior brown, septum white.

2.5 cm. NSW to Fremantle, WA. Lives on dead gastropod shells in the sublittoral zone, usually within the aperture.

▲ GENUS **CLYPEOLA** Gray, 1867
Sigapatella tenuis Gray

Round in outline, conically coiled, apex almost central; septum arcuate; false umbilicus lacking.

Southern Australia and New Zealand. Sometimes treated as a subgenus of the American *Sigapatella* Lesson. There is one Australian species and three in New Zealand. Synonym: *Zegalerus* Finlay, 1927 (*tenuis* Gray).

Clypeola hedleyi (Smith, 1915)

Whorls slightly convex, surface sculptured with weak, concentric growth lines; periostracum light brown, thin, concentrically lamellose. White to light brown.

6 mm. NSW to SA. This southern Australian species is very like the New Zealand *C. tenuis*.

FAMILY **CAPULIDAE**

Cap Limpets and Trichotropes

The Capulidae and Trichotropidae were once regarded as separate families, but are now put together with the former taking priority (Ponder & Warén, 1988, *Malacological Review Supp.* **4**: 300). They are small to minute, turbinate to limpet-like snails.

There are many genera. The group is poorly studied and the provisional classification used here follows Vaught, 1989. Iredale & McMichael (1962) introduced the family names Lippistidae and Siriidae for genera of this group but this has not been followed by other authors.

In the past the genus *Amathina* Gray, 1847 (*tricarinata* Linnaeus) has been assigned to either this family or the Hipponicidae, but Winston Ponder (1987, *Asian Marine Biology* **4**: 1-34) has shown that the anatomical features of the type species are like those of the Pyramidellaceans and that it is not a prosobranch.

▲ GENUS **CAPULUS** Montfort, 1810
Patella hungarica Montfort

Cap-shaped, apex coiled and usually overhanging the posterior end; upper surface smooth or radially ribbed.

Cosmopolitan. The animals live attached to shells of bivalves or gastropods, or sometimes to stones. There are several species in Australia which remain unidentified and may be undescribed. Synonym: *Pileopsis* Lamarck, 1812; *Capulonix* Iredale, 1929 (*calyptra* Martin = *danieli* Crosse); *Tenpetasus* Iredale, 1929 (*liberatus* Pease).

Capulus danieli (Crosse, 1858)

Pl. 22; fig. 7 Spire small, high, not prominently coiled but recurved beyond the posterior margin; sculptured with rough radial ribs; aperture approximately round, margin undulate, interior smooth. White with red-brown streaks; interior red-brown.

1.5 cm. Western Pacific; north Qld. The classification of this species remains uncertain. Iredale (1929, *Memoirs Qld Museum* 9: 277) wrote that it has " little to do with the true *Capulus*" and proposed the new generic name *Capulonix* for it, but this is hardly an adequate generic diagnosis and more study is needed before this is accepted.

Capulus devexus May, 1916

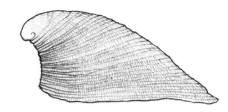

Also like *C. violaceus*; much higher than *C. devotus*, with a larger, smooth apical coil of about 2¹/₂ whorls; body whorl flaring at the margin, curiously flat on the anterior side which bears prominent, transverse growth ridges; mouth arched in lateral view. White.

2.2 mm. Tas. Dredged in about 100 m.

Capulus devotus Hedley, 1904

Like *C. violaceus* but much smaller and less laterally compressed; spire whorls rounded, smooth, with about two tightly and planispirally coiled turns, body whorl flaring out into a rounded, shallow and widely open cup. White.

6 mm. Cape Moreton, Qld to SA. *C. banksi* Cotton, 1939 may be a synonym.

Capulus liberatus Pease, 1868

Pl. 22; fig. 8 Spire whorls moderately large, strongly coiled on the right side, more or less elevated; aperture round, smooth within, margin finely crenulate; sculptured with scaly or granulose radial cords; covered with a finely tufted periostracum. White, cream or yellow; aperture white.

2 cm. South-west Pacific; Torres St. to Capricorn Group, Qld and Norfolk I. Iredale (1929) argued that the apex of this species is quite unlike that of *Capulus* and he introduced the generic name *Tenpetasus* for it, but the matter needs to be reviewed.

Capulus sycophanta Garrard, 1961

Pl. 22; fig. 6 a-c Subcircular, moderately elevated; nucleus smooth and recurved, straight or inclined to either left or right, usually level with the edge of the shell or projecting a little beyond it; sculpture of numerous radiating striae commencing at the beak and extending radially to the margin, crossed by irregular growth lines, often with folds that mimic the sculpture, if present, of the underlying host shell. Apex pink to rose, body whorl fawn with obscure radiating pink rays; interior glossy, white or with a pink edge, often flushed with carmine; periostracum pale brown.

3.5 cm. Qld south to at least Keppel Bay. There is a specimen from Port Hedland, WA in the Australian Museum so that a northern Australian distribution is likely. This species is large for the genus and lives on the shells of bivalves such as *Amusium* and *Pecten*. Garrard commented on the resemblance of his species to *C. dilatatus* Adams of Japan and noted that the host shells are bored by the cap limpets.

Capulus violaceus Angas, 1867

Pl. 22; fig. 10   Elevated, somewhat laterally compressed, outline of the mouth irregular; spire almost planispiral, with $1^1/_2$ turns which may be open or closed; sculptured with numerous spiral striae. Pink, yellow or violet.

1.3 cm. NSW to Albany, WA. For an illustration see Macpherson & Gabriel (1962, p. 128, fig. 154).

▲ GENUS **ICUNCULA** Iredale, 1924
Cingulina torcularis Tenison Woods

Tall-spired, prominently keeled, narrowly umbilicate, columella narrowly reflected.

Eastern Australia. There are three described species.

Icuncula consobrina (May, 1915)

Height almost twice the width; with $4^1/_2$ concave-sided whorls; upper whorls with a single peripheral, very sharp and prominent keel, body whorl with an additional two smaller keels below the periphery. Pale buff.

4 mm. Tas. A Qld species named *zodiacus* Hedley, 1907 (described from Mast Head Reef) also has four keels in the last whorl but they end as prominent teeth on the lip.

Icuncula torcularis (Tenison Woods, 1878)

Height about twice width; with $3^1/_2$ whorls, each with a very strong central spiral keel. Pale buff.

2 mm. Cape Moreton Group, Qld to southern NSW. This was originally described as a species of *Lippistes* but, in a discussion of that genus Iredale (1924, *Proceedings Linneaean Society NSW* 49: 251) showed that there is no close relationship and introduced a new generic name for it.

▲ GENUS **LIPPISTES** Montfort, 1810
Argonauta cornu Gmelin

Shell subdiscoidal, spire low, the last whorl tending to separate from the early coils forming a very wide and open umbilicus; aperture slightly channelled at the base; operculum with an apical nucleus.

High latitudes of both hemispheres. *Separatista* Gray, 1847 (*separatista* Dillwyn = *helicoides* Gmelin) is usually treated as a subgenus in which the last whorl separates almost completely from the earlier whorls.

Lippistes gabrieli Pritchard & Gatliff, 1898

Pl. 22; fig. 1 a-b   Thin, fragile, turbinate; last whorl remaining loosely attached; spire whorls bicarinate, last whorl tricarinate; umbilicus wide, columella vertical, meeting the outer lip at a sharp angle.

1 cm. Vic. to SA. The authors of this species subsequently decided that their name was a synonym of "*S. blainvilleanus* Petit, 1851". Originally this species was associated with the New Zealand *Zelippistes* Finlay, 1926. Synonym: *meridionalis* Verco, 1906.

Lippistes helicoides (Gmelin, 1791)

Pl. 22; fig. 2   Early whorls with channelled suture and steeply sloping, upper sides slightly convex, last whorl completely separating; tricarinate, the upper keel forming a sharp periphery. Fawn to brown.

1.5 cm. Indo-West Pacific; Port Hedland, WA to southern Qld. Synonyms: *separitista* Dillwyn, 1817; *tricarinata* Brazier, 1877. *? fraterna* Iredale, 1936. *Lippistes gracilenta* Brazier, 1877, reported from Torres St. to Swain Reefs, Qld, is very similar but with a taller spire.

▲ GENUS **SIRIUS** Hedley, 1900
Raulinia badia Tenison Woods

Shell thin, tall-spired, turbinate; columella broad, expanded, concave, terminating abruptly without reaching the siphonal notch, fasciolar area also concave, with a keeled edge. Periostracum apparently lacking.

Eastern Australia. Vaught (1989) included as subgenera *Dolichosirius* (*cupiens* Iredale) and *Opposirius* (*idoneus* Iredale) both named in 1931. Other Australian species attributed to this genus are: *badius* Tenison Woods, 1878; *meracus* Iredale, 1936; *desponsus* Iredale, 1936.

Sirius idoneus (Iredale, 1931)

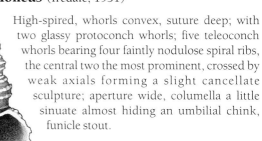

High-spired, whorls convex, suture deep; with two glassy protoconch whorls; five teleoconch whorls bearing four faintly nodulose spiral ribs, the central two the most prominent, crossed by weak axials forming a slight cancellate sculpture; aperture wide, columella a little sinuate almost hiding an umbilial chink, funicle stout.

1.3 cm. Described from Sydney Harbour dredgings.

SUPERFAMILY
XENOPHOROIDEA

T HE origins of this small group are uncertain. There are several late Mesozoic fossil genera which display some similarities, but unquestionably fossil xenophoroids did not appear until the Cretaceous. There is a single family, with a range from the Cretaceous until the present.

which has a paucispiral protoconch indicative of a short pelagic stage or direct development.

The family was thoroughly reviewed by Dr Winston Ponder (1983, *Australian Museum Memoir* 17: 126 pp, 42 figures), recognising only one genus with three subgenera. This account of the Australian species is drawn from Dr Ponder's publication.

FAMILY **XENOPHORIDAE**

Carrier Shells

This small family contains 22 living species found in tropical and temperate seas, often on the outer part of the continental shelf or on the slope beyond. They earn their common name for their habit of attaching pieces of debris such as small shells or pebbles to their own shell, presumably as a form of camouflage. These objects are cemented in rows around the periphery, which is flanged. Objects are selected within a size range peculiar to each species. They are placed in position by the long muscular proboscis or the front edge of the foot.

Xenophorids are detrital feeders that live on sandy or muddy substrates. They scrape edible detritus from the substrate surface with their proboscis. It is reported that some species bury their faeces in small holes in the substrate, which they cover after themselves. The structure of the foot is unusual; it is not broad and flat like most gastropods but stromboid, and locomotion is by a jerky leaping motion like that of strombs.

Little is known of reproduction in this family. However, with one exception, the protoconch is multi-spiral which indicates that the larvae are planktonic. The exception is the southern Australian species *X. flindersi*

▲ GENUS **XENOPHORA** Fischer von Waldheim, 1807
Xenophora laevigata Fischer von Waldheim
(= *conchyliophorus* Born)

Depressed-conical and peripherally flanged; base broad, flattened, often umbilicate; teleoconch whorls usually with foreign particles attached; operculum horny, nucleus lateral; periostracum thin or lacking.

▲ SUBGENUS **XENOPHORA** s.s.

Peripheral flange narrow, simple, its ventral side non-porcelaineous; foreign objects usually covering more than a third of the dorsal surface in adults.

Cosmopolitan in tropical and temperate waters. There are ten species in the Indo-West Pacific region, five of which occur in northern Australia, and one species in the southern Australian region. There are also two species in the Atlantic and one in New Zealand. Synonyms: *Phorus* Montfort, 1810 (*agglutinans* Montfort); *Xenophorus* Philippi, 1849 (error); *Endoptygma* Gabb, 1877 (*umbilicatum* Tuomey).

Xenophora cerea (Reeve, 1845)

Pl. 23; fig. 6   Spire angle 70°-100°; dorsal sculpture of fine, wavy, inclined riblets and prominent rounded rugae giving a warty appearance; base slightly concave to slightly convex, usually with an umbilical depression, with heavy radial lamellae crossed by spiral striae; foreign objects large, covering much of the dorsal surface; operculum subtriangular. White to yellowish or brown.

5.5 cm diam. Indo-West Pacific; North West Cape, WA to Moreton Bay, Qld. This species appears to be closely related to *X. conchyliophora* of the Western Atlantic. Distinguished from *X. mekranensis* by a lower spire, less convex whorls and weaker spirals on the base. Synonyms: *agglutinans* Cuvier, 1849; *corrugata* Philippi, 1855; *trochiformis* Bergh, 1904; *torrida* Kuroda and Ito, 1961.

Xenophora flindersi (Cotton and Godfrey, 1938)

Pl. 23; fig. 5 Spire angle 80°-103°; protoconch paucispiral; surface rough, sculpture obscured by attached foreign objects that cover most of the dorsal surface; base flat to slightly concave or weakly convex, with distinct growth lines and very fine spiral striae; umbilical depression weak in adults, narrowly open in juveniles; operculum elongate, edges serrate. Grey or white, with pale brown spots on the base near the lip.

5 cm diam. Port Lincoln, SA to King George Sound, WA. The paucispiral protoconch and stromboid operculum of this species are unique in the family. In southern WA there is an Upper Pliocene-Lower Pleistocene fossil subspecies named *ludbrookae* Ponder, 1983. Synonyms: *tatei* Verco, 1909 (not Harris, 1897).

Xenophora mekranensis konoi Habe, 1953

Pl. 23; fig. 9 a-b High-spired, spire angle 62°-80°; dorsal sculpture of close, sharp, fine riblets and folds crossed by fine growth lines rendering riblets finely scabrous; base slightly convex, with about four spiral cords, and a narrow umbilicus which may be almost closed by a collumellar callus; foreign objects large, well spaced, usually covering about half of the dorsal surface; operculum suboval, nucleus displaced to the left. White.

5.5 cm diam. Central Indo-West Pacific; Geraldton, WA to Swain Reefs, Qld; 27-292 m. The nominate form of this species is a Pliocene fossil from India but the extant subspecies *konoi* is widely distributed in the Indian and western Pacific. It has a similar distribution to *X. cerea* but the two species usually occupy different habitats. Synonym: *caperta* Schepman, 1909 (not Philippi, 1851).

Xenophora pallidula (Reeve, 1842)

Pl. 23; fig. 8 a-c High-spired, spire angle 65°-92°; dorsal sculpture of distinct riblets which cross folds and growth lines; base flat to slightly concave, with subspiral, wavy striae crossing distinct, curved growth lines, and a small umbilicus which may be closed in adults; foreign objects large, usually elongate, obscure about $^2/_3$ of the dorsal surface; operculum suboval, with weak rugose growth lamellae and a midlateral nucleus. White to pale yellow.

9 cm diam. Indo-West Pacific; Carnarvon, WA to central NSW; 50-1500 m. The foreign objects project like spines from the periphery.

Xenophora peroniana (Iredale, 1929)

Pl. 23; fig. 4 a-b Spire angle 70°-94°; sculpture of fine riblets and irregular rugae giving a warty appearance; outer lip thickened internally; base flat to slightly concave, with fine oblique striae and strongly curved growth lines; umbilicus closed in adults, umbilical depression lacking; foreign objects obscure more than half the dorsal surface; operculum suboval to rectangular. Yellowish white, brown growth lines on the base.

5.5 cm diam. Southern Qld to Bass St.; from the outer part of the continental shelf. Distinguished from *X. cerea* by the flat base and absence of an umbilical depression, but these two species may intergrade in southern Qld. A subspecies named *X. p. kondoi* Ponder, 1983 occurs in Hawaii.

Xenophora solarioides (Reeve, 1845)

Pl. 23; fig. 2 Depressed, spire angle 85°-111°; dorsal sculpture of fine, wavy striae crossed by irregular growth lines; largely obscured by large foreign objects; base flat to convex, with 3-6 gemmate spiral cords over the inner half; deeply and widely umbilicate; operculum oval, smooth. Yellowish white.

3.2 cm diam. Indo-West Pacific; Carnarvon, WA to Yamba, NSW; 1-170m. The small size and wide umbilicus distinguish this species from others. Synonyms: *javanicus* Gray, 1857; *australis* Souverbie & Montrouzier, 1870.

▲ SUBGENUS **STELLARIA** (Schmidt MS) Moller, 1832
Trochus solaris Linnaeus

Peripheral flange wide, with digitations or tubular spines, its ventral surface nonporcelaineous; foreign objects covering less than half of the dorsal surface in adults; umbilicus wide to moderately wide.

Indo-West Pacific and tropical Atlantic Ocean. There are five species, two of them known from Australian waters. The type species is found throughout the tropical Indian Ocean except in Western Australia. Synonym: *Haliphoebus* Fischer, 1879.

Xenophora chinensis (Philippi, 1841)

Pl. 23; fig. 10 Depressed, spire angle about 95°; peripheral flange simple, occupies 30-37% of the base diameter; dorsal sculpture of fine, granular, irregular oblique ridges, crossed by growth rugae; base slightly convex, with close spiral ridges crossed by radial ridges, beaded at the intersections, radials weakly lamellate within the deep and wide umbilicus; row of small foreign objects narrow, most of the dorsal surface exposed. Ranges from yellowish white to pale yellowish brown.

8.2 cm diam. Indo-West Pacific; Dirk Hartog I. to Dampier Archipelago, WA; 4-140 m. The wide umbilicus and heavily spirally ridged base are characteristic. Synonym: *calculiferus* Reeve, 1842.

Xenophora gigantea Schepman, 1909

Pl. 23; fig. 7 a-b Spire angle 81°-84°; peripheral flange simple, occupies about 35% of the total base width; dorsal sculpture of distinct, rather wavy oblique riblets crossed by growth striae; base convex, smooth except for weak growth lines that become lamellate within the deep umbilicus; with a narrow row of foreign objects on all whorls, most of the dorsal surface exposed; operculum rather thin, subopaque, with distinct growth lines. Pale brown to yellowish white.

12 cm diam. Indo-West Pacific; Port Hedland, WA to Tweed Heads, NSW; 60-700m. Differs from *X. chinensis* by the lack of spiral sculpture on the base and a thicker and less oblique operculum.

▲ SUBGENUS ONUSTUS Swainson, 1840

Trochus indicus Gmelin

Ventral side of the peripheral flange porcelaineous; umbilicus narrow to wide, sometimes plugged with callus; foreign objects small and inconspicuous, attached to all or only a few whorls, leaving most of the shell surface exposed.

Indo-West Pacific and Western Atlantic. There are four living species, two in the Western Atlantic and two in the Indo-West Pacific region, both of the latter occurring in northern Australia. The fossil lineage of this subgenus can be traced back to the Eocene of Europe. There is little difference in the anatomy of *Onustus* species. Ponder (1983, *loc. cit.*) noted that they lack a metapodial sole, the opercular lobe is very long, and the mantle edge is smooth; but these characters are not sufficient to justify separation at generic level. The diagnostic character of the subgenus is the porcelaineous under-surface of the shell peripheral flange. Synonyms: *Tugurium* Fischer, 1879 (*indica* Gmelin); *Trochotugurium* Sacco, 1896 (*borsoni* Sismonda - a European fossil).

Xenophora indica (Gmelin, 1791)

Pl. 23; figs. 3 a-b Depressed, spire angle 92°-100°; peripheral flange simple though often ragged, occupying 30-35% of the base diameter; dorsal sculpture of irregular, close-set, oblique riblets, interrupted by intersecting irregular growth lines; base slightly convex to almost flat, with a concave part near the outer edge bordered by a ridged fold bearing 0-4 weakly gemmate spiral ridges, growth lines traverse the entire base and become subfoliate within the umbilicus; umbilicus deep; foreign objects confined to early whorls, the last 2-3 whorls completely free of attachments; operculum oval, thin, sharply pointed at one end, sculptured with heavy radial ridges on the outer surface, rendered minutely foliate by concentric lamellae. Upper surface shiny, yellowish white to pale orange-brown, base often white, sometimes with an orange-brown band at the edge.

5.4 cm diam. Indo-West Pacific; Jurien Bay, WA to northern NSW; 4-150 m. Occurs in the same area as *X. exuta* over much of its wide range but distinguished from that species by the simple peripheral flange and radial sculpture on the operculum. Synonyms: *helvaceus* Philippi, 1852; *wagneri* Philippi, 1855.

Xenophora exuta (Reeve, 1842)

Pl. 23; fig. 1 a-b Resembles *X. indica* ; suture and peripheral flange strongly and evenly undulating, umbilicus bordered by a sharp edge; foreign objects very small and sometimes lacking, when present they are confined to the first 1-2 whorls; operculum smooth except for growth lines and fine radial striae. Uniformly pale yellowish brown, shiny.

10 cm diam. Central Indo-West Pacific; Shark Bay, WA to the Gulf of Carpentaria, Qld; 18-340 m. It appears that this species is not present on the eastern Qld coast although it is sympatric with *X. indica* throughout the rest of its range. The similarity of its shell to that of *X. indica* is superficial judging from opercular and anatomical characters described by Ponder (1983, *loc. cit.*).

VERMETOIDEA

FAMILY **VERMETIDAE**

Worm Shells

The vermetids are sessile gastropods which take their vernacular name from their worm-like, tubed shells which may be regularly or irregularly coiled and are usually cemented to hard substrates. They may be solitary or colonial animals. Some intertidal, colonial species form prominent zones on rocky shores. Their sharp-edged apertures can puncture unwary feet.

The Siliquariidae were once classified together with the vermetids, or at least considered to be close relatives. However, it is now agreed that there is little relationship. The similarity of their shells and feeding strategies is a case of evolutionary convergence.

The taxonomy of the Vermetidae remains poorly known. Keen (1960) published a provisional classification of the family. She recognised five genera, four of which appear to be represented in the Australian fauna although there has been no review and it is not possible to identify most specimens below generic level.

These animals are suspensory feeders; they catch and feed on micro-plankton or detrital fragments suspended in the water. This feeding strategy is characteristic of the bivalves but unusual in gastropods. The vermitid sessile habit is related to this feeding strategy as the animal is not obliged to move about to find its food.

Among the vermetids there are two means of capturing suspended food particles. They may be swept into the mantle cavity with the incurrent water and caught on the gill filaments, wrapped in mucous secreted on the gills and carried along special ciliary tracts to the mouth. Alternatively, mucous produced in a secretory gland near the front edge of the foot (the pedal gland) may be formed into sticky mucous nets or strings and "hung out" into the water until floating particles are entangled

in them and then drawn back to be swallowed. Most vermetids use both methods of food capture but some specialise in one method or the other.

Surprisingly, vermetids have retained a radula. It is of the typical taenioglossate type with a broad central tooth flanked on each side by a lateral and two marginals. The laterals and marginals are erectile and seize the food boluses when they come to the mouth. In the mucous feeding specialists this function is performed by prominent jaws.

Female vermetids brood the eggs within the mantle cavity or in capsules attached to the shell within the aperture.

References

Hadfield, M.G., Kay, E. A., Gillette, M.U. & Lloyd, M.C. (1971). The Vermetidae (Mollusca: Gastropoda) of the Hawaiian Is. *Marine Biology* **12** (1): 81-98.

Keen, A.M. (1960). A proposed reclassification of the family Vermetidae. *Bulletin British Museum (Natural History) Zoology* **7** (3): 183-213, Pls 54-55.

Morton, J.E. (1965). Form and function in the Vermetidae. *British Museum (Natural History) Zoology* **11** (9): 585-630.

Yonge, C.M. (1932). Notes on the feeding and digestion in *Pterocera* and *Vermetus*, with a discussion on the occurrence of the crystalline style in the Gastropoda. *Scientific Reports Great Barrier Reef Expedition* **1**: 259-281.

Yonge, C.M. & Iles, E.L. (1939). On the mantle cavity, pedal gland and evolution of mucous feeding in the Vermetidae. *Annals & Magazine Natural History* (Ser. 11) **3**: 536-556.

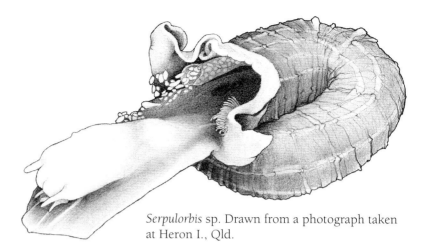

Serpulorbis sp. Drawn from a photograph taken at Heron I., Qld.

▲ GENUS **DENDROPOMA** Mörch, 1861

Siphonium (Dendropoma) lituella Mörch

Shell tube irregularly coiled, scaly, carinate; operculum large, conoidal to discoidal, filling the aperture of the tube, with a button-like process or "mamilla" on the underside, outer side with spirally arranged chitinous plates either compactly welded to form a smooth surface or lamellose and agglutinated with foreign materials; shell partly submerged into the substrate. Colonial or solitary. Specialist ciliary feeders.

Cosmopolitan. The large operculum is diagnostic. Synonyms: *Veritoma* Kuroda, 1928 (*luchuanum* Kuroda); *Veristoa* Iredale, 1937 (*howensis* Iredale).

There are several species of this genus in the Indo-West Pacific. One of them, *D. maximum* (Sowerby, 1825), lives partly embedded in corals, and was the species which Yonge (1932) studied during the Great Barrier Reef Expedition under the name *Vermetus novaehollandiae*. *D. platypus* (Morch, 1861) is another large Indo-West Pacific species which lives completely embedded in massive corals. It may also be present on Australian coral reefs. Iredale (1937) named a vermetid from Middleton and Elizabeth Reefs *Veristoa howensis*. According to Keen (1961) this generic name is a synonym of *Dendropoma* but the status of the species remains uncertain.

▲ GENUS **PETALOCONCHUS** Lea, 1843

P. sculpturatus Lea

Tubes tending to be loosely coiled or irregular; distal parts emerging from the substrate; tube with a spiral internal lamella; operculum small, concave, with an erect spiral lamella. Colonial or solitary. Believed to use both ciliary and mucous net feeding methods.

Cosmopolitan. The majority of described living species belong to the subgenus *Macrophragma* Carpenter, 1857 (*macrophragma* Carpenter). Two species of this subgenus are recorded from northern Australia: *nerinaeoides* Carpenter, 1857; *cochlidium* Carpenter, 1857. An unidentified shell from Shark Bay is chosen to represent the genus (Pl. 15; fig. 6).

▲ GENUS **SERPULORBIS** Sassi, 1827

S. polyphragma Sassi = *arenaria* Linnaeus

Shells usually large, early whorls regularly coiled, later whorls usually detaching but still tending to be coiled; operculum lacking. Colonial or solitary. Mucous feeders.

Cosmopolitan. The absence of an operculum is an obvious identification feature. Two subgenera are recognised; *Serpulorbis s.s.* and *Cladopoda* Gray, 1850 (*grandis* Gray). The latter is characterised by *planorbid* coiling throughout life. *S. (S.) sipho* (Lamarck, 1818) is a well-known species in southern Australia (Pl. 15; figs. 1, 8). It is readily identified by its irregularly coiled later whorls and sculpture of rugose growth striae and rough axial riblets. *S. (S.) novaehollandiae* (Rousseau per Cotton Chenu, 1843) is presumed to be from Australia. Three species are described from NSW under this generic name: *watei* Hedley, 1903; *constrictor* Mörch, 1862; *hedleyi* Finlay, 1927.

▲ GENUS **VERMETUS** Daudin, 1800

V. adansonii Daudin

Tubes loosely coiled or very irregular, the distal parts erect from the substrate; tube without an internal lamella; operculum spiral, concave, small, one half or less the diameter of the aperture. Usually solitary or in clusters. Primarily mucous feeders.

Cosmopolitan. Apart from the presence of the small operculum these animals resemble *Serpulorbis*. There are two subgenera, *Vermetus s.s.* and *Thylaeodus* Mörch, 1860 (*contortus* Carpenter). There is no confirmed record of either from Australia.

CYPRAEOIDEA

*I*N this group there are two families, the true cowries (Cypraeidae) and the egg and spindle cowries (Ovulidae) whose shells are glossy smooth, more or less bilaterally symmetrical, with a long, narrow aperture and terminal anterior and posterior canals. There is no periostracum. An operculum is present only during the veliger larval stage. The mantle is bilobed, thin and contractile, and the lobes may be slid up over the sides of the shell to meet in the midline on top, so completely enclosing the shell.

Some authors have allied the lamellarias, trivias and eratos with the cowries in this superfamily. However, the cowries have a veliger larval stage, while the others have a very distinctive double-shelled larva called an echinospira; and they are now usually treated as quite separate groups.

Both the cypraeoidean families are strongly represented in the Australian marine fauna and they are among the most popular shells with collectors.

FAMILY **CYPRAEIDAE**

Cowries

The true cowries have ovate, beautifully coloured or patterned shells characterised by prominent ridges or "teeth" along the lips on both sides of the narrow aperture. In immature shells the spire is exsert but the last whorl usually encloses it so that in adults the spire is covered.

The character of the apertural teeth is a useful means to distinguish the species. They may be mere denticles along the edges of the lips, or ridges which cross the base to the margins. Deep within the aperture anteriorly the columella may be broad, forming what is known as the fossula, which may be flat or concave. The end tooth on the columellar side usually forms a prominent, angled terminal ridge which is another useful identification character. Colour and pattern are also useful to distinguish between species but there is a great deal of variation in this character within some species.

Immature cowry shells are elongate with a wide aperture and thin outer lip. The shell grows wider and longer by additions to the outer lip, just as most other gastropods do. But when maturity is reached the outer lip turns inward, the shell is thickened, and the apertural teeth are formed.

At one time it was believed that cowry animals could dissolve their shell wholly or in part and make a new, larger one to accommodate the growing animal. But that is now known not to be the case. Once an individual reaches maturity and the outer lip is inturned, further increase in shell length is not possible, although by thickening the shell, especially at the margins, it may become wider.

Cowry shells were at one time used as currency in the African and Asiatic regions. The name is said to have been derived from the Hindu word "kauri". Specimens taken to Europe by early explorers were called "shells of Venus" or "Porcellana" by the first iconographers. The generic name *Cypraea*, introduced by Linnaeus in 1758, is a Greek rendering of the name of the Roman goddess Venus.

Linnaeus named 27 cowry species which are found in northern Australia, although it is unlikely that any of his specimens came from Australian waters. Most of them were named from illustrations in an early book published by Rumphius (1705) which were based on collections from Indonesia. The first cowry to be named from an Australian specimen was probably *C. angustata*

Gmelin, 1791, an endemic south-eastern species, but the source of Gmelin's specimens is not known.

The Cypraeidae originated more than 50 million years ago in the Cretaceous period. There are many fossil species. The exact number of living species depends of course on the taxonomic system followed. Different authorities do not always agree whether a particular name represents a true species, or whether it is merely a synonym of another. Nevertheless, the nomenclature now seems to be fairly settled (in comparison with that of some other groups). In his masterly treatise on the living cowries, Burgess (1985) listed a total of 202 living species. In this account 78 species are listed in the Australian fauna.

Natural history

Most cowries live in rocky or coral habitats. They have a broad, flat foot for crawling on firm substrates. During the day they tend to secrete themselves in crevices or under stones but at night many species emerge to feed in the open. Some species, like the Western Australian *C. (Z.) marginata,* live permanently hidden in underwater caves and are difficult to find. Perhaps the most interesting feature of these animals is the mantle which spreads over the shell and is responsible for creating and maintaining the high polish.

The majority of living species are found in the shallow waters of the tropical and subtropical zones. The largest variety of species ocurs in the Central Indo-West Pacific region, of which the northern coast of Australia is a part. There are a few species in the temperate waters of southern Australia and South Africa.

There is a great range in the size of cowry shells. The smallest Australian species is *C. minoridens,* which reaches a maxiumum length of about 13 mm, while the largest is *C. tigris,* which grows up to 150 mm. The largest of all cowries is the Caribbean species *C. cervus,* which grows up to 190 mm long.

Some species of cowry exhibit sexual dimorphism in shell size, females tending to be longer and more tumid than males. This is true of *C. (Zoila) venusta* and may be a general rule in the family. The mantles of true cowries may be velvet-smooth, or bear projecting pustules or filaments. The filaments may be long and pointed or complexly branched.

The true cowries are omnivores or herbivores, although some feed on sponges and possibly bryozoans. Their radula is taenioglossate with seven teeth per row.

The very common intertidal cowries *C. moneta* and *C. caputserpentis* are known to eat algae. Most cowries will graze on algae on the glass walls of aquaria. It is likely that the majority of tropical cowries are herbivorous. Species of the Western Australian subgenus *Zoila* feed on sponges. They graze on the sponge flesh by rasping with the radula. As sponges are animals these cowries may be regarded as carnivorous. There is some evidence that *C. (Austrocypraea) reevei* and possibly *C. (Umbilia) hesitata* may feed on bryozoans.

In aquaria *C. tigris* and *C. eglantina* have been observed to feed on the flesh of fish or bivalves, while the Californian cowry *C. spadicea* will eat anemones and frozen prawn meat. However, whether any cowries are flesh-eating scavengers in nature is not known. Observant collectors could contribute much useful knowledge by testing the feeding habits of cowries in home aquaria.

Female cowries lay an egg mass consisting of up to 300-400 oval or spherical capsules, each one containing many tiny eggs. The female remains on the egg mass, covering it with her foot, until the larvae hatch. In most species the majority of the eggs develop. The larvae hatch after one or two weeks as free-swimming veligers which feed on microplankton before they settle to the sea floor in preferred habitats, grow their cowry shell, and assume adult benthic life.

A few species, including those in the Australian subgenera *Zoila*, *Notocypraea*, *Austrocypraea* and *Umbilia*, have very different larval development (Wilson, 1985). In these cowries only one egg develops in each capsule. As it grows it consumes the undeveloped eggs. It develops through the veliger stage within the capsule and hatches as a tiny crawling snail, that is, there is no planktonic larval stage. Development within the capsule may take as long as 55 days (as in *C. reevei*) before hatching. The mother remains in her brooding position on the egg mass all that time. Only a few other cowries are known to have this kind of "direct" development, including species of the temperate South African subgenus *Cypraeovula*, and three tropical cousins of the Australian *Zoila* group.

Classification of the cowries

The classification of this family remains very confused. Franz and Maria Schilder published many papers on the matter over several decades during the middle part of this century. They based their classification almost entirely on the characters of the cowry shells. In their last major paper (1971) they split the family into four subfamilies, divided into tribes and genera. The following table is presented so that the reader may see how these genus-level groups were arranged in the Schilders' classification. (Only the groups with Australian representatives are shown.)

Subfamily **Bernayinae**:   *Zoila*

Subfamily **Cypraeinae**:   *Cypraea, Mauritia, Lyncina, Talparia, Chelycypraea, Luria*

Subfamily **Erroneinae**:

    Tribe **Zonariini**:   *Schilderia, Nesiocypraea*

    Tribe **Erroneini**:   *Adusta, Erronea, Notadusta, Palmadusta, Purpuradusta, Blasicrura, Bistolida, Ovatipsa, Cribrarula*

    Tribe **Cypraeovulini**:   *Umbilia, Notocypraea*

Subfamily **Erosariinae**:   *Monetaria, Erosaria, Staphylaea, Nucleolaria Pustularia*

There is no doubt that there is considerable merit in the Schilders' classification. It is clear that there are several distinct evolutionary lineages within the family. For example, the southern Australian *Zoila*, *Notocypraea* and *Umbilia* represent clearly distinct groups, differentiated by shell and anatomical characters, larval developmental type, feeding speciality and discrete fossil lineages traceable far back into the Tertiary.

However, studies on the anatomy of many species have shown that a classification based on anatomical characters does not conform well with the conchological one proposed by the Schilders. For this reason Kay (1960) and many subsequent authors chose to disregard generic classification of the family and to revert to the Linnean generic name *Cypraea* for all species. This practical approach avoided assuming evolutionary relationships for which there is contrary anatomical evidence. However, it also failed to present the available evidence that some groups of species are clearly related and have had a common ancestry.

A compromise arrangement is to treat the Schilder "genera" as subgenera of *Cypraea*. An account of some species groups which may be treated as subgenera was given by Alison Kay in Burgess (1985). She listed ten distinguishable groups: *Cypraea s.s., Cypraeovula, Erosaria, Luria, Lyncina, Mauritia, Notocypraea, Trona, Zoila, Zonaria*. These are all groups recognised as genera in the Schilder classification although the two systems do not exactly match in respect of the distribution of species among them.

More recently Cossignani & Passamonti (1991) took another line. They used the Schilders' system but recognised two subfamilies (Bernayinae and Cypraeinae) with only ten genera, reducing the rest to subgenera. There are parallels between this arrangement and that of Kay, but not exact agreement, even allowing for the different ranks assigned to the categories. Unfortunately Cossignani & Passamonti gave no diagnoses of their classification units and no discussion. There are also some obvious anomalies in their arrangement. For example they include the highly divergent *Nucleolaria* and *Pustularia* as subgenera of *Naria*, along with more typical cowries of the subgenera *Monetaria* and *Erosaria*.

This author concludes that it is not yet possible to propose a classification that represents the relationships among all the species of the family, even though some distinctive groups may be differentiated. However, there are advantages in using subgeneric units. As an interim practical measure the Schilder genera will be used as subgenera in this account of the Australian members of the family, with a few minor amendments.

Species taxonomy

Most tropical cowries have planktotrophic veliger larvae and are geographically widespread, although some have restricted distributions. Some show indications of genetic differentiation in geographically isolated populations. The Schilders and other workers named many geographic variants as subspecies on the basis of statistical analysis of shell form, apertural dentition and colour pattern.

The direct-developing cowries of southern and western Australia exhibit extreme intrapopulation, interpopulation and clinal variation in shell characters, giving the impression that they are presently undergoing evolutionary change. In *C. (Z.) friendii*, for example, the shells vary from pure white to almost black, from slender to obese, and from almost edentulous to moderately toothed along both sides of the aperture. It is very difficult to determine the limits of species in variable complexes like this.

The practice of some contemporary writers to give taxonomic names to selected variants within such polytypic species complexes is regrettable. The use of "form" names for distinctive variants can be helpful to collectors seeking to represent the range of species' variations in their samples, but the use of formal binominal nomenclature is inappropriate for that purpose and causes great confusion. Naming a distinctive form as a species supposes that the living population which it represents is reproductively isolated from other populations of the complex. In extremely polytypic species complexes, local variations in shell form and colour pattern provide very poor evidence for reproductive isolation. In this account of the Australian cowries a conservative approach is taken to species taxonomy and many of the recently introduced names are treated as form names.

Because of the excessive splitting this family has suffered, the synonymies of many species are very long. As two comprehensive checklists have been published in recent years, full species synonymies are not given here. Only names which have been based on Australian material, or are otherwise relevant in the Australian context, are given. For full details the reader should consult the lists of Schilder & Schilder (1971) and Cossignani & Passamonti (1991).

References

There is a vast literature on cowries. The following references contain bibliographies which deal with the most relevant details of the Australian species.

Allan, J. (1956). *Cowry Shells of World Seas*, Georgian House, Melbourne, 170 pp.

Burgess, C. M. (1985). *Cowries of the World*, Gordon Verhoef Seacomber Publications, Cape Town. 289 pp.

Cossignani V. & Passamonti, M. (1991). *Cipraeidae - Systematic Catalogue Synonyms and Values*, Mostra Mondiale Malacologia, Cupra Marittima, Italy, 135 pp.

Iredale, T. (1935). Australian cowries. *Australian Zoologist* **8**: 96-135, pls 8-9.

Iredale, T. (1939). Australian cowries: Part 11. *Australian Zoologist* **9** (3): 297-323, pls 28-29.

Kay, E. A. (1960). Generic revision of the Cypraeinae. *Proceedings Malacological Society London* **33**: 278-287.

Schilder, M. & Schilder, F. A. (1971). A catalogue and bibliography of Triviacea and Cypraeacea, *Memoirs Institut Royal des Sciences Naturelles de Belgique 2nd Ser.*, **85**: 1-246.

Wilson, B.R. (1985). Direct development in southern Australian cowries (Gastropoda: Cypraeidae). *Australian Journal Marine & Freshwater Research* **36**: 267-280.

▲ GENUS **CYPRAEA** Linnaeus, 1758

C. tigris Linnaeus

With characters of the family.

Cosmopolitan in tropical and warm-temperate seas. The subgenera which follow (in alphabetical order) are the genera used by Schilder and Schilder (1971) - see discussion on page 174.

▲ SUBGENUS **AUSTROCYPRAEA** Cossmann, 1903

Cypraea contusa McCoy, 1877

Ovate to globose, tumid; teeth fine, extending deep into the aperture on the columellar side; fossula wide, concave, crossed by the columellar teeth; surface covered with small dents as if lightly hammered.

Southern Australia. This group is an enigma. The only living species is *reevei* but there are several Miocene fossil species, including the type.

Shell dentition and radular dentition of *reevei* resemble the tropical group *Lyncina* and some authors have placed this species with that group. Schilder (1927) introduced the generic name *Prolyncina* for this species. However, the Miocene fossil *contusa*, type species of *Austrocypraea*, is almost identical to the living *reevei* in form and shell dentition and shares the hammered-like surface. Also, living *reevei* has direct development and in this respect it is very different to the tropical *Lyncina* species. This author believes that *C. reevei* is the sole survivor of an ancient south-eastern Australian group and is not related to any tropical species. Separate subgeneric rank is used for it accordingly. It probably deserves full generic rank. Synonym: *Prolyncina* Schilder, 1927 (*reevei* Sowerby).

Cypraea reevei Sowerby, 1832

Pl. 35; fig. 15 a-d   Ovate, sometimes almost globose, thin and light, with rounded sides and narrow aperture; dorsal surface characteristically bearing small dents like hammer marks. Top and sides light brown to yellow, unspotted but sometimes faintly banded; ends pink, base and sides white or pink-tinted.

4.5 cm. Western SA to about Geraldton WA. Occasionally found intertidally but occur right across the continental shelf. Males tend to have smaller and less tumid shells than females. Some deepwater specimens are almost globular. Fossil specimens indistinguishable from living shells occur in the Early Pleistocene sediments of the Roe Plain. Deepwater specimens tend to be globose. Raybaudi, 1980 used four form names for variants of this species (*elongata, globosa, macrospaerica* and *microsphaerica*), none of which has any taxonomic standing, and they may be disregarded. The egg masses and larval development of this species have been described by Wilson (1985).

▲ SUBGENUS **BISTOLIDA** Cossmann, 1920

C. stolida Linnaeus

Ovate; teeth strong and usually extending onto the base on the columellar side; characterised by a distinct and sharply defined dorsal blotch.

Indo-West Pacific.

Cypraea brevidentata Sowerby, 1870

Pl. 32; fig. 9   Oblong-ovate; columellar teeth short and not extending up onto the base. Top blue-grey without spots but with a conspicuous central brown patch; sides with or without spots; pairs of small brown blotches present at each end; base white.

2.5 cm. North West Cape, WA to central Qld. Lives in the intertidal and shallow sublittoral zones. Some authors treat this as merely a form of *C. stolida*, claiming that colour pattern and dentition intergrade. However, this is not the case with the material examined here and this author prefers to treat them as distinct species for the time being. The two species occur sympatrically throughout northern Australia. Synonyms applicable to the *brevidentata* form include *moniontha* Melvill, 1888; *irvineanae* Cox, 1889 (for the WA population); *fluctuans* and *deceptor* Iredale, 1935 (for the Qld population).

Cypraea hirundo Linnaeus, 1758

Pl. 33; fig. 11 a-c   Ovate, subcylindrical; teeth fine, extending across the base almost to the margins. Base white, top pale blue with three poorly defined blue zones separated by white bands, and fine brown spots most numerous on the sides; with a prominent brown dorsal blotch and conspicuous dark brown blotches at the ends.

2 cm. Indo-West Pacific; Shark Bay, WA to northern NSW. The species is moderately common in shallow rocky habitats in northern Australia. Several subspecies are named although the diagnostic characters distinguishing them are very minor. Synonyms relevant to Australian material include *neglecta* Sowerby, 1837 (for the WA population); *rouxi* Ancey, 1882; *abbreviata* and *marteli* Dautzenberg, 1903; *francisca* Schilder & Schilder, 1938; *endela, cameroni* and *peropima* Iredale, 1939 (all introduced for the Qld population). See remarks on *C. kieneri*.

Cypraea kieneri Hidalgo, 1906

Pl. 33; fig. 12 a-c   Slender, subcylindrical; teeth rather coarse, outer lip teeth short anteriorly but extending onto the base posteriorly. Top bluish grey, crossed by two irregular pale bands and flecked with small brown spots; there is no dorsal blotch but the anterior blue-grey area encloses a pale patch; base and sides white, sides spotted.

2.4 cm. Indo-West Pacific; North West Cape, WA to north Qld. Rather uncommon. Lives in the intertidal and shallow sublittoral zones. Iredale (1939) introduced the name *marcia* as a subspecies for the eastern Australian population but there seems no basis for this. The short anterior columellar teeth that do not extend across the base distinguish this species from *C. hirundo* and *C. ursellus*.

Cypraea stolida Linnaeus, 1758

Pl. 32; fig. 8 a-c Oblong-ovate, ends slightly beaked ends; with strong columellar teeth extending over the base. Base yellow-white, sides with indistinct brown spots; light orange-brown blotches present on each side of the anterior and posterior ends, and two blotches of the same colour on each side; top blue-grey with indistinct brown spots and a large irregular brown central blotch that is usually partially connected to the side blotches.

4 cm. Indo-West Pacific; Geraldton, WA to central Qld. Found in rocky and coral habitats in the intertidal and shallow sublittoral zones. There is much variation in the colour pattern. See remarks on *C. brevidentata*.

Cypraea ursellus Gmelin, 1791

Pl. 33; fig. 13 a-b Ovate to pyriform; teeth fine, extending across the base almost to the margins. Base white, top pale blue with three well-defined blue zones of irregular outline separated by white bands, and fine brown spots, brown dorsal blotch small and irregular, pairs of conspicuous dark brown blotches present at the ends.

1.5 cm. Western Pacific including central Qld. Uncommon. Very like *C. hirundo* but usually smaller, a little more globose, and the three pale dorsal bands tend to be more distinct. Nevertheless, it is sometimes difficult to separate these species. The fine teeth extending right across the base and outer lip to the margins and the less cylindrical form distinguish the species from *C. kieneri*.

▲ SUBGENUS **BLASICRURA** Iredale, 1930
C. rhinoceros Souverbie (= *pallidula* Gaskoin)

Small; elongate-ovate to pyriform; spire invert; teeth moderately strong, usually spreading onto the base on the columellar side; fossula poorly developed.

Indo-West Pacific. There are eight living species assigned to this group, four of them occurring in northern Australia.

Cypraea alisonae Burgess, 1983

Pl. 32, fig. 4 Ovate, right margin calloused; columellar teeth not spreading onto the base. Top fawn to grey, crossed by three double bands of squarish brown spots, and with numerous light brown flecks which may tend to form axial streaks; commonly there is a prominent, irregular brown dorsal blotch; base off-white to pinkish; margins milky, with prominent brown spots which may extend onto the sides of the dorsum.

4 cm. Western Pacific and tropical Eastern Pacific Regions; with positive Australian records from Coral Bay and Dampier Archipelago, WA. The shells of *C. alisonae* are very hard to tell apart from those of *C. teres* and specimens are often misidentified as that species. Shells of *C. alisonae* are usually broader and the dorsal blotch, though not always present, is diagnostic. It was originally recognised as a different species by its distinctive animal. The carmine mantle is blotched with orange-white and bears tapering, columnar, transversely banded papillae. The brilliant, carmine mantle of *C. teres* is not blotched and its papillae, when present, are white or grey, sparse and branching. The upper surface of the foot of *C. alisonae* is orange-carmine and spotted while that of *C. teres* is bright carmine like the mantle and may be flecked but not spotted. Further study is needed to determine the distribution of *C. alisonae* in northern Australia. It is likely that it occurs in Qld as well as in WA.

Cypraea pallidula Gaskoin, 1849

Pl. 33; fig. 14 a-c Elongate-ovate to subcylindrical; teeth strong, extending right across the base on the columellar side. Top greenish grey, with many brown freckles and four distinct, interrupted, equally spaced bands, sides white.

3 cm. Central Indo-West Pacific; Cape Naturaliste, WA to central Qld. Intertidal and shallow coral reef and rocky habitats; uncommon. The Cape Naturaliste record is based on a dead specimen and it is doubtful whether the species has a breeding population so far south. Closely resembles *C. interrupta* Gray, 1824 from the Philippines and Indo-Malay Archipelago but distinguished by the evenly spaced dorsal bands. Qld and WA populations are said to show minor differences and have been given subspecies status with the names *rhinoceros* Souverbie, 1865 and *simulans* Schilder & Schilder, 1940 respectively, although these names could perhaps be better relegated to the synonymy.

Cypraea quadrimaculata Gray, 1824

Pl. 33; fig. 15 a-b Elongate-ovate to pyriform; spire slightly umbilicate; columellar teeth spread right across to the margin. Top blue-grey, with three slightly darker bands and profuse fine brown flecking; a pair of nearly black spots are very conspicuous at each end; base and sides uniformly white or yellowish.

3.5 cm. Central Indo-West Pacific; Dampier, WA to central Qld. Moderately common in rocky habitats in shallow water. The name refers to the four terminal spots. The Australian population tends to be more cylindrical and more heavily calloused than others and was given subspecies status as *C. q. thielei* Schilder & Schilder, 1938. *C. coxeni* Cox, 1873 (Pl. 32, fig. 11 a-b) from Melanesia is similar but heavily spotted.

Cypraea teres Gmelin, 1791

Pl. 32; fig. 3 a-b Oblong-ovate, right margin calloused; columellar teeth not spreading onto the base. Top bluish grey, crossed by three double bands of squarish brown spots, often with irregular V-shaped marks between the bands; base white, sides white with large brown spots.

5 cm. Indo-West Pacific; Abrolhos, WA to central NSW. Lives in the intertidal and shallow sublittoral zones. Iredale (1939) named the eastern Australian form as a subspecies which he called *pentella* but the name is regarded as a synonym here. See remarks on *C. alisonae.*

▲ SUBGENUS **CHELYCYPRAEA** Schilder, 1927
C. testudinaria Linnaeus

With the characters of the type species.

Indo-West Pacific. Monotypic.

Cypraea testudinaria Linnaeus, 1758

Pl. 31; fig. 10 a-b   Large, cylindrical; aperture narrow with fine and numerous teeth; columellar teeth short posteriorly, but extending deep into the aperture on the fossula. Base light brown, sides spotted, top fawn with large brown spots, overlain by dense dark brown blotches and scattered, minute, dust-like, pale grey dots.

13 cm. Indo-West Pacific. There are few northern Australian records but the species was collected in 1985 by a WA Museum expedition at Seringapatam Atoll off the north coast of WA. A coral reef species found under slabs in the shallow sublittoral zone but never abundant. There can be no mistaking this large, cylindrical, finely toothed cowry which is unlike any other.

▲ SUBGENUS **CRIBRARULA** Strand, 1929
C. cribraria Linnaeus

Small, oblong-ovate to pyriform; teeth moderate but short; fossula poorly developed; characterised by white base, red-brown to yellow-brown top, and very conspicuous white spots.

Indo-West Pacific. There are seven species in the group. They are among the most striking of all cowries. There are two Australian representatives. The others have restricted distributions in the Pacific and Indian Oceans. *Cribraria* Jousseaume, 1884 was widely used as the generic name for the group but it was preoccupied and Strand provided an alternative.

Cypraea cribraria Linnaeus, 1758

Pl. 35; fig. 19 a-d   Ovate to pyriform; margins rounded or moderately calloused; ends moderately produced. Base and sides milk-white and unspotted; top red-brown with sharply defined white spots.

4 cm. Indo-West Pacific. Albany, WA to southern Qld. Widely distributed but never abundant. The mantle is crimson and velvety but the dorsal spots can be seen through it. This is one of the tropical cowries which is occasionally found far south in the temperate zone on the west coast.

An entirely white form from Qld was named *melwardi* Iredale, 1930. In this author's opinion the white shells are merely aberrations and this name is here treated as a synonym.

In their 1971 catalogue the Schilders recognised several subspecies, although in practice there is considerable difficulty in differentiating between these with any consistent morphological characters. The recognised subspecies relevant to eastern Australia are *orientalis* Schilder & Schilder, 1940 from the New Guinean region, *northi* Steadman & Cotton, 1943 from the Fijian region, and *zadela* Iredale, 1939 from Qld, the type locality of the latter being Hamilton I.

There has been much confusion and debate about the subspecific status of the WA populations of this species and its relationships to *C. fallax* Smith. The Schilders and most other recent authors have used the latter as a subspecies name for the WA *cribraria* but Smith's name applies to a different shell from the south coast. (See remarks on *fallax.*) It is then appropriate to reinstate *exmouthensis* Melvill, 1888 as a subspecies name for the northern WA population. This name is not a *nomen dubium* as some authors have assumed, there being a holotype in the British Museum (well figured by Luigi Raybaudi, 1986, *The Connoisseur of Seashells* **14**: 6-11, fig. 3) and a specified type locality (Exmouth Gulf collected by T.H. Haynes). Raybaudi (1987, *ibid.* **13**: 41-43) introduced another name, *cribraria occidentalis,* to replace Smith's *fallax,* even though he also recognised *exmouthensis* as a valid taxon living in the same area. Raybaudi did not nominate a type or a type locality and the name is both invalid and unnecessary.

Compounding the confused nomenclature of this species, the shells are extremely variable. Raybaudi (1986, *ibid.* **11**: 23-28) illustrated a large number of specimens from localities between Darwin and Albany. Most specimens from the north west coast of WA (e.g. fig. 19 (a) from the Dampier Archipelago) fall well within the morphological range of the species and are hard to differentiate from Qld specimens. However, shells from the west coast (i.e. south of North West Cape) are commonly dark (figs. 19 b) and sometimes globose (fig. 19 d).

Cypraea fallax Smith, 1881

Pl. 35; fig. 20 a-c   Elongately pryiform, margins moderately calloused, ends produced; columellar teeth inclined and extending over the fossula. Base white, top light yellow-brown with scattered white spots.

3.6 cm. Denmark to Cape Naturaliste, WA. This species was believed to be confined to the vicinity of Denmark but the specimen illustrated here from Canal Rocks (just south of Cape Naturaliste) extends the known range a little northwards.

A pale yellow dorsum distinguishes this species from *cribraria,* which it otherwise closely resembles. Were it not for the existence on the south coast of typical *cribraria* it might be concluded that these southern shells are merely an end-of-range variant of that tropical species. The matter needs further study.

Smith's name was based on a type specimen preserved in the British Museum of Natural History (well figured by Raybaudi,

1986, *The Connoisseur of Seashells* **12**: 24-28, fig. 2). The locality was given only as WA. For many years it was assumed that the name referred to the northern *cribraria* but the identity of the type seems now established as belonging to this different species confined to the south-west corner of the state. In the meantime Trenberth (1973) introduced a new name for the south-west shell, *haddnightae*, which is now placed into synonymy.

▲ SUBGENUS **CYPRAEA** s.s

Large, ovate to pyriform, with rounded margins; labial and columellar teeth coarse, short, fossula deeply concave, crossed by columellar teeth; terminal ridge strong, curved.

Indo-West Pacific. There are two living species.

Cypraea tigris Linnaeus, 1758

Pl. 31; fig. 1 a-b   Ovate, inflated; labial and columellar teeth coarse but not extending across the base, fossula deeply concave, crossed by long columellar teeth, posterior columellar teeth short; terminal ridge strong, curved. Base white, sides and top white or pale brown with large brown spots; prominent mantle line present.

13 cm. Indo-West Pacific; Abrolhos, WA to northern NSW. This is a common shell found in coral reef habitats from low tide to 30 m. Unlike most tropical cowries, *C. tigris* is often found crawling in the open, even during the day. Many Qld specimens found in the open on intertidal reefs are heavy and very pale. Living specimens make a spectacular sight with their long mantle filaments extended. When naming this species Linnaeus seems to have been confused about his feline taxonomy - *leopardus* might have been a better choice. Lightfoot (1786) did better in naming the similar Red Sea species *C. pantherina*. There are many synonyms; two relevant are *nephelodes* Lancaster, 1928 and *fuscoapicata* Coen, 1949, both named from Qld.

▲ SUBGENUS **EROSARIA** Troschel, 1863
C. erosa Linnaeus

Ovate, usually marginally calloused and flat-based; teeth strong; fossula poorly to moderately developed; top spotted.

Cosmopolitan in tropical and warm temperate waters. This is a large group of small to medium-sized shells. *Ravitrona* Iredale, 1930 (for *caputserpentis*) is sometimes treated separately but this account follows Schilder & Schilder (1971) in associating it with *Erosaria*.

Cypraea beckii Gaskoin, 1836

Pl. 35; fig. 6 a-b   Pyriform, inflated, spire depressed, ends slightly projecting; labial teeth rather strong, columellar teeth

fine at the centre but strong and spread onto the base posteriorly; anteriorly the first four or five teeth fusing to form an unusual terminal ridge. Top fawn to orange with white spots and reddish spots encircled by white rings; sides prominently spotted; some teeth marked with brown.

2 cm. Indo-West Pacific, north Qld at least as far south as Innisfail. This is a rare little shell and there are few Australian records. It is similar to the very restricted *C. thomasi* Crosse, 1865 now known to come from the Marquesas Is. in eastern Polynesia.

Cypraea caputserpentis Linnaeus, 1758

Pl. 34; fig. 2 a-c   Solid, depressed, with thick, expanded sides and flat base; teeth very strong. Top brown, flecked with white spots and blotches, mantle line indistinct; base and sides chocolate brown.

3.5 cm. Indo-West Pacific; Albany, WA to southern NSW. This is a very common cowry in both coral reef and rocky habitats. A large breeding population occurs on the intertidal rock platforms at the West End, Rottnest I., WA but specimens are not uncommon alive on rocky shores as far south as Albany. It is one of the few species which is frequently found in the open during the day (i.e. not always under stones). It has an unusually wide geographic range extending well into the temperate zone of both hemispheres. Southern Australian specimens (i.e. in temperate waters on both east and west coasts) tend to be more humped and less laterally expanded than tropical specimens (Pl. 34, fig. 2c). The name *kenyonae* Schilder & Schilder, 1938 was introduced for this form from southern WA but it is really only an end-of-range variant and a similar variant occurs in NSW. The Schilders also treated the Qld population as a distinct subspecies under the name *caputanguis* Philippi, 1849. Both names are regarded here as synonyms.

Cypraea cernica Sowerby, 1870

Pl. 32; fig. 12 a-b   Ovate to deltoid, with calloused sides and projecting ends; sides pitted at the upper margins of the callus; teeth moderately strong, labial teeth extend half way across the outer lip, columellar teeth short at the centre. Top olive-green to orange, with many diffuse white spots of varying size; sides with prominent brown spots; base white.

3.5 cm. Indo-West Pacific; Albany, WA across northern Australia to northern NSW and the Kermadec Is. Never common. This is another tropical species which is occasionally found far south in temperate waters. This author has collected the species alive under granitic boulders at Cape Naturaliste. Schilder & Schilder (1971) recognised four subspecies, the nominate form being from the western Indian Ocean. The others are: *viridicolor* Cate, 1962 for WA shells; *tomlini* Schilder, 1930 for eastern Australian and South Pacific shells; *ogasawarensis* Schilder, 1945 for Japanese and northern Pacific shells. The characters distinguishing between them seem trivial and they are regarded by this author as synonyms along with *percomis* Iredale, 1931 and *prodiga* Iredale, 1939.

Cypraea erosa Linnaeus, 1758

Pl. 34; fig. 4 a-b Solid with thickly calloused, often nodulose margins; labial teeth strong, extending almost to the margin, columellar teeth strong but short. Top olive green or fawn, with scattered light brown spots, many small white dots, a prominent mantle line and large squarish brown patches on each side extending onto the base; base and sides yellow-white with scattered brown dots or vertical lines.

5 cm. Indo-West Pacific; Abrolhos, WA to northern NSW. A common species with a wide range. Although quite variable it is always easy to identify by way of the lateral patches. The only similar species is *C. nebrites* Melvill, 1888 of the north-western Indian Ocean, in which the lateral blotches do not extend onto the base. There are many synonyms; two relevant in the Australian context are *purissima* Vredenburg, 1919 and *diaphana* Coen, 1949, both named from Qld.

Cypraea flaveola Linnaeus, 1758

Pl. 32; fig. 10 a-b Elongate-ovate; margins calloused and pitted along their upper edges; teeth strong, those on the outer lip almost to the margin. Base white and brown terminal blotches, sides white with conspicuous dark brown spots above the margins; top fawn or olive-green with round white spots and a prominent mantle line.

3 cm. Central Indo-West Pacific; North West Cape, WA to central NSW. This moderately common species has generally been known as *C. labrolineata* Gaskoin, 1848 but the Linnaean name has priority. Its shell form and colouring are variable and Iredale introduced two names for Qld variants, both of which are relegated to synonymy, viz.: *nashi* Iredale, 1931; *maccullochi* Iredale, 1939.

Cypraea helvola Linnaeus, 1758

Pl. 32; fig. 2 a-b Ovate, sides of adult shells heavily calloused and pitted along their upper margins; teeth strong, extending half way across the outer lip and onto the base on the columellar side. Top pale brown, spotted with white and orange-brown, a weak mantle line is present; sides dark brown above the marginal calluses, base and calluses orange, canals mauve.

3 cm. Indo-West Pacific; Albany, WA to northern NSW. Relatively common throughout its range in coral reef and rocky habitats. It is another of the tropical species which extends into the temperate zones on both sides of Australia. Iredale (1935) introduced the subspecies name *citrinicolor* for WA specimens but they do not differ significantly from others and the name is regarded as a synonym.

Cypraea miliaris Gmelin, 1791

Pl. 32; fig. 13 a-d Pyriform, ends projecting, sides only moderately calloused, pitted along upper edges of the calluses; teeth moderately strong, extending a little way onto the base on the columellar side. Top fawn or olive-green with many small white spots and a distinct mantle line; base and sides white, unspotted.

5 cm. Central Indo-West Pacific; Shark Bay, WA to central NSW. Lives in the intertidal and shallow sublittoral zones of relatively turbid shores. Several subspecies have been described, e.g. *metavona* Iredale, 1935 in eastern Australia and *diversa* Kenyon, 1902 in WA, although the differences seem minor and both names are included here in the synonymy. In the South Pacific and Qld there is an all-white variant (Pl. 32, fig. 13 a-b) which some authors treat as an entirely separate species under the name *C. eburnea* Barnes, 1824. Other synonyms used in the Australian connection are: *diversa* Kenyon, 1902; *nivea* Preston, 1909; *gabrieli* Gatliff, 1916; *gabrielana* Gatliff, 1929; *metavona* Iredale, 1935.

Cypraea poraria Linnaeus, 1758

Pl. 33; fig. 3 a-b Solid, ovate, marginally calloused; sides pitted above the calluses anteriorly; teeth moderately strong, short on the labial lip, extending a little way onto the base on the columellar side. Top light brown with lilac blotches and tiny white spots, mantle line indistinct; base and sides dark lavender.

2.5 cm. Indo-West Pacific; Cape Naturaliste, WA to southern Qld. This is normally a coral reef species which lives in the intertidal and shallow sublittoral zones. Its occurrence at Cape Naturaliste in a granitic rock habitat seems to be an abnormality; breeding populations do not seem to occur south of Shark Bay on the west coast. The Australian population is of the nominate subspecies. The western Indian Ocean form is *poraria mbalavuensis* Ladd, 1945. The names *theoreta* Iredale, 1939 and *wilhelmina* Kenyon, 1897 were introduced for western and eastern Ausralian specimens but both these names are regarded as synonyms.

▲ SUBGENUS **ERRONEA** Troschel, 1863
C. errones Linnaeus

Small to medium-sized; ovate, pyriform or cylindrical; aperture rather wide, teeth strong, fossula narrow and not well developed.

Indo-West Pacific. A large number of species are attributed to this group. *Adusta* Jousseaume, 1884 has been used as a generic name for a group of species (*hungerfordi, pyriformis, onyx, subviridis, walkeri* and *xanthodon*) with prominently coloured, elongated teeth at the centre of the columellar lip. If anatomical studies confirm that these constitute a distinct group then *Adusta* should be ranked equal to *Erronea*.

Cypraea caurica Linnaeus, 1758

Pl. 34; fig. 1 a-c Subcylindrical, with thickened sides and unusually strong teeth. Top greenish fawn, densely flecked with brown spots, crossed by three faint brown bands and with a large central brown blotch; widely spaced brown spots on the sides; base and sides fawn or brown.

6 cm. Indo-West Pacific; Abrolhos, WA to northern NSW. Relatively common in shallow water. Like most of the wide-

ranging cowries this species has been divided into several subspecies. The two named Australian subspecies are *longior* Iredale, 1935 (= *velesia* Iredale, 1939) in the east and *blaesa* Iredale, 1939 in the west, but the differences are trivial and these names are regarded here as synonyms.

Cypraea chinensis Gmelin, 1791

Pl. 32; fig. 1 a-c Ovate; sides calloused; outer lip teeth short, strong and extend half way to the margin, columellar teeth finer and more numerous; fossula shallow or lacking. Base beige or fawn, orange between the teeth, sides spotted with violet; top blue-grey with orange-brown reticulations.

5 cm. Indo-West Pacific; Cape Naturaliste, WA to central NSW. It is not a common species but widespread, living in intertidal and sublittoral zones of coral and rocky habitats. Some authors have regarded the western and eastern Australian forms as separate subspecies under the names *whitworthi* Cate, 1964 and *sydneyensis* Schilder & Schilder, 1938 respectively. The western shells are more solid, broader, with a shallow but heavily toothed fossula, large and dense lateral spots and more brilliant lavender colouring. However, whether such characters warrant subspecies recognition is a matter of opinion. Iredale (1935) cited this species as type of his monotypic genus *Ovatipsa*.

Cypraea cylindrica Born, 1778

Pl. 34; fig. 3 a-b Subcylindrical, sides pinched in at the anterior end; columellar teeth fine, extending deep into the aperture. Top bluish or green, with fine brown freckles, a prominent brown dorsal blotch, and a pair of large dark brown spots at each end; base and sides yellow-white.

5 cm. Central Indo-West Pacific and Micronesia; Shark Bay, WA to central Qld. Moderately common in the intertidal and shallow sublittoral zones. The eastern Australian population is typical of the species but WA shells were said to show subspecific differences and named *sowerbyana* by Schilder, 1932. Iredale (1939) introduced the name *lenella* and Steadman & Cotton (1943) the name *wangga* for variants of the species. All these names are regarded here as synonyms. For a picture of the living animal see p. 358.

Cypraea errones Linnaeus, 1758

Pl. 34; fig. 5 a-b Subpyriform; columellar teeth short and weak. Base and sides creamy white or yellow and unspotted; top green with three faint blue bands overlain by fine diffuse brown freckles, usually with a large central brown blotch; anterior terminal blotches or spots sometimes present.

4 cm. Indo-West Pacific; Shark Bay, WA to northern NSW. A very common intertidal species. It lives unusually high in the intertidal zone. Shell form and colour are both very variable and there are many synonyms. The western and eastern forms have been treated as subspecies by some authors. WA variants have been named *coxi* Brazier, 1872, *proba* Iredale, 1939, and *azurea* Schilder, 1968. Eastern Australian variants have been named *coerulescens* Schroter,

1804, *nimiserrans* Iredale, 1935, and *magerones* Iredale, 1939. All of these names are regarded here as synonyms of a single species.

Cypraea felina Gmelin, 1791

Pl. 33; fig. 1 a-d Small, pyriform-ovate to subcylindrical, flat-based; teeth moderately strong except on the parietal wall, where they are short and weak. Top fawn or bluish grey, with four bands of squarish grey-brown blotches, and crowded brown flecks; there are pairs of prominent brown spots at each end; sides spotted; base white.

3 cm. Indo-West Pacific; Qld, northern NSW and Lord Howe I. A moderately uncommon but very variable species. Cate (1968) reported it from North West Cape but the presence of the species in WA needs verification. The nominate form is ovate and comes from the south-western Indian Ocean. A population in the north-western Indian Ocean characterised by tumid, ovate shells has been given subspecific status as *felina fabula* Kiener, 1843. In the remainder of the species' range, including Australia, the shells are slender and subcylindrical. Several names are available for the slender Australian form, if subspecific distinction is required, the oldest being *listeri* Gray, 1824. The species appears in the NSW list of Iredale & McMichael (1962) as *Melicerona velesia* Iredale, which is a synonym.

Cypraea hungerfordi Sowerby, 1888

Pl. 34; fig. 10 Thin, pyriform and slender; right hand margin projecting beyond the left one at the posterior end; outer lip teeth short and fairly coarse, columellar teeth short and fine, especially posteriorly. Base and sides pale yellow-orange, with a few large brown spots on the sides; top pale orange to fawn with rust-coloured flecks and three transverse bands or zones of darker blotches.

4.5 cm. Western Pacific; southern Japan to Taiwan in the north, and southern Qld in the southern hemisphere. The Qld specimens have come from localities between the Capricorn Channel and Cape Moreton. This kind of bipolar distribution with populations in the waters of Japan and southern Qld is repeated in several species and may have an ancient evolutionary explanation, or it may be simply an artifact of inadequate sampling in the area between. The Qld population was named *C. coucomi* Schilder, 1965, based on specimens trawled off Cape Moreton where it is taken only rarely, but the shells are indistinguishable from specimens from Japan and Taiwan and the name is now regarded as a synonym. See remarks on *C. hirasei* and *C. langfordi*.

Cypraea ovum Gmelin, 1791

Pl. 34; fig. 6 a-b Pyriform; very like and often confused with *C. errones* but has stronger columellar teeth, no trace of anterior blotches, and the aperture is less dilated anteriorly. Orange colouring between the apertural teeth is characteristic.

4 cm. Central Indo-West Pacific and Micronesia; Abrolhos WA to central Qld. Lives in the intertidal and shallow

sublittoral zones. Many specimens identified in collections as *ovum* are actually *errones*. Anatomical and natural history study of the two species is needed to establish the relationship between them.

Cypraea pyriformis Gray, 1824

Pl. 32; fig. 15 a-f Pyriform, anterior end distinctly pinched in at the sides; columellar teeth fine but long, reaching deep into the aperture and often spreading part way onto the base centrally. Fawn or greenish on top, flecked with tiny brown spots and sometimes with an incomplete transverse band; sides fawn, sometimes pinkish, spotted; ends with pairs of brown blotches; base fawn, columellar teeth red-brown.

3.5 cm. Central Indo-West Pacific; Kimberley, WA to north Qld. Lives in the intertidal and sublittoral zones, usually in muddy or sandy habitats. Not a very common species. Shells of the Kimberley population are smaller, more solid and ovate than eastern specimens and were named *smithi* Sowerby, 1881 (Pl. 32, figs. 14 e-f). A series collected by the author at One Arm Point indicates that the differences are significant and consistent and worthy of differentiation at subspecific rank. Kenyon (1897) also introduced a name for the WA shell, calling it *C. kaiseri* which may be relegated to the synonymy of *pyriformis smithi*. Iredale (1935) introduced the name *problematica* for Qld specimens, which is a synonym of *pyriformis s.s.*

Cypraea subviridis Reeve, 1835

Pl. 32; fig. 16 a-e Pyriform; columellar teeth rather weak. Top fawn or faintly green, flecked with tiny brown spots, and crossed by a faint darker band; sides, base and teeth white, fawn, mauve or rose, unspotted.

4 cm. South-western Pacific and northern Australia from Rottnest I., WA to central NSW. The species seems to prefer rocky habitats on intertidal sandflat areas. Shells from WA usually have a prominent dorsal blotch and have been given the subspecies name *dorsalis* by Schilder & Schilder, 1938. The species appears in the NSW list of Iredale & McMichael (1962) as *Solvadusta vaticina* Iredale, 1931 which is a synonym.

Cypraea walkeri Sowerby, 1832

Pl. 32; fig. 14 a-b Ovate to pyriform; teeth strong, those at the centre on the columellar lip tending to spread onto the base. Top grey, fawn or greenish, flecked with minute brown spots and with a broad brown central band marked on each side by thin transverse white lines; ends with prominent brown blotches; sides spotted; base pale orange, conspicuously stained lilac between the teeth.

3.5 cm. Central Indo-West Pacific; Broome, WA to northern NSW. Usually found associated with sponges; intertidal and sublittoral. Not common. The Philippine population is sometimes regarded as a subspecies under the name *surabajensis* Schilder, 1937. Two names were introduced by Iredale for eastern Australian shells, *continens* Iredale, 1935

and *merista* Iredale, 1939, but both should be treated as synonyms. See p. 354 for a picture of the living animal.

Cypraea xanthodon Sowerby, 1822

Pl. 32; fig. 17 a-b Ovate, attenuating and slightly pinched anteriorly; teeth short and moderately strong. Top bluish or greyish green, flecked with minute brown spots and usually with three darker bands, a central brown blotch; and pairs of prominent dark brown terminal spots; base and sides rose or orange-brown, sides prominently spotted, teeth darker colour.

3.5 cm Torres St. to northern NSW. Found in shallow-water rocky habitats along the coast. Moderately common. Preston, 1909 introduced the name *carnicolor,* which is a synonym. This is the only endemic cowry on the tropical east coast of Australia.

▲ SUBGENUS LURIA Jousseaume, 1884
C. lurida Linnaeus

Cylindrical with rounded margins; aperture narrow, with many fine teeth confined to the aperture; fossula deeply concave, crossed by anterior columellar teeth.

Cosmopolitan in tropical and temperate waters. This is a fairly well defined group with six living species, only one of which occurs in Australian waters. Synonym: *Basilitrona* Iredale, 1930 (*isabella* Linnaeus).

Cypraea isabella Linnaeus, 1758

Pl. 32; fig. 7 a-d Top fawn or pale grey with faint transverse bands and irregular, axially aligned black lines; ends with orange blotches; base and sides white, lacking spots.

4 cm. Indo-West Pacific; Abrolhos, WA to central NSW. The species is rather common in coral reef and rocky habitats throughout its range and notable for remarkable size variation even within single populations. The mantle is evenly black. The eastern and western Australian populations have been given subspecies status with the names *lekalekana* Ladd, 1934 and *rumphii* Schilder & Schilder, 1938 respectively but the shells are variable in size and form without consistent distinguishing geographic characters.

▲ SUBGENUS LYNCINA Troschel, 1863
C. lynx Linnaeus

Medium to large sized; ovate to subcylindrical, sides subangulate to rounded, rather inflated; fossula deeply concave, columellar teeth long and extending deep into the aperture throughout its length; terminal ridge strong, curved.

Indo-West Pacific; four northern Australian species are assigned to this group. Some authors have included the southern Australian *reevei* but it has a distinct fossil record and different larval development.

Synonyms: *Ponda* Jousseaume, 1884 (*achatina* Perry); *Callistocypraea* Schilder, 1927 (*aurantium* Gmelin); *Arestorides* Iredale, 1930 (*argus* Linnaeus); *Mystaponda* Iredale, 1930 (*vitellus*).

Cypraea argus Linnaeus, 1758

Pl. 31; fig. 4   Cylindrical, spire whorls visible. Fawn or fawn-grey, top crossed by two pale grey bands, prominently patterned with large irregular brown rings, base with two large reddish brown blotches.

10 cm. Indo-West Pacific; not common but recorded from coral reef habitats between North West Cape, WA and central Qld.

Cypraea carneola Linnaeus, 1758

Pl. 34; fig. 7 a-c   Slightly pyriform, margins subangulate; teeth fine. Top pale orange-brown with four darker transverse bands, sides and base fawn, sides with a granular pattern, teeth deep lilac or purple.

5 cm. Indo-West Pacific; Abrolhos, WA to northern NSW. This is a moderately common cowry in coral reef and rocky habitats. Cate (1968) named a large shell from North West Cape as *C. leviathan gedlingae* but the status of this taxon is doubtful. After examination of the holotype this author believes it is merely a large specimen of *C. carneola*. The extreme variability of this species is reflected in a very long synonymy.

Cypraea lynx Linnaeus, 1758

Pl. 34; fig. 8 a-c   Ovate to pyriform; base rather flat, teeth strong. Top fawn or blue-grey, clouded with overlapping brown and mauve spots and a few larger brown spots; sides fawn with large brown spots, base fawn or white, bright orange between the teeth.

5 cm. Indo-West Pacific; Abrolhos, WA to northern NSW. Moderately common in coral reef and rocky habitats of the intertidal and shallow sublittoral zones. The shells are very variable in size, shape and colouring, and consequently there is a very long synonymy.

Cypraea porteri Cate, 1966

Pl. 34; fig. 13 a-b   Thick-shelled, subglobose; aperture narrow, with distinctly formed but short, fine teeth which ascend onto the narrow fossula anteriorly. Base and sides orange-tan, top pale yellow with many suffused orange-tan spots of varying size and a distinct mantle line.

5.8 cm. Central West Pacific; off the Swain Reefs, Qld. Trawled from deep water. This is a very rare species originally described from the Philippines but now with several records from Taiwan, the Solomon Is. and southern Qld. The species was also named *joycae* Clover, 1970 based on specimens from Taiwan.

Cypraea vitellus Linnaeus, 1758

Pl. 34; fig. 9 a-c   Tumid and rather pyriform; teeth moderately strong. Top pale brown to fawn, usually with two obscure pale transverse bands, sides darker brown, top and sides with prominent white spots of varying size; sides characteristically striated with cloudy vertical white lines; base white to fawn, teeth white.

7 cm. Indo-West Pacific; Cape Leeuwin, WA to central NSW. Moderately common in coral reef and rocky habitats of the intertidal and shallow sublittoral zones. Occasionally found well south of the tropical zone on both sides of the continent. A living population along the Naturaliste-Leeuwin coast of WA seems persistent although this author has not seen evidence of breeding there. In the list of NSW molluscs, Iredale & McMichael (1962), the record of *Mystaponda orcina* Iredale, 1931 was based on a dead specimen of *C. vitellus*.

▲ SUBGENUS **LEPORICYPRAEA** Iredale, 1930
C. mappa Linnaeus

Shells large, ovate, tumid, pinched in at the sides anteriorly; outer lip teeth moderate, short, columellar teeth fine and extend deep into the aperture.

Indo-West Pacific. The position of the two species assigned to this subgenus is problematical. They have been associated with *Cypraea s.s.* and *Mauritia* but do not seem close to either. Iredale's generic name is therefore retained as a provisional subgenus.

Cypraea mappa Linnaeus, 1758

Pl. 31; fig. 2 a-b   Ovate to pyriform and moderately inflated. Base fawn or pinkish; sides spotted; top fawn with crowded irregular brown axial lines interrupted by rounded white spaces; mantle line very prominent, zigzag, with short alternate side branches.

10 cm. Indo-West Pacific; north Qld. This is an uncommon species living in sublittoral coral reef habitats. The colour varies in this beautiful shell. Some Qld shells are rose-tinted.

Cypraea valentia Perry, 1811

Broadly to globosely ovate, anterior end slightly produced and rostrate; terminal ridge on the columellar side pronounced and separated from the fossula by a deep channel. Sides clouded greyish yellow or green with prominent brown spots, top of the same ground colour but not clouded and with numerous brown freckles, an irregular brown central blotch and a vague mantle line; thin brown lines overlie the anterior and posterior ends.

10 cm. Western Pacific; Torres St., Qld. Lives in the deeper sublittoral zone. This handsome shell differs from *C. mappa* by its more inflated form and dorsal colour pattern, namely a much less prominent dorsal line and a conspicuous central

blotch. For many years this was one of the really rare cowries but deep trawling in the Philippines and scuba diving in Melanesia have produced more specimens in the last two decades. However, it is still very rare in Australian waters. Cox (1881, *Proceedings Linnean Society, NSW* **1881**: 539) recorded a specimen from Warrior Reef, Torres St. Searches by divers in that area have never confirmed the occurrence of the species there but in 1990 a CSIRO survey, using a submersible, collected a specimen from a rock wall at 114 m, east of Don Cay at the extreme northern end of the Great Barrier Reef (Loch, 1991, *Australian Shell News* **74**: 1). That specimen is now in the Qld Museum and is the only Australian specimen known at this time. Synonym: *princeps* Gray, 1824.

▲ SUBGENUS MONETARIA Troschel, 1863
C. moneta Linnaeus

Ovate to deltoidal, depressed, with thickened and expanded sides; teeth strong and short; fossula not developed.

Indo-West Pacific. There are three living species. The name *Ornamentaria* was introduced by Schilder & Schilder for *C. annulus* but there is a close affinity between that species and *C. moneta* so that *Monetaria* is preferred here as a subgeneric name for both. *C. obvelata* Lamarck, 1810 from the French Polynesian islands is the third species. Synonyms: *Aricia* (Gray) Broderip, 1837 (*moneta* Linnaeus) - preoccupied; *Aria* Deshayes, 1865; *Mercatoria* Melvill, 1888 (*moneta* Linnaeus); *Ornamentaria* Schilder & Schilder, 1937 (*annulus* Linnaeus).

Cypraea annulus Linnaeus, 1758

Pl. 32; fig. 5 a-b   Sides not nodulose. Sides and base cream, white, yellow or grey, lacking spots; top bluish white with a conspicuous yellow or orange encircling ring.

3 cm. Indo-West Pacific; Rottnest I., WA to southern NSW. Like its relative *C. moneta,* this species is very abundant, living in the open in the intertidal zone. It is often found in tide pools surprisingly high above low water level. A very long synonymy is listed in full by Schilder & Schilder (1971) and Cossignani & Passamonti (1991).

Cypraea moneta Linnaeus, 1758

Pl. 32; fig. 6 a-b   Margins irregularly nodular. Top yellow, orange or greenish grey with three faint darker transverse bands, a faint encircling line sometimes present; base and sides paler than the top and unspotted.

3 cm. Indo-West Pacific; Abrolhos, WA to northern NSW. This is one of the most common of all cowries and lives out in the open high in the intertidal zone of rocky shores. As its name implies it was once commonly used as a form of currency in Indian and Pacific Ocean countries. The

western and eastern Australian populations have been given the subspecies names *rhomboides* Schilder & Schilder, 1933 and *barthelemyi* Bernardi, 1861 respectively, but the species is extremely variable throughout its range and this distinction seems unwarranted. Iredale (1939) introduced three names for Qld shells, all of which can be relegated to synonym here, viz.: *harrisi, isomeres* and *monetoides*. The synonymy is very long and is listed in full in Schilder & Schilder (1971) and Cossignani & Passamonti (1991).

▲ SUBGENUS MAURITIA Troschel, 1863
C. mauritiana Linnaeus

Cylindrical to ovate; spire slightly exsert; margins rounded or calloused; teeth fine to strong, short posteriorly on the columellar side but extending across the deeply concave fossula anteriorly; apertural teeth brown.

Indo-West Pacific. There are eight living species, six of which occur in Australian waters. A group of species known as the "arabic" cowries (*arabica, depressa, eglantina, grayana, histrio, maculifera, scurra*) appear to differ in some respects from *mauritiana*. In their account Cossignani & Passamonti (1991) separate this group as a distinct subgenus, i.e. *Arabica* Jousseaume. However, Schilder & Schilder (1971) and Kay (in Burgess, 1985) lump them all together in *Mauritia* and this procedure is followed here. Synonyms: *Maurina* Jousseaume, 1884 (*mauritiana* Linnaeus); *Arabica* Jousseaume, 1884 (*arabica* Linnaeus); *Mauxenia* Jousseaume, 1884 (*mauritiana* Linnaeus); *Etronsa* Jousseaume, 1886 (*mauritiana* Linnaeus); *Arabia* Jousseaume, 1886 (*arabica* Linnaeus).

Cypraea arabica Linnaeus, 1758

Pl. 31; fig. 8 a-c   Solid, ovate to subcylindrical; margins calloused, especially at the anterior end. Base creamy grey, sometimes with a reddish blotch on the left side, teeth brown; top fawn-grey with a hieroglyphic pattern of brown markings, many tending to be longitudinal lines interrupted by clear spaces; large brown spots present on the sides, dark brown blotches at the ends, but there is no spire blotch (compare with *eglantina*); mantle line prominent.

10 cm. Indo-West Pacific; Shark Bay, WA to central NSW. Moderately common in the intertidal and shallow sublittoral zones of coral reef and rocky habitats. Cate, 1964 introduced the subspecific name *brunnescens* for the WA population, but this seems unwarranted and the name is here regarded as one of the many synonyms.

Cypraea depressa Gray, 1824

Pl. 31; fig. 6 a-b   Ovate, depressed, flat-based, margins thickened and subangular; teeth strong and relatively few. Top chestnut brown with crowded white spots and a narrow mantle line; sides grey with many large dark brown spots; base fawn; teeth brown.

5 cm. Indo-West Pacific; collected from oceanic coral reef habitats of the Rowley Shoals and Scott Reef off WA where it lives in the intertidal zone near the wave-swept reef edge. The thick margins and depressed form distinguish this species. It most resembles the larger Pacific cowry *C. maculifera* Schilder, 1932 but lacks the prominent grey blotch on the base so characteristic of that species.

Cypraea eglantina Duclos, 1833

Pl. 31; fig. 7 a-c Elongate and subcylindrical, with a colour and pattern very like that of *C. arabica* except for the presence of a prominent brown blotch above the posterior canal.

8 cm. Indo-West Pacific; Abrolhos, WA to eastern Qld. Common in shallow water in coral reef and rocky habitats. The name *perconfusa* Iredale, 1935 was introduced for the WA population but should be regarded as a synonym as there are no consistent differences between them and eastern specimens. It is not always easy to distinguish this species from *C. arabica*, although it is usually more slender and less calloused at the margin.

Cypraea histrio Gmelin, 1791

Pl. 31; fig. 5 a-c Ovate with small, short teeth; margins broad, calloused and expanded behind the anterior end. Base white or pale fawn, sides with large dark brown spots; top fawn or blue-grey with indistinct bands and a reticulate pattern of light brown lines; dark brown blotches present at the ends.

9 cm. Indian Ocean; North West Cape, WA to Darwin. The Australian form has been recognised as a subspecies with the name *westralis* Iredale, 1935 although the basis of this seems doubtful.

Cypraea mauritiana Linnaeus, 1758

Pl. 31; fig. 11 a-b Broad and flat-based, with angular margins and strong teeth. Base and sides dark chocolate brown, top slightly paler with large, more or less rounded cream spots; a mantle line is present.

13 cm. Indo-West Pacific Kimberley, WA to north Qld. These large and heavy cowries are found in crevices of rocky headlands exposed to strong wave action. At night they emerge to browse. The mantle is thin and black.

Cypraea scurra Gmelin, 1791

Pl. 31; fig. 9 a-b Elongate and cylindrical, aperture straight, narrow; teeth short, fine, numerous; base and margins rounded. Base and sides fawn, teeth red-brown, sides with large brown spots; top with a reticulate pattern of light brown lines enclosing large round blue-grey spaces; mantle line distinct, spire blotch lacking.

5 cm. Indo-West Pacific; eastern Qld and northern NSW. This species is not frequently taken alive by collectors because it lives deep down among the protective fronds of

living corals. Iredale (1939) unnecessarily introduced the name *antelia* for the Qld population.

▲ SUBGENUS NUCLEOLARIA Oyama, 1959
C. nucleus Linnaeus

Solid; spire depressed and covered in adults; teeth extending across the outer lip and base and over the sides; top heavily nodulose but with a distinct incised axial groove representing the mantle line; fossulate.

Indo-West Pacific. Three living species are assigned to this very different group of cowries, which probably warrants full generic rank. Only one species of the group is found in Australian waters.

Cypraea nucleus Linnaeus, 1758

Pl. 35; fig. 2 a-b Broadly ovate; teeth divide on the base. Top yellow-brown, shiny, nodules ringed with brown; base and sides white, pale fawn or orange; teeth orange or tan.

3 cm. Indo-West Pacific; Albany, WA across northern Australia to southern Qld. Another tropical cowry which is sometimes found in the temperate zone on the south coast of WA. Its colouring distinguishes it from *C. staphylaea* which is also nodulose. The living animal is quite spectacular with its very long, branching mantle filaments.

▲ SUBGENUS NOTADUSTA Schilder, 1935
N. victoriana Schilder

Small, thin-shelled, oblong-ovate to pyriform; teeth fine and short; fossula absent or moderately developed.

Indo-West Pacific. The type species is a Miocene fossil from Victoria. There are five living species, all from the tropical Indo-West Pacific region, and all are rare or uncommon. Only one definitely occurs in Australian waters. Burgess (1985) reports a second species, the very rare *C. martini* Schepman, 1907, from off Lady Elliot I., Qld but the record needs to be substantiated. It is a light brown, tumid, pyriform shell with projecting ends, fine, dark brown teeth and a distinct fossula.

Cypraea punctata Linnaeus, 1771

Pl. 33; fig. 9 Oblong-ovate to pyriform; right margin thickened and slightly angulate, left margin rounded; fossula lacking. Milky white with medium-sized brown spots on the top and margins.

2.5 cm. Indo-West Pacific; North West Cape, WA to Qld. This is an uncommon shell occasionally found in the lower intertidal zone but most Australian records coming from deeper water. Schilder & Schilder (1938) gave the south western Pacific form subspecific status with the name *iredalei*. Iredale, 1939 named the eastern and western Australian shells *persticta* and *carula* respectively. All these names may be relegated to synonymy.

▲ SUBGENUS **NOTOCYPRAEA** Schilder, 1927
C. piperita Gray

Ovate to pyriform; spire impressed; teeth fine, short; fossula concave, usually narrow; mantle thin with short papillae or pustules.

Southern Australia. There is little consensus about the taxonomy of these cowries. In this rather conservative account five polytypic species are recognised. Most of them may be found under stones in the intertidal zone but also occur across the continental shelf.

In this group females lay egg masses consisting of a cluster of oval capsules loosely connected together by jelly strings. Only one egg develops in each capsule and it hatches as a crawling snail. There is no pelagic veliger larval stage, which may account for the extreme variability between isolated populations (Wilson, 1985). Most of the species are very variable and in several there are distinctive local variants whose taxonomic status is uncertain. Some of the variants may be the result of either local inbreeding or phenotypic response to different environmental conditions. The deepwater forms tend to be more tumid and less coloured than the shallow water ones, which further confuses the taxonomy. This is a case where more sophisticated methods than analysis of shell morphology are needed to establish relationships between populations.

The group has a long fossil record in southern Australia. The south-eastern Australian and New Zealand Middle Tertiary group *Notoluponia* Schilder, 1935 (*elegantior* Schilder) appear to be their ancestors. Their nearest living relatives are the South African cowries of the subgenus *Cypraeovula*.

Griffiths (1962, *Memoirs National Museum Victoria* **25**: 211-231) summarised the taxonomy of the group and provided details on the radulae of most of the species.

Synonyms:*Guttacypraea* Iredale, 1935 (*pulicaria*); *Thelxinovum* Iredale, 1931 (*molleri* = *angustata*).

Cypraea angustata Gmelin, 1791

Pl. 35; fig. 16 a-c   Ovate to subpyriform and humped; labial teeth moderately large, extending part way across the outer lip; fossula narrow, crossed by regular columellar teeth which are often divided into upper and lower parts with a central trough running along the fossula and the columellar sulcus. Top brown, tan or grey, usually lacking in either spots or bands; base off-white, sides heavily spotted; ends with prominent spots on the sides of the canals. Mantle red.

3 cm. Eden, NSW to St. Vincent Gulf, SA including northern Tas. The shells are more slender than those of *C. declivis* and the dorsum is rarely spotted. Griffiths described the radula as having large, almost semicircular central teeth with lateral ridges rather than basal cusps like thos of *C. piperita*.

N. verconis Cotton and Godfrey, 1932 is a synonym. Iredale (1931) introduced the name *molleri* for specimens taken from deep water off Twofold Bay, NSW. Griffiths (1962) provided a more detailed description of the holotype and figured the radula. The shell has an exsert spire which led Iredale to introduce the generic name *Thelxinovum* for it but it is surely only an aberration. The dorsum is faintly banded but otherwise the shells fall within the range of *C. angustata*. Griffiths noted that the radula of *C. molleri* had ridges instead of basal cusps on the central teeth, and for that reason *molleri* is treated here as a synonym of *C. angustata*. An albinistic specimen from Tas. named *C. albata* Beddome, 1897 has been allied to *molleri* by Griffiths and probably falls into the synonymy of *C. angustata* as well.

Cypraea comptoni Gray, 1847

Pl. 35; figs. 13 a-f; 17   Subpyriform, rather slender; fossula not well developed, crossed by columellar teeth which may be divided into upper and lower parts by a central depression. Top plain orange-brown, with two faint, continuous transverse bands close together near the centre, sometimes with faint bands near the ends; base off-white; sides sparsely spotted, brown blotches usually present at the ends. Mantle red or orange.

2.5 cm. Southern NSW to Cape Leeuwin, WA. The slender shape and two discrete dorsal bands will usually suffice to distinguish this species. Griffiths described and illustrated the radula as having a central tooth shaped like a truncated triangle and with oblique, outward-pointing basal cusps in the lower corners.

In most populations the dorsum of the shells is light orange-brown but in some, e.g. in Westernport Bay and Tumby Bay, it is dark brown. At Port MacDonnell, SA some specimens are completely white; Schilder & Summers, 1963 named this albinistic form *N. casta*. Griffith (1963 - in the same publication as the Schilder & Summers description) showed that there is a gradation from coloured to colourless in the Port MacDonnell population and that the radulae of pure white shells do not differ from the others. He concluded that *casta* is merely a colour form of *C. comptoni*. Other likely synonyms: *subcarnea* Beddome, 1896; *mayi* Beddome, 1897; *trenberthae* Trenberth, 1961; *hartsmithi* Schilder, 1967.

Cypraea declivis Sowerby, 1870

Pl. 35; fig. 18 a-c   Broadly ovate and humped; fossula shallow, crossed by columellar teeth which are undivided. Top milky white, fawn or sepia, profusely spotted with brown, transverse bands lacking except in juveniles; sides strongly spotted, large brown spots present on the ends. Mantle yellow-grey.

2.5 cm. Northern Tas. and Vic. The humped form and distinctive spotted, unbanded dorsal pattern of the shells of this species make them relatively easy to identify. Griffiths described and illustrated the radula as having a central tooth shaped like a truncated triangle and with oblique, outward-pointing basal cusps in the lower corners.

Iredale (1935) named a population at Geographe Bay, WA (actually Cape Naturaliste) characterised by a reticulate dorsal pattern as a subspecies *N. declivis occidentalis*. However, that population relates to *piperita,* not to *declivis*. There are no authentic records of *C. declivis* west of Spencer Gulf, SA. Synonyms: *lentiginosa* Coen, 1949; *eugeniae* Cate, 1975.

Cypraea piperita Gray, 1825

Pl. 35; fig. 12 a-g   Subpyriform; teeth fine, columellar teeth high at their ends, depressed at their centres, thus partially divided into upper and lower parts; fossula only moderately developed, upper and lower parts of the columellar teeth may not exactly align at the centre of the fossula. Top cream or pale tan, crossed by four interrupted bands and usually profusely spotted and flecked with brown, sometimes with a delicate reticulate pattern; base cream or fawn; sides heavily spotted. Mantle mottled yellow-grey.

2.5 cm. Southern NSW to Cape Naturaliste, WA. This is the most widely distributed and most variable species of the group. Griffiths (1962) described and figured the radula as having squarish central teeth, slightly narrower on top, with basal cusps situated near the centre line of the tooth, and pointing downwards parallel to it.

Variations in the prominence of the dorsal bands and the degree of dorsal reticulation has led to the naming of several populations. A reticulate form of *C. piperita* from south-western Australia was named *occidentalis* Iredale, 1935 (see remarks on *C. declivis*). It is typical of the population north of Cape Leeuwin. Most specimens of that population are without bands but in some there are diffuse dorsal blotches that may tend to align transversely.

An unspotted form with uninterrupted bands from the south coast of WA was named *N. reticulifera* Schilder, 1924. There seems to be an abrupt change from the *reticulifera* form to the *occidentalis* form at Cape Leeuwin but this needs to be verified before any taxonomic conclusions are drawn.

Some Victorian and SA populations have shells with four very prominent dorsal bands. The central pair are usually more prominent than the terminal bands and may fuse to obscure the dorsum. The name *C. bicolor* Gaskoin, 1848 was introduced for a strongly banded form but it clearly falls within the range of variation of the species and must be relegated to synonymy.

A colour form with an unbanded, yellow or golden dorsum found in the Flinders and San Remo areas of central Victoria, was named *N. wilkinsi* Griffiths, 1962 but that too is here regarded as a synonym. The radula is like that of *C. piperita*.

Iredale (1931) named two species from deep water off southern NSW, viz. *N. emblema* and *N. dissecta*. The holotype of the former is an ovate shell with a bluish white, unbanded dorsum and spotted sides. The shell of *dissecta* is a little more cylindrical with a faintly banded dorsum and spotted sides. Both forms fall within the range of variation of *C. piperita*. Griffiths noted that the radula taken from the holotype of *C. emblema* had very large central teeth with prominent basal cusps, similar to the central teeth of *C. piperita* except for the

large size. The central radular teeth of *C. dissecta* were like those of *C. piperita* but with small basal cusps. Mainly because of these differences in their radulae, Griffiths retained these two names as representing distinct species. For the moment these names are treated by this author as synonyms of *C. piperita* but further study of these NSW animals is needed.

Cypraea pulicaria Reeve, 1846

Pl. 35; fig. 14 a-d   Elongate, subcylindrical; teeth very fine; fossula wide and deeply concave with its inner edge quite angular and projecting into the aperture; columellar teeth extending well down into the aperture, including across the fossula, but are divided at their centres to form two ranks. Pale orange, rose or pink, sides with rather large brown spots which usually partly merge to form four interrupted transverse bands. Mantle orange.

2 cm. Southern WA from the Great Australian Bight to Rottnest I. The distinct fossula led Iredale (1935) to separate this species into its own genus *Guttacypraea,* but its close affinity to the *Notocypraea* group is readily apparent. Griffiths (1962) noted that the radula resembles that of *N. piperita.*

Shells of the south coast population, east of Cape Leeuwin, differ a little from those of the west coast population. The shells are cream, often with a weak reticulate dorsal pattern but showing only traces, if any, of transverse bands. Also the lateral spots are finer and the fossula does not project so far into the aperture. This WA south coast population was named *N. euclia* Steadman & Cotton, 1946 and as the characters seem to be consistent the name is retained here as a geographic subspecies.

▲ SUBGENUS **PALMADUSTA** Iredale, 1930
C. clandestina Linnaeus

Small, oblong-ovate to pyriform; teeth usually short, moderately well formed; fossula poorly developed.

Indo-West Pacific. Many species are assigned to the group, which seems to be something of a ragbag for small shells which don't fit anywhere else. Schilder & Schilder (1971) treat *Palmadusta* and *Purpuradusta* as distinct genera.

Synonyms: *Evenaria* Iredale, !930 (*asellus*); *Palmadusta* Iredale, 1930 (*clandestina* Linnaeus); *Cupinota* Iredale, 1939 (*macula*); *Opponaria* Iredale, 1939 (*minoridens*); *Purpuradusta* Schilder, 1939 (*fimbriata* Gmelin).

Cypraea asellus Linnaeus, 1758

Pl. 33; fig. 4 a-b   Pyriform to ovate; teeth short except on the posterior end of the columella where they spread on to the base. Porcelaineous white except for three broad bands of black or dark brown across the top.

2 cm. Indo-West Pacific; Coral Bay, WA to central NSW. Intertidal and shallow sublittoral. Schilder (1930) proposed the subspecies name *latefasciata* for NSW specimens, which are said to be more dilated at the ends and to have shorter posterior columellar teeth, but the name is now included in the synonymy. The mantle is black. This is not a very common species but very widespread.

Cypraea clandestina Linnaeus, 1767

Pl. 33; fig. 16 a-c   Stout and ovate; teeth extending part way across the outer lip and onto the base on the columellar side. Base creamy white, top ivory usually crossed by two slightly darker bands and with thin zigzag brown lines; spots lacking.

2.5 cm. Indo-West Pacific; Abrolhos, WA to central NSW. Relatively common in intertidal and shallow sublittoral rocky habitats. The names *whitleyi* and *extrema* were introduced by Iredale (1939) for Qld and NSW forms. The southern shells (from Shellharbour) were said to be "larger, less solid, less pyriform, anteriorly broader, clouded but zoning indistinct, and zig zag lines notable...; the teeth less pronounced, coarser". These characters seem to be within the variation of the species throughout its range, with no basis for differentiating the NSW population. Both names are included in the synonymy. The black mantle and white shell of this little species make an unusual contrast.

Cypraea contaminata Sowerby, 1832

Pl. 33; fig. 10 a-b   Elongate-ovate, ends slightly projecting, spire impressed; teeth short. Top pinkish fawn with three pale bands, a few indistinct brown spots and a prominent red-brown central blotch; anterior ends with small spots at the sides and tips, posterior end unspotted; base and sides yellow or pale tan and strongly spotted.

1.8 cm. Indo-West Pacific; North West Cape, WA to Qld. A rare little cowry with few confirmed records in Australia. Collected by the WA Museum at the Dampier Archipelago. Reports of the species from North West Cape and Qld need confirmation.

Cypraea fimbriata Gmelin, 1791

Pl. 33; fig. 7   Subcylindrical, with slightly beaked ends; teeth rather coarse. Top pale orange-brown or grey, with numerous minute brown freckles and a central dark band which is usually divided; conspicuous lilac spots present on each side of the anterior and posterior ends, spire blotch brown to slate; base white, unspotted, sides usually faintly spotted above the lateral callus.

2 cm. Indo-West Pacific; Cape Naturaliste, WA to central NSW. Not very common but widely distributed. It lives in intertidal and shallow sublittoral rocky and coral reef habitats. Iredale (1939) gave the subspecific name *blandita*

to NSW specimens, with no sound reasons. There are several species like this, i.e. *microdon*, *minoridens* and *hammondae*, which the collector may have difficulty telling apart. They all have carmine mantles. *C. fimbriata* is most like *C. minoridens* but has coarser teeth, a more curved aperture, and the spire blotch is brown or slate and not bright purple to carmine as it is in the other species. *C. hammondae* differs in having prominent lateral and basal spots, while *C. microdon* is distinguished by its pyriform shape and very fine teeth.

Cypraea gracilis Gaskoin 1849

Pl. 33; fig. 2 a-d   Rather solid, pyriform to ovate; teeth short. Base and sides yellowish or white; sides with conspicuous brown spots; top blue-grey, with numerous small brown freckles and some larger spots and usually a large central brown blotch; a pair of brown spots present at the anterior end and a brown blotch in the spire.

2.5 cm. Indo-West Pacific; Cape Naturaliste, WA to northern NSW. This is a moderately common little shell which lives in the intertidal and shallow sublittoral zones of coral and rocky habitats. There are several named subspecies. The Schilders proposed use of the names *gracilis macula* Angas, 1867 (= *interpunctata* Henn, 1896) for the Qld shells and *gracilis irescens* Sowerby, 1870 for the western population. Iredale (1939) regarded the eastern *macula* as distinct from *gracilis* and named the western population *macula hilda* on the grounds that shells from Shark Bay were "smaller and paler". However, there seems to be no consistent differences between eastern and western populations, or between them and *C. gracilis* from elsewhere in the species-wide range. Use of the available subspecies names is not recommended.

Cypraea hammondae Iredale, 1939

Pl. 33; fig. 5 a-c   Subcylindrical to slightly pyriform; teeth moderately coarse and short. Base yellowish white, sides and base conspicuously spotted; top fawn or brown with brown spots but not freckled; four faint transverse bands usually evident; black to brown spire blotch prominent.

2 cm. Central Indo-West Pacific; Abrolhos, WA to northern NSW. Not common. Lives in the intertidal and shallow sublittoral zones. Iredale's original material came from the Clarence River, northern NSW and Yirrkala, NT. The eastern Australian shells tend to be cylindrical while those from WA and the Philippines are more pyriform. The Philippine and WA populations have been recognised as a subspecies under the names *h. raysummersi* Schilder, 1960 and *h. dampierensis* Schilder & Cernohorsky, 1965. As well as the pyriform shape the WA *dampierensis* are usually smaller than eastern specimens, with the anterior end attenuated, the right side less margined, the left side rather rounded, and the labial teeth fewer and more widely spaced. Further study of NT specimens is needed to determine whether the differences are maintained in the intermediate populations.

Cypraea humphreysii Gray, 1825

Pl. 35; fig. 10 a-b Pyriform, teeth moderately strong. Top light mustard colour, densely spotted with orange-brown, crossed by three pale bluish white bands which may be obscure; base and sides orange-brown and prominently spotted.

2 cm. South-west Pacific; Qld to Long Reef, NSW. Rather uncommon. Gray introduced this name as a subspecies of *C. lutea* and that affinity is still recognised by some authors, although the differences in colour pattern are very pronounced and consistent. This author prefers to treat them as distinct but related species. Gray's description and illustration have sometimes been interpreted as referring to the banded form of *C. lutea*, but the type specimen in the British museum clearly ties the name *humphreysii* to the densely spotted, triple-banded shells from the south-west Pacific. It follows that the name *yaloka* Steadman and Cotton, 1943, introduced for specimens from Fiji, is a synonym of *humphreysii*.

Cypraea lutea Gmelin, 1791

Pl. 33; fig. 17 a-c Ovate to pyriform. Base and sides rich orange-brown with distinct dark brown spots on the sides. Top deep orange or fawn with two thin white spiral lines and often with scattered brown spots.

2 cm. Central Indo-West Pacific; North West Cape to the Kimberley, WA. Moderately uncommon. Lives in the intertidal and shallow sublittoral zones. Curiously there seems to be no record from Qld. The heavily spotted, triple-banded species *C. humphreysii* may be its replacement there. Specimens from WA typically have two thin and very distinct white bands across the dorsum, leading Iredale, (1935) to name them *C. bizonata*, but most contemporary authors regard this form as within the range of the species.

Cypraea microdon Gray, 1828

Pl. 33; fig. 8 a-b Tiny, pyriform to oblong-ovate, ends projecting in mature shells, especially the left side of the posterior canal; teeth numerous and very fine. Top white or pale orange-brown, with two faint dorsal bands which are often bordered by brown dots, surface with minute brown freckles; lateral spotting sparse, base and sides white; tips of canals lilac, spire with a bright purple blotch.

1.5 cm. Indo-West Pacific; north Qld. This appears to be a coral reef species and is not uncommonly taken by collectors. The pear-shape, fine teeth and bright purple spire blotch separate *microdon* from other like species but it is very close to *minoridens*.

Cypraea minoridens Melvill, 1901

Pl. 33; fig. 6 a-b Tiny, subcylindrical to ovate; teeth fine; ends slightly projecting. Top fawn, profusely freckled with brown and crossed by a single broad band; sides unspotted, ends conspicuously spotted with rose-brown and there is a prominent red-brown spire blotch.

1.3 cm. Indo-West Pacific; eastern Qld south to Port Jackson, NSW. Lives among dead coral and coral rubble. This is the smallest of all the cowries. Iredale (1939) introduced *blandita* as a subspecies name for NSW shells on the grounds of their "larger size, less projecting posterior extremities, less pronounced columellar ledge and also their greater proportional breadth". Further study is needed to demonstrate whether these characters are sufficiently consistent to warrant taxonomic separation. Lack of lateral spots distinguishes *minoridens* from *fimbriata* although the species are very close.

Cypraea saulae Gaskoin, 1843

Pl. 33; fig. 19; Pl. 35; fig. 11 a-b Pyriform and rather tumid, anterior end rostrate, spire depressed; teeth short, moderately strong. Yellow or cream, darker at the ends with a large irregular brown central blotch and usually scattered brown spots on the top, sides strongly spotted; base white, usually with pale yellow between the teeth.

3 cm. Central Indo-West Pacific; Dampier, WA to southern Qld. An uncommon species, rarely found in the intertidal zone. The shells are variable and several doubtful subspecies names have been introduced. Iredale (1935) proposed *saulae nugata* for Qld shells on the ground that they are more elongate and graceful than the Asian shells. Cate (1968) used the name *saulae crakei* for large, broad, anteriorly attenuate specimens from WA.

Cypraea ziczac Linnaeus, 1758

Pl. 33; fig. 18 a-c Light, pyriform, with numerous small short teeth. Base orange-brown with dark brown spots which extend to the sides; top fawn or light brown with bands of brown spots at each end and three transverse white bands containing V-shaped light brown lines.

2.5 cm. Uncommon Indo-West Pacific; Shark Bay, WA to northern NSW. This very distinctive cowry is rather uncommon. It lives in the intertidal and shallow sublittoral zones. The only species like it is *C. diluculum* Reeve, 1845 from East Africa. Sometimes the V-shaped marks join to form zigzag axial lines from end to end. At one time Schilder treated eastern Australian specimens as a separate subspecies under the name *vittata* Deshayes, 1831. Iredale (1939) introduced the name *signata* as a subspecies of *ziczac*, based on a specimen from the mouth of the Clarence River in northern NSW but without giving any differentiating characters. Both these names can probably be relegated to synonymy.

▲ SUBGENUS **PUSTULARIA** Swainson, 1840
C. cicercula Linnaeus

Small globular shells with rostrate and projecting ends; back pustulose or smooth; teeth fairly strong and spreading at least part way onto the base; fossula deep and concave.

Indo-West Pacific. There are five living species. They live in coral reef habitats, deep within crevices and among the fronds of corals. A distinctive group probably warranting full generic rank. Synonyms: *Globularia* Sowerby, 1842 (*globulus*); *Epona* Adams & Adams, 1854 (*cicercula*).

Cypraea bistrinotata Schilder & Schilder, 1937

Pl. 35; fig. 7 a-b   Like *cicercula* but more globose; spire covered with callus; back pustulose and has an incised axial sulcus; columellar teeth fairly short at the centre but spreading across the base in the anterior and posterior thirds, reaching up onto the side anteriorly. Top spotted light yellow-brown with three pairs of darker blotches bordering the sulcus; base yellow-brown, with four prominent brown blotches.

2 cm. Western Pacific; eastern Qld. Uncommon. The four basal blotches and three pairs of obscure dorsal blotches distinguish this species from C. *cicercula*. and C. *globulus*.

Cypraea childreni Gray, 1825

Plate 35; fig. 4 a-b   Ends moderately beaked and projecting; apertural teeth extending over the margins forming closely-spaced, sharp, transverse ribs over the back, interrupted by an axial sulcus. Pale yellow-brown.

1.2 cm. Indo-West Pacific; Rowley Shoals, WA (Hugh Morrison collection). This rare little shell probably lives on the Great Barrier Reef as well but I have not seen specimens.

Cypraea cicercula Linnaeus, 1758

Pl. 35; fig. 8 a-b   Ends beaked and prominently projecting; back pustulose, with a deeply incised sulcus (representing the mantle line) terminating posteriorly in a pit beside the sunken spire; teeth long and extending across the base to the margins on both sides. Top pale yellow with faint brown spots but no blotches; base fawn or pale yellow, teeth darker.

2.5 cm. Indo-West Pacific; North West Cape, WA to southern Qld. The lack of basal blotches distinguishes this species from C. *bistrinotata* and the deep dorsal sulcus from C. *globulus*. Some authors consider shells from Tahiti to be a distinct species named C. *margarita* Dillwyn, 1817.

Cypraea globulus Linnaeus, 1758

Pl. 35; fig. 9 a-d   Like *cicercula* but the back is smooth and there is no sulcus; teeth relatively short. Top pale orange-brown, usually spotted; base usually with four brown blotches.

2.5 cm. Indo-West Pacific; North West Cape, WA to southern Qld.

Cypraea mariae Schilder, 1927

Plate 35; fig. 5 a-b   Ends not beaked; surface smooth; teeth fine and short. Pale yellow, with a pattern of darker yellow or pale orange spots enclosed within dark yellow or orange rings.

2 cm. Western Pacific; Scott Reef, NT (Hugh Morrison collection). A coral reef species rarely collected alive.

▲ SUBGENUS **SCHILDERIA** Tomlin, 1930
C. utriculata Lamarck

Pyriform, sides rounded, aperture narrow, teeth fine, fossula well developed.

Central Atlantic and western Pacific. The type species is a Pliocene fossil from the Mediterranean region. (Cossignani & Passamonti, 1991, give it subgeneric status within the genus *Schilderia*.) The group has an interesting distribution with Middle and Upper Tertiary representatives in the European area and living survivors in the eastern Atlantic and western Pacific. Two of the western Pacific living species occur in deep water off the southern coast of Qld.

Cypraea hirasei Roberts, 1913

Pl. 34; fig. 11   Tumid, with markedly projecting and sharply margined anterior end; teeth fine and numerous. Top pale yellow with a large central brown blotch which has irregular and variable outline, and crowded yellow-brown spots which may form a netlike pattern; base and sides white, sides bear rather large but faint brown spots.

6 cm. Southern Japan and off Cape Moreton, southern Qld. Dredged or trawled on the outer part of the continental shelf. When the first Qld specimens were discovered they were named C. *queenslandica* by Schilder, 1966, but most authors now regard them as a subspecies at best of the Japanese C. *hirasei*.

Cypraea langfordi Kuroda, 1938

Pl. 34; fig. 12 a-c   Base and sides light orange-brown, usually unspotted; top paler, more or less evenly mottled with brown and sometimes crossed by three darker bands or zones of blotches.

6 cm. Southern Japan and southern Qld. Most of the Qld specimens have come from the outer shelf between the Capricorn Group and Cape Moreton but the species has been taken as far north as off Leopard Reef. Dredged or trawled on the outer part of the continental shelf. When first discovered Qld shells were named *moretonensis* by Schilder, 1965, but more specimens have shown that they do not differ consistently from Japanese shells. The Qld shells from more northerly locations tend to be more tumid than those from the southern end of the range.

▲ SUBGENUS **STAPHYLAEA** Jousseaume, 1884
C. staphylaea Linnaeus

Ovate; labial teeth extending across the outer lip and onto the marginal callus and projecting slightly onto the sides, columellar teeth extending at least part way over the base; fossulate; top smooth or nodulose.

Indo-West Pacific. There are three living species, two of them widespread and common in northern Australia.

The group seems to be close to *Nucleolaria*. Synonyms: *Purperosa* Iredale, 1935 (*facifer* Iredale = *limacina* Lamarck); *Eustaphylea* Steadman & Cotton, 1943 (*semiplota* Mighels).

Cypraea limacina Lamarck, 1810

Pl. 35; fig. 1 a-b   Elongate-ovate to pyriform, spire evident, ends produced and ridged above; right margin calloused and pitted; columellar teeth spreading only part way across the base; top smooth or only weakly nodulose. Top brown or bluish grey with white spots and a weak mantle line; base and sides white, ends and teeth orange.

3.5 cm. Indo-West Pacific; Cape Leeuwin, WA to northern NSW. Moderately common in the intertidal and shallow sublittoral zones. The more elongate form, relatively short teeth and lack of dorsal nodules easily distinguish this species from *C. staphylaea*. The rare *C. semiplota* Mighels, 1845 from Hawaii is a close relative.

Iredale (1935) noted the variation in shell form and colouring of this species in Qld. He described a specimen from Lindeman I. as having a smooth upper surface without raised spots, contrasting it with a series from the Capricorn Group in which the upper surfaces were rendered pustulose by raised spots. This led him to name both kinds as a new species and subspecies respectively, the former as *S. facifer* and the latter as *S. facifer monstrans*. The larger series now available indicates that both forms fall within the range of variation of *C. limacina* and contemporary authors relegate both names to the synonymy of this species.

Cypraea staphylaea Linnaeus, 1758

Pl. 35; fig. 3 a-b   Ovate; ends projecting; top nodulose and with a weakly incised mantle line; teeth on the columellar side bifurcate and spreading across the base to the edge of the left margin. Back fawn or grey, nodules white; base and teeth orange, ends darker orange.

3 cm. Indo-West Pacific; Rottnest I., WA to southern Qld. Not very common but widespread. Iredale (1935) named a variant from the Capricorn Group, Qld *S. staphylaea descripta* and Coen, 1949 used *staphylaea nitida* for Qld shells, but the alleged diagnostic characters fall well within the range of variation exhibited by the species and these names are treated as synonyms of *staphylaea* here. The mantle is black with mostly short pointed filaments and a few dendritic ones intermixed.

▲ SUBGENUS **TALPARIA** Troschel, 1863
C. talpa Linnaeus

Cylindrical; aperture narrow, teeth fine and numerous; fossula short, deeply channelled anteriorly; columellar teeth not extending far into the aperture.

Indo-West Pacific. There are only two living species, one of them widespread in the region.

Cypraea talpa Linnaeus, 1758

Pl. 31; fig. 3 a-b   Base and sides chocolate brown without spots; top pale fawn or yellowish brown with four transverse bands of slightly creamy brown.

9 cm. Indo-West Pacific; North West Cape, WA to northern NSW. Found in coral reef habitats and never common. The mantle of this species is quite unusual, having short, thick, bubble-like papillae instead of filaments. *C. exusta* Sowerby, 1832 is a similar species from the western Indian Ocean.

▲ SUBGENUS **UMBILIA** Jousseaume, 1884
C. umbilicata Sowerby (= *hesitata* Iredale)

Pyriform, spire depressed (umbilicate); anterior end projecting, more or less rostrate and flat on the underside and with a conspicuous distortion on the upper side; teeth fine and short; fossula lacking or very narrow and concave, smooth, its inner edge angulate.

Eastern and southern Australia. It remains uncertain whether there is a single, extremely variable living species in this group, or three geographically separated species or subspecies. In this account, three species are recognised. The group has a long fossil record in southeastern Australia. It has no relatives outside the region and may warrant generic rank. The Schilders (1971) accept the Victorian Miocene groups *Rhynhocypraea* Cossmann, 1898 (*leptorhyncha* McCoy) and *Palliocypraea* Cossmann (*gastroplax* McCoy) as subgenera. The giant Miocene fossil cowry *Gigantocypraea* Schilder, 1927 (*gigas* McCoy, 1867) may also belong here, or taxonomically near by.

Cypraea armeniaca Verco, 1912

Pl. 30; fig. 12-16   Globosely pyriform, anterior end weakly rostrate, aperture strongly arched; labial teeth moderately fine, numbering about 35, fossula lacking. Cream, ivory to apricot, usually with suffused light brown spots and a large goldenbrown dorsal blotch, sides unspotted, base sometimes with a darker patch on the columellar side.

11.4 cm. South coast between at least Cape Nuyts, SA to Albany, WA - possibly extending further east and west of these limits. Trawled on the outer part of the continental shelf. This is one of Australia's rarest and most sought after shells. It may be only a bulbous western variant of *hestitata* although the dorsal blotch seems to be consistent in the few specimens examined and a calloused anterior rostrum is poorly developed. There appear to be no intermediate specimens between Bass St. and Port Lincoln. Species status is appropriate on the available information.

Precise distributional limits are not known. South Australian trawlers operating from Ceduna have been taking this species in the Head of the Bight but there are unconfirmed reports of specimens from as far east as Port Lincoln. The original material came from the Eucla-Esperance area and additional specimens have been taken there in recent years. During the last few years specimens have been taken from as far west as Albany.

Raybaudi (1980) recognised two forms, a typically large and light-coloured western form from the outer shelf between Esperance and Eucla, and a smaller, darker coloured eastern form from the Head of the Bight off Yalata. Raybaudi dubbed the western form *westralica* and the eastern form *southstralica*. These names were introduced as varietal names and they have no taxonomic validity. There is also some doubt whether the western and eastern populations are consistently different. In a more recent article Raybaudi (1992, *World Shells* **1**: 15-21) provided additional information, noting that large shells are also found in the eastern population. The Albany specimen illustrated here (Pl. 30, fig. 12) shows that small shells also occur at the western limit of the species range. There seems little ground for maintaining the names *westralica* and *southstralica* even as "form" names.

The name issue has been further complicated by Raybaudi (1992, *loc. cit.*) listing specimens as *minima, major, maxima* and *saturata*, in the form of formal taxonomic names. There is no description or designation of types and it clearly was not his intention to introduce these names as taxa. They should be disregarded for the purposes of taxonomy.

Cypraea capricornica Lorenz, 1989

Pl. 30; figs. 1-6, 8 Inflated to globular with strongly rostrated extremities; anterior canal extended, bearing a prominent dorsal tubercle; aperture narrow, columella with a distinct fossula, labral teeth fine, about 30, extending only slightly onto the lip, columellar teeth strong, extending onto the base, numbering about 22; dorsal surface dull, minutely granulate. Dorsum and margins greyish brown, the dorsum paler and profusely spotted; base darker, sometimes blotched with reddish brown.

8 cm. Swain Reefs and Capricorn Channel, Qld. Trawled at depths of 120-360 m. This species was discovered some years ago. Lorenz first used the name *capricornica* as a "varietal name" in 1988 (*La Conchiglia* **230**: 14-20) but formalised it the following year (*Schriften zur Malako-zoologie* **2**: 2-29). Although the Capricorn population was initially regarded as a form of *C. hesitata* by many cowry specialists, the shell characters are convincingly different, especially the strong columellar teeth and the presence of a distinct fossula. As Lorenz has shown, this living population is more like some of the Miocene fossil representatives of *Umbilia* than it is to *C. hesitata*.

There are two variants within the Capricorn population that appear to inhabit different depth zones in slightly different areas. One form is light coloured, relatively small and heavy, with a nearly straight aperture and comes from the edge of the continental shelf (150-225 m) off Lady Musgrave and Lady Elliot Is. (Pl. 30; figs. 3, 8). This is the form that Lorenz cited as "typical *capricornica*". A darker, larger and more bulbous form has been trawled in deeper water (270-360 m) off the Swain Reefs (Pl. 30; figs. 1, 2, 4, 5, 6). It often has a dark base.

Cypraea hesitata Iredale, 1916

Pl. 30; fig. 7, 9-11 Pyriform; surface usually granular; anterior end pronouncedly rostrate, with a dorsal callus; aperture arched, labial teeth fine, numbering about 33, columellar teeth fine, short posteriorly where they do not descend into the aperture; fossula lacking or very narrow. Creamy white with suffused tan spots on the top and larger more sharply defined spots along the lower sides; base white.

12 cm. Capricorn Group, Qld to Bass St.Trawled on the outer continental shelf. *C. umbilicata* Sowerby, 1825 and *C alba* Cox, 1879 are prior names but both had been used previously for different species.

This species is common on the trawling grounds of south-eastern Australia but less so (or less commonly collected) in the northern part of its range. The specimens illustrated at Pl. 30, figs. 10, 11 represent the typical south-eastern form. It is an extremely variable species and several distinct forms occur. A summary of the variations was published by Felix Lorenz Jr. (1988, *La Conchiglia* **230**: 14-20).

A large white form from Bass St. was named *howelli* Iredale, 1931 but is merely a variant (Pl. 30, fig. 9). Shells from Moreton Bay to central NSW are much smaller and darker than those from further south. A dwarf form from northern NSW was named *beddomei* Schilder, 1930 (Pl. 30, fig. 7). Most contemporary workers treat this as a regional variant so that the name *beddomei* falls into the synonymy of *hesitata*.

More recently, similar but thinner dwarf specimens taken off Cape Moreton were illustrated by Luigi Raybaudi (1979) together with another variant from the Capricorn Channel. Raybaudi introduced the name *nordica* for these shells, without distinguishing between the Cape Moreton and Capricorn specimens. The name was introduced as a form name without designation of a type and it has no taxonomic validity.

▲ SUBGENUS **ZOILA** Jousseaume, 1884
C. scottii Broderip

Ovate to pyriform; aperture narrow, fossula well developed and concave; labial teeth short, columellar teeth short or lacking posteriorly, not extending onto the fossula.

Western and South Australia. There are eight living species but several of them have distinctive geographical variants which have been named. The group was revised, with anatomical and radular details given of the species known at that time, by Wilson & McComb (1967, *Indo-Pacific Mollusca* **1** (8): 457-484).

The living species appear to be survivors of a Tertiary lineage which flourished in the Sea of Tethys. A Tethyan fossil group known as *Bernaya* Jousseaume, 1884 was a relative. The species of *Zoila* are sponge-eaters. They have a thin mantle on which papillae are small and simple or lacking. The females lay typical cypraeid egg masses but, unlike most cowries, only one egg in each capsule develops and there is no pelagic veliger stage (Wilson, 1985).

Cypraea decipiens Smith, 1880

Pl. 28; figs. 7-11 Ovate to subpyriform, narrowing anteriorly, very humped with a steep posterior slope, spire just evident; base rather flat though inclined inwards, margins subangulate, not flanged anteriorly, ends not very projecting, sides of canals rounded, right side of the posterior channel curved and projecting anteriorly; aperture rather straight and little dilated anteriorly; outer lip deep, labial teeth weak, extending deep (vertically) but not extending across the outer lip, numbering about 22; columellar teeth weak but usually present along the whole length of the aperture. Base and sides chocolate brown and unspotted; top ivory cream with a distinctive hammer pattern of obscure rectangles, covered with merging brown spots and blotches.

7 cm. North West Cape to West Kimberley, WA. Lives on sponges in the sublittoral, usually below 15 m. This is a tropical member of the *Zoila* group but it has direct development like its southern relatives. It is quite common on the pearling grounds between Broome and Port Hedland and pearl divers began bringing in specimens many years ago.

Although the base and sides are typically unspotted, in some specimens faint lateral spots may be seen beneath the heavy brown colouring. Federico Bodini (*World Shells* **2**: 32-34) reported "golden" specimens of this species collected north of Broome. Similar golden shells, i.e. lacking brown pigments, are known in *friendii, venusta, rosselli* and now *decipiens*. It seems to be a feature of the brown-shelled species of this subgenus that they occasionally produce these aberrant yellow or golden shells. See also remarks on *C. perlae* and *C. mariellae*.

Cypraea eludens Raybaudi, 1991

Pl. 28; figs. 12-15 Broadly ovate to subpyriform, solid and heavy; base flat, dorsum high and humped with a steep posterior slope, spire slightly protruding but covered with callus, body whorl suture impressed; right margin rounded centrally, left margin subangulate and slightly flanged anteriorly, ends slightly projecting, sides of channels not elevated; aperture almost straight, weakly dilated anteriorly; labial teeth strong, deep vertically but not extending across the outer lip, numbering 18-21; columellar teeth weaker than the labials but nevertheless moderately strong, even-sized and extending along the entire length of the inner lip. Base orange; ends brown, sides orange at the margin, becoming marbled and clouded bluish grey dorsally, with prominent, large, dark brown spots; dorsum ivory, overlain to the point of being almost obscured by merging dark chocolate brown spots and maculations, an indistinct mantle line usually present.

6 cm. Western end of the North West Shelf between North West Cape and Onslow. Lives on sponges at depths from 20 m. On the limited information available at present, this newly described species appears to have a restricted distribution. It is not yet clear whether its range overlaps with that of *C. decipiens*. Trevor Sutcliffe (pers. comm. November, 1992) has informed me that he has collected *decipiens* in the same general area as *eludens* (i.e. within a few kilometres) but that the two species live in different types of habitat. In shell form and apertural dentition, these two species are quite similar, although *eludens* has a much heavier shell with a more prominent marginal callosity. The colour patterns, however, are quite different. The clouded, marbled, bluish grey, heavily spotted sides recall *C. venusta* (the dark form) and the flat-based *friendii jeaniana* from Quobba, and are very unlike the uniform, unspotted chocolate sides of *decipiens*. There can be little doubt that this very beautiful shell represents a distinct species.

Cypraea friendii Gray, 1831

Pl. 26; figs. 1-18; Pl. 27; figs. 12-19 The species complex has populations from St.Vincent Gulf, SA west and north to at least Point Maud on the north west coast of WA. There are also poorly provenanced records of the species from the outer part of the North West Shelf. North of the Perth metropolitan area there are no shallow-water populations (i.e. less than about 10 m). Conversely, there appear to be no deepwater populations (i.e. on the outer edge of the continenal shelf) east of the Great Australian Bight.

This is an extremely variable species. The shells vary from dark chocolate to white, from slender to globose, from almost edentulous to moderately toothed. Wilson & Summers (1966, *Journal Malacological Society Australia* **1** (1): 3-24) described the geographic variation in the group, interpreting it as a single variable species, with the SA form representing an isolated subspecies (*friendii thersites*). Since that paper the known species range has been extended, several new variants have been discovered and new names have been proposed. The following account represents this author's present view of the species complex, recognising four geographic subspecies.

Wilson & Summers (1966) described a statistical trend for progressively increasing shell width (relative to length) in shallow-water populations from Fremantle south and east to SA exhibit. This was interpreted as a clinal change. The most slender specimens are from the shallow reefs immediately north of Fremantle. On average, specimens from Geographe Bay are wider. South coast WA shells, i.e. from east of Cape Leeuwin, are like those from the west coast in every respect except that they are much wider. SA shells are, on average, wider still. Since this clinal trend was described, midshelf populations of the species complex have been discovered in Shark Bay and further north which are characterised by wide, flat-based shells. Some of these northern shells have the proportions of south coast specimens, suggesting a reversal of the clinal trend north of Fremantle.

The clinal trend in relative shell width is more complicated than this, however. There is a matching increase in width from shallow water to deep water. Shells from the outer part of the shelf off the central west coast are broader than shallow water specimens from the same latitude and as broad as shells from shallows on the south coast. Similarly, shells from the shelf edge off Esperance on the south coast are wider than shells from the adjacent shallows and as broad as shells from shallows in SA. The cline is therefore two-dimensional.

It is possible that the relative width of the shell is influenced by temperature, i.e. the colder the water the more inflated the shell grows. There is a progressive decline in water temperature from north to south on the west coast and from west to east on the south coast and, presumably, a matching decline from shallow to deep water. The reversal of the trend north of Fremantle seems to discredit this interpretation although study of bottom water temperatures may provide an explanation.

Another variable character is shell colour. Shallow water shells throughout the species' range have a chocolate base and sides and usually a pale dorsum overlain with suffused brown spots. (Entirely dark brown specimens occur.) There are also large, round brown spots on the sides. But along the outer edge of the shelf the shells tend to be white or pale tan with mid tan spots. This trend is independent of other characters. It is interpreted as a phenotypic response without a genetic basis. These cowries feed on sponges. The sponges found along the inner part of the shelf are usually strongly pigmented with brown and yellow whereas the outer shelf sponges are mostly white or pale yellow. The whiteness of the deepwater shells is likely to be simply because brown colour pigments are not available. If this is true then lack of colour in the shells is not a suitable character for distinguishing species or subspecies.

Some other shell characters are correlated with relative width; for example, wide shells are also high and humped. Others appear to be local population variants probably relating to local inbreeding. For example, the population from the Marmion Marine Park just north of Fremantle has shells with very prominent spires (Pl. 26, fig. 9). However, there are several characters which appear to be regional and useful for taxonomic purposes, most notably aperture dentition.

Taking account of this extraordinary variability and the interpretation of the variable characters, this author recognises four geographic subspecies within the *friendii* complex.

Cypraea friendii jeaniana (left); note the presence of small denticles along the entire inner lip. *C. friendi friendiii* (right) with edentulous lip except anteriorly.

1. C. friendii friendii Gray, 1831

Pl. 26; figs. 7-12   Elongate and narrowing anteriorly; spire projecting, sides of anterior and posterior canals sharp-edged and extended; labial teeth short but moderately strong,

columellar teeth represented by only a few small anterior denticles. Top cream with about four bands of blue-grey bars or rectangular blocks, overlain by large merging brown spots, sometimes with an indistinct mantle line; sides chocolate with large, round, darker spots; base chocolate, the colour extending to the sides of the aperture. Mantle black, velvety, without papillae.

13 cm. Geographe Bay to Fremantle, WA. Lives in the sublittoral zone from the low tide level to at least 180 m. This beautiful species lives out in the open, usually on the sponges on which it feeds. The type specimens were from the "Swan River Colony" and most probably came from Cockburn Sound, where there was a large shallow-water population until the Sound was industrialised and polluted. The species is still fairly common elsewhere in the Fremantle area.

In some specimens the dark chocolate of the sides extends over the dorsum as well, obscuring the ground colour and the lateral spots. In such shells there is a thin mantle line. An even rarer colour form has the density of the chocolate spots much reduced, the spots orange instead of brown, and the ground colour a pale pink. These are sometimes referred to as "golden" *friendii*. (See also comments on "golden" *venusta*, *rosselli* and *decipiens*.)

The very handsome shells with prominently exsert spire (Pl. 26, fig. 9) from the Marmion Marine Park are another very striking variant of this subspecies. These shells are quite localised, even within the park, and are another example of inbreeding effects in a direct-developing species.

Midshelf specimens (Pl. 26, fig. 11) tend to be wider than shallow-water shells (Pl. 26, fig. 9) from the same latitude. Also, shallow-water Geographe Bay shells (Pl. 26, fig. 8) tend to be wider than those from similar depths in the Fremantle area. In the Cape Naturaliste to Cape Leeuwin region the western *friendii s.s.* changes into the broader *friendii vercoi*, and along the coast immediately north of Fremantle it grades into *friendii jeaniana*, which is typically toothed along the columellar side of the aperture.

The names *nana, maxima,* and *dorata* used by Raybaudi (1992, *World Shells* **1**: 91) were introduced as form names and have no taxonomic standing. Synonym: *scottii* Broderip, 1831.

2. C. friendii vercoi Schilder, 1930

Pl. 26, figs. 13-16; Pl. 27, figs. 16-19   Like the west coast *f. friendii* in every respect except that the shells are, on average, broader and the aperture is wider.

10 cm. Eucla to Cape Leeuwin, WA. The habitat is the same as that of the west coast population. Recognition of the south coast population as a geographic subspecies assumes that there is a step in the cline at Cape Leeuwin, or thereabouts.

Shallow-water and midshelf shells (Pl. 26, figs. 13-16) have the same range of colour and colour pattern as the west coast nominate subspecies. Deep-water shells, i.e. from the outer part of the continental shelf between Albany and the SA border, are wider than the shallow-water shells of the same longitude and are very weakly pigmented (Pl. 27, figs. 17-18). Some deepwater shells are pure white or cream although most specimens have at least a pale tan base and tan spots on the sides and sometimes bluish grey rectangles beneath light brown spots on the dorsum. There are intermediate specimens on the midshelf (e.g. Pl. 26, fig. 16).

The pale, deepwater shells were named *thersites contraria* Iredale, 1935. Iredale associated them with the SA *thersites* because of their width but they clearly intergrade in that respect with the adjacent shallow *vercoi*. In those specimens in which the base has pigmentation, the colouring extends to the edge of the aperture as it does in the adjacent shallow-water shells and unlike the condition of the SA *thersites*.

The most easterly record of this subspecies in shallow water known to this author is a single shell from Eucla on the WA-SA border (Pl. 26, fig. 16). That specimen is subadult but shows a slight pale patch on the side of the aperture, suggesting an intermediate condition between *vercoi* and *thersites*. Synonym: *thersites contraria* Iredale, 1935.

3. C. friendii thersites Gaskoin, 1849

Pl. 26, figs. 17, 18   Broad, flat-based and humped; ends less channelled than in the WA shells; with a few anterior teeth on the columella and 2-4 small denticles on the parietal wall. Colouring as in *friendii friendii* except for a white patch on each side of the aperture centrally.

10 cm. Central SA. The greater width of the SA shells seems to be merely a continuation of the clinal trend in this character and the other distinguishing characters are relatively minor and unworthy of full species status. Nevertheless, the different basal colouring and columellar dentition are consistent. The central SA population appears to be geographically isolated and it is interpreted as an eastern subspecies of the *friendii* complex.

4. C. friendii jeaniana Cate, 1968

Pl. 26; figs. 1-6; Pl. 27; figs. 12-15   Elongate to broad, attenuating anteriorly; columellar side of the aperture with a continuous row of very short denticles, base tending to be flat with an angulate margin. Colour patterns as in *friendii friendii*.

10 cm. Fremantle to Point Maud, WA and along the outer North West Shelf possibly as far north as Broome (pale form).

Some recent authors have accorded these shells full species status but they clearly relate to the *friendii* complex and subspecies seems more appropriate, given the extreme variability of the complex. The presence of weak columellar teeth along the whole length of the aperture is a consistent character in specimens from north of Fremantle.

Along the midshelf (rarely in less than 10 m) between Fremantle and Point Maud, the colour pattern of the shells is typical of the species complex with a brown base, spotted sides and irregular brown spots and blotches on the dorsum, often overlaying a pale ground with bluish grey rectangles. Some localised populations show distinctive colour patterns but, for the most part, these are minor variations on the *friendii* theme.

Midshelf populations of this subspecies from between Fremantle and Shark Bay (e.g. Pl. 26, fig. 6) tend to be slightly more tumid than those of the nominate subspecies from the Fremantle region. In the Shark Bay to Point Maud area the shells become wide and flat-based.

The type specimen of *jeaniana* came from near Koks I. at the mouth of Shark Bay. It is of the wide and flat-based form. Since the original description a number of similar shells have been taken from that vicinity, mostly west of Bernier and Dorre Is. (e.g. Pl. 26, fig. 5). On shape alone these shells could be mistaken for the south coast *friendii vercoi* but the columellar dentition immediately distinguishes them.

A population in the Quobba area represents the most northerly known locality of the dark form of this subspecies. The shells from this population have similar form and dentition to the Shark Bay shells but are smaller and have a darker base with spreading and calloused margins (Pl. 26, figs. 1-4). Raybaudi (1990) introduced the name *jeaniana sherylae* for these shells, although he did not designate either a type or type locality and his description does not meet the requirements for a valid introduction of a new name. In any case, this author believes that the Quobba shells are merely a localised variant not worthy of taxonomic distinction, and the population is included within the concept of *jeaniana*.

Specimens from the outer edge of the shelf along the central west coast between Fremantle and Geraldton are pale, sometimes white, with only a few light orange-brown spots like those from similar depths on the south coast. Few of these deepwater west coast shells have been taken (see Wilson & Gillett, 1971, Pl. 25, fig. 11) but those examined consistently have the form and apertural dentition of *jeaniana* and are regarded here as colour morphs of that subspecies, just as "*contraria*" is regarded as a deep colour morph of the southern *vercoi*.

In the north, west of Cape Maud, there is a remarkable population of pale shells on the outer part of the continental shelf. Many of them are pale orange rather than tan. The shells are large and, like the inner-shelf population at the same latitude, they tend to be flat-based, sometimes with spreading, calloused margins. They also have the typical *jeaniana* apertural dentition. These shells have been named *aurata* Raybaudi, 1978 but are regarded here as colour morphs of *jeaniana*.

There are no records of any member of the *friendii* species complex on the inner part of the North West Shelf but pale shells of the "*aurata*" colour form have been taken from the outer edge of the continental shelf between North West Cape and Broome (Pl. 27, figs.12, 15). These localities need confirmation as they are based on specimens taken by trawlers and are not accurately provenanced.

Synonyms: *aurata* Raybaudi, 1978; *sherylae* Raybaudi, 1990.

Cypraea marginata Gaskoin, 1849

Pl. 29; figs. 1-21   Ovate, spire slightly projecting; with a dorsal hump, flat-base and sharply angulate, irregularly indented margins; aperture narrow, only slightly arched; teeth relatively strong along both sides of the aperture, labial teeth rather small and sharp, numbering 24-29, columellar teeth numbering 16-23. Top white, with varying degrees of brown spotting, sometimes almost completely brown; base white with brown spots in the indentations; sides heavily spotted, the larger spots corresponding with the marginal indentations. Mantle pale grey, velvety, almost transparent, papillae lacking.

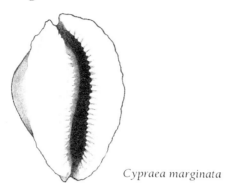

Cypraea marginata

7 cm. Central SA to Monte Bello Is., WA. Usually lives in the darker recesses of underwater caves at depths from 5-200 m, where it feeds on sponges. There is some evidence of sexual dimorphism in the species, with males being smaller and narrower than females.

At one time this species was considered to be one of the world's rarest cowries. Until 1961 only the unprovenanced type specimen was known. The Australian habitat of the species was confirmed only when trawlers brought specimens into the port of Albany in the early 1960s and a little later when specimens turned up in lobster pots off the central west coast. In recent years skin divers have collected it rather commonly in moderately shallow water.

This species exhibits similar patterns of geographic variation to those of *C. (Zoila) friendii*. Deep water shells tend to be more tumid than shallow water shells of the same region, and south coast shells are more tumid than those from the west coast. There are also many distinctive local population variants, no doubt due to the lack of a planktonic larval stage and local inbreeding. Nevertheless, it is possible to distinguish several regional forms which may warrant recognition as subspecies.

If subdivision into subspecies is to be applied it is necessary to establish which should be regarded as the nominate population. Some recent authors have assumed that the west coast population is the nominate form. The locality of the original specimen was not given by Gaskoin. When first reporting the "rediscovery" of the species, on the basis of a specimen trawled off the south coast, Cate (1961, *Veliger* **3**: 76-78) restricted the type locality of *marginata* to "Albany Bay". However, the details given by Cate were inaccurate - there is no such place as Albany Bay and his quoted

geographic coordinates would position the type locality in the town of Albany itself. In fact the Albany trawlers at that time were operating much further east, mostly in the western part of the Great Australian Bight, and Cate's specimen probably came from that area. Because of these ambiguities Cate's designation of a type locality might be disregarded. Later (1964, *Veliger* **7**: 23-24) he inappropriately "corrected" the type locality to the Houtman Abrolhos.

The type specimen is preserved in the British Museum of Natural History (figured Wilson & McComb, 1967, *Indo-Pacific Mollusca* **1** (8): Pl. 342). Although it is unprovenenced it appears to be of the broad southern form and not like the west coast shells. Biraghi (1993) stated that the type "incontrovertibly shows to be a typical specimen of *marginata raybaudii* sensu Lorenz" (i.e. from the South Australian population). In this author's opinion, having inspected the type specimen, a South Australian type locality (rather than a southern Western Australian one) is not "incontrovertible".

Thus, *Cypraea (Zoila) marginata* may be arranged into four subspecies, from north to south and east. In the plate list (Pl. 29) they are named, provisionally, according to the conclusions of Biraghi (1993) although some aspects of that arrangement are clearly unsatisfactory. Further study of the type specimen and its history is needed to settle the question of the type locality before the correct subspecies names can be confirmed.

1.   In the northern part of the species' range, i.e. between the Monte Bello Is and the Abrolhos (Pl. 29, figs 1 & 2) the shells are typically stoutly ovate, very strongly flanged, deeply indented along the margins, dorsally spotted and orange-based. The subspecies name *marginata ketyana* Raybaudi, 1978 was introduced for the northern population. An unusual variant from the Abrolhos is cylindrical with rounded sides and protruding spire (Pl. 29, figs 6 a-b). Most specimens from Shark Bay and further north fit the subspecies diagnosis well but from the Abrolhos and southwards the characters merge with those of lower west coast shells (Pl. 29, figs 3, 4, 7). Bodoni (1986, *Connoisseur of Seashells* : 20-22) noted and illustrated these intermediate forms and actually named them *"intermedia"* although the name was not validly introduced and has no taxonomic standing. Because of the intermediate forms and the lack of a clear geographic boundary between *m. ketyana* and the lower west coast population it is not always possible to assign specimens to one subspecies or the other.

2.   A slender, shallow water form of the species is moderately common in the Fremantle area. On the dorsum the shells are suffusely spotted or mottled with brown over a pinkish background. The degree of dorsal spotting is variable (Pl. 29, figs 5, 8, 9, 10, 11, 12, 14, 15). Shells of this form are found in depths from 5 to 30 m along the lower west coast. The subspecies name *marginata consueta* Biraghi, 1993 was introduced for the west coast population. However, the diagnostic characters said to distinguish the subspecies are not common to all populations. In some specimens from off Rottnest I. (Pl. 29, fig. 20) the dorsal spots are relatively discrete, resembling the colour patterns of the northern

m. ketyana and South Australian shells. Specimens from the outer part of the continental shelf (Pl. 29, fig. 5) are usually much more tumid than the shallow water shells and approach the form of the southern subspecies.

3. The WA south coast population ranges between Cape Leeuwin and the Recherche Archipelago. The shells are tumid and broad-based. They are usually pale with only weak, suffused spotting on the dorsum (Pl. 29, fig. 17) but the colouring ranges from unspotted white or pink (Pl. 29, fig. 13) to heavily mottled (Pl. 29, fig. 16). Raybaudi (1985) named this population *marginata albanyensis*. This name would become a synonym if a south coast type locality were confirmed.

4. SA shells are easily recognised by their inflated form and distinctly and evenly spotted dorsum (Pl. 29, figs 18, 19, 21). It is not yet clear whether there are intermediate forms in western South Australia and the Great Australian Bight. The South Australian population has been named *marginata raybaudii* Lorenz, 1992. That name would fall into the synonymy of *marginata* if the SA population were confirmed as the nominate one. In his cards and catalogues Raybaudi referred to the South Australian shells as *marginata meridionalis* and later as *marginata orientalis* These names have no taxonomic standing.

Cypraea mariellae Raybaudi, 1983

Pl. 28a; figs. 1 a-b Thin, bulbous, humped, pyriform, spire evert and only lightly calloused; posterior channel inclined, with unequal sides, right side more prominent and curved around anteriorly to the left side; aperture sinuous, outer lip thick and deep with a near-vertical face, labial teeth numbering 20-24, strong and vertical, the anterior teeth slightly finer; columellar teeth numbering 16-19, fine, extending down the parietal wall into the aperture but not onto the wide fossula. Dorsum greyish white, incremental growth lines evident, crossed by faint spiral lines producing an obscure hammer pattern; sides light orange-brown sometimes with faint darker spots; base ivory.

5.6 cm. Exact locality unknown; believed to be the outer edge of the continental shelf off the Kimberley coast, WA, at depths of 100-200 m. The original material was collected by Taiwanese trawlers and accurate locality information is not available. The illustrated specimen is the holotype kindly made available by Raybaudi Massilia. Additional specimens have been illustrated by Raybaudi (1993, *World Shells* 3: 10-15).

Most recent authors have considered these shells to be a deep-water, pale form of *C. decipiens* which is found along the inner shelf nearby. This may be the case. Although the bulbous shape of *mariellae* is distinctive, deepwater populations of the *Zoila* complex all exhibit a trend towards greater width and degree of humping. The labial dentition and nature of the anterior and posterior channels clearly resemble shells of *decipiens*. However, this author prefers to maintain the status quo and retain *mariellae* as a distinct species until more material and better locality information are available.

Cypraea perlae Lopez & Chiang, 1975

Pl. 28a; fig. 2 a-b Bulbous, humped, slightly pyriform, spire evert and only thinly calloused; channels slightly extended but with rouded sides; anterior end marginated, with a lateral sulcus above the margination; aperture straight centrally, wider and turning left posteriorly and anteriorly; labial lip moderately thick, with about 23 teeth, teeth even-sized throughout; columellar teeth fine, numbering about 17, not extending deep into the aperture; base flat but slightly inclined inwards. Dorsum ivory-white to biege with scattered, irregular light-brown to chestnut spots; sides marbled with a few round brown spots; ends brown; base buff to orange.

4.7 cm. Exact location unknown; believed to be on the outer edge of the North West Shelf, north of North West Cape, WA. The original material was collected by Taiwanese trawlers and accurate locality information is not available. In the original description Lopez & Chiang quoted 18-25 degrees of latitude (i.e. Broome to Carnarvon) but added "approximately 21°0' S; 114° 10' W" which narrows the range to the North West Cape area. They also quoted a depth range of 182-274 m.

Subsequent authors have tended to treat this population as a colour form of *C. decipiens*. However, the distinctly spotted sides and dorsum are quite unlike the colour pattern of that species and are more akin to the pale, outer shelf form of *friendii jeaniana* found a little further south (i.e. the colour form called "aurata" by collectors). Although the shape of *perlae* shells resembles that of *decipiens*, the labial dentition is much more like that of the *friendii* complex. The labial lip is not deep and the teeth do not extend vertically into the aperture as they do in both *decipiens* and *mariellae*. With such little material and such ambivalent locality information the status of these shells cannot be properly assessed. This author prefers to maintain the status quo and retain *perlae* as a species until the matter can be properly studied.

Cypraea rosselli Cotton, 1948

Pl. 28; figs. 16-19 Broadly ovate, narrowing anteriorly, flat-based, spire impressed and rarely visible; margins angulate, especially anteriorly where the sides are flanged, elevated centrally, often indented; dorsal hump steep posteriorly, maximum height just behind centre; anterior channel almost vertical; aperture arched, only slightly dilated anteriorly; teeth relatively strong along both sides of the aperture, labial teeth relatively strong, numbering 25-31, columellar teeth weaker, even-sized, numbering 18-26. Base and sides dark chocolate brown; top cream but usually almost covered by brown spreading up from the sides leaving only a cream dorsal patch; unspotted. Mantle black, with a few small and simple papillae.

6 cm. Esperance to Carnarvon, WA. This is another of the *Zoila* group which was once very rare but lately collected frequently, by divers. Specimens are rarely taken in less than 20 m.

The typical colour form has a dorsal cream patch but in some shells dark chocolate spreads up the sides onto the dorsum and completely obscures the dorsal patch.

Zoila rosselli albosignata Raybaudi, 1985 is a form name referring to a pale colour form and has no taxonomic standing. *Zoila rosselli edingeri* Raybaudi, 1990 was introduced as a subspecies name for a "golden" form of the species which comes from a population (said to be in Shark Bay) near the northern end of the species' range. The holotype was figured by Raybaudi (*The Connoisseur of Cowries* **24**: 11-13, fig. 4). It is believed that the population is small and localised. Subspecies status is inappropriate. All of the brown-shelled *Zoila* occasionally produce golden-shelled individuals. (See also remarks on *C. (Z.) decipiens*.)

Synonym: *rosselli edingeri* Raybaudi, 1990.

Cypraea venusta Sowerby, 1846

Pl. 27; figs. 1-11; Pl. 28; figs. 1-6   Ovate to subpyriform and tumid, maximum height posterior to the centre; spire depressed; sides of canals rounded; margins rounded centrally, becoming subangulate anteriorly; aperture dilated anteriorly, arched posteriorly; labial teeth short but moderately strong, numbering from 18-27; columella with 5-7 short but moderately strong teeth anteriorly and 3-4 weak denticles posteriorly, sometimes with denticles along the entire length of the columella; fossula deep and concave. Colour extremely variable, ranging from dark brown to chestnut, pink, yellow and white. Mantle thin and velvety grey, papillae lacking.

Cypraea venusta: holotype of *bakeri* (left) compared with the holotype of *venusta* (right).

8 cm. Great Australian Bight to Shark Bay, WA. Lives on sponges on the sea floor and in underwater caves at depths from 5 m to the outer edge of the continental shelf. Sexual dimorphism in shell form is evident, females tending to be tumid and males tending to be narrow.

There is considerable interpopulation variation in size and especially colour. The most common colour form has: an off-white or yellowish base, the teeth often striped with orange-brown; grey and clouded sides becoming darker dorsally and at the ends and with a few large dark brown spots; a cream or pale tan dorsum overlain by merging brown spots or diffuse maculations, sometimes almost uniformly brown across the entire dorsum. The brown colouring of the spots ranges from chocolate to light chesnut.

As in the *friendii* complex, specimens from the outer part of the continental shelf tend to be pale, the brown pigments greatly reduced in intensity. This seems to be an environmental effect rather than a genetic one. However, pale shells also occur in relatively shallow water throughout the species range. These may be cream, off-white to yellow, or pink, with the sides suffused with orange-brown and spotted, and with or without discrete brown or orange-brown spots on the dorsum.

Pink or cream shells like the type of *venusta* are often toothed along the entire columellar side of the aperture while dark shells of the "*episema*" form only occasionally have central denticles on that side. However, this is not a consistent character. Some authors regard the two forms as distinct species (i.e. *venusta* the pink or pale form and *episema* the dark form).

Cate (1962, *Veliger* **5**: 6-14, pls 1-4) discussed the complex nomenclatural history of the species. The original description of *venusta* was based on a type specimen of the pink variety with teeth along the whole length of the inner lip (Cambridge University Saul Collection). The origin of that specimen was uncertain. Similar pale shells are now known from the Great Australian Bight to the central west coast and from depths ranging from 5 m to 200 m. The name *thatcheri* Cox was founded on specimens originally said to have been collected in the Dampier Archipelago (now in the collection of the Institut Royal Sciences Naturelle de Belgique; figured by Cate, 1962). That locality has since been corrected to Cervantes on the central west coast, which is now taken as the type locality of *venusta*. Similar shells have since been collected in quite shallow water in that area (e.g. Pl. 27, fig. 4) but also from a number of localities on the west and south coasts (Pl. 27, figs. 1-9). The type of *bakeri* was from an unspecified WA locality. It lacks central and posterior columellar teeth and resembles unspotted specimens recently taken from the Esperance area.

Iredale's name *episema* was based on two large brown types from Cape Naturaliste (like the specimen from the type locality figured here; Pl. 28, fig. 1). Similar but smaller shells from the Marmion Marine Park near Perth were the basis of Schilder's name *sorrentensis* (Pl. 28, figs. 4). The type of *catei* from the Abrolhos (figured Cate, 1962) is an aberrantly shaped specimen of the dark colour variety. All of these names are considered here to be synonyms of *venusta*.

In spite of an existing surfeit of names for populations of this complex, more have been introduced recently, as form names or to designate subspecies. A number of form names used in his catalogues by Luigi Raybaudi may be disregarded for taxonomic purposes. *Zoila venusta orientalis* Raybaudi, 1992 was intended as a subspecies for a deepwater, pale colour form of *venusta* from the Great Australian Bight. The published illustrations indicate specimens with an unusual colour pattern but, in view of the extreme colour variation within this species, it is doubtful whether this warrants subspecies status.

Synonyms: *thatcheri* Cox, 1869; *roseopunctata* Melvill, 1888; *bakeri* Gatliff, 1916; *episema* Iredale, 1939; *sorrentensis* Schilder, 1963; *catei* Schilder, 1963; *venusta orientalis* Raybaudi, 1992.

FAMILY **OVULIDAE**

Egg and Spindle Cowries

Egg and spindle cowries are relatives of the true cowries but generally lack the regular, strong apertural teeth of that group. There are also anatomical differences.

Shell form is very variable in this family, ranging from globose to extremely long and slender with long, drawn-out ("produced") anterior and posterior canals. In most species the upper surface is smooth and polished but in a few it is striate. As in the true cowries the retractable mantle may spread over the dorsal surface. Form of the shell, dentition of the outer lip and the shape and form of the columella are useful characters for distinguishing between the species. Although the inner lip is toothless, there is often a thickening (the funiculum) at the posterior end beside the posterior canal that may bear nodules or ridges (rugae) and this is a very helpful character in some groups.

All ovulids live in association with coelenterates, feeding on the polyps. The polyps are bitten off by the pair of strong jaws, aided by the median and lateral radular teeth.

The hosts include true corals, fleshy soft corals such as *Sarcophyton*, gorgonians (sea whips or sea fans), and hydrozoans. However, particular species are usually host-specific. In some cases the coelenterates may come in several colours and the ovulids take on the colour of the host. The mantles of these elegant snails very often mimic both the colour and the surface structure of the host, with mantle pustules taking the form of the polyps. They may be very hard to see nestled among the coelenterate fronds.

Egg masses are laid on the surface of the host. There is a veliger larval stage in those species whose development has been studied. It has four long velar lobes for swimming. It can be assumed that ovulid larvae may be widely distributed by ocean currents.

There are more than 100 species and about 16 genera in the family but a thorough review is needed to clarify relationships. The generic nomenclature used here follows that of Schilder & Schilder (1971). More recently Cate (1973 a, 1973 b, 1974) introduced a number of new generic names, using shell form as the sole character. In this author's opinion this procedure is inappropriate in the absence of any habitat, anatomical or other biological data, and all of Cate's generic names are treated here as synonyms until more reliable information is available.

The author gratefully acknowledges assistance with specimens and information provided by Lorraine Rutherford of Qld for this section.

References

Cate, C.N. (1973 a). Systematic revision of the Recent Cypraeid family Ovulidae. *Veliger* **15** (Supp.): 1-116.

Cate, C.N. (1973 b). The Ovulidae: Replacement names for four ovulid hononyms. *Veliger* **16** (2): 238-239.

Cate, C.N. (1974). The Ovulidae: A key to the genera, and other pertinent notes (Mollusca: Gastropoda). *Veliger* **16** (3): 307-313.

Iredale, T. (1935). Australian cowries. *Australian Zoologist* **8**: 96-135, pls 8-9.

Schilder, F.A. & Schilder, M. (1971). A catalogue of living and fossil cowries. *Memoirs Institut Royal Sciences Naturelle de Belgique, 2nd Series* **85**: 1-246.

▲ GENUS **CALPURNUS** Montfort, 1810
Bulla verrucosa Linnaeus

With the characters of the type species.

Indo-West Pacific. The genus has only one representative, although some authors treat *Procalpuruus* as a subgenus.

Calpurnus verrucosus (Linnaeus, 1758)

Pl. 24; figs. 26 a-b   Thick, depressed, ovate, with sub-angulate shoulder forming a dorsal hump; anterior and posterior canals not projecting; posterior canal slightly reflected; aperture wide, curved; outer lip dentate; fossula narrow; terminal ridge weak; dorsum minutely striate; characterised by button-like knobs above the anterior and posterior canals. White, pink at the ends. Animal white, with even-sized brown spots on the mantle and foot (figured Wilson and Gillett, 1971, plate 6).

3.5 cm. Common. Indo-West Pacific; Qld. Lives on the fleshy soft coral *Lobophytum*. Although Cate (1973) claimed that his illustrated specimen was collected "living on a spreading unidentified green algae", that habitat seems very unlikely.

▲ GENUS **CYMBOVULA** Cate, 1973
Ovula acicularis Lamarck

Oblong, subcylindrical, usually rather thin; without an adapical umbilication.

Indo-West Pacific, Panamic and Caribbean regions. This is a replacement name for *Cymbula* Cate, 1973 a (not H. & A. Adams, 1854). The very brief diagnosis is taken from the original description of the genus in which Cate placed three American species and the Indo-West Pacific *C. deflexa.* He later (1984) described another Australian species, *C. queenslandica,* in the genus. Although the generic diagnosis is hardly adequate, the two Australian species have unusual anterior and posterior canal structure and Cate's generic placement of them is retained provisionally until the matter has been reviewed.

Cymbovula deflexa (Sowerby, 1848)

Lenticular; posterior end tapering to a point, concave on the left side; anterior end tapering but terminally obliquely truncate, concave on the left side, convex on the right; base of body whorl calloused and conspicuously keeled at the outer margin, flat and inclining inward; outer lip also inclining inward, weakly crenulate, indented anteriorly; anterior canal wide, truncate, slightly flaring, posterior canal shallow, reflected, spooned, pointed; fossula well developed; funiculum represented by a thickening which partly occludes the posterior canal; dorsum with very fine, incised transverse striae. Beige or pale yellow, with orange tips, the orange sometimes spreading as a thin line along the marginal sutures anteriorly.

2 cm. Indo-West Pacific; central Qld. Lorraine Rutherford reports that in Qld this species lives on "pink shrub-like gorgonians".

Cymbovula queenslandica Cate, 1974

Elongate and subcylindrical, peripheral keel lacking; outer lip inwardly inclined, smooth except for about three weak denticles at the posterior end and one or two anteriorly plus a prominent terminal ridge bordering the anterior canal; anterior canal well-formed and U-shaped, posterior canal divided into left and right channels by a terminal nodule; funiculum a thickened spiral ridge on the left side of the posterior canal; fossula wide, concave, with a conspicuously angulate inner margin; left side of base calloused, the callus meeting that of the outer lip at each end thus encircling the shell; dorsum transversely striate. Maroon; peripheral callus orange to brown.

1cm. Central Qld. The mantle is maroon, like the sea whips on which the animal lives.

▲ GENUS **DIMINOVULA** Iredale, 1930

D. verepunctata Iredale = *alabaster* Reeve)

Ovate or pyriform; posterior end projecting, anterior end attenuate, drawn out; outer lip thickened, dentate along the inner side; aperture curved, narrow, widening slightly behind the anterior canal; anterior canal deeply notched, bordered on the columellar side by a distinct, steep, terminal ridge; fossula wide and concave, with an inner marginal keel; funiculum consisting only of a short thickened ridge bordering the posterior canal; dorsum transversely striate and with three pairs of spots.

Indo-West Pacific. Cate (1973) treated this group of small species as a subgenus of *Pseudosimnia* Schilder, 1927.

Diminovula alabaster (Reeve, 1865)

Pyriformly ovate; funiculum bearing three or four crenulations. Milky white or pale grey, with a distinct golden line encircling the shell above the margin and usually three pairs of brown spots on the back.

8 mm. Indo-West Pacific; Exmouth Gulf, WA to Moreton Bay, Qld. Cate (1973, *loc. cit.,* fig. 49) figured the holotype. The cited type locality was Senegal but Cate noted that the species is not present in West Africa and designated Exmouth Gulf, WA as the "new type locality".

Diminovula aurantiomacula Cate & Azuma, 1973

Elliptical to pyriform; funiculum smooth, without crenulations; outer lip weakly dentate; dorsal striae widely spaced and weak at the centre. Fawn or pale yellow with three obscure encircling golden bands within which lie three pairs of darker, coalescing, irregular blotches.

6.3 mm. Southern Japan and Central Qld. This tiny species closely resembles *P. punctata* but is distinguished by the less bulbous shape and the smooth funiculum.

Diminovula nielseni Cate, 1976

Elliptical to pyriform; funiculum smooth, without crenulations; outer lip teeth moderately strong; dorsal striae confined to the ends. White, with three pairs of small orange spots.

7 mm. Central Qld. Lorraine Rutherford says this tiny species lives on the same kind of alcyonarians as *D. punctata* and *P. cavanaghi.* From the Qld material used for this study it is difficult to separate *P. nielseni* from *P. aurantiomacula* as in both species the funiculum is smooth. Further series and field data are needed to determine whether the outer lip dentition and colouration are consistent diagnostic characters.

Diminovula punctata (Duclos, 1831)

Pl. 24; fig. 22 a-b Bulbous, with constricted, projecting ends; funiculum weakly and irregularly crenulate; terminal ridge weak. Yellow or white, with a thin yellow line around the

marginal suture on the right side and three pairs of prominent yellow-brown or mauve spots on the back. Mantle cream with reddish brown and white pustules; siphon and eye tentacles red-maroon, foot white with red-maroon lines and edges. (Animal figured Wilson and Gillett, 1971, Pl. 40.)

11 mm. Central Indo-West Pacific; Exmouth Gulf, WA to central Qld. Lives on a spiky alcyonarian.

Diminovula whitworthi Cate, 1973

Elongate, ovate-elliptical; posterior end pointed and projecting; funiculum short and bearing three conspicuous nodules; inner lip sharply angulate deep within the aperture; basal conspicuously striate, especially near the ends. Pink to fawn, with three rose transverse bands and three pairs of diffuse, obscure darker blotches.

8 mm. Southern Japan and central Qld. Little is known of the habitat of this tiny species. The diffuse dorsal spots and distinctive funiculum are characteristic.

▲ GENUS **HIATAVOLVA** Cate, 1973
Ovulum depressum Sowerby

Elongately lenticular to cylindrical, anterior and posterior ends evenly tapering; aperture straight, narrow, except at the anterior end where it widens abruptly; anterior and posterior canals obliquely truncate, wide open and notched; base of body whorl calloused; outer lip thickened, inturned, flat lower side inclined, may be obscurely crenulate; fossula wide, shallow, bordered on its inner edge by a curved keel which protrudes into the aperture; funiculum may be represented by a small thickening which may partly occlude the posterior canal.

Indo-West Pacific. There are several species. This is a replacement name for *Hiata* Cate, 1973 (not Zetek & McLean, 1936).

Hiatavolva brunneiterma (Cate, 1969)

Pl. 24; figs. 7 a-b Elongate-lenticular; base calloused, flat and inclined inwards, keeled on its outer margin; outer lip smooth; funiculum obsolete; anterior and posterior canals terminally wide and slightly flaring; dorsum smooth. Milky white, the tips of the canals dark brown with a tiny fleck of tan behind.

2.1 cm. Western Pacific; central Qld. The figured specimens are from Keppel Bay. This species appears to be closely related to the western *H. depressa*.

Hiatavolva depressa (Sowerby, 1875)

Pl. 24; fig. 6 Elongate cylindrical; base of body whorl rounded, thinly calloused; outer lip smooth, its inner edge slightly sinuous; funiculum obsolete; anterior and posterior canals terminally wide and slightly flaring; dorsum smooth. White, pink or beige; tips orange or brown.

2.6 cm. Central Indo-West Pacific; Exmouth Gulf, WA to the NT. Lives on multistemmed sea whips growing on sandy substrates just below low-tide level.

▲ GENUS **MARGOVULA** Iredale, 1935
Ovulum pyriform Sowerby

Pyriform; posterior end projecting, curved on the right side, notched by the reflexed posterior canal on the left; outer lip thickened, dentate along its inner margin; aperture curved, wide; anterior canal deeply notched, bordered on the columellar side by a very prominent, spiralling, steeply angled terminal ridge which is continuous with a keel along the inner edge of the fossula; fossula wide but short, concave, with a curved and keeled inner margin; funiculum a prominent, angular nodule on the parietal wall beside the posterior canal; dorsum striate.

Indo-West Pacific. Although Iredale claimed "this group is easily separated from *Diminovula* or *Prionovolva* by its shape and columella features" the differences in shell characters are not great. *Diminovula* is sometimes treated as a subgenus of the Atlantic genus *Pseudosimnia* Schilder, 1927 (*carnea* Poiret) and perhaps *Margovula* and *Prionovolva* belong there too. Cate (1973) introduced *Inflatovula* (*marginata* Sowerby) as a subgenus of *Pseudosimnia* but it is treated here as a synonym of *Margovula*. This confusion illustrates the need for a proper revision using anatomical data.

Margovula aboriginea Cate, 1973

Sides evenly attenuate anteriorly, not constricted; outer lip wide, teeth fine and numerous; aperture moderately curved; funiculum elongate, sharp-edged, faintly crenulate; dorsum finely striate. Milky white, with three rows of large, pale-yellow, merging dorsal spots.

1 cm. Exmouth Gulf, WA. The biology and distribution are not known. The lack of an anterior constriction distinguishes this species. Drawn from the holotype.

Margovula bimaculata (Adams, 1854)

Pl. 24; fig. 23 a-b Pyriform-ovate; ends less produced than in *M. pyriformis* and the anterior canal less constricted; fossula shallow; dorsum finely striate. Ivory, with three tan bands on the back. Mantle grey, mottled with brown spots; siphon and

eyestalks brown, foot flesh-grey with transverse brown streaks. (Animal figured by Cate, 1973, figs. 96 c, 30 c.)

1.4 cm. Central Qld. Lorraine Rutherford reports that this species lives on the same grey anemone as *pyriformis*.

Margovula marginata (Sowerby, 1828)

Pl. 24; fig. 25   Thin-shelled, tumid, pyriform-ovate; funiculum merely a small, oblique, weakly crenulate ridge, without any callus spreading over the base; outer lip teeth weak; columella callus weak; fossula shallow; dorsum conspicuously striate, at least at the ends. White, sometimes with a thin golden line above the outer lip thickening.

2.8 cm. Indo-West Pacific; Broome, WA to central Qld. More inflated and less pyriform than *P. pyriformis*, and characterised by weak development of the funicular and columellar callus. This is the type species of *Inflatovula* Cate 1973.

Margovula pyriformis (Sowerby, 1828)

Pl. 24; fig. 24 a-b   Markedly pyriform, with a long, constricted anterior end and slightly upturned anterior canal; fossula rather deep; funiculum heavy and boss-like; dorsum finely striate. Fawn to brown, usually obscurely double-banded on the back, outer lip and base darker. Animal black, with large warty mantle pustules.

2.2 cm. Torres St. to southern Qld. Lorraine Rutherford reports that this species "lives on a grey anemone".

▲ GENUS **OVULA** Bruguière, 1789
Bulla ovum Linnaeus

Ovate-globular, centrally bulbous, thick-shelled; posterior end projecting, notched on the left side, convex on the left; anterior end broad, drawn out; aperture wide, curved; outer lip weakly and irregularly dentate; fossula lacking; funiculum represented by weak spiral cords on the side of the posterior canal; dorsum glossy, smooth.

Indo-West Pacific. There are several living species. Synonyms: *Amphiperas* Herrmannsen, 1846 (*ovum* Linnaeus; *Ovulus* Montfort, 1810; *Ovulum* Sowerby, 1828; *Ovularia* Link, 1830; *Parlicium* Iredale, 1935 (*costellata* Lamarck).

Ovula costellata Lamarck, 1810

Pl. 25; figs. 2 a-b   Like *O. ovum* but the posterior canal is shorter and more deeply notched on the left side; exterior milky white, interior pink. Foot and head white, mantle mottled orange with numerous small filaments (figured by Coleman, 1981, *Shells Alive* p. 44).

4 cm. Indo-West Pacific; eastern Qld and northern NSW.

Ovula ovum (Linnaeus, 1758)

Pl. 25; figs. 1 a-b   Posterior canal drawn out with a straight or weakly concave left side. Exterior milky white, interior orange-brown. Animal black with a few white pustules on the mantle.

10 cm. Indo-West Pacific; Abrolhos Is., WA to northern NSW. Lives on the large fleshy soft coral *Sarcophyton*. This is the "egg cowry" of renown.

▲ GENUS **PHENACOVOLVA** Iredale, 1930
Phenacovolva nectarea Iredale = *rosea* Adams

Elongate, spindle-shaped, with drawn-out posterior and anterior canals; at each end the inner lip ending abruptly at an angular corner but the outer lip merely attenuating to nothing so that the anterior and posterior canals are asymmetrical in end-on view, terminating obliquely in plan view; outer lip thickened, edentulous; columella usually straight, rounded, terminal ridge lacking; fossula lacking or very narrow; funiculum represented by a low spiral cord on the inner lip a short distance before the end of the posterior canal, or lacking; exterior smooth or finely transversely striate.

Indo-West Pacific. Where known the habitat appears to be gorgonian sea fans or sea whips. There are many species in the region but some names may apply to variants; supposed diagnostic characters are often merely variations in shell form such as length of the canals. The type species of *Pellasimnia* Iredale, 1931 (*angasi* Adams) is one of several species with relatively short canals and there seems no basis for separating them even at subgeneric level; this genus-level name is treated here as a synonym of *Phenacovolva* although the Schilders (1971) ranked it as a subgenus. Similarly *Aclyvolva* Cate, 1973 (*lanceolata* Sowerby) was diagnosed as "unusually long, narrow and lanceolate", *Calcarovula* Cate, 1973 (*longirostrum* Sowerby, 1828) was proposed for species with particularly long canals, while the subgenus *Turbovula* Cate, 1973 (*brevirostris* Schumacher) refers to centrally inflated shells; these names are regarded here as synonyms of *Phenacovolva*.

Phenacovolva angasi (Reeve, 1865)

Pl. 24; figs. 11 a-b   Ends moderately elongate, gradually tapering; fine transverse striae cross the back but may be obsolete centrally; fossular lacking. Ivory white, interior slightly pink, the tips of the canals may be mauve inside and orange or brown outside. Animal white.

3 cm. Western Pacific; Cape York, Qld and central NSW. Lorraine Rutherford reports that this species lives on white gorgonians.

Phenacovolva brevirostris (Schumacher, 1817)

Pl. 24; fig. 9   Rather sold, fusiform to lenticular, sometimes bulbous or with a subangular shoulder just behind the centre, sides slightly concave at the posterior end; canals short; aperture curved; funiculum weak, obscurely crenulate. Fawn, beige, rose or white; often with a pale central band on the dorsum.

3 cm. Indo-West Pacific; the illustrated specimen is from Cape York. The short, stout extremities distinguish this species from *P. rosea*, with which it has been confused.

Phenacovolva dancei Cate, 1973

Pl. 24; fig. 8   Spindle-shaped; body whorl inflated, with moderately projecting, curved canals; thin-shelled and almost transparent; aperture wide; columella straight; funiculum on a moderately stout spiral cord; ends striate, dorsum smooth. Dorsum pale pink, with three white bands; outer lip white; ends orange.

2 cm. Indian Ocean; Dampier to Broome, WA; dredged from 40 to 90 m. These are distinctive shells unlikely to be confused with any other Australian species. Cate (1973) placed the species in his subgenus *Turbovula* along with the stout *P. brevirostris*.

Phenacovolva haynesi (Sowerby, 1889)

Pl. 24; fig. 2   Slender and elongate, ends upward-curving in side view; canals very long and narrow; base of body whorl thinly calloused; fossula lacking; funiculum represented by a slight, wide, obscurely crenulate swelling about half way down the posterior canal. Cream, pale pink or brown within the aperture.

3.8 cm. Northern WA. The type locality is Exmouth Gulf. The figured specimen was dredged at 60 m off Dongara, WA. Sowerby reported that the type was collected on the gorgonian *Melitodes*. This species resembles *P. philippinarum* but is more slender and has much longer and narrower canals. *Aclyvolva framea* Cate, 1973 (from Broome) WA may be a juvenile of *P. haynesi*.

Phenacovolva improcera Azuma & Cate, 1971

Pl. 24; fig. 10   Cylindrical to lenticular, rather thick-shelled, with stout anterior and posterior canals; base thinly calloused and marginally ridged; aperture sinuous, very narrow posteriorly, widening on both sides behind the short anterior canal; columella thick, bent, with a very narrow fossula; inner lip thickened with callus, posteriorly terminating in a projecting angular ridge; funiculum not evident; dorsum smooth, ends transversely striate. Milky white, sometimes faintly olive at the tips.

2 cm. Japan; Port Hedland, WA to Darwin, NT. Closely resembles *P. angasi* but the sinuous aperture seems to be a consistent differentiating character. Also resembles *P. philippinarum* and may be merely a stout form of that species. In northern WA this species may be confused with *H ioatovolva depressa* but the very weak fossula and form of the canals are more characteristic of *Phenacovolva* and the ends lack colour.

Phenacovolva philippinarum (Sowerby, 1848)

Pl. 24; figs. 3 a-b   Body whorl slender, cylindrical, with the long canals upward-curving in side view; posterior end evenly tapering to a point though usually slightly concave on the left side; anterior end straight on the left side, angled on the right; base of body whorl thinly calloused in adults; outer lip thickened, abruptly angled near the anterior end; fossula poorly developed; funiculum absent; dorsum glossy smooth, finely striate near the ends. Yellow, apricot or rose in life, fading to ivory in collections. Mantle the same colour as the shell with white pustules.

3.5 cm. Indo-West Pacific; Esperance, WA around western and northern Australia to central Qld. Lives on the gorgonian *Mopsella*. The figured specimens from King George Sound, WA and Keppel Bay, Qld are indistinguishable. Cate placed *philippinarum* Sowerby in the synonymy of *birostris* Linnaeus, 1758 and this may be correct.

Phenacovolva rosea (Adams, 1854)

Pl. 24; figs. 5 a-c   Body whorl narrowly ovate, ends drawn out to long spine-like canals, right margin abruptly bent inwards anteriorly at the base of the anterior canal, posterior end more or less evenly tapering; surface finely striate; aperture outward-curving at the centre, wide anteriorly behind the abrupt canal constriction; funiculum represented by a small wart-like nodule on the parietal wall at the base of the posterior canal. Apricot, fawn or mauve, often with a pale transverse central band, outer lip white.

4.2 cm. Western Pacific; Qld and northern NSW.

Phenacovolva tokioi Cate, 1973

Pl. 24; fig. 4   Spindle-shaped, with an elongate-ovate body whorl and long, pointed canals; posterior canal sometimes turned to the left, anterior canal shorter; funiculum represented by a weak spiral cord at the base of the posterior canal; dorsum transversely striate although the central part may be smooth. Pink with darker tips and usually a central white band.

4 cm. Central Indo-West Pacific; Broome, WA to central Qld. The shape and length of the spines is quite variable and this may turn out to be merely a variant of *P. rosea*. Cate associated this species with a group of long-spined, Indo-West Pacific species for which he erected the subgenus *Calcarovula* but it is difficult to separate it from the type species of *Phenacovolva*.

Phenacovolva weaveri pseudogracilis Cate & Azuma, 1973

Pl. 24; fig. 1 Lenticular, spindle-shaped; posterior end evenly tapering though slightly concave on the left side; base of the body whorl thinly calloused; anterior end short, concave on the left side, steeply angled on the right; aperture narrow and gently curved in the posterior half, widening anteriorly; fossula very narrow and shallow but with a keeled inner edge; funiculum represented by a spirally twisted cord; dorsum striated only at the ends. White or beige, with a yellow-orange line at the outer lip suture and an irregular purplish axial band on the left side.

3.1 cm. Western Pacific; central Qld. Tom Nielsen trawled the figured specimen in Keppel Bay; it was living on black coral. The nominate form of this species is from Hawaii. Both forms may turn out to be merely variants of *P. gracilis* (Adams and Reeve, 1848). *P. carneopicta* Rehder and Wilson, 1975 is another species in this complex, dredged on hydrozoans in the Marquesas.

▲ GENUS **PRIMOVULA** Thiele, 1925
Amphiperas beckeri Sowerby

Elongate-ovate to cylindrical, with a subangular shoulder posteriorly; posterior end projecting with steeply sloping sides, concave on the left, anterior sides more acutely sloping; posterior canal slightly reflexed or straight, sometimes divided into two separate channels; outer lip thickened, dentate; aperture narrow posteriorly, wide behind the anterior canal; anterior canal deeply notched, with a low, steeply angled terminal ridge on the columellar side; fossula present, with a keeled margin anteriorly; funiculum a thick spiral rib or a heavy boss-like thickening beside the posterior canal; dorsum transversely striate or smooth.

The subgenus *Adamantia* Cate, 1973 (*Ovulum concinnum* Adams & Reeve) was diagnosed as "roughly diamond-shaped", rhomboid, with a general outline of rounded extensions of the shell periphery." The genus *Crenavolva* Cate, 1973 (*Ovulum striatulum* Sowerby) was diagnosed as having "shells .... narrow, centrally angled dorsally, sub-rhomboid; terminal ends blunt, terminals usually of equal, tapering projection." These subjective characters intergrade among the species and both taxa are treated here, provisionally, as synonyms of *Primovula*. Gowlett-Holmes & Holmes (1989) figured the radulae of two southern Australian species.

Primovula concinna (Adams & Reeve, 1848)

Ovate to subrhomboid, with broad subangulate shoulder; posterior end thick and projecting, anterior end broad, attenuate, concave on the left side; outer lip with strong teeth; funiculum wide and thick, strongly lirate; dorsum striate. White.

8 mm. Western Pacific; central Qld. The broad diamond shape of this species led Cate (1973) to introduce the subgeneric name *Adamantia* with *concinna* as type. The type locality is in the Philippines. Although Cate listed Proserpine, Qld as a locality for the species, no authentic Australian specimen was available for illustration here. For an illustration see Cate (1973, *loc. cit.*, fig. 90).

Primovula cruenta (Gowlett-Holmes & Holmes, 1989)

Elongate, both ends produced, base concave; both canals U-shaped, bordered by slightly projecting terminal ridges; outer lip flat, inward sloping, bearing rounded transverse ridges which become weak anteriorly, its outer margin with four slight terminal nodules posteriorly and one strong and three weak nodules anteriorly; funicle weak, smooth; fossula well developed. Orange-red, ends bright red, with a light yellow central band, a paler diffuse band posteriorly, and a light yellow patch on the dorsum anteriorly; fossula pinkish white. Animal mantle lobes translucent, pale greyish pink with dark red spots and large white papillae; foot ruby, tentacles, siphon and proboscis ruby, tentacles with white tips.

1.3 cm. Spencer Gulf, SA. Found on the gorgonid *Echinogorgia* in 15 m. The authors of this recently described species noted that it is most like *P. striatula* and *P. verconis*, from which it is distinguished by its narrower, more elongate form, more produced ends, and the colour pattern. Originally described as a species of *Crenavolva*.

Primovula heleneae Cate, 1973

Pl. 24; figs. 15 a-b Elongate-ovate to lenticular, boat-shaped; shoulder rounded; both ends tapering; funiculum a thick crenulate swelling; aperture gently curved, only slightly swelling anteriorly; posterior lip swollen, dividing the posterior canal into two diverging portions; outer lip with about 20 thick teeth; fossula hardly developed; terminal ridge low, vertical; dorsum striate, strongly so posteriorly. White, brown-tinted at the ends.

1.3 cm. SA. The holotype was originally a paratype of *P. verconis* (Cotton & Godfrey) but recognised as distinct by Cate. The more lenticular form, heavy crenulate funiculum and narrow aperture readily distinguish it. In many respects the shell resembles *P. platysia* but the colouration is very different. This species was recently rediscovered by Karen Gowlett-Holmes and N. Holmes off Port Douglas, St.Vincent Gulf, where it was living on the gorgonid *Euplexaura* in areas of strong tidal movement. The mantle lobes lack papillae and their colour is transluscent red to dark orange, with irregular reddish brown patches and white dots, closely mimicking the polyps of the host. An egg mass contained 57 colourless, transparent, jelly-like capsules layed sequentially on a frond of the host and fused together; each capsule contained many white-shelled embryos (figured Gowlett-Holmes & Holmes, 1989, *Transactions Royal Society SA*, 113 (4): 205-214, fig. 4).

Primovula platysia Cate, 1973

Pl. 24; figs. 14 a-b Periphery moderately angular, located at the $^2/_3$ position; whorl steeply angled behind the shoulder on the left side; outer lip smooth except for a few teeth posteriorly; aperture straight; funiculum thick, transverse, distorting the posterior canal which protudes more than in most species and is divided into two wide, diverging parts. Mauve with yellow ends and yellow spots on the peripheral keel. Body white but with black spots on the mantle, black siphons and black tentacles.

1.2 cm. Central Qld. Lives on green sea whips. *P. howlandae* Cate, 1974, also found on green sea whips in central Qld, was differed mainly on variations in shell form. Further study is needed to determine the status of this name.

Primovula renovata (Iredale, 1930)

Narrow, moderately keeled at the $^2/_3$ position; outer lip thickly toothed posteriorly, smooth anteriorly; aperture straight, sometimes even twisted to the right at the rear end, widening behind the anterior canal; funiculum poorly developed, may be represented by one or two small nodules on the thickened left side of the posterior canal. Maroon in life, fading to mauve; outer lip, columella and ends orange.

1.5 cm. Central Indo-West Pacific; Broome, WA to central Qld. Lives on sea whips. Placed in *Crenavolva* by Cate (1973).

Primovula rutherfordiana (Cate, 1973)

Pl. 24; fig. 21 a-b Rather thick, with broad, subangular shoulders, the periphery located at about the $^3/_4$ posterior position; whorl sharply angled behind the keel; posterior end spout-like; aperture rather wide, recurved to the left posteriorly; outer lip dentate throughout its length, though the anterior teeth are weak; funiculum thick, triangular, crenulate. White or pink in life, depending on the colour of the host gorgonian, but even the pink shells fade to white after death. Animal ivory, with brown spots on the mantle and brown siphons and tentacles.

1.6 cm. Cape York to central Qld. Lives on hollow-stemmed alcyonarians. Placed in the subgenus *Adamantia* by Cate (1973). The thick, broad shell distinguishes this from other Australian species.

Primovula striatula (Sowerby, 1828)

Pl. 24; fig. 17 a-c Cylindrical to fusiform, periphery sub-angulate; aperture bent posteriorly; outer lip inclined inwards, with several strong teeth at the posterior end, moderate teeth centrally becoming obsolete anteriorly; funiculum stout, irregularly crenulate; posterior canal with four or five rather strong denticles on the inner side, edentulous elsewhere; surface finely transversely striate. Back pink, rose or orange, usually with a pair of pale spots at the periphery and a single pale spot at each end; outer lip and tips darker coloured, usually with two central blotches.

1.2 cm. Indian Ocean; Exmouth Gulf to Broome, WA. The more slender Qld shells identified here as *P. traillii* may be merely a form of this species.

Primovula tinctura (Garrard, 1963)

Peripheral keel pronounced at the $^2/_3$ position, whorl steeply angled behind the keel; outer lip teeth strong posteriorly, becoming progressively smaller anteriorly; columellar keel pronounced; funiculum a thick, crenulate boss. White or fawn with a white transverse band at the periphery; outer lip white or yellow; extremities red-orange; fossula pink or mauve.

9 mm. Yeppoon to Moreton Bay, Qld. Lives on the gorgonian *Mopsella*. Treated as a subspecies of *P. striatula* (in *Crenavolva*) by Cate (1973) but its form is distinctive.

Primovula traillii (Adams, 1855)

Pl. 24, fig. 16 a-b Cylindrical, periphery rounded, located at the $^2/_3$ position; aperture almost straight, widening anteriorly behind the anterior canal; outer lip strongly dentate on the posterior half, teeth obsolete anteriorly; underside calloused between the funiculum and the columella in adults; funiculum short, crenulate. Mauve or pink, with a paler band around the periphery; outer lip, columella and ends orange.

1.3 cm. Central Indo-West Pacific; central Qld. Treated as a subspecies of *P. striatula* (in *Crenavolva*) by Cate (1973). The material examined from Keppel Bay is quite variable in form and a close affinity wth *P. striatula* certainly seems likely, although the spots are lacking and most specimens have a characteristic peripheral band.

Primovula verconis (Cotton & Godfrey, 1932)

Elongate-ovate, shoulder rounded, aperture almost straight, slightly reflected posteriorly, widening abruptly anteriorly behind the broadly open anterior canal; outer lip inturned, flat and inclined, weakly dentate; funiculum smooth, merely a calloused area posteriorly; fossula narrow. White, dorsum rose, with reddish tint at the ends.

1.1 cm. SA. Biology and distribution unknown. Verco (1908, *Transactions Royal Society SA* **32**: 338-361) collected three specimens from SA which he identified as *Ovula formosa* Adams & Reeve, 1848. Recognising that Verco's specimens are not of the Adams & Reeve species, Cotton & Godfrey renamed them after Verco. The specimen labelled as the holotype of *verconis* carries the locality St.Francis Is, Nutys Archipelago. This is the only known locality of the species but it may be widely distributed along the southern coasts of SA and WA. The holotype was figured by Cate (1973, *loc. cit.*, fig. 109). Cate (1973) noted that the type material contained two species and nominated the remaining paratype as the holotype of a new species, *C. heleneae*.

▲ GENUS **PRIONOVOLVA** Iredale, 1930
Ovulum breve Sowerby

Helmet-shaped, ovate to globular; posterior end of the outer lip curved, slightly projecting, notched on the left side by the strongly reflexed posterior canal; outer lip thickened, edentulous or weakly dentate; aperture curved, widening anteriorly; anterior canal shallow or deeply notched, bordered on the columellar side by a short, oblique terminal ridge; fossula present but shallow; funiculum a prominent, angular nodule, often continuous with a callus forming an angular margin along the outer left side of the base and nearly meeting a similar columellar callus anteriorly.

Indo-West Pacific. The generic names *Galerovolva* (*hervieri* Hedley, 1899), *Globovula* (*spatiosa* Cate, 1973), and *Testudovolva* (*orientis* Cate, 1973) all introduced by Cate (1973) are synonyms of *Prionovolva*. See *Margovula*.

Prionovolva brevis (Sowerby, 1828)

Pl. 24; figs. 20 a-b Broadly ovate, narrowing anteriorly; outer lip toothed, teeth transverse across the width of the inclined lip; base demarcated on the left side by an axial keel (except at the centre); dorsal surface minutely transversely grooved at the anterior and posterior ends. Calluses, columella and outer lip white, dorsum flesh and usually with two faint paler bands. Mantle pink with large white spots at the base of the filaments (figured Keppel Bay Tidings, April - May, 1987, 26 [1]).

1.5 cm. Townsville, Qld to Newcastle, NSW. Lives on alcyonarians.

Prionovolva cavanaghi (Iredale, 1931)

Pl. 24; fig. 18 a-b Globular to pyriform; aperture narrow, outer lip edentulous; fossula rather deep with a sharp edge within the aperture anteriorly; base not keeled on the left side. Flesh or pink, outer lip, columella and ends white. Mantle flesh-coloured, with irregular red blotches outlined with dark red; siphon and tentacles red, foot white with a few light red blotches (figured Wilson and Gillett, 1971, Pl. 1).

18 mm. Carnarvon, WA to Sydney, NSW.

Prionovolva cottesloensis (Cate, 1973)

Globular; aperture wide, outer lip broad, semicircular, faintly crenulate; funiculum low but thick, crescent-shaped; dorsum glossy smooth. Pale grey to mauve.

1.2 cm. Cottesloe, WA. Nothing is known of the biology or distribution of this species. Cate (1973) placed it in *Globovula*. Drawn from the holotype.

Prionovolva manifesta (Iredale, 1936)

Ovate; aperture wide, curved; outer lip rather strongly dentate along its whole length; funiculum thick, high, elongate; dorsum smooth centrally, striate towards the ends. White.

9 mm. Central NSW; dredged in 90 to 108 m. Nothing is known of the biology of this species. Placed by Cate (1973) in his new genus *Galeravolva*.

Prionovolva margarita (Sowerby, 1828)

Broadly ovate, attenuate anteriorly; aperture curved; outer lip broad, dentate along its inner edge; funiculum wide and crenulate; terminal ridges weak, fossula narrow; dorsum striate. White.

9 mm. South-west Pacific; eastern Qld. Placed in *Globovula* by Cate (1973).

Prionovolva pudica (Adams, 1854)

Pl. 24; fig. 19 a-b Elongate-ovate, base only moderately keeled on the left side; lip thickly varicose and finely toothed. Dorsum pink or mauve with two transverse pale bands or blotches.

1.6 cm. Indo-West Pacific; recorded from Keppel Bay, Qld. Resembles *P. brevis* but more elongate and with weaker margination on the left side. The Australian population seems typical of the species.

Prionovolva spatiosa (Cate, 1973)

Pyriform, tumid but attenuate anteriorly; aperture semicircular, widening anteriorly; posterior canal strongly recurved; outer lip faintly crenulate; funiculum represented by a rounded mound of smooth callus; with an angular margin bordering the posterior canal; dorsum smooth. Milky white, lip and ends beige.

2.1 cm. Cottesloe, WA. Nothing is known of the biology or distribution. Cate (1973) cited this species as the type of *Globovula*. Drawn from the holotype.

▲ GENUS **PROCALPURNUS** Thiele, 1939
Ovula lactea Lamarck

With characters of the type species.

Indo-West Pacific. With only one species. Except for the lack of teeth on the labial side, these shells look like cowries. The rounded form and lack of the terminal knobs readily distinguish the genus from *Calpurnus*, although Thiele (1939), Schilder & Schilder (1971) and Cate (1973) treated it as a subgenus.

Procalpurnus lacteus (Lamarck, 1810)

Pl. 24; fig. 27   Ovate to pyriform; aperture narrow, curved, widening immediately behind the anterior canal, posterior canal reflected; outer lip dentate; funiculum represented by a weak, calloused swelling beside the posterior canal; fossula short, narrow, with a sharply keeled inner margin; terminal ridge lacking; dorsum smooth or finely striate. Uniformly milky white.

2 cm. Central Indo-West Pacific; Rowley Shoals, WA to Moreton Bay, Qld.

▲ GENUS **PROSIMNIA** Schilder, 1927
Ovula semperi Weinkauff

With the characters of the type species.

Indo-West Pacific. Apparently there is only one highly variable species in this genus.

Prosimnia semperi (Weinkauff, 1881)

Pl. 24; figs. 13 a-b   Elongate, subcylindrical, with nodulose and subangular shoulder; ends moderately projecting; aperture straight, widening behind the anterior canal, slightly reflected to the left posteriorly; columella smooth, fossula poorly developed but there is a sharp keel bordering the inner lip deep within the aperture anteriorly; funiculum toothed, outer lip with numerous irregular teeth which may project beyond the outer margin; whole shell conspicuously striate and finely granulose.

1.2 cm. Western Pacific; Cape Leeuwin, WA to Sydney, NSW. Lives on the sea fan *Mopsella*, the shell taking the red, orange or yellow colouring of the host. The form of this species is highly variable; the posterior end may be short or projecting and there may be a ridge of irregular transverse teeth on the base between the funiculum and the columella. The variability of shell form is confusing and there may be more than one species in the complex. The two subspecies *P. s. draconis* and *P. s. boshuensis* were described by Cate, 1973 from the Caroline Is. and Japan respectively.

▲ GENUS **VOLVA** Röding, 1798
Bulla volva Linneaus

Body whorl inflated, subglobular; anterior and posterior canals exceedingly long, curved, of about equal length; both canals reflected; aperture wide, crescent-shaped; columella smooth, terminal ridge fossula and funiculum lacking; outer lip thin, edentulous; dorsum smooth or fine striate, canals finely striate.

Indo-West Pacific. Monotypic. Synonyms: *Radius* Montfort, 1810 (*volva* Linnaeus) *Birostra* Swainson, 1840 (*volva* Linnaeus); *Volvula* Pilsbry, 1895 (*volva* Linnaeus).

Volva volva (Linnaeus, 1758)

Pl. 24; fig. 12   Cream or flesh-pink, outer lip white; interior light brown. Foot and mantle flesh-coloured, mantle with large irregular brown mound-like pustules bearing black terminal papillae.

10 cm. Indo-West Pacific; North West Cape, WA to northern NSW. Lives on large sea whips. For a picture of the living animal see p. 341.

SUPERFAMILY
LAMELLARIOIDEA

*T*HERE are two families in this interesting group, the lamellarias and the bean cowries. Although at one time placed in the Cypraeoidea they are distinguished by a helicoid echinospira larva, quite unlike the veliger larvae of cypraeids and ovulids, and there are major differences in the structure of the nervous system and sensory organs. The echinospira larvae have large ciliated vela for swimming and strange, double-hulled shells consisting of an inner layer of calcified conchiolin and a much larger, uncalcified outer layer, with a fluid-filled space between. This unusual larval type is peculiar to the group and is thought to be an adaptation for long larval life in the plankton.

References

Allan J. (1956). *Cowry Shells of World Seas*. Georgian House, Melbourne, 170pp.

Cate, C. (1979). A review of the Triviidae. *San Diego Society of Natural History Society, Memoir* **10**: 1-126.

Iredale, T. (1935). Australian cowries. *Australian Zoologist* **8** (2): 96-135, pls 8-9.

Schilder, F. A. (1933). Monograph of the subfamily Eratoinae. *Proceedings Malacological Society London* **20**: 244-284.

Schilder, M. & Schilder, F. A. (1971). A catalogue of the living Cowries. *Memoires Institut Royal des Sciences Naturelles de Belgique, 2nd series* **85**: 246 pp.

FAMILY **TRIVIIDAE**

The two subfamilies in this group, the trivias and the eratos, have strong shells very different to the fragile, enclosed shells of the slug-like lamellarias. In form and apertural dentition the shells of trivias, also known as bean cowries, are quite like those of the cypraeids, while the shells of eratos are more elongate and resemble marginellids except for the lack of strong columellar plaits. It has been customary to treat the two groups as separate families but they are kept together in this account because there is very little difference in their anatomy.

SUBFAMILY **TRIVIINAE**

Bean Cowries

In this subfamily the shell is very like that of true cowries but characterised by strong transverse ribs across the back, some of which may be bifurcate on the sides. In many species there is an axial groove, or sulcus, on the dorsum. The ribs may be interrupted by the sulcus or they may cross it. The aperture is straight and toothed on both sides. There is usually a deeply concave fossula (an interior widening of the columellar wall within the aperture) which normally is crossed by ribs. There is no terminal ridge and the posterior and anterior canals are simple U-shaped folds. In most forms the spire is covered and the posterior ribs are laid over it.

The mantles of trivias are pustulose and brightly coloured and may be spread over the shell. The anterior siphon is much longer than it is in true cowries.

Trivias feed on compound ascidians and are usually

found alive in association with encrusting colonies of these fleshy animals. Thus, they may be regarded as carnivorous. They bite off pieces of their prey by means of their horny jaws. They also lay their eggs in capsules embedded in the tissue of the host ascidian. The capsules are vase-shaped with funnel-like openings fashioned by the side of the female's foot from capsule material secreted from her genital tract.

It is odd that these little molluscs, with such an unusual life style, have received so little attention in Australia. Some species are common in beach litter but they are rarely collected. Even in museum collections they are seldom correctly sorted and identified.

The taxonomy of the group is still very confused. The family has not been reviewed adequately. A paper by Crawford Cate (1979, *loc. cit.*) introduced a number of new species and generic names but there is doubt about their status. In this account the classification of Schilder & Schilder (1971, *loc. cit.*) is followed, except that their genera are treated as subgenera of *Trivia*, and one of Cate's generic names is added. Six subgenera are represented in the Australian fauna. The following key to the subgenera as used here may assist identification.

1.	Spire slightly exsert, shell thin	2
	Spire invert, shell solid	3
2.	Ribs fine; with pink dorsal blotch & extremities	ELLATRIVIA
	Ribs moderately coarse	SEMITRIVIA
3.	Shell globular, ends blunt	CLEOTRIVIA
	Shell ovate to cylindrical, ends subrostrate to rostrate	4
4.	Ends markedly rostrate, sides thick	DOLICHUPIS
	Ends subrostrate, sides round	TRIVIROSTRA

▲ GENUS **TRIVIA** Gray, 1832
Cypraea europaea Montagu

The subgenus *Trivia* s.s. is an Atlantic and Mediterranean group with several species characterised by subglobular form, blunt ends and a weak (or lacking) dorsal sulcus. The oldest fossil is from the upper Eocene of Europe.

▲ SUBGENUS **FOSSATRIVIA** Iredale, 1931
Trivia caelatura Hedley

Spire slightly exsert, outer lip rounded, shell thin but with prominent ribs.

Central Indo-West Pacific and south-eastern Australia. There are two living and two fossil species in the south-eastern Australian and New Zealand Tertiary faunas. The group has been treated sometimes as a subgenus or synonym of *Semitrivia* Cossmann, 1903 (*erugata* Tate).

Trivia caelatura Hedley, 1918

Dorsal area smooth, but not excavate, opaque; sides with about 37 ribs interrupted dorsally by the smooth area; aperture rather wide, outer lip narrow. White.

4.5 mm. NSW to Tas. Dredged on the outer shelf. Cate (1979, *loc. cit.*, fig. 1) figured the holotype.

▲ SUBGENUS **CLEOTRIVIA** Iredale, 1930
Cypraea pilula Kiener = *globosa* Sowerby

Globose with blunt ends; spire invert; dorsal sulcus deep, the ribs terminating at its sides; fossula weakly developed.

Cosmopolitan. There are two living species in the Indo-West Pacific and three in the Central American region. In addition to these older names Cate (1979) introduced three names for southern Australian shells (*meridionalis, dorsennus, corallina*) which he assigned to *Cleotrivia* but the status and generic position of these names requires further study. Shells of *Cleotrivia* are usually quite small or tiny and are said to be almost "as round as a pea". Schilder & Schilder (1971) treated the group as a subgenus of *Niveria* Jousseaume, 1884.

Trivia globosa (Sowerby, 1832)

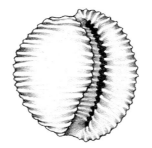

Ribs fine; dorsal sulcus sharply cut and reaches almost to the extremities; aperture narrow; quite globular. White.

7 mm. Indo-West Pacific; circum-Australian. Unusually variable in size. There are several named forms. The Pacific form is sometimes called *T. pilula* Kiener, 1843 and the species has been recorded from eastern Australia under that name. Cate (1979) named a South Australian shell *pilula euclaensis*. Other names used in this connection, which may be synonyms, are *sphaerula* Mighels, 1845; *pisum* Gaskoin, 1846; *brevissima* Sowerby, 1870; *acutisulcata* Kenyon, 1900; *cosmoi* Dautzenberg, 1921; *pisulum* Schilder, 1932; *bathypilula* Iredale, 1935.

Another species, *T. (C.) vitrea* Gaskoin, 1849, has been listed from Queensland by Hedley. It is distinguished by coarser ribs, rounded base, straight aperture and glassy surface.

▲ SUBGENUS DOLICHUPIS Iredale, 1930
Cypraea producta Gaskoin

Rostrate, spire invert, sides thickened, dorsal sulcus lacking, with many ribs continuous across the dorsum, fossula deep and crossed by the ribs.

Central Indo-West Pacific. There are two living members of the group and several Tertiary fossils. Although the projecting ends of these shells make them quite distinctive it is really only a matter of degree and the separation of this group from *Trivirostra* is questionable. Schilder & Schilder (1971) treated *Dolichupis* as a synonym of *Pusula* Jousseaume. Synonym: *Trivellona* Iredale, 1931 (*excelsa* Iredale = *producta* Gaskoin).

Trivia producta (Gaskoin, 1836)

Pl. 24; figs. 29 a-b Body whorl bulbous, slightly pyriform; base rather flattened because of the calloused sides; aperture narrow, posterior and anterior canals wide; with about 22 strong ribs crossing the labial lip, about 20 on the columellar side, about 10 ribs are continuous across the back. White.

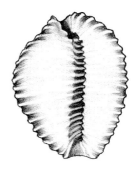

1.8 cm. Central Indo-West Pacific. A form from NSW named *Trivellona excelsa* by Iredale, 1931 is regarded as a synonym. The thickly calloused margins, flat base and slightly projecting ends will usually suffice to distinguish this moderately common species.

▲ SUBGENUS ELLATRIVIA Iredale, 1931
Triviella merces Iredale

Ovate; ends blunt to subrostrate; spire slightly evert; dorsal sulcus very weak or lacking, dorsum usually smooth or crossed by weak ribs; fossula well developed and crossed by ribs; extremities red-tinged.

Southern Australia and New Zealand. There is only one living species but several southern Australian and New Zealand late Tertiary fossils are assigned to this group. Schilder & Schilder (1971) treat it as a subgenus of *Niveria*.

Trivia merces (Iredale, 1924)

Pl. 24; fig. 32 White with rose or red-brown blotches on the dorsum as well as red-brown extremities, but this colouration may be lacking in deepwater specimens.

1.3 cm. Southern Qld to Geraldton, WA. Common in the intertidal zone, especially on the south coast; also dredged across the continental shelf. The species was known as *T. australis* Lamarck, 1822 in older literature but this name was preoccupied and Iredale provided a replacement. As well as lacking colouration, deepwater specimens may have ribs crossing the dorsum; Iredale (1931) introduced the subspecies name *addenda* for shells of this kind from NSW. The species occurs in New Zealand where it is given subspecies status as *T. merces memorata* Finlay, 1927.

▲ SUBGENUS **TRIVIROSTRA** Jousseaume, 1884
Cypraea oryza Lamarck

Ovate to cylindrical, ends subrostrate, sides rounded; spire invert; dorsal sulcus weak to strong, ribs strong; fossula well formed and crossed by ribs.

Indo-West Pacific. Schilder & Schilder (1971) list 14 living species and several subspecies under this name. The group needs much more study because the shells are very variable and some of the supposed species may be only variants.

In addition to the Australian species illustrated here, Cate (1979) has described a shell from St.Vincent Gulf as *Trivirostra cydarum*. The holotype is thin-shelled and has a moderately well formed dorsal sulcus but the spire is evident, recalling the condition of *Ellatrivia*.

Trivia corinneae Shaw, 1909

Oblong-ovate; ends almost rostrate; ribs rather coarse. White, often suffused with pink.

5 mm. Pacific; NSW. See Cate, 1979, *loc. cit.* figs. 112, 112a. Synonyms: *insecta* Mighels, 1845; *hidalgoi* Vredenberg 1920.

Trivia edgari Shaw, 1909

Subglobular; dorsal sulcus weakly impressed, crossed by many ribs; aperture narrow, fossula deeply concave, teeth fine. White. Animal described by Kay (1979: 186-187).

7 mm. Indo-West Pacific; Cape Jaubert, WA to Qld. Synonyms used in the Australian connection are *grando* Gaskoin, 1849 and *oryzoides* Iredale, 1935.

Trivia hordacea (Kiener, 1843)

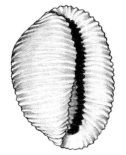

Elongate-ovate; dorsal sulcus narrow, moderately impressed; ribs interrupted by the sulcus, some bifurcating on the sides. White.

4 mm. Indo-West Pacific; eastern Australia. The elongate form distinguishes this small species. Synonym: *koroensis* Ladd, 1934.

Trivia obscura (Gaskoin, 1849)

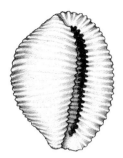

Pl. 24; fig. 31   Ovate, almost rostrate; dorsal sulcus deeply cut; ribs end abruptly at the sulcus, about 24 ribs cross the labial lip, about 17 on the columellar side, plus a few small denticles in the canals. White; margins rosy pink, reddish brown blotches border the dorsal sulcus.

9 mm. Abrolhos to Cape Leveque, WA. This little endemic WA species is readily identified by the deep sulcus and distinctive colouring. Synonyms: *bipunctata* Odhner, 1917; *pargrando* Iredale, 1935; *exmouthensis* Cate, 1979.

Trivia oryza (Lamarck, 1810)

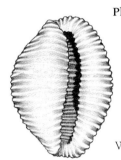

Pl. 24; fig. 30   Ovate; dorsal sulcus well formed and reaches almost to the extremities; ribs strong, almost parallel on the sides, stopping at the sulcus, some ribs originating in the interspaces on the sides, about 27 crossing the labial lip and 20 crossing the columellar lip; interspaces minutely transversely lamellate. White.

8 mm. Indo-West Pacific; Dongara, WA to southern Queensland. This is the most common trivia in northern Australian shallow waters. Its ovate form, well formed dorsal sulcus and white shell will usually distinguish it from others. Synonyms: *intermedia* Kiener, 1843; *pediculus* Roberts, 1870; *minor* Schepman, 1909; *zzyzyxia* Cate, 1979.

Trivia scabriuscula (Gray, 1827)

Ventricose; spire evident; ribs coarse; inner lip ends in a sharp edge.

1.1 cm. Indo-West Pacific; recorded from Qld and NSW. Holotype figured by Cate (1979, *loc. cit.*, fig. 149). Synonym: *gaskoini* Roberts, 1869.

Trivia pellucidula (Reeve, 1846)

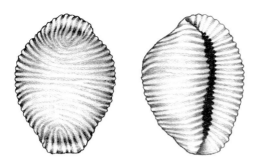

Ovate; dorsal sulcus lacking, ribs continuous over the dorsum. White.

7 mm. Indo-West Pacific; recorded from Qld by Hedley as *T. pellucida* Gaskoin, 1846, which is an unpublished manuscript name. The differences between this and *T. edgari* are minor and the species may be the same.

SUBFAMILY **ERATOINAE**

Eratos

This is a small subfamily of small shells. Superficially they are more like marginellas than cowries. Usually they are conical in form with an exsert spire, although the sides of the aperture are toothed. Anatomically there is very little to distinguish eratos from trivias. Like trivias they feed on ascidians and have an echinospira larva.

In their 1971 catalogue Schilder and Schilder listed 23 living species in the world, the majority occurring in the Indo-West Pacific region. There is still much work to do before the Australian species are sorted. There are at least six in Australian waters.

▲ GENUS **PROTERATO** Schilder, 1927
Erato neozelanica Suter

Conical or pyriform, attenuate anteriorly; spire exsert; aperture long, narrow, toothed on both sides; terminal ridge similar and parallel to subsequent columellar teeth; posterior canal shallow; fossula narrow but distinctly concave.

Indo-West Pacific, Southern Australia and New Zealand. *Proterato* s.s. is a fossil group from the Tertiary of south-eastern Australia and New Zealand. There are three subgenera with living representatives according to the classification of Schilder and Schilder (1971), all of them with Australian representatives. Synonym: *Lachryma* Sowerby, 1832 (*lachryma* Sowerby, 1832).

▲ SUBGENUS **CYPRAEERATO** Schilder, 1932
Erato bimaculata Tate

Pyriform, distinctly constricted anteriorly; dorsal sulcus lacking or represented by a weak anterior impression, surface smooth; fossula deeply concave, projecting.

Indo-West Pacific and south-eastern Australia. With five living species plus the Victorian Miocene fossil *P. minor* (Tate, 1878).

Proterato angistoma (Sowerby, 1832)

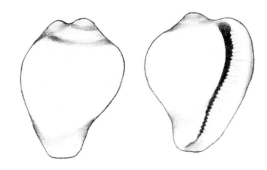

Yellow or cream, sometimes with an obscure, narrow rose band at the suture and rose extremities; outer lip thickened and a little projecting posteriorly, teeth rather short.

4 mm. Indo-West Pacific; south to Caloundra, Qld. The Qld shells are sometimes regarded as a distinct subspecies under the name *angulifera* Sowerby, 1859 on the basis of more prominent teeth than in the Indian Ocean form. Shells from NSW named *P. bisinventa* Iredale, 1931 may be a subtropical form of this group. Synonym: *pellucida* Reeve, 1865.

Proterato bimaculata (Tate, 1878)

Dorsal sulcus absent or represented by a very weak anterior impression. Pale pink, extremities purple or darker pink.

5 mm. Vic. and Tas. to southern WA. Not figured.

Proterato gallinacea (Hinds, 1844)

Outer lip angular and projecting posteriorly; teeth almost cross the outer lip; extremities granulate. White, ends brown.

5 mm. Central Indo-West Pacific; Qld.

▲ SUBGENUS **ERATOENA** Iredale, 1935
Ovulum corrugata Hinds

Conical to ovate; dorsal sulcus deeply excavated, surface granulate, at least posteriorly; fossula concave but hardly angular, collumellar teeth distinct throughout, terminal ridge transverse.

Indo-West Pacific. There are seven living species, one of them present in Qld.

Proterato sulcifera (Sowerby, 1832)

Ovate; finely granulate. White or pale green, sometimes with three obscure bands or a coalescent dorsal notch; usually spotted with pink.

5 mm. Indo-West Pacific; recorded from Qld. *P. corrugata* (Hinds, 1844) may be merely a more coarsely granulate, unspotted form of this species. Synonyms: *nana* Sowerby, 1859; *schmeltziana* Crosse, 1867; *capensis* Schilder, 1933.

▲ SUBGENUS **SULCERATO** Finlay, 1930
Erato (Eratopsis) illota Tate

Conical; dorsal sulcus lacking, surface smooth; fossula concave but hardly angular, columellar teeth usually distinct throughout (except *in lachryma*), terminal ridge rather oblique.

Indo-West Pacific. Schilder & Schilder (1971, *loc. cit.*) listed eight living species and two Pliocene fossils in Victoria.

Proterato denticulata (Pritchard & Gatliff, 1901)

Very like *lachryma* but the central band is obsolete and there are usually short columellar teeth posteriorly.

7 mm. Vic. to WA. Schilder & Schilder (1971, *loc. cit.*) treated this as a subspecies of *lachryma*.

Proterato lachryma (Sowerby, 1832)

Columellar teeth weak anteriorly, lacking posteriorly. Greyish or yellowish white, sometimes green at the ends, with three reddish or purple bands, the posterior band the widest.

8 mm. Qld to northern NSW. Synonym: *trifasciata* Cotton & Godfrey, 1932.

FAMILY **LAMELLARIIDAE**

Lamellarias

At first sight you might take these animals to be opisthobranch slugs but closer inspection will show otherwise. They have a greatly reduced, fragile shell which is permanently enveloped by the fleshy mantle lobes. The general aspect of the anatomy is quite like that of the bean cowries and they have a similar life history and feeding habits.

Even though they are so fragile, dead shells are sometimes cast up on the beach. In most species the shell is auriform, very thin, translucent, coiled and depressed with a very wide aperture (Pl. 24, fig. 28 a-b). The shells are so lacking in characters as to be of little use for classification or identification. The animal is brightly coloured and intricately patterned but very variable. In short, this is a very difficult group for the taxonomist. Consequently, the family has been little studied.

Lamellarias are common under stones on the southern shores of Australia and there are tropical species as well, but it is not clear how many species there are. A much abbreviated account of them is given here pending further study. Names used in the Australian context are:

Caledoniella Souverbie, 1869 (type species *montrouzieri* Souverbie): *labyrinthina, pulchra* and *testudinis*, all described from SA by Basedow, 1905, *Transactions Royal Society SA* **29**: 183-185, Pl. 28, figs.1-4.

Chelynotus Bergh,1853 (type species *nigra* Blainville): *tonganus* Quoy & Gaimard, 1832, an Indo-West Pacific species recorded from northern Australia.

Lamellaria Montagu, 1815 (type species *perspicua* Linnaeus): *australis* Basedow, 1905, from SA and southern WA; *ophione* Gray, 1849, from Bass St. to Cape Naturaliste, WA.

Marseniopsis Bergh, 1886 (type species *pacificus* Bergh): *innominatus* Iredale, 1936, from NSW.

Mysticoncha Allan, 1936 (type species *contusiformis* Baselow = *wilsoni* Smith): *wilsoni* Smith, 1886, from Bass St.

NATICOIDEA

*T*HERE is only one living family in the superfamily. The naticids are predatory snails which crawl in sand or mud substrates in search of their prey. The meandering trails they make are a feature of intertidal sand flats.

FAMILY **NATICIDAE**

Sand and Moon Snails

This is a moderate-sized family, well represented in the coastal and estuarine faunas of both temperate and tropical Australia. Three subfamilies are recognised, the Naticinae, Polinicinae and Sininae. There has never been a comprehensive review involving consideration of anatomical features and there is some evidence of convergent evolution among the genera, so that the classification of the family remains problematical.

Naticids feed on bivalves, as far as is known (Kabat, 1990). They catch their prey buried in the substrate and hold it wrapped in the foot. Using a fold in the front edge of the foot as a steadying support, the proboscis is applied to the bivalve shell and drills a hole in it. There is still argument about whether this is achieved only by mechanical abrasion by rotation of the radula, or whether chemical secretions from an accessory gland under the lip assist. In any event, the end result is a neat, round hole through which the proboscis may attack the body of the prey. Bivalve shells with such holes are common on our sand flats and beaches.

Females lay egg masses of stiffened jelly in which very many tiny eggs are embedded. In many species the jelly becomes impregnated with sand giving extra support. The sandy egg collars so common on Australian sandflats are the egg masses of naticids. The eggs hatch as shelled veligers and there is a planktotrophic larval stage.

References

Hedley, C. (1924). Some naticoids from Queensland. *Records Australian Museum* **14** (3): 154-162, Pl. 22.

Cernohorsky, W. O. (1971). The family Naticidae in the Fiji Is. *Records Auckland Institute & Museum* **8**: 169-208.

Kabat, (1990). Predatory ecology of naticid gastropods with a review of shell boring predation. *Malacologia* **32** (1): 155-193.

SUBFAMILY **NATICINAE**

The Naticinae are characterised by a more or less globular, umbilicate shell with a flat semicircular calcareous operculum. Usually there is a callus on the parietal wall which may spread over and cover at least the posterior part of the umbilicus. The columella is straight or slightly arched and often strengthened by a rib on its back side which spirals into the umbilicus and is called the funicle. The operculum has a corneous layer on its under surface. It is rounded at the anterior end and pointed at the other and its upper surface may be smooth or spirally grooved. The animal of many Naticinae is beautifully coloured and patterned. Although the foot is large, the whole animal may be quickly withdrawn into the shell behind the closed operculum. The generic classification of the subfamily is based almost entirely on the characters of the umbilicus and operculum. However, there is a great deal of intergradation. In this account several genus-level groups in the Australian fauna are treated as subgenera of *Natica*.

▲ GENUS **NATICA** Scopoli, 1777

Nerita vitellus Linnaeus

Umbilicate, with a funicle on the back of the columellar lip and a parietal callus which may merge with the funicle and occlude the umbilicus posteriorly; operculum calcareous, semicircular, round anteriorly, pointed posteriorly, outer surface smooth or spirally grooved.

Cosmopolitan. Three subgenera are recognised (see Cernohorsky, 1971). *Notocochlis* is commonly used for some Australian species, the presence of a single opercular groove being the defining character; because this is a common condition among Indo-West Pacific naticids this generic name is regarded here as a synonym. The New Zealand *Tanea* Marwick, 1931 (*zelandica* Quoy and Gaimard) is defined as having two weak spiral cords on the operculum; although this character fits within the range of *Natica* s. s. Cernohorsky (1971) reported that the type species has a very different radula. Thus it would be unwise to synonymise *Tanea* with *Natica*. However, some Australian species have been placed in *Tanea* solely on opercular characters, and this seems inappropriate. Synonyms: *Cochlis* Röding, 1798 (*vitellus* Linnaeus); *Notocochlis* Powell, 1933 (*migratoria* Powell).

In addition to the species illustrated here Hedley has listed the following names from Qld: *buriasiensis* Reeve, 1844; *deiodosa* Reeve, 1855; *variabilis* Récluz in Reeve, 1856; *limpida* Smith, 1884. The status of these names is not known to this author.

▲ SUBGENUS **NATICA** s.s

Operculum with 1-3 marginal spiral grooves; funicle prominent or merging with the umbilical wall; umbilicus sometimes partly closed by the parietal callus.

Cosmopolitan.

Natica arachnoidea (Gmelin, 1791)

Pl. 36; fig. 17 a-b   Globose, spire low; parietal callus large, covering the funicle and most of the umbilicus; operculum with a serrate margin and one primary spiral rib near the edge, marginal serrations sometimes running back a short distance as spiral cords. White, cream, yellow or orange, irregularly patterned with brown; protoconch purple-brown; callus white.

2.5 cm. Indo-West Pacific; Torres St. to Mossman, Qld. Synonym: *raynoldiana* Récluz, 1844.

Natica areolata Récluz, 1844

Globose, rather thin, spire low; parietal callus small; funicle prominent, spiralling into the umbilicus and almost filling it in some specimens; operculum with one or two marginal ribs. White, with three broad bands of curved or V-shaped orange markings which may merge together.

1.5 cm. Indo-West Pacific; Qld south to Caloundra. Compare with *N. sagittata*.

Natica euzona Récluz, 1844

Pl. 36; figs. 21 a-b   Globose, thin, spire low; usually sculptured with fine, arcuate axial plicae at the suture; umbilicus open but occluded by a broad funicular pad separated from the parietal callus by a narrow U-shaped hole descending into the umbilicus; operculum with a single marginal spiral groove. Cream with wavy axial orange-brown lines and 3-4 spiral bands of alternating white and orange-brown chevron marks, those below the suture and around the base often prominent.

2.5 cm. Indo-West Pacific; Dampier, WA to Sydney, NSW. Widespread but not very common anywhere within its range. There is difficulty distinguishing this species from the highly variable *N. sagittata* complex because the colour patterns of the two are similar. However, mature *euzona* are considerably larger then *sagittata*. Another available name is *N. picta* Récluz, 1844 (= *elegans* Récluz, 1850 - not Sowerby, 1836) which may be a distinct species from the same area.

Natica fasciata (Röding, 1798)

Pl. 36; fig. 12 a-b Globose, spire low; umbilicus open but covered posteriorly by a tongue-like callus folding into it from the parietal wall; operculum with a single spiral marginal groove and a few weak striae, the straight edge finely serrated. Dark brown to orange, usually with a pale subsutural band and base, callus dark brown. The operculum is illustrated here.

3 cm. Indo-West Pacific; Dampier Archipelago, WA to Solitary Is., NSW. The colouring is usually sufficient to indentify this species. Synonyms: *melanostoma* var. *e* Gmelin, 1791; *solida* Blainville, 1825; *cinnamomea* Menke, 1830; *lupinus* Deshayes, 1838.

Natica gualtieriana Récluz, 1844

Pl. 36; fig. 19 Globose; spire low with subsutural costae; parietal callus not reflecting over the umbilicus which is deep posteriorly, but there is a massive funicle running more or less vertically into the umbilicus and occluding it anteriorly; operculum smooth, convex, with a single marginal groove and smooth edges. Greenish grey or cream, with two spiral bands of short, irregular, brown axial lines and usually a peripheral band of spots or crescentic marks around the base; columella and funicle white.

2 cm. Indo-West Pacific; Albany, WA around the northern coast to southern Qld. Very common in sandy shallows. Its meandering tracks are a feature of many sandy shores at low tide. *Cochlis migratoria* Powell, 1927 is a New Zealand shell believed to be related. A NSW form of this same group has been named *vafer* Finlay, 1930 and may be the same species. Synonyms: *marochiensis* Quoy & Gaimard, 1833 (not Gmelin, 1791); *tesselata* Philippi, 1849.

Natica lineata (Röding, 1798)

Pl. 36; fig. 25 Globose and relatively thin; spire slightly elevated; funicle strong and spiralling into the deep umbilicus, parietal callus thin, small and separated from the funicle by a U-shaped space; operculum concave, with two deep spiral grooves. Fawn or cream with numerous thin, red or orange axial lines; umbilicus and funicle white.

4 cm. Central Indo-West Pacific; Kimberley, WA to Torres St., Qld. According to Cernohorsky (1971) this species has a radula similar to that of the New Zealand genus *Tanea* but here it is retained in *Natica* pending further study.

Natica luculenta Iredale, 1929

Pl. 36; fig. 11 Globose, thin, spire short, mouth large; umbilicus open but narrow with a distinct but narrow funicle behind the columella and a spreading callus on the parietal wall; operculum not known. Buffy cream, regularly spotted with splashes of orange-brown.

2.4 cm high. Trawled on the outer continental shelf from southern Qld to eastern Vic.

Natica phytelephas Reeve, 1855

Pl. 36; fig. 13 Globose, spire moderately high with convex whorls and well impressed suture; umbilicus small but deep, without a funicle but there is a thick posterior callus which spreads onto the parietal wall; operculum not seen by this author. White.

2.3 cm. Central Indo-West Pacific; Kimberley coast, WA and NT. The columellar and umbilical features of this species do not fit the generic pattern very well and it may turn out to be better placed in the Sininae when more material is available for study.

Natica pseustes Watson, 1881

Pl. 36; fig. 18 a-b Globose, suture lacking axial plicae; parietal callus covers the funicle leaving only a deep groove entering the umbilicus; operculum with a single weak marginal spiral rib and serrate, straight edge. Cream with four indistinct spiral bands of irregular chevron marks, those below the suture most prominent, and an irregular reticulate pattern of fine bown lines; callus and columella brown or lavender.

1.5 cm. Indo-West Pacific; Lagrange Bay, WA to Moreton Bay, Qld. Synonyms: *telaaranae* Melvill, 1901; *lavendula* Woolacott, 1956; *roscoei* Kilburn, 1976.

Natica sagittata Menke, 1843

Rather thin, globose; protoconch small, planate, not differentiated from the teleoconch whorls; umbilicus narrow, occluded anteriorly by a massive funicle which does not connect with the small parietal callus posteriorly; operculum smooth except for a single marginal rib between two shallow grooves, edge of the straight side not serrated. Fawn to white, with light brown curved axial lines in three spiral bands, tending to form acute V-marks in the centre, and three spiral rows of darker brown (sometimes orange) spots at the suture, periphery and base, the peripheral ones usually V-shaped; columella and umbilicus white.

1.5 cm. Circum-Australian. Common intertidally. In colour pattern and umbilical features this shell is like a miniature *N. euzona* and has in fact been confused with that species. *N. areolata* has a different colour pattern. *N. cothurnata* Iredale, 1936 from NSW, is probably a synonym.

Natica sertata Menke, 1843

Small, globose, thin; parietal callus weak or lacking, umbilicus open except for the plug-like funicle behind the columella; operculum concave, smooth except for a wide spiral marginal rib. White or cream, with a faint tan subsutural band and several V-shaped rows of obscure tan marks.

7 mm. Vic. to Fremantle, WA. Although the operculum is typical of the subgenus, the open umbilicus is not and the subgeneric position of the species remains problematical. Synonym: *colliei* Récluz, 1844.

Natica stellata (Hedley, 1913)

Pl. 36; fig. 24 More elongate than *N. vitellus* with a higher spire and less flaring lip anteriorly; parietal callus thick and with a tongue-like portion spreading over the posterior part of the umbilicus leaving only a narrow opening anteriorly; funicle lacking or small; operculum with two or three marginal ribs. Orange or yellow with a darker central band and two spiral rows of white blotches; aperture and umbilicus white.

4 cm. Indo-West Pacific; Torres St. to Swain Reefs, Qld. Not a common species. When he used this name for specimens he had collected in the Torres St., Hedley attributed it to Martyn, 1786 but that author's names are unavailable and so Hedley himself inadvertently becomes the author of the species. The rather bright colouring is unusual - compare with *N. vitellus* in which the parietal callus is much less prominent.

Natica suffusa Reeve, 1855

Pl. 36; fig. 16 a-b Height greater than width; parietal callus thick, covering the funicle and almost closing the umbilicus; operculum with a single marginal spiral groove. White with a broad, pale mauve, central spiral band and often several axial growth zones of darker mauve; interior, outer lip, callus and columella mauve.

1.2 cm. Indo-West Pacific; north Qld. The colouring of this species is quite distinctive and it is very unlikely that it could be confused with any other.

Natica tigrina (Röding, 1798)

Pl. 36; fig. 10 Rather thin with a relatively tall spire and deeply impressed suture; parietal callus weak or lacking, umbilicus very narrow, partly filled by a small funicle. White with 14-16 spiral rows of reddish brown spots which may be spirally elongate; columella and funicle white; operculum not seen by this author.

3 cm. Indo-West Pacific; apparently rare in Australia but recorded from a sand cay off Carwell, Qld. Synonyms: *maculata* Perry, 1811; *javana* and *maculosa* Lamarck, 1822; *pellistrigina* Dunker, 1882.

Natica vitellus (Linnaeus, 1758)

Pl. 36; fig. 28 Globose, spire low, anterior end slightly flaring; parietal callus thick but low, partly covering the posterior part of the deep umbilicus; funicle lacking or small; operculum with two marginal ribs. White or yellowish, with one or two broad brown bands.

4 cm. Indo-West Pacific; Exmouth Gulf, WA to Caloundra, Qld. This common, widespread species has many synonyms. *N. stellata* was misidentified as *N. vitellus* for many years and care must be taken when using the older literature. Cernohorsky (1971, *loc. cit.*, p. 176) compared the two. Synonyms: *rufa* Born, 1778; *leucozonias* and *spadicea* Gmelin, 1791; *albula* Röding, 1798; *fuscata* Link, 1807; *helvacea* Lamarck, 1822; *forscalii* Sowerby, 1825; *pallens* Philippi, 1849; *chemnitzii* Philippi, 1852; *globosa* Philippi, 1855.

▲ SUBGENUS NATICARIUS Duméril, 1806
Nerita canera Linnaeus

Operculum with many spiral grooves; funicle prominent; umbilicus open.

Cosmopolitan. Although the radula is like that of *Nattica* s.s., most contemporary authors regard the multiple grooves on the operculum as sufficient to warrant subgeneric distinction. Synonyms: *Naticus* Montfort, 1810 (*canrena* Linnaeus); *Quantonatica* Iredale, 1936 (*subcostata* Tenison Woods).

Natica alapapilionis (Röding, 1798)

Pl. 36; fig. 31 Globose, spire of moderate elevation; outer lip flaring and angulate anteriorly; umbilicus wide and deep with a narrow central funicle, parietal callus small, reflecting over the umbilicus; operculum with seven or eight spiral ribs. Fawn or blue-grey, body whorl with four white spiral lines containing rectangular brown spots.

4 cm. Indo-West Pacific; Torres St., Qld to northern NSW. The shell is very distinctive. The beautiful animal is red-brown with radiating white lines. Synonyms: *zonaria* Lamarck, 1816; *taeniata* Menke, 1828; *articulata* Philippi, 1852; *crenata* Récluz, 1853.

Natica colliei Récluz, 1844

Pl. 36; fig. 23 a-b Globose, spire low, with subsutural axial costae; parietal callus small, separated by a U-shaped notch from the massive funicle almost filling the deep umbilicus; operculum with 6-8 prominent spiral ribs and two sharp, narrow, outer spiral cords. White or cream, with a pale orange subsutural zone crossed by paler costae and two wide bands of rectangular brown blotches.

2 cm. Indo-West Pacific; Geographe Bay, WA to southern Qld. *Although* a different shade, the shell is similar to *N. gualtieriana*; the multiribbed operculum is different. See note on *N. zonalis*. Synonym: *seychellianum* Watson, 1886.

Natica lineozona Jousseaume, 1874

Height greater than width; parietal callus narrow, thin, only partly overlapping the umbilicus, funicle indistinct or lacking; operculum with many spiral grooves. White with four brown spiral lines, the central two sometimes enclosing a tan spiral band, and irregular brown axial lines which stop at the periphery without extending onto the base.

1 cm. Indo-West Pacific; north Qld. For illustrations of the shell, radula and operculum, see Cernohorsky (1971, *loc. cit.* figs. 36-38). This is an uncommon species. A very similar Indo-West Pacific species, *N. insecta* Jousseaume, 1874, is also found in north Qld. Synonyms: *gaidei* Souverbie, 1874; *notata* Sowerby, 1883.

Natica onca (Röding, 1798)

Pl. 36; fig. 26   Subglobose, aperture wide; umbilicus wide and deep; funicle prominent and thick, bordered by an arched groove which enters the umbilicus; callus small and confined to the parietal wall; operculum with 6-8 grooves. White with 2-5 rows of brown spots.

2.5 cm. Indo-West Pacific; NT to Sydney, NSW. Synonyms: *pavimentum* Röding, 1798; *litterata* Link, 1806; *chinensis* Lamarck, 1822; *candida* Wood, 1825.

Natica orientalis (Gmelin, 1791)

Pl. 36; fig. 22 a-b   Globose spire low, suture impressed; columella projecting to meet the outer lip at a pronounced, spatulate, anterior angle; parietal callus small but thick, funicle large, separating a deep umbilical pit posteriorly from a shallow, narrow, anterior spiral groove; operculum with about five prominent spiral ribs and one or two thin cords near the outer edge, margin toothed. Uniformly cream or pale yellow; columella and funicle white.

3 cm. Indo-West Pacific; recorded from Exmouth, WA but probably present also in north Qld. Like *N. alapapilionis* in form but the anterior end is narrower and the characteristic colour pattern of that species is lacking. Synonyms: *explanata* Röding, 1798; *eburnea* Deshayes, 1838.

Natica subcostata (Tenison Woods, 1878)

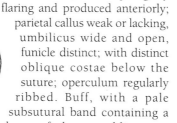

Pl. 36; fig. 20   Globose with weak subsutural costae, outer lip slightly flaring and produced anteriorly; parietal callus weak or lacking, umbilicus wide and open, funicle distinct; with distinct oblique costae below the suture; operculum regularly ribbed. Buff, with a pale subsutural band containing a spiral row of obscure, oblique tan streaks and one or more white spiral lines on the body whorl containing short tan dashes.

1 cm. Southern Qld around the southern Australian coast as far north as Fremantle, WA.

Natica zonalis Récluz, 1850

Pl. 36; fig. 15   Globose; subsutural zone with fine, oblique plicae; parietal callus moderate, funicle prominent and oblique, umbilicus deep posteriorly; operculum with a raised callus over the nucleus and 7-8 strong spiral ribs. White to light tan, with two broad spiral bands of darker brown on the body whorl, bands sometimes becoming confluent.

2.8 cm high. Indo-West Pacific; Broome, WA to Bunker Group, Qld. This may be a synonym of *N. sertata*. Compare with *N. seychellianum*.

▲ SUBGENUS **TECTONATICA** Sacco, 1890
Natica tectula Sacco

Small to medium-sized shells; umbilicus nearly filled with callus leaving only a narrow lateral groove; operculum with a single marginal rib.

Indo-West Pacific. The type species is a Miocene-Pliocene fossil from Italy. Cernohorsky (1971 and later) has assigned several Indo-West Pacific species to this subgenus.

Natica bougei Sowerby, 1908

Pyriform-ovate, penultimate whorl longer than usual in the genus; parietal callus usually completely filling the umbilicus; operculum with a single weak groove. White, with wavy dark brown axial lines which may coalesce, sometimes the lines are concentrated in a spiral band.

1 cm. Indo-West Pacific; recorded from Ningaloo Marine Park and the Rowley Shoals, WA in coral reef lagoons.

Natica robillardi Sowerby, 1843

Pl. 36; fig. 27   Subglobose, spire moderately high, with radial riblets below the suture; parietal callus completely filling the umbilicus, bordered by an arched groove; operculum with a single weak spiral groove. Milky white to fawn, sometimes with a broad, darker cream or fawn band on the body whorl; operculum with a prominent, crescent-shaped brown spot over the nucleus.

2 cm. Indo-West Pacific; Rowley Shoals, WA to Swain Reefs, Qld. The spotted opercular nucleus is distinctive.

Natica shorehami (Pritchard & Gatliff, 1900)

Tiny, but quite strong, globose; large protoconch of one whorl, then two teleoconch whorls; weak oblique costae below the suture; with a thick funicle almost completely plugging the umbilicus; operculum smooth, with a slight prominence after the initial coils and a very obscure groove near the margin. Mauve, with a white subsutural band.

5 mm. Twofold Bay, NSW to Bass St. The generic placement of this species is problematical. Some authors have placed it in the New Zealand genus *Proxiuber* Powell (*australis* Hutton). *N. elkingtoni* Hedley & May, 1908, described from off Cape Pillar, Tas., may be the same species.

Natica simplex Schepman, 1909

Pl. 36; fig. 14   Globose, spire moderately low; parietal callus small, spreading over a small but distinct funicle, umbilicus narrow but deep; operculum not seen. Cream to fawn with a distinct white subsutural band and white base; callus and funicle brown.

2 cm. Central Indo-West Pacific; dredged off the Kimberley coast of WA in 90 m on muddy bottom.

Natica violacea Sowerby, 1825

Solid, pyriformly ovate to globose, spire low; suture grooved, body whorl sculptured with fine axial striae; aperture semi-ovate, parietal callus almost completely covering the umbilicus. White with five spiral rows of large, irregular, rectangular spots which may tend to merge axially, one row of spots on the penultimate whorl; parietal callus and base rose-violet.

2 cm. Central Indo-West Pacific; there are a few uncertain records from north Qld. No specimen could be obtained for illustration. Synonyms: *glabra* Wood, 1828; *rhodostoma* Philippi, 1842.

Natica zonulata Thiele, 1925

Ovate, spire moderately low; parietal callus thick, umbilicus almost filled by a thick funicle; operculum with a spreading nuclear callosity and a single, deep spiral groove near the outer edge, margin minutely serrate. Cream, subsutural band white, followed by a wide pale fawn band and then one or two spiral bands of brown spots around the basal periphery; columella cream, funicle light brown.

1 cm. Two Peoples Bay to Onslow, WA. The brown funicle renders this little endemic WA species quite distinctive.

▲ GENUS **TASMATICA** Finlay & Marwick, 1937
Natica schoutanica May

With characters of the type species.

South-eastern Australia. There is one, possibly two species in this genus. The shell form and lack of a funicle are akin to *Polinices* but as the operculum is unknown it seems best to retain *Tasmatica* for the time being.

Tasmatica schoutanica (May, 1912)

Ovate; protoconch of one large whorl, then 2½ rapidly expanding teleoconch whorls; body whorl obliquely inflated; suture moderately impressed; aperture angled posteriorly, rounded anteriorly; columella inclined, straight, funicle lacking but there is a thick parietal callus occluding but not filling the open umbilicus. White, with two spiral rows of irregular, tan, axial bars.

8 mm. North West I., Qld to the Great Australian Bight. Iredale (1936) gave the NSW form the subspecies name *diatheca* and in the absence of an operculum, associated the species with *Notocochlis* (= *Natica*). The SA species *Natica sticta* Verco, 1909 differs only in having a more prominent colour pattern, including a subsutural row of spots, and it may be the same thing.

▲ GENUS **FRIGINATICA** Hedley, 1916
Natica beddomei Johnston

With characters of the type species.

Southern Australia. Monotypic.

Friginatica beddomei (Johnston, 1884)

Small, globose, rather solid; suture deeply channelled; aperture ovate, not sharply angular posteriorly, peristome complete, outer lip slightly produced and pointed anteriorly; parietal callus thick but not covering the wide and open umbilicus; columellar lip concave on the inner side, funicle lacking; operculum unknown. White.

7 mm. North West I., Qld to southern WA. The deeply channelled suture distinguishes this species within the family, although there is a similar but much larger, undescribed species in deep water off south-eastern Australia. Synonym: *effossa* Watson, 1886.

SUBFAMILY **POLINICINAE**

The shells of the Polinicinae are thick and either globular, ovate or flattened. They possess a thin corneous (horny) operculum and have similar umbilical characters to the Naticinae, there being a parietal callus and a funicle in most species. However, in this group the foot is very greatly enlarged and may envelop the shell. It is so large that it may not be withdrawn within the shell quickly. A large anterior portion, the propodium, forms a thick muscular shield, like a monster bulldozer blade, which covers the front of the animal as it pushes through the substrate. The several genus-level groups in the Australian fauna are treated as subgenera of *Polinices* because of the intergradation of the diagnostic characters.

▲ GENUS **POLINICES** Montfort, 1810
Nerita alba Montfort = *mammilla* Linnaeus

Ovate, solid, smooth, glossy; body whorl inflated; aperture semicircular, inner lip oblique, calloused, callus partially or completely filling the umbilicus; operculum thin, horny, completely filling the aperture.

Cosmopolitan. Several subgenera are sometimes used, as in this account. They are distinguished by features of the columella and umbilicus although the groups intergrade somewhat in these respects. Synonym: *Uber* Humphreys, 1797 (nonbinominal).

▲ SUBGENUS **POLINICES** s.s

High-spired, conical; umbilicus completely filled by callus.

Cosmopolitan. In addition to the species illustrated here Gabriel, 1961 described *P. ayresi* from Victoria.

Polinices aurantius (Röding, 1798)

Pl. 36; fig. 29   Solid, conically ovate, spire moderately high; columella oblique, very slightly arched. Exterior pale yellow or orange, aperture and callus white.

4.5 cm. Central Indo-West Pacific; NT to central Qld. More ovate and less common than *P. mammilla*. Synonyms: *straminea* and *sulphurea* Récluz, 1844; *mellosum* Hedley, 1924.

Polinices mammilla (Linnaeus, 1758)

Pl. 36; fig. 30   Glossy, solid, conically ovate, spire high; columella oblique, straight. White.

5 cm. Indo-West Pacific; Fremantle, WA to southern Qld. A very common species in sandy habitats, often associated with coral reefs. Synonyms: *albula* Récluz, 1851; *cygnea* Philippi, 1852; *ponderosa* Philippi, 1852; *pyriformis* Récluz, 1844; *tumidus* Swainson, 1840.

▲ SUBGENUS **CONUBER** Finlay & Marwick, 1937
Natica conicus Lamarck

High-spired, conical; callus pad posterior, ending abruptly leaving a narrow umbilicus and half the funicle exposed.

Southern and northern Australia.

Polinices conicus (Lamarck, 1822)

Pl. 36; fig. 34 a-b   High-spired, elongate-ovate, suture impressed. Cream, fawn or grey-blue, with a brown or orange subsutural band, base often paler or even white; columella, callus and umbilicus chocolate brown.

4 cm. Circum-Australian. Abundant in sand substrates; its wriggly trails are a feature of intertidal sand flats. The northern forms (Pl. 36, fig. 34b) are smaller, less tumid, have a more prominent sutural band and may not be the same species. Synonyms: *pyramis* Reeve, 1855; *ustulata* Sowerby, 1883.

Polinices melastomus (Swainson, 1822)

Pl. 36; fig. 35   Ovate, rather flat-based; spire low, with a shallow channel below the suture. Fawn or grey-blue; columellar callus orange; aperture red-brown.

4 cm. Southern Qld to eastern Vic.

Polinices putealis Garrard, 1961

Pl. 36; fig. 32   Ovate, solid, glossy, spire small and acute; aperture wide, semilunate, umbilicus wide, columella thickly calloused posteriorly, the callus spreading onto the parietal wall. White, sometimes with traces of light fawn.

2.7 cm. Lady Musgrave I., Qld to Jervis Bay, NSW. Dredged on the outer continental shelf.

Polinices sordidus (Swainson, 1821)

Pl. 36; fig. 39   Conically ovate, body whorl globose. Tan or blue-grey, with an orange subsutural band; columellar callus and umbilicus orange or brown; aperture chocolate-brown.

5 cm. North Qld to SA. This is a very common species on muddy sand flats. Synonyms: *plumbea* Lamarck, 1822; *microstoma* Quoy & Gaimard, 1833; *leucophaea* Reeve, 1855; *strangei* Reeve, 1855.

▲ SUBGENUS **GLOSSAULAX** Pilsbry, 1929
P. reclusianus Deshayes

Depressed; umbilicus and funicle partly or entirely occluded by a callus on the back side of the columella.

Cosmopolitan. The type species is from the eastern Pacific. The relationship to *Neverita* needs to be reviewed.

Polinices incei (Philippi, 1851)

Pl. 36; fig. 41 a-b   Very depressed, spire low; button-like callus completely filling the umbilicus leaving only a narrow arched marginal groove. Upper surface white, pale yellow, grey, brown or even purple; lower surface usually white; callus sometimes brown.

3 cm. Central Qld to SA. Lives on muddy substrates. Synonym: *fibula* Reeve, 1855.

Polinices didyma (Röding, 1798)

Pl. 36; fig. 40 Depressed-ovate, spire low; umbilicus deep, with a small spiral funicle, both almost closed by a deeply grooved, thick, tongue-like reflection of the parietal callus at the posterior end of the columella. Blue-grey or fawn, with a narrow orange subsutural band and sometimes faint spiral lines around the body whorl; parietal wall and callus brown, umbilicus and base white, aperture brown.

7 cm. Central Indo-West Pacific; Kimberley, WA to southern Qld. Lives on muddy substrates in shallow water. *P. ephebus* Hedley, 1915 described from the Hawkesbury River, NSW is very similar and is probably a southern variant of this species. There are many synonyms.

▲ SUBGENUS **MAMMILLA** Schumacher, 1817
Natica melanostoma Gmelin

Thin-shelled; ovate to conical, spire low; whorls oblique, rapidly enlarging; aperture oblong; columella narrow, reflected; umbilicus open, not funiculated.

Indo-West Pacific. The open umbilicus distinguishes the group from *Polinices* s.s. in which the umbilicus is filled by a callus plug.

Polinices fibrosa (Eydoux & Souleyet, 1852)

Surface dull, finely spirally striate; umbilicus narrow, almost closed by the columellar reflection. White, with two spiral bands of brown blotches and a spiral row of brown axial dashes below the suture; columella and umbilicus brown.

5 cm. Indo-West Pacific, southern Qld. This species has often been misidentified and illustrated as *P. mammatus* (Röding, 1798) which is a synonym of *P. melanostomus* (Gmelin, 1791). The thin shell and dull surface distinguish *P. fibrosa* from *P. melanostomus*. Synonym: *filosa* Reeve, 1855; *mammata of* authors not Röding, 1798.

Polinices maurus (Lamarck, 1816)

Pl. 36; fig. 42 Broad, moderately thin; surface dull; columellar reflection not fully covering the umbilical slit. Spire whorls white, body whorl tan or brown; columella and umbilicus dark brown; aperture white.

5 cm. Indo-West Pacific; Qld south to the Bunker Group. Synonym: *nuxcastanea* Hedley, 1913.

Polinices melanostomoides (Quoy & Gaimard, 1833)

Like *P. melanostomus* but very thin and higher-spired, almost pyriform; columellar reflection narrow, bearing a narrow groove-like umbilicus. White, with three spiral bands of irregular brown spots; columella and umbilicus dark brown.

3 cm. Indo-West Pacific; north Qld. See Cernohorsky, 1971, *loc. cit.*, figs. 66-67.

Polinices melanostomus (Gmelin, 1791)

Shiny, minutely spirally striate; reflection of the columella partly covers the umbilicus. Off-white with pale fawn or grey bands, spire and subsutural area milky white; columella and umbilicus dark brown.

5 cm. Indo-West Pacific; Abrolhos, WA to southern Qld. Synonyms: *mammata* Röding, 1798; *opaca* Récluz, 1851; *putamen* Mörch, 1852; *succineoides* Reeve, 1855. See Wilson & Gillett, 1971, *loc. cit.*, Pl. 23, figs. 6, 6a.

Polinices sebae (Récluz, 1844)

Surface dull, finely spirally striate; umbilicus moderately narrow. White, often with two indistinct, interrupted brown spiral lines around the body whorl; columellar inner margin white, reflected part brown.

5 cm. Qld south to Moreton Bay. See Wilson & Gillett, 1971, *loc. cit.*, Pl. 23, fig. 7.

Polinices simiae (Deshayes, 1838)

Pl. 36; fig. 36 Dull and finely spirally striate; reflection of columella covering or almost covering the umbilicus. Off-white with three broad bands of red-brown, jagged axial lines, subsutural zone white; columella and umbilicus dark brown.

4 cm. Indo-West Pacific; Shark Bay, WA to central NSW. Found in coral reef habitats. Synonyms: *propesimniae* Iredale, 1929 and *plumatilis* Iredale, 1936, both from Sydney Harbour, appear to be variants of this species.

▲ SUBGENUS **NEVERITA** Risso, 1826
P. josephinia Risso

Depressed; parietal wall heavily calloused, columella with a central thickening which spirals into the deep and wide umbilicus as a thick cord or funicle.

Cosmopolitan. Synonym: *Mammillaria* Swainson, 1840 (*lactea* Swainson = *peselephanti* Link).

Polinices albumen (Linnaeus, 1758)

Pl. 36; fig. 37 Ovate, slightly compressed; columella very slightly arched, umbilicus almost occluded by a massive funicle. Cream or pale yellow, funicle and umbilicus white.

5 cm. Indo-West Pacific; Dampier Arch., WA to central Qld.

Polinices peselephanti (Link, 1807)

Pl. 36; fig. 33 Solid, globose, polished; spire more or less conical; columella straight; umbilicus partly occluded by a broad funicle. White, cream, yellow or pale orange; columella and umbilicus white.

4 cm. Central Indo-West Pacific; Gulf of Carpentaria, Qld to Sydney, NSW. The funicle is narrower than *P. albumen*. Synonyms: *alba* Gray, 1827; *columnaris* Récluz, 1850; *lactea* Swainson, 1840.

Polinices powisiana (Récluz, 1844)

Pl. 36; fig. 38   Solid, conical-ovate, spire moderately elevated; columella slightly arched, funicle thick but the exposed area of the umbilicus is wide and deep. White, usually with a central orange-brown band.

6 cm. Indo-West Pacific; Geraldton, WA to southern Qld. This distinctive species is usually placed in *Polinices* s.s. but the columellar characters seem closer to *Neverita*. However, the condition of the columellar callus or funicle varies among the species and this example suggests that the current subgeneric divisions of *Polinices* need to be reviewed, and other characters taken into account. A similar species recorded from northern WA and Qld is *P. flemingianus* Récluz, 1844 (= *virginea* Philippi, 1852 and *jukesii* Reeve, 1855) which is distinguished by its taller spire, approaching the form of *P. tumidus* but with the lower part of the umbilicus open. Synonym: *cumingianus* Récluz, 1844.

SUBFAMILY **SININAE**

Species of the subfamily Sininae have thin shells, which may be globular or depressed and the operculum, when present, is thin corneous. They are the least well known of the subfamilies and there are many ambiguities about its classification.

▲ GENUS **SINUM** Röding, 1798
Helix haliotideum Linnaeus

Depressed, discoidal to ear-shaped; spire low, with one or two whorls, followed by two rapidly increasing adult whorls; aperture wide and shallow; umbilicus lacking; operculum lacking.

Cosmopolitan. The animals are large, white and slug-like and, although rarely common, they are a feature of intertidal sandflats. The taxonomy of the tropical representatives of the genus remains confused. The account given here is provisional. Synonyms: *Sigaretus* Lamarck 1799 (*haliotiodeum* Linnaeus); *Cryptostoma* Blainville, 1818; *Ectosinum* Iredale, 1931 (*pauloconvexum* Iredale).

Sinum eximium (Reeve, 1864)

Pl. 36; fig. 2   Thin, depressed, upper surface convex, arched in side view, with three postnuclear whorls; spire moderately low and marginal; basal periphery rounded, columella arched with a faint marginal sulcus at the junction with the outer lip; sculpture of crowded, minute spiral striae. Cream with a paler subsutural band.

3 cm. Indo-West Pacific; Kimberley, WA to Keppel Bay, Qld.

Sinum haliotoideum (Linnaeus, 1758)

Pl. 36; fig. 5   Elongately discoidal, much depressed, slightly arched in side view, with nearly three postnuclear whorls; spire almost flat, subcentral; columella strongly arched, base wide and almost flat, periphery angulate; upper surface sculptured with weak growth lines, spirals lacking. White.

3.5 cm. Indo-West Pacific; North West Cape, WA to Keppel Bay, Qld. Synonym: *planulatus* Récluz, 1843.

Sinum incisum (Reeve, 1864)

Pl. 36; fig. 4 a-b   Discoidal, upper surface and spire flat, with three whorls, spire subcentral; columella strongly arched, base wide and almost flat, periphery angulate; upper surface sculptured with fine but conspicuous, undulating, flat-topped spiral cords. White.

3.5 cm. Indo-West Pacific; Torres St., Qld to Port Stephens, NSW. Synonym: *undulatus* Lischke.

Sinum javanicum Gray in Griffith & Pidgeon, 1834

Pl. 36; fig. 1 a-b   Depressed, with three postnuclear whorls, upper surface convex, slightly arched in side view, spire low and submarginal; columella arched, base wide and flat, periphery subangulate, upper surface sculptured with prominent spiral cords. Cream, spire apex purple.

3.5 cm. Central Indo-West Pacific; Qld south to Innisfail.

Sinum zonale (Quoy & Gaimard, 1833)

Pl. 36; fig. 3   Depressed, with two postnuclear whorls, upper surface convex with a weak subsutural sulcus, arched in side view, spire moderately low and submarginal; columella arched, basal periphery rounded; parietal callus small; sculpture of growth lines crossed by obscure spiral striae. White with a thin yellow periostracum.

3 cm. Southern Qld to about Geraldton, WA. Cernohorsky (1971) has noted that *S. laevigatus* Lamarck, 1822 may be an earlier available name for this species. A small NSW form named *Ectosinum pauloconvexum* Iredale, 1931 is probably only a variant.

▲ GENUS **EUNATICINA** Fischer, 1885
Nerita papilla Gmelin

Ovate, thin; sculptured with spiral grooves crossed by axial growth lines; columella sinuous, sharp-edged, anteriorly thickened and folded over the umbilicus posteriorly, continuous with a narrow parietal callus; umbilicus open; operculum horny.

Indo-West Pacific. In addition to the species listed, the following names in the Australian literature probably belong here: *nitida* (Reeve, 1864) from Vic.; *pictum* (Reeve, 1864) from SA.

Synonyms: *Naticina* Gray, 1847 (*papilla* Gmelin); *Propesinum* Iredale, 1924 (*umbilicata* Quoy & Gaimard); *Pervisinum* Iredale, 1931 (*dingeldeii* Iredale = *linnaeana* Récluz).

Eunaticina albosutura (Verco, 1909)

Pl. 36; fig. 9   Ovate, moderately thin, spire moderately high; umbilicus narrow, columella thick and overhangs the umbilicus posteriorly, parietal callus small; sculpture of minute, wavy spiral striae. White.

2 cm. SA to Geographe Bay, WA. Little is known of this snail. In form it resembles *Polinices* and its position in *Eunaticina* will need to be reviewed.

Eunaticina linnaeana (Récluz, 1843)

Pl. 36; fig. 6   Spire moderately high with the penultimate whorl large and exposed. White or cream, umbilical area fawn or faintly mauve.

2 cm. Indo-West Pacific; Shark Bay, WA to Townsville, Qld. The higher spire and shallowly impressed suture distinguish this species from *E. papilla* (see Majima, 1989, *Bulletin American Paleontology* 331: 67-69). Synonyms: *dingeldeii* Iredale, 1931; *flindersi* Cotton & Godfrey, 1931.

Eunaticina papilla (Gmelin, 1791)

Pl. 36; fig. 7   Suture of the body whorl rather deep, spiral cords flat-topped, 40-60 on the body whorl. White.

3.3 cm. Indo-West Pacific; Fremantle, WA to Gunnamatta Bay, NSW. Synonyms: *tranquebarica* Röding, 1798; *costulata* Quoy & Gaimard, 1833; *cancellata* Swainson, 1840.

Eunaticina umbilicata (Quoy & Gaimard, 1833)

Pl. 36; fig. 8   Ovate, thin, spire low; umbilicus wide, columella thickened posteriorly below its meeting with the parietal wall, parietal callus narrow; sculpture of fine spiral striae crossed by fine growth lines. Cream, sometimes axially streaked with light brown and with pale subsutural and central bands.

2 cm. Central Qld to Fremantle, WA.

▲ GENUS **GENNAEOSINUM** Iredale, 1929
G. peleum Iredale

Globose; aperture wide; umbilicus large, open, perspective, with a spiral funicle on the inner side of the columella; columella straight, slightly flattened and recurved anteriorly, with a callus spreading onto the parietal wall; sculptured with spiral ribs or threads; operculum unknown.

Eastern Australia. Iredale described two species in the genus.

Gennaeosinum intercisum Iredale, 1931

Spire a little elevated; sculpture of fine, flattened spiral threads, closely packed, overridden by fine growth lines; funicle small. White.

1.3 cm high. NSW.

Gennaeosinum peleum Iredale, 1929

Spire short and somewhat flattened; body whorl with about 30 spiral lirae, tending to become finer toward the base. White, rather glossy, with three spiral rows of yellow spots, lines, or V-shaped marks.

1.4 cm high. North Qld.

TONNOIDEA

*T*HERE are eight families in this group. They are mostly medium to large-sized animals which prey on other invertebrates. A feature of these gastropods is their large-sized and long-lived veliger larvae, which have been termed "teleplanic". The larval shells of some species are so large that they were once assumed to be planktonic adults and given distinctive generic names.

An important paper on the comparative anatomy and classification of the superfamily has been published by Anders Warén and Philippe Bouchet (1990), Laubierinidae and Pisanianurinae (Ranellidae), two new deep-sea taxa of the Tonnoidea (Gastropoda: Prosobranchia) *Veliger* 33: 56-102).

FAMILY **BURSIDAE**

Frog Shells

Frog shells or bursids take their vernacular name from the warty, toad-like appearance of many species.

Their shells resemble those of tritons and trumpets but may be distinguished by the presence of a deeply notched posterior canal produced to form a spout at the posterior end of the aperture. Like the tritons and trumpets they are strongly varixed and have a horny operculum. In most bursids the varixes are aligned up the side of the spire, at least on the early whorls. When present the periostracum is thin.

It seems that bursids are carnivores which prey on polychaete worms (Houbrick & Fretter, 1969). They have an extendible proboscis with, in many species, a broad flat tip and lateral lips. The salivary glands are large and believed to produce a substance which anaesthetises the worms in their tubes so that they may be extracted by the proboscis and swallowed. There are three accessory salivary glands. The radula is very similar to that of the Tonnidae but the jaws are reduced.

Females lay egg masses consisting of stiff, urn-shaped capsules on the undersurface of stones or corals and there is a pelagic larval stage. The veliger larvae of some western Atlantic bursids have been described by Laursen (1981).

Species of the genus *Bursa* are found in shallow water, rocky or coral reef habitats of the tropical and warm temperate zones. Species of *Tutufa* are also firm substrate creatures in both tropical and subtropical shallow waters. *Bufonaria* are sand or mud substrate dwellers in tropical and temperate seas, often living in deeper water.

The nomenclature of this group has been terribly confused. This book follows reviews by Alan Beu (1981, 1987). Full synonymies are given in those papers and are not repeated here except in cases of particular interest in the Australian context. See also references cited in the introductory section for the Ranellidae.

References

Beu, A.G. (1981). Australian gastropods of the family Bursidae. Part 1. The families of Tonnacea, the genera of Bursidae, and a revision of some species previously assigned to *Tutufa* Jousseaume, 1881. *Records Australian Museum* **33** (5): 248-324.

Beu, A.G. (1987). Taxonomy of gastropods of the families Ranellidae (= Cymatiidae) and Bursidae. Part 2. Descriptions of 14 new modern Indo-West Pacific species and subspecies, with revisions of related taxa. *N.Z. Journal Zoology* **13**: 273-355.

Houbrick, J. R. & Fretter, V. (1969). Some aspects of the functional anatomy and biology of *Cymatium* and *Bursa*. *Proceedings Malacological Society London* **38**: 415-429.

Laursen, D. (1981). Taxonomy and distribution of teleplanic prosobranch larvae in the North Atlantic. *Dana Report* **89**: 1-43.

▲ GENUS **BUFONARIA** Schumacher, 1817

Bufonaria spinosa Schumacher (=*echinata* Lamarck)

Dorso-ventrally compressed, relatively thin-shelled; sculptured with numerous rows of small granules and in some species, spines or nodules at shoulders of the varixes; varixes aligned along the spire sides, at least on the early whorls; posterior canal short to long, when long canals of successive whorls aligned with the previous varixes on the spire; operculum fan-shaped with the nucleus central or subcentral near the axial margin.

Indo-West Pacific. There are about 12 described species. Synonyms: *Chasmotheca* Dall, 1904 (*foliata* Broderip); *Bursina* Oyama, 1964 (*nobilis* Reeve).

Bufonaria crumena (Lamarck, 1816)

Pl. 43; fig. 2   Squat and broadly ovate, much depressed; later teleoconch whorls with subangulate shoulders bearing small spiky nodules, body whorl with two primary spiral cords bearing small pointed nodules and gemmate secondary and tertiary cords; varixes wide and thin, with three short spines (sometimes obsolete); outer lip weakly crenulate at the edge, with about 15 strong inner denticles; posterior canal relatively wide and shallow, adherent to the previous whorl and not spinose; parietal wall with several small ridges and a wide callus, columella with a free edge and strongly plicate anteriorly, finely plicate centrally; fasciole compressed, with a nodulose and angular edge. Cream with spiral rows of red spots and yellow-pink varixes.

6 cm. Indo-West Pacific; Qld. The nominate form of this species is from the northern Indian Ocean. The Qld and Western Pacific form is the subspecies *B. c. cavitensis* (Beck in Reeve, 1844). So far there is no record of either form from northern WA

Bufonaria foliata (Broderip, 1825)

Pl. 43; fig. 4   Ovate-fusiform; protoconch conical of three convex whorls; early teleoconch whorls angulate at the periphery, body whorl convex; sculptured with strong, close, raised, weakly nodulose spiral ribs, the peripheral rib bearing short but prominent spinose nodules; varixes weak; outer lip sharp-edged, flaring, lirate, with pairs of sharp denticles along the inner margin; parietal wall lirate, columellar shield wide, reflected, lirate; posterior canal a deep channel extending well up the side of the previous whorl; anterior canal short; fasciole weak. Fawn or cream exterior; aperture and columella orange with white lirae.

7 cm. Western Indian Ocean. Recently reported from the North West Shelf but this locality needs confirmation. The specimen illustrated here for reference comes from Mozambique.

Bufonaria margaritula (Deshayes, 1832)

Pl. 43; fig. 3 a-c   Squat; shoulder nodules prominent and compressed; varixes wide; outer lip reflexed anteriorly, with a crenulate edge and strong inner denticles; posterior canal a wide channel projecting over the previous whorl; columellar-parietal callus low, adherent, plicate; fasciole very wide. Colour variable, cream to dark brown, usually blotched with red-brown; columella sometimes yellow.

5 cm. Central Indo-West Pacific; Qld. Resembling *B. crumena* but the spiral cords do not form nodules where they cross the varixes. A unique specimen from Sydney Harbour named *Gyrineum pacator* by Iredale, 1931 appears to be this species. If that is so then the species range could be extended to central NSW.

Bufonaria rana (Linnaeus, 1758)

Pl. 43; fig. 1 a-b Ovate-fusiform; protoconch of three convex whorls; teleoconch whorls angulate, sculptured with minutely gemmate threads becoming cords on the anterior half of the body whorl, shoulders bearing short, sharp spines, with a second row of much smaller spines near the centre of the body whorl; varixes narrow, with three prominent spines, the posterior one at the suture largest, forming one side of the posterior canal; anterior canal long, slightly upturned; outer lip reflexed anteriorly, with a toothed edge, plus inner denticles anteriorly; columellar callus poorly developed, plicate anteriorly; fasciole wide but low. Fawn with pale tan blotches on the back and bands on the varixes; interior white.

6 cm. Western Pacific; Onslow, WA to central Qld. Probably the most common species of the genus, taken mostly by trawlers. Development of the spines is variable.

Bufonaria subgranosa (Sowerby, 1836)

Very like *B. rana* but more elongate and coarsely granulose.

10 cm. Western Pacific; north Qld. This species occurs in the same habitats as *B. rana* and care must be taken to distinguish between them. See Springsteen & Leobrera (1986, *Shells of the Philippines* Pl. 33, fig. 9).

Bufonaria thersites (Redfield, 1846)

Squat, almost biconical, spire short; last two whorls with an irregular, compressed tubercle at the shoulder; varixes thick and low, weakly nodulose where crossed by the spiral cords; outer lip wide, flaring anteriorly, with crenulate margin and pointed denticles; columellar callus thin, narrow, weakly plicate at the anterior margin. Cream, speckled with red-brown.

1.5 cm. South West Pacific; north Qld. Uncommon. For illustrations see Beu (1987, figs. 245-256).

▲ GENUS **BURSA** Röding 1798
Murex bufonia Gmelin

Dorso-ventrally compressed; spire moderately tall, thick-shelled and coarsely sculptured, varixes aligned up the sides of the whorls; operculum lanceolate with abapically terminal nucleus.

Cosmopolitan in tropical seas. Some authors have recognised two subgenera, *Bursa* s.s. and *Colubrellina* Fischer, 1884 (*candista* Lamarck = *condita* Gmelin) but the characters are gradational and they are regarded as synonymous here. Synonyms: *Lampadopsis* Jousseaume, 1881 (*rhodostoma* Sowerby); *Pseudobursa* Rovereto, 1899 (*bufonis* Gmelin); *Bufonariella* Thiele, 1929 (*scrobiculator* Linnaeus); *Dulcerana* Iredale, 1931 (*granifera* Lamarck); *Annaperenna* Iredale, 1936 (*verrucosa* Sowerby); *Tritonoranella* Oyama, 1964 (*ranelloides* Reeve).

Bursa bufonia (Gmelin, 1791)

Pl. 43; fig. 9 Outer lip expanded, with 4-5 pairs of heavy denticles on its inner margin; parietal wall and columella heavily calloused and coarsely, transversely lirate. Cream or white, profusely spotted with brown; aperture white.

8 cm. Indo-West Pacific; Rowley Shoals and Scott Reef, WA to Qld. Synonyms: *mammata* Röding, 1798; *dunkeri* Kira, 1961.

Bursa condita (Gmelin, 1791)

Elongate, with a very tall turreted spire and convex whorls; varixes low, almost aligned; spiral cords variable in size, numerous, finely beaded; aperture elongate-ovate, outer lip varixed, finely denticulate within; columellar shield finely plicate anteriorly and centrally, granulose at the centre. Fawn or yellow, many cords with brown spots which become blotches on the varixes; aperture white.

11 cm. Central Indo-West Pacific; central Qld. There are few Australian records. See Springsteen & Leobrera (1986, *Shells of the Philippines* Pl. 33, fig. 5).

Bursa cruentata (Sowerby, 1835)

Pl. 43; fig. 8 Very strong sculpture of three spiral ribs bearing a pair of granulose spiral cords, ribs elevated at intervals forming double nodules, those at varixes particularly high; outer lip expanded, irregularly scalloped with four or five pairs of solid denticles on its inner margin; former posterior canals on the spire barely discernible; a thin wide reflected shield present over the parietal wall and columella, the whole shield and columellar margin covered by strong nodules and irregular lirae. Exterior white or cream with a few purple-brown blotches mainly on the tubercles; aperture white, parietal wall usually with conspicuous red-brown spots between the lirae.

4 cm. Common Indo-West Pacific; Rowley Shoals and Scott Reef, WA to Qld.

Bursa granularis (Röding, 1798)

Pl. 43; figs. 11 a-b, 12 Slightly compressed dorso-ventrally; spire tall with convex whorls; varixes rounded though channelled and buttressed by the spiral cords behind; sculptured with more or less equal spiral cords bearing small nodules, five cords on the penultimate whorl, nine on the body whorl plus several on the stout fasciole; outer lip with about 14 small denticles on its inner margin, the posterior two larger; parietal ridge narrow and deep, columella plicate throughout. Orange-brown or red-brown, with darker spiral lines and often a central cream band; interior white.

6 cm. Indo-West Pacific, Rottnest I., WA to northern NSW. A very common and variable intertidal species. *B. cubaniana* Orbigny, 1842 of the Caribbean is a close relative. A form with tall spire and angulate whorls bearing prominent tubercles has been called *Bursa affinis* (Broderip, 1832) by some authors (Pl. 43, fig. 12). The two forms occur in the

same areas in WA and there may be good reason for distinguishing between them; but Alan Beu informs me that the Broderip types are actually typical *granularis* so that even if they are distinct species the name *affinis* cannot be used for the tuberculose form.

Bursa humilis Beu, 1981

Pl. 43; fig. 10 a-b   Stout, ovate; spire relatively short; whorls with angulate shoulders and wide, rounded varixes; sculptured with six primary spiral ribs bearing prominent nodules and finely nodulose secondary cords in the interspaces; outer lip reflected, wide, sharp-edged, with about 10 fine inner denticles; parietal ridge strong, columellar callus adherent, strongly lirate; fasciole thick. Cream, some nodules and varixes buff-pink; aperture white but with a conspicuous red-brown blotch on the parietal wall between the plicae.

4.5 cm. Eucla to Jurien, WA. A deepwater species from 130 m to 240 m. Beu described this as one of several subspecies of the widespread *B. ranelloides* (Reeve, 1844) but the differences warrant species status in this author's opinion. The nominate form of *B. ranelloides* is from the western Pacific while *B.r. tenuisculpta* Dautzenberg and Fischer, 1906 is an Atlantic population. The shells of the WA *B. humilis* are smaller, broader, have shorter spires, very inflated body whorls, and very short anterior canals. Although these distinctive characters are quite consistent, a close relationship to *B. ranelloides* is assured.

Bursa lamarckii (Deshayes, 1853)

Pl. 43; fig. 7 a-b   Whorls with two thick prominent tubercles between each pair of varixes and unusually long, projecting, previous posterior canals; other sculpture of three pairs of low spiral ribs becoming enlarged and gemmate where they form three bifid nodules on the varixes; outer lip with a crenulate edge, six pairs of inner denticles; parietal ridge moderate, parietal-columellar callus low and adherent, finely lirate throughout but with several thick, short plicae anteriorly; fasciole broad. Grey or fawn but with dark purple-brown blotches between the tubercles, dark grey on the varixes; aperture throat (including the inner lip callus) dark purple-brown between the orange or fawn denticles and lirae.

6 cm. Central Indo-West Pacific; Rowley Shoals, WA and coral reef habitats on the Great Barrier Reef, Qld. A truly remarkable species.

Bursa latitudo Garrard, 1961

Pl. 43; fig. 13   Tall-spired; body whorl with angulate shoulder and three rows of semipointed, widely spaced nodules and fine interstitial spiral cords; outer lip reflected, nodulose along its inner edge; inner lip with wide parietal-columellar shield, parietal wall irregularly plicate, columella plicate; fasciole strong. Yellow-brown; interior white, with a red-brown patch on the columella which may be obliterated by plicae in old specimens.

9.5 cm. There are two geographically separated subspecies. The nominate form was described from the continental slope of southern Qld. (*B. l. wolfei* Beu, 1980, described from Hawaii, is a synonym.) The second subspecies *B. l. natalense* Coelho and Mathews, 1970 is from the central Western Atlantic.

Bursa rhodostoma (Beck in Sowerby, 1835)

Pl. 43; fig. 5   Small, ovate; varixes low, excavated and buttressed behind by the three primary ribs; body whorl sculptured with one single and two bifid primary ribs bearing small bifid nodules, plus tiny gemmate secondary threads in the interspaces; aperture nearly round, outer lip crenulate, with four pairs of tiny inner denticles; posterior canal short, open not tubular; parietal ridge strong, callus weak and adherent, bearing a few tiny posterior denticles and short plicae posteriorly. Fawn, ribs with purple spots, nodules yellow; throat of aperture purple, denticles and plicae yellow.

3 cm. Central Indo-West Pacific; Ashmore Reef, NT to eastern Qld. There is a central Atlantic subspecies known as *B. r. thomae* Orbigny, 1842. The Australian population belongs to the nominate form. This is the smallest species in the genus.

Bursa rosa (Perry, 1811)

Pl. 43; fig. 6 a-b   Surface with weak granulose cords, two spiral ribs which fuse to form two or three double knob-like protuberances between the varixes, and a smaller nodulose spiral rib nearer the anterior end; outer lip rather narrow, weakly scalloped, with 4-5 pairs of small denticles on its inner margin; posterior canal semitubular and very long; parietal wall and columella thickly calloused but a reflected shield not formed; strong nodules present on the columella anteriorly, and long but weak irregular lirae centrally and posteriorly on the parietal wall. Exterior cream faintly blotched with purple-brown, deep interior white, throat of aperture coloured deep wine-red.

4 cm. Indo-West Pacific; North West Cape, WA to northern NSW. Synonym: *siphonata* Reeve, 1844.

Bursa verrucosa (Sowerby, 1825)

Stout, with a spiral rib bearing very prominent, rounded nodules around the spire whorls and the shoulder of the body whorl, and two additional lesser ribs on the body whorl; columella strongly lirate. Cream, nodules dark brown to black, spaces between the columellar lirae orange or red.

4.5 cm. Southern Qld (Caloundra) to northern NSW and New Zealand. Common at the Kermadec Is. This distinctively coloured species has a more restricted geographic distribution than most members of the genus. See Beu (1985, *Conchologists of America Bull.* **13** (4): fig. 46).

▲ GENUS **TUTUFA** Jousseaume, 1881
Murex bubo Linnaeus

Medium-sized to very large and tall-spired; not dorso-ventrally compressed; posterior and anterior canals short;

varixes at every $^2/_3$ whorl; operculum with nucleus at or slightly below half height and slightly towards centre from the axial margin.

Indo-West Pacific. The nomenclature of this genus has been terribly confused. There are two subgenera.

▲ SUBGENUS TUTUFA s.s.

Very large; columella weakly plicate or smooth.

Indo-West Pacific. There are five species, three of them found in Australian waters.

Tutufa bubo (Linnaeus, 1758)

Pl. 43; fig. 16   Whorls with strong spiral ribs, those at the shoulders and at the centre of the body whorl bearing large protruding tubercles which may be paired and granulose, other ribs smaller and merely heavily nodulose; outer lip expanded, scalloped, with pairs of denticles on the inner margin; parietal wall calloused, columellar callus raised and reflected to form a shield; columella strongly transversely lirate. White, heavily blotched and spotted with brown; aperture white.

30 cm. Moderately common. Indo-West Pacific, Pt. Cloates, WA to Qld. There has been much confusion about the correct name for this species, but the matter has been clarified by Beu (1980) who has shown that the name *Bursa lampas* has been misapplied to it. This is a coral reef species which may be collected in relatively shallow water by divers, or even in the intertidal zone. It is less widely distributed than *T. bufo*.

Tutufa bufo (Röding, 1798)

Pl. 43; fig. 17   Whorls with weak spiral ribs, 2-3 bearing tubercles, others with weak nodules, 4-5 ribs swell to form thick ridges where they cross the varixes; interspaces between the ribs patterned with fine oblique striations; outer lip expanded, scalloped, bearing 5-6 pairs of denticles on its margin; a thin but wide reflected shield formed over the parietal wall and columella, polished and smooth except for a few teeth or short lirations at the anterior end. White, cream or pink; deep interior white, shield and outer lip margin pink or white, throat of aperture rusty red.

18 cm. Indo-West Pacific; Cape Naturaliste, WA to central NSW. This species is widely distributed, extending well into the subtropical margins of the Indo-West Pacific region. It has been reported from northern New Zealand. It is not a shallow-water species but is generally taken at depths below 40 m. *T. bufo* resembles *T. bubo* but may be distinguished by the smooth parietal shield. Synonym: *rubeta lissostoma* Smith, 1914.

Tutufa oyamai Habe, 1973

Pl. 43; fig. 14 a-b   Small to medium-sized; anterior canal protruding; body whorl with three narrow major spiral cords, the peripheral one bearing large, compressed nodules which are large on the varixes, interspaces crowded with secondary cords, entire surface finely gemmate; outer lip narrowly flaring, its outer edge made digitate by the ribs, its inner edge bearing a row of 12-14 small paired nodules; inner lip flaring forming a wide, plicate, parietal-columellar shield, its inner part bearing a row of nodules; fasciole well developed, with a small umbilical hollow below. Pale fawn to red-brown; aperture white.

7.5 cm. Central Indo-West Pacific from Japan south to the Abrolhos in WA and northern NSW in eastern Australia, usually trawled on sandy substrates to depths of 100 m. The white aperture readily distinguishes this species from *T. rubeta*. Another species found in the South China Sea, *T. tenuigranosa* Smith, 1814 is similar but much larger.

Tutufa tenuigranosa (E.A. Smith, 1914)

Pl. 43; fig. 15   Spire high; whorls with angulate shoulders and finely granulate spiral cords, body whorl with three spiral cords bearing pointed nodules, the shoulder nodules relatively large, the central and anterior rows smaller and becoming obsolete toward the lip; outer lip fluted, with 12-14 paired denticles along its inner side; shield thin, strongly plicate, with a narrow parietal ridge; anterior canal moderately long and upturned, fasciole narrow, leaving a false umbilical chink. Cream or fawn, flecked with brown; aperture and shield white.

18 cm. Central Indo-West Pacific. Trawled in moderately deep water. Taller-spired and less coarsely sculptured than *T. bubo*. There is a shell very like this from the North West Shelf, WA but its identity is uncertain and specimens were not available for illustration so that a Taiwan shell is figured for reference.

▲ SUBGENUS TUTUFELLA Beu, 1981
Lampas hians Schumacher (= *Murex rana rubeta* Linnaeus)

Shell small to large; columella closely plicate.

Indo-West Pacific. Beu introduced this subgeneric name as a replacement for *Lampas* Schumacher, 1817, a name used earlier by Montfort (1808) for a different group.

Tutufa rubeta (Linnaeus, 1758)

Pl. 43; fig. 18   Body whorl with three major spiral ribs bearing rounded nodules and with weak secondary, finely nodulose cords in the interspaces, giving an overall evenly nodulose appearance; anterior canal short, reflected to the right; outer lip narrowly flaring, bearing rows of 12-14 small nodules on its inner edge and a second row of elongate ridges deeper within the aperture; parietal and columellar pad lirate and nodulose. Exterior mottled red; inner lip red, lirae paler; outer lip red, nodules pink or white.

11 cm. Indo-West Pacific; Abrolhos, WA to northern NSW. Lives in shallow water, usually associated with coral reefs.

FAMILY **CASSIDAE**

Helmet Shells

Some species of this family are large and very decorative and are popular in the shell trade. Their vernacular name is derived from their heavy, helmet-shaped shells. One species, *Cypraecassis rufa*, has been used for centuries for the manufacture of cameos.

The shells of cassids are typically globular to cylindrical with an inflated body whorl and narrow aperture. There is usually a shield-like callus on the parietal wall and columella and the outer lip is thickened and reflected. Varixes of earlier lips are sometimes present on the spire whorls. There is often an open umbilicus and also a second or "false" umbilicus formed near the end of the siphonal canal.

Cassids are sand-dwelling, carnivorous animals which feed on echinoderms, primarily sea urchins. The giant helmet, *Cassis cornuta*, eats the Crown of Thorns starfish and for that reason is totally protected in Qld. Hughes and Hughes (1981) and Hughes (1985) gave information on prey and feeding behaviour of cassids. Warén & Bouchet (1990) illustrated the radula of several cassids.

Sexual dimorphism is common in the family, males being smaller than females. Female cassids lay egg capsules containing many eggs. In some species the egg mass is a symmetrical tower-like structure but in others, e.g. *Phalium glaucum* and *Semicassis labiata,* it is an irregular mass which is the work of several females spawning together. Within the capsules development may be direct or there may be a planktotrophic veliger stage after hatching. The larval shells are calcified, round and smooth (Laursen 1981).

A world-wide review of this family was published some years ago by Tucker Abbott (1968). There have been many changes since then. Alan Beu (1981) distinguished two subfamilies, the Cassinae and Phaliinae, mainly on the basis of different opercular and siphonal canal structure, and gave brief details of some anatomical characters.

References

Abbott, R.T. (1968). The helmet shells of the world. Part 1. *Indo-Pacific Mollusca* **2** (9): 7-201.

Beu, A.G. (1981). Australian gastropods of the family Bursidae. Part 1. The families of Tonnacea, the genera of Bursidae, and a revision of some species previously assigned to *Tutufa* Jousseaume, 1881. *Records Australian Museum* **33** (5): 248-324.

Hughes, R.N. and Hughes, H.P. I. (1981). Morphological and behavioural aspects of feeding in the Cassidae (Tonnacea), Mesogastropoda. *Malacologia* **20**: 385-402.

Hughes, R. N. (1985). Notes on the feeding behaviour of *Tonna zonatum* (Tonnidae) and *Phalium bisulcatum* (Cassidae) (Tonnacea) in *Proceedings Second International Workshop on the Macrofauna of Hong Kong and Southern China*, pp. 623-626, Hong Kong University Press.

Laursen, D. (1981). Taxonomy and distribution of teleplanic prosobranch larvae in the North Atlantic. *Dana Report* **89**: 1-43.

Warén, A. & Bouchet, P. (1990) Laubierinidae and Pisanianurinae (Ranellidae), two new deep-sea taxa of the Tonnoidea (Gastopoda: Proso-branchia). *Veliger* **33** (1): 956-102.

SUBFAMILY **CASSINAE**

Operculum oval, the nucleus situated near the anterior end; siphonal canal straight or weakly twisted.

▲ GENUS **CASSIS** Scopoli, 1777
cornuta Linnaeus

Large, globose, with a broad parietal shield, strongly upturned anterior canal, and a thick tuberculose outer lip; operculum small, oblong; umbilicus sealed.

Indo-West Pacific and tropical Atlantic. There are seven large species, three of which occur in Australia.

▲ SUBGENUS **CASSIS** s.s

Parietal shield very wide and flat, bounded on the left side by a large former varix; with a deep gutter behind the anterior canal.

Pan-tropical. Synonyms: *Cassidea* Bruguière, 1789 (*cornuta* Linnaeus); *Cassinia* Rafinesque, 1815 (misspelling); *Goniogalea* Mörch, 1857 (*madagascariensis* Lamarck = *cornuta* Linnaeus); *Fimbriola* Scudder, 1882 (*nomen nudum*); *Galeodocassis* Sacco, 1890 (*anceps* Sacco = *saccoi* Rovereto); *Cassisoma* Rovereto, 1899 (nonbinominal).

Cassis cornuta (Linnaeus, 1758)

Spire very short, nucleus usually corroded; surface of whorls pitted; body whorl with three or four thick nodulose spiral ribs; aperture narrow, outer lip broad and flat with an upturned shelf-like rim, and 5-7 large teeth on its inner edge; columella bears strong, irregular plicae. Dorsal side and spire grey or light brown, upper edge of outer lip with 6-7 brown patches; base, including outer lip and parietal-columellar shield glossy cream or orange, teeth white.

35 cm. Indo-West Pacific; Onslow, WA to Capricorn Group, Qld. This is the largest living cassid and one of the largest gastropods. The colour, shape of the parietal-columellar shield and strength of the dorsal knobs are variable. Male shells are smaller than females and have fewer and larger dorsal knobs. This species is totally protected in Qld. Synonyms: *caput-equinum* and *hamata* Röding, 1798; *labiata* Dillwyn, 1817; *amboinensis* and *brevirostrum* Tryon, 1885.

▲ SUBGENUS HYPOCASSIS Iredale, 1927
Cassis fimbriata Quoy & Gaimard, 1833

Parietal shield thin and less well developed than in *Cassis* s.s.; gutter behind the anterior canal shallow.

Southern and eastern Australia. The case for retaining this subgenus is weak as the diagnostic characters may relate to small size. Synonym: *Nannocassis* Iredale, 1927 (*nana* Tenison Woods).

Cassis fimbriata (Quoy & Gaimard, 1833)

Pl. 38; fig. 8 a-c   Ovate to globose; nuclear whorls bulbous; postnuclear spire whorls axially ribbed and spirally striate, body whorl with numerous irregular axial ribs, spiral cords anteriorly, and 2-3 spiral rows of tubercles; shield broad in the columellar region but often thin or lacking on the parietal wall; columella smooth except for 4-7 marginal lirae anteriorly; outer lip smooth or weakly toothed. Exterior shiny, cream, usually with axially elongated brown patches and interrupted brown spiral lines; interior light brown.

12 cm. Western Vic. to the Abrolhos, WA. Synonyms: *bicarinata* Jonas, 1839; *decresensis* Hedley, 1923.

Cassis nana (Tenison Woods, 1879)

Pl. 38; fig. 9 a-b   Small, rather thin; wide at the shoulder, narrowing anteriorly; spire smooth, nuclear whorls semibulbous; body whorl smooth except for two spiral rows of short, sharp nodules on the shoulder and 1-3 spiral rows of weak nodules on the anterior half; shield thin, sharp-edged, indented near the centre giving a gnarled appearance; outer lip, shield and parietal wall strongly toothed. Exterior shiny, yellow-brown with faint brown spots between the whitish nodules; faint brown axial rays on the spire; teeth white; deep interior yellow-brown.

6 cm. Southern Qld and northern NSW. Usually dredged in deeper water.

▲ GENUS CYPRAECASSIS Stuchburry, 1837
Buccinum rufum Linnaeus

Heavy, oblong shells, lacking former varixes but possessing many fine teeth on the columella and a large apex of 4-5 translucent whorls; aperture narrow; operculum small, oval, with few whorls.

Pan-tropical. Synonym: *Cypraeicassis* Cossmann, 1903 (amended spelling).

Cypraecassis rufa (Linnaeus, 1758)

Pl. 37; fig. 12   Spire short, with five nuclear whorls; postnuclear whorls sculptured with fine nodulose ribs; body whorl with rounded shoulders bearing three posterior spiral rows of rounded knobs, one central spiral row of axially elongate knobs, spiral rows of raised axial bars anteriorly and with 2-3 nodulose spiral ribs between them; outer lip very thick with heavy teeth; columellar and parietal areas covered by glazed callus forming a thick rim along the left side of the base, strongly lirate, columellar margin with an axial swelling bearing heavy teeth. Dorsal side orange-brown or reddish with grey to white patches, base creamy orange, parietal wall, columella and interior dark orange-red.

18 cm. Indo-West Pacific; ? north Qld. This is the shell from which cameos are cut. There is doubt whether it occurs in Australia although there are anecdotal accounts of it. Synonyms: *pullum* Born, 1778; *ventricosum* and *pennatum* Gmelin, 1791; *rufescens* and *tuberosa* Röding, 1798; *labiata* Perry, 1811.

▲ GENUS GALEODEA Link, 1807
Buccinum echinophora Linnaeus

Globose to ovate; spire shouldered, high, previous varixes normally lacking; anterior canal produced, slightly upturned; aperture wide, outer lip varixed, smooth or weakly dentate within, posterior canal notch present; parietal-columellar shield thin, narrow, but well formed; umbilicus and false umbilicus open; protoconch small, with few whorls, markedly deviated; operculum ovate with a marginal nucleus.

Cosmopolitan. The genus has a very long Tertiary fossil history and several living species, most of them found in deep water. Reynell (1905, *Proceedings Malacological Society London*, 6 (5): 292-299; 1906, *ibid*, 7 (1): 67) has described the anatomy of the type species.

▲ SUBGENUS GALEODEA s.s.

Small and chunky; globose to ovate; outer lip varixed.

Cosmopolitan. Synonyms: *Morio* Montfort, 1810; *Echinora* Schumacher, 1817; *Cassidaria* Lamarck, 1812.

Galeodea echinophorella Habe, 1961

Pl. 38; fig. 6 a-b   Elongate-ovate; spire high, conical, suture impressed; protoconch of two whorls; teleoconch whorls spirally corded, with angulate shoulders and concave subsutural ramps; shoulders bearing a row of pointed nodules, body whorl with two further but weaker rows becoming obsolete toward the lip; outer lip varixed, with obscure denticles along its inner margin and a weak posterior nodule beside the poorly defined anal notch; columella obscurely lirate anteriorly. White but with a pale orange patch on the back.

3.5 cm. Southern Japan and recently discovered in deep water on the North West Shelf near the Rowley Shoals, WA.

Galeodea maccamleyi Ponder, 1983

Pl. 38; fig. 5 a-b   Globose; suture deeply channelled; body whorl inflated, finely spirally corded and with four spiral rows of prominent nodules; penultimate whorl weakly carinate; outer lip sharp-edged but thickened within, with weak denticles along its inner side and a large posterior denticle bordering the anal notch; parietal callus thin, transparent; columellar shield narrow, erect, weakly lirate along its margin; anterior canal rather short, twisted dorsally; umbilicus narrow, open to closed. Orange-brown to fawn, nodules white.

3.7cm. Central Qld. Described from 220 m to 238 m in the Capricorn Channel.

▲ SUBGENUS GALEOOCORYS Kuroda & Habe, 1957
Galeodea leucodoma Dall

Large, thin-shelled, body whorl inflated; outer lip reflected, flaring and expanded anteriorly, smooth within.

Indo-West Pacific.

Galeodea nipponica (Sakurai & Habe, 1961)

Pl. 38; fig. 7 a-b   Protoconch conical, of two smooth whorls; teleoconch whorls spirally corded, with a steep subsutural ramp and angulate shoulders bearing nodules which are pointed on the early whorls becoming progressively obsolete and usually lacking on the back of the body whorl; outer lip and columella smooth. White.

11 cm. Southern Japan, and recently discovered on the North West Shelf, WA in 400-500 m.

SUBFAMILY PHALIINAE

Operculum slender, fan-shaped, with the nucleus half way down the columellar edge, sometimes radially ribbed; siphonal canal strongly twisted.

▲ GENUS CASMARIA H. & A. Adams, 1853
Buccinum vibex Linnaeus (= *erinacea* Linnaeus)

Elongate-ovate, smooth, glossy; columellar shield poorly formed, smooth; outer lip with 1-2 rows of sharp prickle-like teeth anteriorly; umbilicus closed; operculum fan-shaped to ovate.

Caribbean, Eastern Pacific, Indo-West Pacific. There are three living species. Synonym: *Casmaria* Jousseaume, 1888 (*torquata* Reeve = *ponderosa* Gmelin).

Casmaria erinacea (Linnaeus, 1758)

Pl. 38; fig. 2 a-c   Outer lip smooth along its inner edge, bearing several prickle-like teeth on the outer edge at its anterior end; parietal wall glazed with callus, columellar shield with a small swelling centrally, one strong spiral fold anteriorly, and several weak marginal spiral lirae behind that. Whorls white, grey, cream or pale tan, sometimes with irregular bands of diffuse brown blotches which may align axially, and usually with axial rows of minute brown spots; outer lip white with large dark brown spots; a dark brown blotch lies in the base of the siphonal notch; columella white; interior brown.

7 cm. Eastern Pacific and Indo-West Pacific; Pt. Cloates, WA to Qld. Two forms of *C. erinacea* occur throughout its range. One of these is rather thin-shelled and has smooth shoulders, the other is heavy and has nodulose shoulders. The Australian population belongs to the nominate subspecies while *C. e. kalosmodix* (Melvill, 1883) is found in the Central Pacific, and *C. e. vibexmexicana* (Stearns, 1894) in the tropical Eastern Pacific. Synonyms: *vibex* Linnaeus, 1758; *denticulata* Röding, 1798; *edentulum* Link, 1807; *ventricosa* Rigacci, 1866. For a picture of the living animal see p. 366.

Casmaria perryi Iredale, 1912

Elongate-ovate, smooth, relatively high-spired, without shoulders; lip varix weak, anterior lip denticles lacking. Pale brown to tan, with brown blotches below the suture and three or four wide but indistinct light brown spiral bands on the body whorl which contain squarish brown blotches, varix with blotches but no posterior spots.

5 cm. Subtropical waters of the south-west Pacific, including eastern Australia from southern Qld to northern NSW; also found in the Kermadec Is. and NZ. Illustrated by Abbott, *loc. cit.*, Pl. 187, who treated it as a subspecies of *C. ponderosa.*

Casmaria ponderosa (Gmelin, 1791)

Pl. 38; fig. 1 a-c   Outer lip with one or two rows of widely spaced tooth-like denticles along its entire length; columellar shield with many fine irregular lirae. White, cream or tan with a spiral row of brown spots below the sutures and another behind the anterior canal; outer lip bears large dark brown spots, a dark brown blotch lies in the base of the siphonal notch; lip and columella white; interior brown.

5 cm. Indo-West Pacific and the Caribbean; Barrow I., WA to NSW. *C. ponderosa* resembles *C. erinaceus* but lacks axial rows of minute spots and has instead prominent subsutural spots. Two forms of this species occur throughout its range, one thin, smooth and without shoulder nodules, the other heavy with shoulder nodules. The nominate subspecies is widely distributed in the tropical Indo-West Pacific including northern Australia. The other subspecies are *p. unicolor* (Dautzenberg, 1926) from the Red Sea; *p. nipponensis* Abbott, 1968 from southern Japan; *p. atlantica* Clench, 1944 from the Caribbean. Some authors regard these as distinct species. Synonyms: *nodulosum* Gmelin, 1791; *quadratum* Link, 1807; *biarmatum* Dillwyn, 1817; *pantherina* "Solander" Dillwyn, 1817; *tenuilabris* Menke, 1828; *torquata* and *turgida* Reeve, 1848; *cernica* Sowerby, 1888.

▲ GENUS **ECHINOPHORIA** Sacco, 1890
Buccinum intermedium Brocchi

Thin-shelled, tuberculate, high-spired; shield weakly developed or lacking; outer lip reflected, sometimes weakly toothed; characteristically having a sharp spiral cord on the fasciole ending at the top of the anterior canal notch; umbilicus small or lacking.

Pan-tropical. The group has a long fossil record at least as far back as the Eocene and there are several synonyms based on fossil species. Synonyms: *Trachydolium* Howe, 1926 (*dalli* Dickerson); *Euspinacassis* Finlay, 1926 (*pollens* Finlay); *Shichiheia* Hatai & Nisiyama, 1949 (*etchuensis* Hatai & Nisiyama); *Bathygalea* Woodring & Olsson, 1957 (*coronadoi* Crosse); *Miogalea* Woodring & Olsson, 1957 (*dalli*).

Echinophoria carnosa (Kuroda & Habe, 1961)

Pl. 38; fig. 11 a-b   Early spire whorls spirally threaded and axially ribbed and with two spiral rows of nodules; body whorl with six rows of nodules (although the sixth may be obsolete) and fine cords anteriorly; outer lip slightly reflected and bearing weak, widely spaced denticles, usually along its entire length; columella smooth or weakly granulose. White or yellow-white; outer lip pale tan on the upper side.

9 cm. Western Pacific from southern Japan to northern WA; recently discovered by trawlers in deep water on the North West Shelf off Port Hedland, WA. See remarks on *E. kurodai*.

Echinophoria kurodai (Abbott, 1968)

Pl. 38; fig. 12 a-b   Early spire whorls spirally threaded, with a single central row of pointed nodules; body whorl with five spiral rows of nodules and weak spiral threads anteriorly; outer lip slightly reflected, typically with 7 small prickle-like denticles anteriorly; columella smooth, parietal wall glazed. Yellow-white but with a tendency to have faint tan bands; outer lip with brown spots.

7 cm. Western Pacific from southern Japan to northern WA; recently discovered by trawlers in 300-400 m on the North West Shelf, north of Port Hedland, WA. There seems to be an overlap in shell characters with *P. carnosum* and further study would confirm whether the two are different species.

Echinophoria wyvillei (Watson, 1886)

Pl. 38; fig. 10   Globose; early spire whorls spirally threaded and with a spiral row of erect pointed nodules; body whorl with a steep subsutural ramp and prominent shoulder nodules; usually with 1-2 additional rows of nodules centrally and numerous spiral threads anteriorly; outer lip reflected, bearing 5-6 very weak lirae on the inner edge; columella swollen in the centre, sometimes lirate anteriorly; with a deep U-shaped gutter above the wide and deep siphonal notch; umbilicate. Yellow to pale tan.

10 cm. Western Pacific from southern Japan to northern Australia. Recorded from the outer continental shelf off NSW and recently by trawlers working on the North West Shelf off Port Hedland. Abbott (1968) treated this as a subspecies of the Caribbean *P. coronadoi* (Crosse, 1867).

▲ GENUS **PHALIUM** Link, 1807
Buccinum glaucum Linnaeus

Medium-sized, spire tall and pointed; former varixes retained at the growth pauses; parietal shield moderately developed; axial and spiral sculpture of about equal strength; outer lip thickened, with one row of denticles and sometimes spines anteriorly; shield plicate or granulose; outer lip strongly toothed; umbilicus deep, false umbilicus a small chink or closed; operculum chitinous, fan-shaped.

Indo-West Pacific. Synonyms: *Bezoardica* Schumacher, 1817 (*glaucum* Linnaeus); *Cassidea* Swainson, 1840 (*areola* Linnaeus); *Bezoardicella* Habe, 1961 (*decussata* Linnaeus).

Phalium areola (Linnaeus, 1758)

Pl. 37; fig. 3 a-b   Ovate; spire with concave sides, a fine pointed apex, and a small nucleus of two whorls; postnuclear whorls sculptured with axial and spiral ribs and varixes; body whorl glossy, smooth except for a few weak spiral cords anteriorly; shoulders smooth and rounded; outer lip thickened, with strong sharp teeth along its inner edge, outer edge smooth; columellar shield thick, moderately wide, crossed by strong lirae. White with five spiral bands of large squarish brown spots on the body whorl and along the upper edge of the lip and varixes; anterior canal and aperture white.

9 cm. Indo-West Pacific; Broome, WA to northern NSW. Lack of shoulder nodules and the much more distinct spots readily distinguish this species from *P. bandatum*. Synonyms: *alea* Röding, 1798; *clathratum, extinctum* and *sulcatum* Link, 1807; *areolata* Schumacher, 1817; *aureola* Bruguière, Lienard, 1877; *agnitum* Iredale, 1927.

Phalium bandatum (Perry, 1811)

Pl. 37; fig. 2 Elongate-ovate; spire nucleus of two whorls; early postnuclear whorls rounded and sculptured with fine, beaded, axial and spiral ribs; later whorls more angulate, with nodulose shoulders; body whorl smooth except for shoulder nodules; outer lip thickened, with sharp teeth along its inner edge and 2-3 poorly developed spines on the outer edge anteriorly; columellar shield moderately broad and flaring, crossed by many strong wavy folds. Cream to white, five spiral bands of rather indistinct squarish yellow-brown blotches in front of the shoulder, and a band of yellow-brown flames behind; lower surface of lip with six tan or orange blotches; anterior canal dark brown; interior light brown.

12 cm. Central Indo-West Pacific; Abrolhos, WA to northern NSW. Resembles *P. glaucum* but the yellow-brown blotches, higher spire and more elongate shape distinguish it from that species. Synonyms: *coronulata* Sowerby, 1825; *?muricata* Menke, 1828.

Phalium glaucum (Linnaeus, 1758)

Pl. 37; fig. 13 a-b Ovate to globular; spire nucleus of 2½ whorls; early postnuclear whorls rounded and sculptured with fine beaded spiral and axial ribs; later spire whorls with more angular, nodulose shoulders and weaker sculpture; body whorl smooth except for weak shoulder nodules and rows of shallow dents; outer lip thickened, with sharp teeth along its inner edge and 3-4 sharp projecting spines anteriorly on its outer edge; columellar shield broad, flared, and crossed by many strong but irregular lirae. Ash-grey, varixes whitish with tan or pale orange blotches; anterior canal white with a brown tip; interior brown.

12 cm. Indo-West Pacific; possibly found in north Qld. Common within the species range but the Australian records are doubtful. Philippine specimens are illustrated for reference. Synonyms: *vulgaris* Schumacher, 1817; *bezoar* Gray, 1839; *strigata* Shirley, 1911.

▲ GENUS SEMICASSIS Mörch, 1852
Cassis japonica Reeve = *bisulcata* Schubert & Wagner

Globular to ovate; varixes rarely retained; usually with spiral threads or incised lines on the body whorl, axial sculpture obsolete; outer lip finely denticulate, often with anterior spines; umbilicus usually deep, false umbilicus small or sealed; operculum fan-shaped and either smooth or radially sculptured.

Cosmopolitan. The genus has a long fossil history as well as a wide distribution. Beu (1976, *Journal Royal Society N.Z.* 6 (4): 413-432) discussed the taxonomy of the genus and its geological history. There are several subgenera, including two which are represented in the modern Australian fauna. *Semicassis* s.s. appears to have had its origin in the Miocene of Europe and to have colonised the temperate waters of southern Australia and New Zealand during the early Pleistocene.

▲ SUBGENUS SEMICASSIS s.s.

Cosmopolitan.

Synonyms: *Faurotis* Jousseaume, 1888 (*faurotis* Jousseaume); *Xenophalium* Iredale, 1927 (*hedleyi* Iredale = *royanum* Iredale); *Xenogalea* Iredale, 1927 (*pyrum royanum* Iredale).

Semicassis angasi (Iredale, 1927)

Pl. 37; fig. 5 a-b Spire of medium height, nucleus of 3 whorls; early postnuclear whorls with beaded spiral and axial threads, later spire whorls smooth; body whorl with faint dents; outer lip narrow, with widely spaced, fine, sharp teeth along its inner edge; columellar shield narrow, margin indented, strongly lirate; false umbilicus and true umbilicus of almost equal size, deep and round; with a pointed projection of the columella bordering the anterior inner edge of the false umbilicus. Translucent white, pale tan or pale pink, sometimes with a few small brown blotches near the sutures.

6 cm. Kimberley, WA to northern NSW. Some authors have regarded this as a subspecies of the uncommon central Indo-West Pacific *S. glabratum* (Dunker, 1852). Synonyms: *denisi* Salmon, 1948; *knighti* Thornley, 1954. For a picture of the living animal see p. 364.

Semicassis bisulcata (Schubert & Wagner, 1829)

Pl. 37; fig. 7 a-c Thin or thick-shelled, ovate to globose; spire of medium height, with a very small nucleus of about three whorls; early postnuclear whorls rounded and sculptured with beaded spiral and axial ribs, axial ribs becoming obsolete on later spire whorls; body whorl smooth or with strong flattened spiral ribs or fine striae; outer lip moderately thick, with teeth along its inner edge but the outer edge smooth; columellar shield narrow, crossed by strong lirae; false and true umbilicus deep with a deep spiral channel behind the anterior canal. White, cream, blue-grey or pink, often with 5-6 spiral rows of reddish or yellow-brown spots which are usually rather square; upper edge of lip varix spotted with brown, inner edge of lip and columellar shield white; interior white.

7 cm. Indo-West Pacific; Shark Bay, WA to southern Qld. Synonyms: *tessellatum* Wood, 1825; *saburon* "Lam." Schubert & Wagner, 1829; *pila* and *japonica* Reeve, 1848; *minor* and *nucleus* Küster, 1857; *pfeifferi* Hidalgo, 1871; *booleyi* Sowerby, 1900; *diuturna* Iredale, 1927; *nashi* Iredale, 1931; *persimilis* "Kuroda" Kira, 1955.

Semicassis labiata (Perry, 1811)

Pl. 37; fig. 1 a-b Oblong-ovate, spire of medium height; whorls shiny smooth, shoulders rounded or weakly nodulose; outer lip usually toothed along the inner side; columella thick, weakly lirate along the margin; umbilicus narrow. Exterior bluish cream, with purple-brown blotches which may be U-shaped and sometimes joined to make axial zigzag lines; a spiral row of brown spots at the suture and an axial row on the outer lip; anterior canal brown; columella orange-yellow, interior light brown.

8 cm. Circum-Southern Ocean; southern Qld to south coast of WA. Synonyms: *achatina* Lamarck, 1816; *insperata* Iredale, 1927 (the latter is a nodulose form from NSW).

Semicassis paucirugis (Menke, 1843)

Pl. 37; fig. 9 a-b   Ovate, spire of medium height; early spire whorls rounded, axially ribbed, spirally striate; penultimate whorl nodulose; body whorl with weak shoulder nodules becoming obsolete toward the lip, otherwise smooth; outer lip thick, usually weakly toothed along the inner edge; columella thick, with short lirae on the inner margin; umbilicus almost closed. White or light yellow with four spiral bands of indistinct yellow spots and yellow-brown flames below the suture; anterior canal tip brown; interior yellow-brown.

7 cm. Esperance to Shark Bay, WA. Moderately common on the inner shelf but rarely seen in shallow water. This is a distinctive species unlike any other in the area.

Semicassis pyrum (Lamarck, 1822)

Pl. 37; figs. 8, 11 a-e   Typically with a spire of medium height; early spire whorls strongly spirally striate and finely axially striate, later whorls with a single nodulose spiral rib on the shoulders; body whorl lacking nodules but with a distinct incised spiral subsutural groove; outer lip narrow, toothless; columella smooth except for a few weak marginal lirae. Fawn, with spiral bands of diffuse brown spots or stains; anterior canal tipped with brown; interior brown; outer lip with brown spots on its outer side.

7 cm. NSW to Fremantle, WA; also found in NZ and South Africa. A specimen from the North West Shelf (fig. 8) also seems referable to this species. Usually taken by trawler. There are several named forms which Abbott (1968) considered to be variants of this species complex. All of them occur throughout the species range and there are intergrades between them. In this account they are treated simply as variants of *pyrum* and the "form names" and are listed in the synonymy.

Elongate-ovate and relatively high-spired shells with smooth shoulders, obsolete subsutural groove and strong oblique anterior lirae along the columellar margin were named *stadiale* Hedley, 1914. Shells like this occur occasionally off NSW and as far west as the south coast of WA. In another common variant the shells have an angulate shoulder on the body whorl bearing two rows of nodules. Such shells were named *niveum* by Brazier, 1872. They may also be found from NSW to Bremer Bay, WA. Specimens intermediate between typical *pyrum* and the *niveum* form were named *spectabilis* Iredale, 1929. Synonyms: *niveum* Brazier, 1872; *stadiale* Hedley, 1914; *spectabilis* Iredale, 1929; *mawsoni* Cotton, 1945; *denda* Cotton 1945; *lalli* Cotton, 1954.

Semicassis royanum Iredale, 1914

Large, tall-spired, with subangulate, nodulose shoulders; columellar shield thick and rounded, bears 5-7 short plicae.

NSW, New Zealand and Kermadec Is. This form was regarded as a subspecies of *pyrum* by Abbott but Alan Beu informs me that it is common and coexists with typical *pyrum* in New Zealand and is clearly a distinct species. It is rare in NSW. Illustrated by Abbott, 1968, *loc. cit.,* Pl.13, fig. 9; Pl. 158, 159. Synonym: *hedleyi* Iredale, 1927.

Semicassis sophia (Brazier, 1872)

Very close to *P. bisulcata* but with more sharply carinate shoulders, smoother surface and broader columellar shield.

8 cm. Southern Qld and NSW. Treated by some authors as a subspecies of *P. bisulcata.*

Semicassis thomsoni (Brazier, 1875)

Pl. 37; fig. 4 a-b   Spire high and pointed, body whorl globose; spire whorls shouldered, axially and spirally striate; body whorl shouldered with 1-3 spiral rows of nodules, spirally striate; outer lip strong, toothed on its inner side; columellar shield narrow, plicate, marginally indented. White, cream or tan, usually with 3-5 spiral rows of red-brown spots; outer lip and columella white.

9 cm. Southern Qld to Bass St; also found in northern NZ. The synonym *palinodia* Iredale, 1931 is a nodulose dwarf form of this species from Sydney Harbour.

Semicassis whitworthi (Abbott, 1968)

Pl. 37; fig. 6 a-b   Spire tall and pointed, body whorl ovate; spire whorls shouldered, axially threaded and spirally ribbed; body whorl with 12-14 wide and nodulose spiral ribs; outer lip thick, toothed on its inner side; columellar shield short, narrow, rugose. Cream or pale tan, with a few light brown spots, outer edge of lip varix and tip of canal dark brown; interior yellow-brown.

8 cm. Bunbury to Shark Bay, WA. Taken by trawlers and in lobster pots in deep water. The strong ribbing is a characteristic of this handsome shell and it could not be confused with any other species.

▲ SUBGENUS **ANTEPHALIUM** Iredale, 1927
Cassis semigranosa Lamarck

False umbilicus slit-like or lacking.

Southern Australia. The group has a long Tertiary fossil history in southern Australia.

Semicassis adcocki (Sowerby, 1896)

Pl. 38; fig. 3 a-b   Ovate, spire moderately high; spire whorls axially and spirally ribbed; short axial folds present on the body whorl posteriorly; outer lip toothless, columellar shield thick, small; umbilicus closed, false umbilicus lacking. White with six spiral rows of small red-brown spots.

4.5 cm. Bass St. to about Esperance, WA.

Semicassis semigranosum (Lamarck, 1822)

Pl. 37; fig. 10 a-b   Elongate-ovate, moderately high-spired; shoulders with beaded spiral cords, body whorl smooth or spirally striate; outer lip smooth, columellar shield small and smooth except for weak marginal lirae; umbilicus small or sealed, false umbilicus lacking. Cream or pink; interior white.

6 cm. Bass St. to Fremantle, WA.

Semicassis sinuosum (Verco, 1904)

Pl. 38; fig. 4 a-b   Small, elongate-ovate; spire whorls rounded, shoulders lacking; sculptured with flat spiral threads; aperture constricted posteriorly, outer lip only slightly inturned and thickened, weakly toothed on its inner side; columellar shield small, finely wrinkled; siphonal canal short; umbilicus very narrow, false umbilicus slit-like. White or cream with five spiral rows of subdued, rather square yellow-brown spots which may be elongate and merge axially.

2.5 cm. SA to about Bunbury, WA. A rare shell taken in relatively deep water. Synonym: *angustatum* Cotton & Godfrey, 1931.

FAMILY **FICIDAE**

Fig Shells

There are only two living genera and a handful of species in this family. The family name comes from the shape and name of the best-known genus, *Ficus*. The second genus, *Thalassocyon*, was discovered recently and, although its shell form is more like the American genus *Busycon*, the radula is taenioglossate and the anatomy like that of *Ficus* (see Beu, 1981, *Records Australian Museum* **33** (5): 248-324; Warén & Bouchet, 1990, *Veliger* **33** (1): 56-102).

Figs are sand or mud dwellers. *Ficus eospila* is found on intertidal sand flats in association with small "biscuit urchins" which are probably their prey. This would be consistent with the echinoderm eating habits of their cousins the cassids.

The animals of fig shells are colourful. The foot is large, the thin mantle spreads over the shell, and the head tentacles and siphon are long. There is very little information on their larval development - see Warén & Bouchet (1990, *loc. cit.*) for a summary.

▲ GENUS **FICUS** Röding, 1798

F. *ficus* Gmelin

Shells thin to moderately thin, without varixes, pyriform; spire low conical to depressed; body whorl inflated, with rounded shoulder; aperture elongate, wide posteriorly, narrowing in a curve to a very long anterior canal; outer lip simple, not thickened; columella not calloused; sculpture cancellate; operculum lacking.

Circumtropical. Synonyms: *Pyrula* Lamarck, 1799; *Ficula* Swainson, 1840; *Sycotypus* H. & A. Adams, 1853.

Ficus eospila (Péron, 1807)

Pl. 40; fig. 1 a-c   Spire nearly flat but with slightly channelled suture and convex whorls; protoconch deviate, consisting of a single, bulbous, smooth whorl; sculpture of flat-topped primary cords with a single, slightly weaker secondary cord plus axial bars in the interspaces. Cream or fawn, typically with scattered prominent brown spots tending to be in spiral rows, plus tiny pale brown dots on some ribs, occasionally also with faint tan spiral bands.

6 cm. Esperance to the North Kimberley, WA. The typically spotted form (Pl. 40 , figs. 1 a-b) is found in shallow water, commonly intertidally, only as far south as North West Cape. In deep water off the mid-west and south coasts as far across as Esperance there is a very delicate and slender form

which is spotless (Pl. 40 , fig. 1c). This has the same kind of protoconch and sculpture as the spotted northern form and it is provisionally identified as the same species. This species has been known as *F. tesselata* Kobelt but in his narrative of the Baudin expedition of 1801-2 François Péron named specimens from Shark Bay and his prior name should be used. Synonym: *tesselata* Kobelt, 1881. For a picture of the living animal see p. 370.

Ficus filosa (Sowerby, 1892)

Pl. 40; fig. 3   Relatively solid, with an emergent spire and wide shoulder; suture impressed, whorls convex, protoconch small, of two smooth whorls; sculpture of strong primary cords made slightly nodulose by the equally strong axials, plus one weaker secondary spiral in each interspace. Cream, primary cords light brown giving a conspicuously striped appearance.

9 cm. Central Indo-West Pacific; North West Shelf, WA to northern NSW. Trawled in deep water. There are few Australian records of this distinctive species.

Ficus subintermedia (Orbigny, 1852)

Pl. 40; fig. 2   Spire moderately elevated, with weakly channelled suture and convex whorls; protoconch deviate, of two smooth whorls; sculpture of primary spiral cords made faintly nodulose by slightly wider crossing axials, and usually three lesser spirals in each interspace. Pale brown with 5-6 obscure paler spiral bands and diffuse faint small brown spots

8 cm. Indo-West Pacific; Onslow, WA to northern NSW. See Wilson & Gillett (1971) for a colour picture of a living animal. The shell of the common Indo-West Pacific species *F. ficus* (Linnaeus, 1758) is similar but has a single secondary cord in each interspace and is a darker colour than *F. subintermedia*. There is a record of *F. ficus* from Barrow I., WA but the presence of that species in Australia needs confirmation. Synonyms: *ficoides* Lamarck, 1822 (not of Brocchi, 1814); *reticulata* Reeve, 1847 (not of Lamarck, 1822); *margaretae* Iredale, 1931.

▲ GENUS **THALASSOCYON** Barnard, 1960
T. bonus Barnard

With the characters of the type species.

South Africa, New Zealand and south-eastern Australia. Monotypic. Anatomical details are given by Warén & Bouchet (1990, *loc. cit.*).

Thalassocyon bonus Barnard, 1960

Shells moderately thick, without varixes; spire moderately high, whorls carinate; aperture subquadrate, outer lip simple, constricted anteriorly; anterior canal long, open, spike-like; columellar and parietal lips with a reflected callus over a narrow umbilical chink; spiral sculpture dominant; operculate; peripheral keel with broad, laterally compressed spines, body whorl otherwise sculptured with numerous spiral cords and weak axial folds. Light brown.

7 cm. South Africa, North I., New Zealand, Kermadec Is, and recently trawled in 570-950 m off the eastern Australian coast from southern Qld to southern NSW. Lives on mud or ooze substrates in very deep water. The New Zealand population was named as *T. tui* Dell, 1967 but Warén & Bouchet (*loc. cit.*) showed it to be the same as that from South Africa. For illustrations see *Australian Shell News*, 1987, No. 59-60: p.1; Warén & Bouchet, 1990, figs. 1-2.

FAMILY **PISANIANURIDAE**

This family was established only recently by Warén & Bouchet (1990, *Veliger*, **33** (1): 56-102) for the genus *Pisanianura*, until then placed in the Buccinidae. They introduced it as a subfamily of the Ranellidae and Beu (pers. comm. April, 1992) believes is worthy of full family rank and that view is followed here. Anatomically these animals conform with the Tonnoidea but there are distinctive features.

The shells are of medium size, ovate to biconical, with poorly developed siphonal canal and curved columella giving the aperture a rounded appearance. There are no varixes. The operculum has an apical nucleus (corroded). The proboscis is very short, broad and muscular; the osphradium is bipectinate with the left row of leaflets less developed than the inner ones.

▲ GENUS **PISANIANURA** Rovereto, 1899
Murex inflatrus Brocchi

Ovate with a tall, conical spire; protoconch multispiral, with a large-meshed reticulate sculpture; teleoconch with rounded to angulate whorls, without varixes; columella curved, aperture wide and round, siphonal canal short. Operculum with an apical nucleus.

Cosmopolitan. The type species is a Pliocene fossil from Italy and there are several other fossil species dating from as early as the Oligocene. There are two living species living in abyssal habitats of tropical and subtropical regions. The anatomy of the living species has been described by Warén & Bouchet (1990, *loc. cit.*). Specimens of neither of the two Australian species were available for illustration and the reader is referred to the same publication, where they are well figured. Synonyms: *Anura* Bellardi, 1873 - not Hodgson, 1841 (*inflatus* Brocchi); *Kaiparanura* Laws, 1944 (*spiralis* Marshall); *Laminilabrum* Kuroda & Habe, 1961 (*brevialaxe* Kuroda & Habe); Nawenia Ladd, 1977 (*bartholomewi* Ladd).

Pisanianura brevialaxe (Kuroda & Habe, 1961)

Whorls convex, base rounded, columella weakly recurved; sculpture of fine spiral threads; outer lip internally bevelled, sharp-edged. White.

Western Pacific; Qld. Dredged at depths from 150 to 837 m.

Pisanianura grimaldii (Dautzenberg, 1889)

Whorls slightly convex, base rounded, columella recurved; sculpture of nodulose axial ribs crossed by sharp spiral cords; outer lip with a slightly fluted edge. White.

3 cm. Central Atlantic and Indo-West Pacific; Qld. Dredged at depths from 680 to 2165 m. Synonym: *clathrata* Dautzenberg & Fischer, 1906.

FAMILY **LAUBIERINIDAE**

This is a very small family only recently distinguished (Warén & Bouchet, 1990, *Veliger*, **33** (1): 56-102). It is diagnosed as having: low-spired, globular or weakly biconic fragile shells with poorly developed siphonal canals; monopectinate osphradium, very large in the larvae; proboscis short with a large buccal mass. The radulae of the group are similar to those of the ranellids, which gives the clue to its relationships with the Tonnoidea.

There are two genera, *Laubierina* Warén & Bouchet, 1990 (*peregrinator* Warén & Bouchet) and *Akibumia* Kuroda & Habe, 1959 (*flexibilis* Kuroda & Habe) both widely distributed and with representatives in abyssal habitats off the eastern Australian coast. Warén & Bouchet described but did not name a species of *Laubierina* from 1000 m off Broken Bay, NSW. They also recorded *Akibumia orientalis* (Schepman, 1909) from 1106-1143 m off Sydney, and *A. schepmani* Habe, 1962 from 550 m off the Gold Coast of southern Qld. Specimens were not available for illustration here and the reader is referred to Warén & Bouchet for details and figures.

FAMILY **PERSONIDAE**

Distorsios

Until recently this small group was regarded as a subfamily of the Ranellidae but Beu (1988) has shown that the anatomy and operculum of the animals are more like those of the Ficidae. The proboscis is very long and slender and is coiled when retracted. The radula is distinctive in having a crescentic central tooth with downturned corners, unlike that of the ranellids. For these reasons Beu elevated the group to full family rank. The shell is varixed and there is a moderately developed posterior canal but no extended siphon.

Only the genus *Distorsio* represents the family in the Australian fauna. Very little is known of the natural history of the group although the anatomy has been described by Lewis (1972) and the larvae by Laursen (1981).

References

Beu, A. G. (1980). Australian gastropods of the family Bursidae, Pt 1. *Records Australian Museum* **33** (5): 248-324.

Beu, A. G. (1985). A classification and catalogue of living world Ranellidae (= Cymatiidae) and Bursidae. *Conchologists of America Bulletin* **13** (4): 55-66.

Beu, A. G. (1987). Taxonomy of gastropods of the families Ranellidae (= Cymatiidae) and Bursidae, Pt. 2 Descriptions of 14 new modern Indo-West Pacific species and subspecies with revisions of related taxa. *N.Z. Journal Zoology* **13**: 273-355.

Beu, A. G. (1988). Taxonomy of the gastropod families Ranellidae (= Cymatiidae) and Bursidae, Pt. 5 ..... and recognition of the family Personidae. *Sato Ho-on Kai Special Publication* **2**: 69-96.

Laursen, D. (1981). Taxonomy and distribution of teleplanic prosobranch larvae in the North Atlantic. *Dana Report* **89**: 1-43.

Lewis, H. (1972). Notes on the genus *Distorsio* with descriptions of new species. *Nautilus* **86**: 27-50, 48 figs.

Parth, M. (1989). Brief notes on the genus *Distorsio* Roeding, 1798 and descriptions of a new species. *La Conchiglia* **21** (233-236): 52.

▲ GENUS **DISTORSIO** Röding 1798

Murex anus Linnaeus

Ovate, solid, whorls moderately to extremely distorted; outer lip wide flaring, heavily toothed; anal notch deep, bordered by a strong parietal ridge; inner lip excavate centrally, bearing heavy lirae anteriorly and along the sides of the anterior canal.

Circumtropical. The genus gets its name from the distorted coiling of the whorls. Synonyms: *Distortrix* Link, 1807 (*anus* Linnaeus); *Persona* Montfort, 1810 (*anus* Linnaeus); *Distorta* Perry, 1811 (*anus* Linnaeus); *Rhysema* Clench & Turner, 1957 (*clathratus* Lamarck).

Distorsio anus (Linnaeus, 1758)

Pl. 40; fig. 21 a-b   Solid, ovate, grossly distorted; outer lip and ventral callus expanded to form a flattened base with a broad, frilled rim; spire of medium height, anterior canal short and nearly vertical, varixes weak; body whorl sculptured with heavy nodules, strong spiral ribs and fine spiral striae; posterior and anterior ends of the columella projecting into and partially occluding the aperture. Dorsum cream or greyish with brown patches and brown spiral band, ventral surface polished, mottled white and fawn, rim patched alternately brown and white.

10 cm. Western Pacific; Rowley Shoals & Scott Reef, WA to Qld. This is a shallow-water, coral reef species.

Distorsio decipiens (Reeve, 1844)

Rather tall-spired, with a nodulose spiral cord at the suture below a concave subsutural slope and angulate shoulder bearing two sharp, nodulose spiral cords, giving the shell a distinctly hunched appearance, centre and anterior surface of the body whorl with 6-7 weakly nodulose cords, the nodules aligned axially producing imperfect axial ribs; shield callus thin, columella with about 15 sharp denticles, separated by a deep furrow from the varix; siphonal canal long, slightly recurved. White with yellow-gold blotching.

6.5 cm. Western Pacific; recorded from Qld as far south as the Bunker Group but not common. This species is distinguished from *D. reticulata* by the subsutural concavity and nodulose rather than reticulate sculpture. For illustrations see Springsteen & Leobrera (1986, *loc. cit.*, Pl. 32, fig. 6).

Distorsio euconstricta Beu, 1987

Extremely distorted and almost triangular; like *D. constricta* in general form but much smaller and with the inner base of the outer lip constricted to a thin, sharp ridge, almost pressed to the base of the columella so forming a narrow slit; spiral sculpture of ill-defined primary cords and sharp threads, axial sculpture of widely spaced folds forming peripheral nodules; outer lip narrow, bearing narrow transverse ridges; inner lip deeply excavate centrally. Cream, buff or pale pinkish orange.

3.5 cm. Indian Ocean; dredged on the North West Shelf, WA. This is the smallest member of the genus. Compare with *D. graceielae*. For illustrations see original description (Beu, 1987, *loc. cit.*, figs. 131-143, 145).

Distorsio graceiellae Parth 1989

Sculpture of eight main spiral ribs, the two most prominent forming sharply angulate nodules where they cross the axials, and secondary cords in the interspaces; aperture small, similar to that of *D. euconstricta* in having a narrow slit at the base formed by compression of the edge of the outer lip, the latter with nine denticles, the third and fourth fused and enlarged, and the eighth and ninth fused to form the border of the slit; prominent parietal tooth present, parietal and columellar zones separated by an oblique nodulose ridge; columella with 9-10 marginal denticles. Light brown, darker at the suture; aperture yellowish brown, darker than the exterior.

3.2 cm. Western Pacific; Qld, NSW. The distinction between this and *D. euconstricta* is in doubt. For an illustration see original description (1989, *La Conchiglia* 21: 233-236).

Distorsio habei Lewis, 1972

Pl. 40; fig. 20 a-b   Shell exhibiting extreme distorstion; nodulose spiral cords dominate the sculpture, the subsutural cord separated by a distinct gap from the next at the subangulate shoulder; outer lip with 8 stout teeth, the third very prominent; inner lip with a single tooth on the parietal wall, deeply excavated centrally, heavily lirate anteriorly; siphonal canal short. Light brown with golden brown blotches.

5 cm. Western Pacific from southern Japan to the Kermadec Is. and New Zealand and recently discovered off the coast of northern WA. A related species is *D. constricta* (Broderip, 1833) from the west coast of America. In fact Lewis originally described *habei* as a subspecies of *constricta*.

Distorsio reticulata (Linnaeus, 1758)

Pl. 40; fig. 22 a-b   Rather light, fusiform, moderately distorted; spire high, anterior canal moderately long and up-turned, varixes weak; sculptured with fine spiral and axial ribs with a small nodule at each intersection; outer lip flat and expanded, crossed by rib-like teeth; parietal ridge prominent, plus several irregular lirae or nodules; columella bearing strong teeth anteriorly; ventral side of body whorl glazed. Dorsal surface cream or yellow with yellow-brown spiral bands, ventral side light brown, teeth white; periostracum brown with a single long bristle from each nodule on the dorsal surface.

9 cm. Indo-West Pacific; Shark Bay, WA to southern Qld. Commonly trawled. Synonym: *reticulata* Röding, 1798.

FAMILY **RANELLIDAE**

Tritons and Trumpets

Many will know this family by the name Cymatiidae, but after much debate about the complex nomenclatural history, Beu & Cernohorsky (1986) concluded that Ranellidae must be used. There are two subfamilies, Ranellinae and Cymatiinae. (A third group, Personinae, is now given full family status.)

A characteristic of the family is a prominent, thick and hirsute periostracum. Most ranellids have thick shells with strong varixes and teeth or lirae along the inner side of the outer lip and usually on the inner lip as well. There is usually an anal notch at the posterior end of the aperture formed between an outer lip nodule and a ridge or nodule on the parietal wall. The columella is usually glazed, the glaze often forming an adherent or raised shield. The anterior canal is produced to form a short spout or a long, projecting limb.

Females lay eggs in capsules clustered in masses attached to the substrate. The larvae of most ranellids hatch as free-swimming veligers and there is a long planktotrophic larval stage. Scheltema (1966, 1971) *Deep Sea Research* **13**: 83-95) reported that *Cymatium parthenopeum* has a planktonic larval stage lasting as long as nine months. Such long larval life may account for the very wide geographical distribution of many species. An exception is the egg mass of *Sassia (Austrotriton) subdistorta*; it has wineglass-shaped capsules often laid in dead bivalve shells and there is no planktonic larval stage. This is unusual in the family but appears to be the rule in *Sassia (Austrotriton)*.

Warén & Bouchet (1990) summarised what is known of larval shell structure and sculpture in this family and gave references. In *Sassia* the larval shells are globular, calcareous, with cancellate sculpture. In contrast, the larval shells of *Cymatium*, *Charonia* and *Ranella* are made of periostracal material, only partly or not at all calcified. In these latter genera the periostracal shell is filled with calcium carbonate after settlement, as it is in the tonnids, and the original material wears or dissolves away leaving an internal cast to form the protoconch.

Ranellids have a variety of diets including bivalves, gastropods and echinoderms. The trumpet shell *Charonia tritonis* feeds on starfish, including the spiny Crown of Thorns (*Acanthaster planci*). *Ranella australasia*, *Cymatium parthenopeum*, and *Cabestana spengleri* eat ascidians. Observations on feeding of tropical and temperate species occurring in Australia have been published by Houbrick & Fretter (1969) and Laxton (1971) respectively.

The taxonomy of this family, especially in the case of species with very wide geographical distribution, is very difficult and confused. There are many examples of circumtropical species showing very little morphological difference between tropical Atlantic and tropical Indo-West Pacific populations. In such cases opinions may differ on whether the separated populations deserve recognition as distinct species or subspecies, or not at all. In this account I have generally followed the conservative approach of Beu, who has so thoroughly researched this family. Some of his important papers are cited in the references below.

References

Beu, A. G. (1981). Australian gastropods of the family Bursidae, *Records Australian Museum* **33** (5): 248-324.

Beu, A. G. (1985). A classification and catalogue of living world Ranellidae (= Cymatiidae) and Bursidae. *Conchologists of America Bulletin* **13** (4): 55-66.

Beu, A. G. & Cernohorsky, W. O. (1986). Taxonomy of gastropods of the families Ranellidae (= Cymatiidae) and Bursidae, Pt. I., *N.Z. Journal of Zoology* **13**: 241-266.

Beu, A. G. (1987). Taxonomy of gastropods of the families Ranellidae (= Cymatiidae) and Bursidae, Pt. 2., *ibid.* **13**: 273-355.

Beu, A. G. & Kay, E. A. (1988). Taxonomy of gastropods of the families Ranellidae (= Cymatiidae) and Bursidae, Pt. 4. *Journal Royal Society N.Z.* **18** (2): 185-223.

Beu, A. G. (1988). Taxonomy of gastropods of the families Ranellidae (= Cymatiidae) and Bursidae, Pt. 5. *Saito Ho-on kai Special Publ.*: 69-96

Houbrick, J. R. & Fretter, V. (1969). Some aspects of the functional anatomy and biology of *Cymatium* and *Bursa*. *Proceedings Malacological Society London* **38**: 415-429.

Laxton, J.H. (1971). Feeding in some Australasian Cymatiidae (Gastropoda: Prosobranchia). *Zoological Journal Linnean Society* **50**: 1-9.

Warén, A. & Bouchet, P. (1990). Laubierinidae and Pisanianurinae (Ranellidae), two new deep-sea taxa of the Tonnoidea (Gastropoda: Prosobranchia). *Veliger* **33**: 56-102.

SUBFAMILY **RANELLINAE**

The Ranellinae usually have a low, turbinate protoconch and a smooth and furry periostracum, or a periostracum with evenly spaced bristles lacking axial blades; the central radular tooth is approximately equidimensional with only a few denticles on each side.

▲ GENUS **ARGOBUCCINUM** Bruguière, 1792
argus Gmelin (= *pustulosum* Lightfoot)

Solid, spire of moderate height, body whorl short and ventricose; anal notch well defined, anterior canal short, varixes low and rounded; with strong spiral sculpture; aperture wide.

Middle to high latitudes of the southern hemisphere. Synonyms: *Gondwanula* Finlay, 1927 (*tumida* Dunker); *Mediargo* Terry, 1968 (*mediocre* Dall).

Argobuccinum pustulosum (Lightfoot, 1786)

Pl. 41; fig. 8 Ovate; whorls convex with impressed suture and very broad, low, rounded varixes; body whorl wide; early spire whorls with both axial and spiral cords, the former becoming obsolete on the later whorls where the primary cords are weakly nodulose, with crowded secondary threads in the interspaces; outer lip with eight double inner denticles; parietal wall with a strong bifid ridge, columella reflected, plicate anteriorly. Fawn, primary ribs red-brown, some secondary threads bearing thin tan lines; aperture white.

10 cm. Temperate shores of South Africa, South America, New Zealand and south-eastern Australia. The Australian and New Zealand form is the subspecies *A. p. tumidum* (Dunker, 1862). *A. vexillum* Sowerby, 1835 has sometimes been incorrectly applied to the Australasian form.

▲ GENUS **FUSITRITON** Cossmann, 1903
Triton cancellatus Lamarck (= *magellanicus* Röding)

Thin, fusiform, spire moderately high with convex whorls and impressed suture; varixes small; anterior canal long; outer lip and columella edentulous; sculpture cancellate; operculum thick.

Middle latitudes of the Southern Ocean and the North Pacific. Synonym: *Cryotritonium* von Martens, 1904.

Fusitriton magellanicus (Röding, 1798)

Pl. 41; fig. 10 With six teleoconch whorls, sculptured with widely spaced axial and spiral cords forming small knobs at the interspaces; columella narrow, sinuous, thinly calloused; parietal nodule small. White.

12 cm. South Africa, South America, New Zealand and south-eastern Australia. The Australian form is the subspecies *A. m. retiolus* (Hedley, 1914) and is found on the outer continental shelf off NSW and eastern Vic. A recent record from the North West Shelf of WA needs confirmation.

▲ GENUS **GYRINEUM** Link, 1807
Murex gyrinus Linnaeus

Shell solid, dorso-ventrally compressed; varixes wide, two per whorl, aligning obliquely along the sides of the spire; anterior canal short with an open slit; aperture round, with a raised rim, outer lip strongly dentate within.

Circumtropical. Synonyms: *Apollon* Montfort, 1810 (*gyrinus* Linnaeus); *Gyrinella* Dall, 1924 (*pusilla* Broderip).

Gyrineum bituberculare (Lamarck, 1816)

Pl. 40; fig. 7 Squat, spire moderately high, body whorl short and constricted at the base, varixes wide; sculpture of low axial folds crossed by spiral primary cords and numerous threads, the folds becoming obsolete on the body whorl except for two large dorsal tubercles; outer lip raised, forming an almost entire peristome with the raised parietal-columellar callus, with about eight sharp inner denticles; columella plicate; anterior canal sharply bent and slightly upturned.

4 cm. Indo-West Pacific; Torres St., Qld. The figured specimen from the Torres Strait (Australian Museum collection) is one of the few Australian records.

Gyrineum gyrinum (Linnaeus, 1758)

Pl. 40; fig. 6 Protoconch of two smooth, convex whorls; teleoconch of six whorls, sculptured with numerous spiral threads and axial rows of prominent, rounded nodules, about six rows between varixes; outer lip with seven strong inner denticles, parietal wall with two weak parietal ridges, columella irregularly plicate. Yellow or white, with a red-brown spiral band on the spire whorls, two on the body whorl and at the tip of the canal, some nodules may be orange.

3 cm. Indo-West Pacific; Qld.

Gyrineum lacunatum (Mighels, 1845)

Pl. 40; fig. 8 a-c Protoconch of 2½ smooth, convex whorls; teleoconch of seven whorls, strongly sculptured with spiral and axial ribs forming a lattice pattern, weakly nodulose at the intersections; outer lip raised, with seven inner denticles; parietal ridge thick, oblique; columella plicate. Brown and white; aperture violet.

3 cm. Indo-West Pacific; Dampier Archipelago, WA to northern NSW. This species has generally been misidentified as *pusillum* (Broderip, 1833), which is a different species from the Tuamotus. Synonyms: *cuspidatum* Reeve, 1844; *facetus* Iredale, 1936; *deliberatus* Iredale, 1936.

Gyrineum roseum (Reeve, 1844)

Small, short, with 6-7 convex teleoconch whorls bearing six spiral cords extending onto the varixes producing nodules there and in the intervarixeal areas where they cross 4-5 axial ribs; columella lirate except centrally, outer lip dentate. Pink with paler nodules, orange and white bands on the varixes, and cream aperture.

2 cm. Indo-West Pacific; Qld south to the Capricorn Group.

▲ GENUS **BIPLEX** Perry, 1811
Argobuccinum (Biplex) perca Perry

Dorso-ventrally depressed, with expanded, wing-like varixes; anterior canal long, opening only by a very narrow slit; outer lip weakly toothed or edentulous; aperture encircled by continuous outer and inner lips; columella raised, smooth.

Indo-West Pacific. Sometimes ranked as a subgenus of *Gyrineum*. There are three species. The shells are known to Australians as "kookaburras" because of their profile.

Biplex perca (Perry, 1811)

Suture only slightly channelled, whorls greatly flattened; varixes wing-like, relatively narrow, bearing short, triangular spines; intervarixeal areas nodulose; surface rather glossy. Light orange-brown, nodules and spine tips white.

4 cm. Indo-West Pacific; uncommon in Australia but with positive records from off Port Hedland, WA and Keppel Bay, Qld.

Biplex pulchellum (Sowerby, 1825)

Pl. 40; fig. 4 a-b   Suture deeply channelled; varixes fin-like and marginally toothed; shoulder rim gemmate, other sculpture of very fine axial threads and spiral cords forming minute nodules at the intersections, the spiral cords running out onto the varixes; aperture oval. Cream or fawn.

2.5 cm. Central Indo-West Pacific; Onslow, WA to central Qld. Dredged on muddy substrates. Synonym: *jucundum* Adams, 1853.

Biplex pulchrum (Gray in Sowerby, 1836)

Pl. 40; fig. 5 a-b   Suture channelled; varixes bearing three prominent spines; sculpture of gemmate primary spiral ribs, numerous spiral threads and weak axial cords connecting the gemmae, about six primary ribs on the body whorl; columella smooth or with very weak plicae. Cream or fawn.

4.5 cm. Central Indo-West Pacific; Shark Bay, WA to Qld. The larger size and prominent varixal spines immediately distinguish the species from *B. pulchellum*. Synonyms: *perca* var. *aculeata* Schepman, 1909; *microstoma* Fulton, 1930.

▲ GENUS **RANELLA** Lamarck, 1816
R. gigantea Lamarck (= *olearia* Linnaeus)

Ovate to fusiform; varixes prominent at about ½ whorl intervals but not aligned; aperture wide, outer lip toothed within; columella weakly plicate; anterior canal short, open; periostracum thick and furry; operculum oval, nucleus subcentral.

Cosmopolitan. Synonyms: *Gyrina* Schumacher, 1817 (*maculata* Schumacher = *olearia* Linnaeus); *Eugyrina* Dall, 1904 (*gigantea* Lamarck = *olearia* Linnaeus); *Mayena* Iredale, 1917 (*australasia* Perry); *Gyrinopsis* Dall, 1925 (*cowlitzi* Dall = *washingtoniana* Weaver).

Ranella australasia (Perry, 1811)

Pl. 41; fig. 11 a-b   Fusiform, spire high with impressed suture, body whorl wide; sculptured with many spiral threads and usually 2-3 spiral ribs bearing nodules, shoulder nodules sometimes quite pronounced; parietal nodule strong, anterior columella plicae weak. Colour variable, usually red-brown but sometimes cream, with alternating dark brown and white bands on the varixes; aperture white.

9 cm. NSW to Abrolhos, WA. Common from the intertidal zone to the edge of the continental shelf. The NZ population of this has been called *vossi* Powell, 1952 and *blacki* Powell, 1954; both names may be regarded as synonyms of *R. australiasia*. The South African *gemmifera* Euthyme, 1889 is treated as a subspecies of *R. australasia* by Beu.

SUBFAMILY **CYMATIINAE**

In the Cymatiinae the protoconch is usually tall and conical and the periostracum is luxuriantly bristled, each bristle with thin axial blades; the central radular tooth is markedly wider than high, with many small denticles on each side of the main cusp.

▲ GENUS **CABESTANA** Röding, 1798
Murex cutaceus Linnaeus

Large, fusiform, anterior canal short; protoconch turbinate, with four whorls; varixes narrow; columella smooth, with a small, bifid parietal nodule and angulate edge anteriorly, reflected over a small umbilicus.

Cosmopolitan. There are four living species, two of them in temperate Australian waters. Synonyms: *Aquillus* Montfort, 1810 (*cutaceus* Linnaeus); *Dolarium* Schluter, 1838 (*caduceus* err. = *cutaceus* Linnaeus); *Neptunella*, Gray, 1854 (*cutaceus* Linnaeus); *Cymatilesta* Iredale, 1936 (*spengleri* Perry).

Cabestana spengleri Perry, 1811

Pl. 41; fig. 13 Shoulders nodulose, with seven primary spiral ribs on the body whorl and secondary ribs or cords in the interspaces, numerous axial ridges crossing the ribs but divided by an incised line on the primaries; outer lip with about eight inner denticles, the central ones double. Yellow-brown, usually with darker spiral lines in the interstices; aperture and columella white.

15 cm. Southern Qld to SA. Common in shallow water among ascidians. Synonyms: *bolteniana* Adams, 1854; *barthelemyi* Bernardi, 1857.

Cabestana tabulata (Menke, 1843)

Pl. 41; fig. 12 Relatively thin; early whorls prominently shouldered and nodulose; varixes narrow; with 7-10 bifid, primary ribs on the body whorl and additional thin spiral cords, all crossed by fine axials, first and second primary ribs weakly nodulose. Uniformly yellow-brown, darker brown between the ribs on the varixes.

10 cm. NSW to Fremantle, WA; also found in New Zealand. Synonyms: *waterhousei* Adams & Angas, 1864; *segregata* Powell, 1933; *otagoensis* Powell, 1954.

▲ GENUS **CHARONIA** Gistel, 1848
Murex tritonis Linnaeus

Large, spire tall, base and anterior canal short; sculpture dominantly spiral, varixes well developed; inner lip broad with one or two callus ridges near the posterior canal, columella irregularly plicate; periostracum, if present, thin, smooth or lightly wrinkled, deciduous; radula with a broad and low central tooth bearing a narrow basal plate and downwardly curved extremities.

Cosmopolitan in tropical and temperate waters. The radula is distinctive. These large and colourful shells are highly decorative and are popularly known as tritons. For a discussion of the nomenclature and a full synonymy see Beu, 1970, *Transactions Royal Society New Zealand, biol. ser.* **11** (16): 205-223.

Charonia lampas (Linnaeus, 1758)

Pl. 41; fig. 15 a-b Solid, body whorl swollen, spire height about half total shell length, varixes moderate; sculptured with low spiral ribs, the shoulder rib strong and nodulose, strong spiral threads in the interspaces; outer lip expanded, denticulate, columellar shield wide and covering much of the ventral side of the body whorl, bearing a single strong fold beside the posterior canal and several irregular lirae anteriorly. Yellow-brown with an overlay of darker red-brown blotches, apex pink, ribs white with dark red-brown spots, shoulder nodules white, lip varix white with dark red-brown bands which extend across the inner face; aperture white, columellar shield brown, lirae white.

15 cm. Swain Reefs, Qld to Jurien Bay, WA. Lives in rocky areas from the intertidal zone to at least the edge of the continental shelf. The *C. lampas* species complex is extremely variable and widespread in temperate waters. There are several named subspecies. The Australian and New Zealand population is named *lampas rubicunda* Perry, 1811 (fig. 15a). Other subspecies are: *lampas lampas* in the eastern Atlantic and Mediterranean; *lampas pustulata* Eurthyme, 1889 in South Africa; *lampas sauliae* Reeve, 1844 in Japan. Even within the Australian population there is great variation. Deep water shells tend to be pale, high-spired and narrow. Synonyms of the Australasian subspecies: *australe* Reeve, 1844; *nodiferus* Tryon, 1881; *euclia* Hedley, 1914; *capax* Finlay, 1927; *euclia instructa* Iredale, 1929; *powelli* Cotton, 1957.

Charonia tritonis (Linnaeus, 1758)

Pl. 41; fig. 14 Body whorl swollen, spire high and pointed, varixes broad but low; sculptured with smooth, wide, rather flat spiral ribs and a single, narrow, axially-plicate rib at the suture; outer lip bearing 9-12 pointed teeth; columella with strong bifid lirae along its whole length. Fawn or light brown with crescent-shaped markings of purple-brown; interior orange-yellow or reddish, columella dark brown in the interstices between white lirae.

45 cm. Indo-West Pacific; Dongara, WA to southern Qld and Lord Howe I. Also present in north-eastern New Zealand. A shallow-water, coral reef species. It is the largest member of the family and one of the largest of all gastropods. It is a known predator of the coral-eating Crown-of-Thorns starfish and is totally protected in Qld.

▲ GENUS **CYMATIUM** Röding, 1798
Murex femorale Linnaeus

Solid shells with 3 strong varixes and spiral ribs; anterior canal short to long, usually bent to the right; anal notch present; outer lip dentate; protoconch multispiral, conical or turbinate, smooth beneath a sometimes bristly horny covering resembling the teleoconch periostracum which is thin or thick, often bristly.

Cosmopolitan. The nominate subgenus is not represented in the Australian fauna. There are several subgenera. Synonyms: *Lotorium* Montfort, 1810 (*lotor* Montfort); *Tritocurrus* Lesson, 1842 (*amophiyridis* Lesson = *tigrinum* Broderip); *Nyctilochus* Gistel, 1848 (*tigrinum* Broderip).

▲ SUBGENUS **GELAGNA** Schaufuss, 1869
Murex succincta Linnaeus

With characters of the type species.

Indo-West Pacific and W. Africa. Monotypic. Beu (1986) treated this as a subgenus of *Linatella*. More recently he has advised (pers. comm. April, 1992) that *Linatella* itself is worthy of subgeneric rank (of *Cymatium*). Synonyms: *Lagena* Mörch, 1852 (not Schumacher, 1817) (*clandestina* Lamarck = *succincta* Linnaeus); *Paralagena* Dall, 1904 (replacement for *Lagena* Mörch).

Cymatium succinctum (Linnaeus, 1771)

Pl. 41; fig. 3 Fusiform, with impressed suture and convex whorls lacking any shoulder angulation; the single varix moderately thick and rounded; with 16-20 spiral ribs, widely spaced, very distinct, flat and smooth on top, interspaces smooth; outer lip slightly raised and sharp-edged beside the varix, bearing 11 inner denticles; parietal wall with a prominent callus, columella thin and narrow, made lirate by the ribs on the whorl below it centrally, but with raised lirae anteriorly; anterior canal of medium length; periostracum smooth but frilled on the varix. Yellow-brown, with darker brown ribs; outer lip denticles brown; columellar lirae white.

6 cm. Indo-West Pacific and West Africa; Barrow I., WA to central Qld. This species is easily recognised by its rounded spiral ribs. The disjunct distribution with Indo-West Pacific and Eastern Atlantic populations is unusual. Synonyms: *doliata* Röding 1798; *clandestinum* Lamarck, 1816; *confinis* Brancsik, 1896; *cynocephala* Habe & Kosuge, 1967.

▲ SUBGENUS **GUTTURNIUM** Mörch, 1852
Cymatium tuberosum Lamarck (= *muricinum* Röding)

With characters of the type species.

Indo-West Pacific and Western Atlantic. Monotypic.

Cymatium muricinum (Röding, 1798)

Pl. 41; fig. 6 Solid and stout, spire moderately high, stepped; anterior canal narrow but moderately short and upturned; varixes broad and heavy; sculptured with 4-5 nodulose axial folds, crossed by strong, irregularly granulose, spiral ribs and striae; outer lip with denticles (sometimes divided) on the inner margin; columella thickly calloused, the callus spreading onto the parietal wall, and reflected over the base, with a low nodule beside the posterior canal and a few weak folds anteriorly. Exterior usually blue-grey or fawn-grey flecked or blotched with brown, sometimes dark brown with a central white spiral band; outer lip and columella glazed, glossy white or yellow; interior red or purple-brown.

6 cm. Indo-West Pacific; Shark Bay, WA to eastern Qld. Also in the West Indies. Synonym: *tuberosum* Lamarck, 1822.

▲ SUBGENUS **LINATELLA** Gray, 1857
Cassidaria cingulata Lamarck (= *caudatum* Gmelin)

Medium-sized, thin, short, inflated, resembling species of Tonnidae; protoconch tall and narrow, multispiral; with a weak shoulder angulation; varixes lacking except for the last which is low and narrow; sculpture of low, flat-topped to weakly convex spiral cords and weak axial threads in the interspaces; periostracum thin, with widely spaced bristles along the cords.

Circumtropical. There are two living species. Beu & Cernohorsky (1986, *N.Z. Journal of Zoology* 13: 241-266) reviewed the complex nomenclatural history of this group and treated it as a distinct genus, but more recently Beu has advised that the opercular characters (terminal nucleus) are typical of *Cymatium* and he now regards the group as worthy of only subgeneric rank (pers. comm. April, 1992).

Cymatium cutaceum (Lamarck, 1816)

Pl. 41; fig. 7 Spire stepped, with impressed suture, early whorls weakly cancellate; body whorl inflated, with 14 spiral cords may be weakly nodulose; outer lip thin, slightly flaring, with 16 low lirae leading into the aperture; columellar callus not extending onto the parietal wall, anteriorly it is smooth and reflected, covering a small umbilical chink beside the low fasciole.

8 cm. Circumtropical; trawled off southern Qld and NSW. This species has been known as *Linatella caudata* (Gmelin, 1791). It was first named as *Buccinum caudatum* Gmelin, 1791, a different species, from *Murex caudatum* Gmelin, 1791 which belongs to *Ranularia*. With the downgrading of both *Linatella* and *Ranularia* to subgeneric rank a homonymy is established between these two names. This can best be resolved by designating *Buccinum caudatum* as the junior homonym and replacing it with the next available name for that species, which is *cutaceus* Lamarck, 1816. This is the course of action taken here on the advice of Dr Alan Beu. Synonyms: the most notable in the Australian context are: *Buccinum caudatum* Gmelin, 1791 (not *Murex caudatum* Gmelin, 1791; *cingulata* Lamarck, 1822; *neptunia* Garrard, 1963.

▲ SUBGENUS **LOTORIA** Emerson & Old, 1963
C. (Lotoria) perryi Emerson & Old

Solid, varixes strong and nodulose; anterior canal of moderate length, slightly upturned.

Indo-West Pacific. With three living species, one of them found in Australian waters.

Cymatium lotorium (Linnaeus, 1758)

Pl. 41; fig. 1 Very thick, solid and of gross, distorted form; varixes thick; penultimate and body whorls with high and thick nodules, crossed by ribs (usually four between the varixes of the body whorl); outer lip varix with six thick folds where the ribs cross and two internal rows of denticles, seven per row, the denticles may be joined to form short lirae; columella with a thick, glazed callus which may bear several weak folds; anterior fasciole thick and bent so that the anterior canal is like a crooked knee. Yellow or orange-brown; varixes dark brown with white or pale folds; upper row of inner apertural teeth brown, inner row and aperture wall white.

15 cm. Indo-West Pacific; Qld. An extraordinarily gross but handsome shell. It is never abundant. During the day it hides under stones and coral slabs.

▲ SUBGENUS **MONOPLEX** Perry, 1811

M. australasiae Perry (= *Murex parthenopeus* Salis)

Medium to large-sized; protoconch large and tall. The periostracum is thick, dark brown and prominently bristled.

Circumtropical. There are many species in this diverse genus. Inclusion of *Cabestanimorpha* in the synonymy here rather than with that of *Turritriton* is on the advice of Beu (pers. comm. April, 1991). Synonyms: *Lampusia* Schumacher, 1817 (*pilearis* Linnaeus); *Cymatriton* Clench & Turner, 1957 (*nicobaricum* Röding); *Cabestanimorpha* Iredale, 1936 (*exaratum* Reeve).

Cymatium aquatile (Reeve, 1844)

Pl. 41; fig. 2   Pyriform; varixes moderately strong, broad and slightly flaring anteriorly; sculptured with spiral ribs tending to be doubled and made weakly gemmate by crossing axial cords, and 6-7 axial rows of moderate nodules; varixes with deep pits between the ribs on the upper side; posterior canal deep, anterior canal slightly reflected; outer lip wide anteriorly, with seven pairs of bifid nodules on the inner side forming thick but short lirae descending into the aperture; columellar callus heavily lirate, lirae irregular on the parietal wall, regular on the side of the canal. Yellow-orange, with two darker zones on the varixes and canal; aperture light orange to apricot, denticles and lirae white.

11 cm. Circumtropical; Barrow I., WA to Capricorn Group, Qld. Broader and more roughly sculptured than *C. pileare*. The denticles on the inner side of the outer lip are bifid and short, not extending deep into the aperture.

Cymatium comptum (Adams, 1854)

Like *C. vespaceum* but smaller, finely and crisply sculptured, with deeper suture and more prominent denticles inside the outer lip, upper two cords weakly bifid over the varix but not on the rest of the surface. Orange, brown or reddish, usually with a pale spiral band on the body whorl, often most evident where it crosses the varixes.

2.6 cm. Indo-West Pacific and both eastern and western Atlantic; eastern Qld. For an illustration see Springsteen & Leobrera (1986, *loc. cit.*, Pl. 31, fig. 3).

Cymatium exaratum (Reeve, 1844)

Pl. 42; fig. 1 a-b   Rather solid; spire high, protoconch usually missing, with 6-7 strongly shouldered adult whorls; suture deeply impressed; shoulder ramp steep; sculptured with weak axial folds and spiral cords, body whorl bearing four strong spiral ribs, the first and second bifid and nodulose, the nodules tending to connect with axial folds, entire surface with fine spiral cords crossed by axial threads; anterior canal of moderate length, straight or slightly sinuous, bent to the right; anal notch weak, outer lip with six thick inner nodules; columella erect, with one strong parietal ridge and several anterior folds. Fawn with tan and purple-brown blotches and weak tan axial lines; aperture white.

6 cm. Indo-West Pacific; Geographe Bay, WA to southern Qld. This is the species incorrectly figured as *C. tabulata* Menke by many Australian authors. The Australian form is the nominate subspecies, while *C. e. durbanense* Smith, 1899 comes from South Africa and *C. e. kiiense* Sowerby, 1915 comes from Japan. Synonym: *zimara* Iredale, 1929.

Cymatium intermedium (Pease, 1869)

Like *C. aquatile* and *C. pileare* but yellow-brown with darker bands, aperture orange-red, denticles white, columellar lirae white with black in the interspace; apertural denticles usually single and short, not descending into the aperture.

6 cm. Indo-West Pacific; recorded from Trinity Bay, north Qld. For illustrations see Kay (1979, *loc. cit.*, fig. 77 c; Beu & Kay (1988, *loc. cit.*, figs. 33-40).

Cymatium mundum (Gould, 1849)

Pl. 42; fig. 11   Small, thin for the genus, fusiform, varixes weak, spire pointed, whorls convex; sculptured with narrow spiral ribs and axial threads, axial ribs on the spire whorls forming small nodules where they cross the spirals, becoming obsolete on the body whorl except on the shoulder; posterior canal small but distinct, anterior canal moderately long, almost straight; outer lip with seven small denticles on the inner side; parietal wall with a small ridge beside the posterior canal, columellar callus narrow, crossed by more or less regular lirae. White; periostracum thin and yellow.

3 cm. Indo-West Pacific; Abrolhos, WA to Qld. Quite common in the intertidal zone. *Septa gemmata* (Reeve, 1844) is a name which has been used for this in Australian literature, but that is a different western Pacific species which appears to be present in north Qld, though rare. *S. gemmata* is smaller, has a taller spire, narrower spiral cords and narrower protoconch, and is of a pale tan colour.

Cymatium nicobaricum (Röding, 1798)

Pl. 41; fig. 9   Heavy and solid, varixes thick, spire high, early whorls with angulate shoulders; sculptured with heavy nodules and spiral ribs and cords, separated by narrow, deeply incised grooves, body whorl with 3-4 nodules between the varixes and six primary ribs, thickening where they ascend the last varix; posterior canal small but distinct, anterior canal short; outer lip with seven heavy nodules on its inner side, plus a matching row deeper within the aperture; parietal wall with a strong fold beside the posterior canal, followed by a series of rather straight lirae which cross the columellar callus. Grey, blotched or flecked with brown; aperture orange or yellow, darker near the lip and on the columella, denticles and lirae white.

9 cm. Indo-West Pacific; Rottnest I., WA to northern NSW. Also in the West Indies. Common in the intertidal zone.

Cymatium parthenopeum (von Salis, 1793)

Pl. 41; fig. 5 Broadly fusiform; varixes moderate, anterior canal almost straight; sculptured with broad, rounded, spiral ribs, minute axial striae, two ribs on the spire whorls, six on the body whorl plus many cords on the canal and fasciole, ribs sometimes weakly nodulose; last varix with broad nodules formed by terminal expansion of the ribs; outer lip with five double teeth and two single teeth; columella irregularly lirate. Light yellow-brown; lip teeth and columellar lirae white, interstices dark brown.

14 cm. Cosmopolitan in middle latitudes; northern NSW to Lancelin, WA. Japanese and Eastern pacific subspecies are *C. p. echo* Kuroda & Habe in Kira, 1961 and *C. p. keenae* (Beu, 1970) respectively. Synonym: *australasiae* Perry, 1811.

Cymatium pileare (Linnaeus, 1758)

Pl. 41; fig. 4 Elongate, fusiform; varixes moderate; whorls sculptured, flat spiral ribs and narrow cords in the interspaces, crossed by axial threads, ribs thicken where they cross the varixes, usually also with axial rows of low nodules; posterior canal a distinct U-shaped notch; anterior canal rather straight and upturned, outer lip with strong denticles, usually in seven pairs, extending into the aperture as strong, straight lirae; columellar callus heavily lirate, lirae irregular on the parietal wall, regular on the side of the canal. Cream, blue-grey, or fawn with brown axial streaks, and spiral bands; interior and columella scarlet to crimson, teeth and lirae white.

10 cm. Indo-West Pacific; Rottnest I., WA to southern Qld. Commonly found under stones and corals in the intertidal and shallow sublittoral zones. Beu & Kay (1988, *loc. cit.*) made a detailed comparison of this common species with *C. aquatile* and *C. intermedium*. The three species have a similar geographic range in the Indo-West Pacific. *C. pileare* is the largest and is readily distinguished by its crimson or scarlet colouring and prominent ridges within the aperture.

Cymatium vespaceum (Lamarck, 1822)

Pl. 42; fig. 3 a-b Very like *C. exaratum* but the axial folds are strong and project over the suture, spiral ribs narrow and not bifid, the last varix broad and flaring, anterior canal long and usually terminally narrow. White or pale tan or pink, with bluish blotches and tan spots and a prominent dark brown spiral band on the anterior canal; aperture white.

6 cm. Indo-West Pacific; Geographe Bay, WA to southern Qld.

▲ SUBGENUS **RANULARIA** Schumacher, 1817
Murex clavator (= *gutturnium* Röding)

Anterior canal long and more or less sinuous; spire low or very low, body whorl short; operculum nucleus near the centre of the columellar edge

Circumtropical. This is a distinctive group characterised by shells with low spires and long, more or less sinuous siphonal canals. Many authors regard it as a distinct genus. For a review of some species of the subgenus see Beu (1987, *loc. cit.*). Synonyms: *Retusum* Jousseaume, 1892 (*retusum* Lamarck = *oboesus* Perry); *Tritonocauda* Dall, 1904 (*caudatum* Gmelin).

Cymatium caudatum (Gmelin, 1791)

Pl. 42; fig. 6, 12 Thin; spire moderately high with a deeply channelled suture; body whorl inflated; anterior canal long, narrow and straight; sculptured with fine spiral and axial threads and narrow spiral ribs, two on early whorls, three on the penultimate, eight on the body whorl; ribs on the early whorls gemmate, those on the penultimate and body whorls with regular, small nodules; outer lip narrow and erect, with eight bifid marginal denticles and a matching row internally; parietal wall with 2-3 narrow spiral ridges or lirae, columellar callus thin, with several narrow lirae anteriorly. Fawn, varixes white; aperture, lip and columella white.

7 cm. Indo-West Pacific; Shark Bay, WA to central NSW. Usually taken by trawlers in deeper water. Beu & Cernohorsky (1986, *loc. cit.*) reviewed the complex history of this species and provided a long synonymy. It was first named as *Murex caudatum* by Gmelin, a different species to *Buccinum caudatum* Gmelin, 1791 which belongs to the subgenus *Linatella*. In placing both species in *Cymatium* a homonymy is avoided by replacing the latter with its next available name (see remarks on *Cymatium cutacea*). Iredale (1936) introduced the subspecies name *vulticula* for a NSW form with low spire and short canal.

Cymatium cynocephalum (Lamarck, 1816)

Pl. 42; fig. 5 Spire height variable, suture weakly channelled; sculptured with axial folds on the shoulders, wide primary spiral ribs which are nodulose on the spire whorls, nodules sometimes obsolete on the anterior ribs of the body whorl, sharp spiral threads present in the interspaces; outer lip with seven thick nodules along the inner margin; parietal fold weak, columellar callus reflected over the base, with only a few weak lirae anteriorly. Yellow-brown or light tan, varixes white; outer lip and nodules apricot, aperture apricot within, white near the lip; columella apricot, with a dark red-brown blotch, sometimes divided on the parietal wall.

7 cm. Tropical seas of the Atlantic and Indo-West Pacific; Shark Bay, WA to southern Qld. Beu & Cernohorsky (1986, *loc. cit.*) showed that there is a single species in the Indian, Pacific and Atlantic Oceans and clarified the confused nomenclature. Synonyms: *moricinctum* Reeve, 1844; *caribbaeum* Clench and Turner, 1957.

Cymatium gutturnium (Röding, 1798)

Pl. 42; fig. 9 a-b   Large, solid, with three varixes; spire rather short, protoconch papilliform, smooth, with about $2^{1}/_{2}$ convex whorls; first teleoconch whorl convex with axial and spiral cords, second and fourth with angulate shoulders bearing compressed and pointed nodulose and axial folds crossed by strong spiral cords; body whorl with six thick primary ribs crossed by axial cords and several axial folds; anterior canal long, sculptured with weakly nodulose spiral cords; outer lip thick and slightly recurved over the varix, bearing seven thick inner nodules descending as lirae into the aperture; parietal wall thickly calloused, the callus continuing onto the columella to form a shield over the neck of the body wall; three spiral lirae on the parietal wall and several weak ridges anteriorly on the columella. Cream or fawn, with brown lines on the axial folds, orange blotches on the varixes; lip and callus yellow-orange.

7 cm. Western Pacific; Qld. *C. encausticum* (Reeve, 1844) is a similar species but with lower axial folds and a much larger callus which usually spreads up to the suture of the penultimate whorl; it was recorded by Hedley from Qld but the locality needs confirmation.

Cymatium oblitum (Lewis & Beu, 1976)

Pl. 42; fig. 13   Solid, with rounded whorls and a single varix; spire low, suture impressed; sculpture of low, wide, rounded spiral cords, 5-6 on the spire whorls, about 20 on the body whorl and canal, five secondary cords in the interspaces, crossing low axial folds at the periphery, forming 2-3 rows of small nodules; outer lip narrowly reflected, with 9-10 inner nodules extending into the aperture, anal notch well defined; anterior canal long, upturned over a tiny umbilicus. Reddish brown, mottled with white; varix white but with red-brown blotches; aperture, lip and callus white.

6 cm. North West Cape, WA to Southern Qld.

Cymatium pyrum (Linnaeus, 1758)

Pl. 42; fig. 8   Exceptionally thick and heavily nodulose; spire moderately high; spire whorls with two spiral rows of flattened and pointed nodules, three on the penultimate whorl; body whorl with six spiral ribs and five axial folds connecting prominent nodules; whole surface finely spirally threaded, the threads crossed by axial striae making the surface minutely gemmate; anterior canal spirally corded, moderately long, sinuous; outer lip with two rows of seven thick denticles; parietal wall with a thick fold beside the posterior canal, callus irregularly lirate. Bright orange, sometimes with white blotches on the varixes; aperture white, apricot or orange near the lip and on the columellar callus, denticles and lirae white.

10 cm. Indo-West Pacific; Qld.

Cymatium sarcostomum (Reeve, 1844)

Pl. 42; fig. 7 a-b   Spire moderately high, stepped; spire whorls with weak, oblique axial folds on the angulate shoulders, sculptured with minutely granulose spiral threads, two spiral rows of sharp, flattened nodules and axial striae; body whorl with finely granulose axial and spiral cords and six spiral ribs, shoulder rib with five high, angulate nodules, others weakly nodulose; also with weak, narrow spiral cords on the long, narrow, sinuous anterior canal; outer lip with seven bifid marginal denticles extending into the aperture as thick folds; columellar callus smooth at the centre, with a strong parietal fold beside the posterior canal, and 6-7 strong lirae anteriorly. Light orange or tan, blotched with white; aperture apricot or pale orange, denticles and lirae paler.

8 cm. Indo-West Pacific; Dampier Arch., WA to Qld.

Cymatium sinense (Reeve, 1844)

Pl. 42; fig. 10   Relatively thin, with three varixes; spire tall, early teleoconch whorls delicately latticed by equal axial and spiral cords, on later whorls the spirals becoming dominant, with primary ribs on the body whorl, at least some bifid, plus many spiral cords, all crossing relatively weak axial costae making them nodulose; anterior canal long and sinuous; outer lip with seven thick inner nodules; parietal-columellar callus thin, heavily lirate. Exterior fawn; lip and callus white.

7 cm. Indo-West Pacific; Onslow, WA to central NSW. *C. s. arthuri* Beu, 1986, is a Red Sea subspecies.

▲ SUBGENUS SEPTA Perry, 1810
S. scarlatina Perry (= *Murex rubecula* Linnaeus)

Solid, fusiform; protoconch small, short; varixes thick, anterior canal short; columella strongly lirate; with strong spiral ribs.

Circumtropical. There are eight living species. See Beu (1986, *loc. cit.*) for a comprehensive review.

Cymatium hepaticum (Röding, 1798)

Pl. 42; fig. 16 a-b   Sculpture as in *C. rubeculum*; anterior canal relatively long and straight; outer lip denticles short and not extending into the aperture. Orange or yellow-brown, with dark brown stripes in the interstices, and two white ribs on the varixes; lip denticles touched with orange, columella orange between the white lirae.

6.5 cm. Rowley Shoals and Scott Reef, WA to Qld. A moderately common intertidal species on coral reefs.

Cymatium occidentale (Mörch, 1877)

Pl. 42; fig. 14 a-b   Sculptured with about nine spiral ribs, both ribs and interspaces crossed by crowded axial cords; outer lip with eight strong inner denticles; varixes wide; aperture small. Uniformly yellow-brown or faintly reddish brown, with two white splashes on the varixes.

6 cm. Western Atlantic and Indo-West Pacific; Marmion Marine Park, WA to northern NSW. Synonyms: *blacketi* Iredale, 1936; *beui* Garcia-Talavera, 1985.

Cymatium rubeculum (Linnaeus, 1758)

Pl. 42; fig. 15 a-b Sculptured with strong, beaded spiral ribs, 7-9 on the body whorl, plus several beaded cords on the canal and fine axial threads in the interspaces; outer lip with 8-10 denticles on the inner side. Colour variable, usually dark crimson, sometimes brown or orange, with a pale spiral band at the centre of the body whorl and two white or yellow splashes on the varixes; aperture white, columella crimson, lirae white.

4.5 cm. Indo-West Pacific; Dampier Arch., WA to Qld. The most common species of the subgenus. Intertidal and shallow sublittoral zones.

▲ SUBGENUS **TURRITRITON** Dall, 1904
Cymatium gibbosum Broderip

Stout; protoconch multispiral conical, with smooth whorls; anterior canal moderate to long, attenuate, bent to the right; primary ribs with three riblets; periostracum sparsely bristled.

Circumtropical. *Cabestanimorpha* Iredale, 1936 and the several species associated with it, have previously been synonymised here but are removed to *Monoplex* on the advice of Beu (pers. comm. April, 1992). This leaves only two species in the subgenus, one from the Eastern Pacific and Atlantic (*gibbosum* Broderip, 1833) and the other from the Indo-West Pacific. Synonyms: *Tritoniscus* Dall, 1904 (*gibbosum* Broderip); *Particymatium* Iredale, 1936 (*strangei* Adams & Angas = *labiosum* Wood).

Cymatium labiosum (Wood, 1828)

Pl. 42; fig. 4 a-b Solid and broad; protoconch conical, with four smooth convex whorls; teleoconch with four whorls sculptured with weak axial folds that are strongest on the shoulder slope, low spiral ribs that are weakly or strongly nodulose, and numerous finely gemmate spiral cords which cover both ribs and interspaces; last varix wide; anterior canal rather short, upturned; outer lip with six inner denticles; a small parietal ridge, columella lirate anteriorly. Fawn or yellow, sometimes with brown spots on the varixes.

4 cm. Indo-West Pacific; Albany, WA to central NSW; also found in the tropical Atlantic. This is an extraordinarily widespread species with variable sculpture. Synonym: *strangei* Adams & Angas, 1864.

▲ SUBGENUS **RETICUTRITON** Habe & Kosuge, 1966
Triton pfeifferianum Reeve

Tall-spired with a long anterior canal and finely cancellate sculpture.

Indo-West Pacific. There are two living species, one of which is moderately common in northern Australia.

Cymatium pfeifferianum (Reeve, 1844)

Pl. 42; fig. 2 Thin, elongately fusiform; whorls rounded, sculptured with many finely gemmate spiral cords, crossed by axial threads; varixes narrow and angulate, crossed by the spiral cords; anterior canal very long, slightly upturned and turned right; outer lip finely dentate along the inner margin; columellar callus thin, narrow, evenly lirate. White or fawn, with darker spiral bands below the suture, at the centre and anteriorly, aperture white.

7 cm. Indo-West Pacific; Kimberley, WA to north Qld. Trawled in moderate depths.

▲ GENUS **SASSIA** Bellardi, 1873
Triton apenninicum Sassi

Protoconch small to moderately large, turbiniform, sometimes dehiscent; teleoconch sculpture of simple spiral cords and axial costae forming nodules at the intersections and fine spiral threads; aperture nearly circular, outer lip flared and nodulose along the inner margin; parietal tubercle present; anterior canal narrow, weakly twisted open.

Cosmopolitan. *Sassia* is the most diverse, the most widespread and longest-ranging genus of the Ranellidae (Beu, 1988, *loc. cit.*, p. 85). It originated in the Cretaceous. For a discussion of the taxonomy of this genus and species synonymies see Beu (1987; 1988, *loc. cit.*).

▲ SUBGENUS **SASSIA** s.s.

Protoconch moderately large turbiniform, equidimensional, whorls rounded; anterior canal short to moderately long, open.

Cosmopolitan. There are at least four Australian species. Synonyms: *Semiranella* Gregorio, 1880 (*apenninicum* Sassi); *Monocirsus* Cossmann, 1889 (*carinulatus* Cossmann); *Cymatona* Iredale, 1929 (*kampyla* Watson); *Charoniella* Powell & Bartrum, 1929 (*arthritica* Powell & Bartrum); *Austrosassia* Finlay, 1931 (*parkinsoniana* Perry); *Phanozesta* Iredale, 1936 (*remensa* Iredale); *Proxicharonia* Powell, 1938 (replacement name for *Charoniella*).

Sassia apenninica (Sassi, 1827)

Pl. 40; fig. 19 Tall-spired; protoconch of $3\frac{1}{2}$ cancellate whorls; teleoconch of about six whorls, slightly distorted and angulate at the periphery, varixes at every $\frac{2}{3}$ whorl; sculptured with fine spiral striae and 3-4 spiral cords, the two at the periphery bearing prominent nodules and weak axial costae connecting the nodules; anal canal deep and well formed; anterior canal moderate to long and slightly upturned; outer lip thin, reflected, with about six strong inner denticles; inner lip forming a raised shield (unusually well developed in the illustrated specimen). Pale fawn, with darker spiral lines and blotches, nodules yellow-white.

4 cm. Temperate waters of southern Japan, South Africa, southern Australia and NZ; dredged in 200-400 m off NSW. The nominate form of this species is a European Miocene-Pliocene fossil. There are two living subspecies, *S. a. remensa* Iredale, 1936 from south-eastern Australia, NZ, and southern Japan (= *semitorta* Kuroda and Habe, 1961), and *S. a. nassariformis* (Sowerby, 1902) from South Africa.

Beu (1978) described another species from the Kermadec Is. as *S. marshalli*, distinguished from *S. apeninica* by the lack of spiral cords, narrow axial costae, larger aperture, and more convex whorls; since then *S. marshalli* has also been recorded off south-eastern Australia.

Sassia kampyla (Watson, 1885)

Pl. 40; fig. 11 a-b Elongate, spire tall, suture impressed, whorls convex; protoconch of three smooth convex whorls; teleoconch of six whorls, sculptured with prominent axial folds bearing small, pointed nodules where the primary spiral cords cross, other sculpture of minute crowded spiral threads; varixes small; outer lip sharp-edged, smooth within, anal notch lacking; columella raised and curved, smooth; anterior canal long, narrow, upturned at an angle of about 45°. Fawn.

5 cm. NSW and NZ. Dredged on the continental slope. There is a distinct living subspecies at Macquarie I., known as *S. k. tomlini* (Powell, 1955).

Sassia parkinsonia (Perry, 1811)

Pl. 40; fig. 10 a-b Fusiform, solid and chunky; spire moderately high; protoconch of three smooth, convex whorls; teleoconch with large, somewhat irregular, axially elongate tubercles, the whole surface covered with faintly beaded, crowded spiral threads; varixes low but strong; outer lip sharp-edged, smooth or with 4-5 denticles within; collumellar lip raised, reflected, smooth, except for a parietal nodule which defines the anal notch; anterior canal short, upturned. Yellow-brown to orange-brown, spiral threads of alternating white and brown spots; aperture glossy white.

4.5 cm. Northern NSW, eastern Vic. and Tas. Common in the shallow sublittoral. Synonym: *basilica* Iredale, 1924.

Sassia ponderi Beu, 1987

Spire, tall, gradate; varixes high, thin, wide, located at each $^2/_3$ turn of the whorls; protoconch of four tall, spirally striate, weakly convex whorls; body whorl short, base contracting rapidly to a moderately long anterior canal, spiral sculpture of wide, flat, smooth cords of varying size, axial sculpture of short folds, rendering the shoulder cords prominently nodulose; aperture oval, with a raised rim and nine low ridges on the inner side of the outer lip, columellar lip smooth except for a very low parietal ridge. Yellow-brown to grey-brown, with indistinct pale red-brown areas between nodules, on the cords, and at the centre of each varix.

1.4 cm. Southern Qld to central NSW. Trawled on the outer continental shelf. It is the smallest member of the subgenus. The tall protoconch is unusual for the genus.

▲ SUBGENUS **AUSTROTRITON** Cossmann, 1903
Triton radialis Tate

Protoconch lacking or, when present, irregularly coiled, usually with a terminal spike, apparently chitinous; larval development apparently direct.

Eastern and southern Australia. This endemic subgenus has a fossil record confined to southern Australia. Murray (in Watson, 1971, *A preliminary account of the benthic flora and fauna of southwestern Western port, Victoria*. Underwater Research Group of Victoria, Melbourne) described the direct larval development of *S. subdistorta* and how the larval shell is lost after hatching. Synonyms: *Negyrina* Iredale, 1929 (*subdistorta* Lamarck); *Charoniella* Thiele, 1929 (*subdistorta* Lamarck).

Sassia bassi (Angas, 1869)

Pl. 40; fig. 16 Ovate; spire moderately short, body whorl broad, anterior canal short; adult whorls sculptured with numerous spiral, faintly gemmate threads and a row of prominent, rounded nodules at the subangulate shoulders and 1-2 rows of smaller nodules on the posterior part of the body whorl; outer lip with about 13 small denticles along the inner side; parietal ridge defining a deep anal notch, columellar callus weak and adherent, bearing short denticles anteriorly. Cream or fawn, with red-brown spots and blotches; aperture slightly violet.

3 cm. Bass St. to Gulf of St Vincent, SA.

Sassia epitrema (Tenison Woods, 1877)

Ovate, slightly dorso-ventrally compressed; protoconch lacking; suture deeply excavate, shoulders sharply angulate and weakly nodulose, sides of whorls convex; varixes not aligning along the spire, slightly hooked at the posterior ends, whorls with thin spiral ribs of varying size crossing the varixes; anterior canal short, columella-parietal shield narrow, adherent, smooth; anterior fasciole stout, bent. Cream, ribs and varixes yellow-orange.

2 cm. Southern NSW and Bass St. This rare and unusual shell lacks a protoconch and apparently has direct development, which is the reason for its placement in *Sassia (Austrotriton)*.

Sassia garrardi (Beu, 1970)

Pl. 40; fig. 18   Stoutly fusiform; protoconch deciduous; spire moderately tall, with heavily nodulose teleoconch whorls, nodules prominent and pointed, one spiral row on spire whorls, two rows at the centre of the body whorl; subsutural ramp steep and slightly concave, other sculpture of low, sometimes lightly beaded spiral threads; outer lip smooth on the inner side except for a few small denticles (up to nine); columellar lip smooth, reflected over a narrow umbilical chink, with a small parietal pad, and several sharp anterior plicae; anterior canal moderately long, slightly upturned. Fawn to red-brown.

6 cm. Capricorn Channel, Qld to NSW. Dredged on the outer part of the continental shelf and slope. The Capricorn Channel specimens are twice the size of that of the NSW form.

Sassia mimetica (Tate, 1893)

Pl. 40; fig. 9 a-b   Ovate; spire low and stepped, early whorls sharply keeled at the periphery, with concave subsutural ramps; body whorl inflated, shoulder subangulate, heavily nodulose, steeply constricted at the base; sculpture of many narrow spiral cords; outer lip stoutly varixed, with about 11 weak inner nodules, parietal nodule weak, columellar callus narrow, adherent; siphonal canal rather narrow, long, attenuate, straight or slightly turned at the end, open only by a narrow slit. White.

2.5 cm. SA.

Sassia petulans (Hedley & May, 1908)

Pl. 40; fig. 12   Fusiform; protoconch lacking; spire very tall, whorls convex, slightly distorted, with low rounded varixes every $^2/_3$ whorl; sculptured with crowded spiral threads, sometimes obscurely nodulose at the shoulders; aperture oval, outer lip with about nine weak inner nodules; columellar callus very weak or lacking, columella with weak lirae anteriorly; anterior canal short, open. Cream, with obscure reddish blotches, especially on the varixes.

3.5 cm. Bass St. to SA. This may be only a deepwater form of *S. subdistorta*.

Sassia subdistorta (Lamarck, 1822)

Pl. 40; fig. 17 a-c   Solid, fusiform; spire high, with impressed suture and somewhat distorted convex whorls; sculptured with nodulose primary spiral ribs and numerous secondary spiral threads; outer lip sharp-edged, with a strong posterior nodule bordering the anal notch, followed by a series of weak nodules along the inner side; parietal ridge moderate, columella raised, smooth except for a few short anterior plicae; anterior canal short. Cream or fawn, mottled with red-brown; aperture white.

7 cm. NSW to Esperance, WA. Quite common in shallow water. Synonym: *delecta* Cotton, 1945.

▲ SUBGENUS **CYMATIELLA** Iredale, 1924
Triton quoyi Reeve (= *verrucosa* Reeve)

Small, tall-spired; protoconch small; sculpture usually finely cancellate; anterior canal short.

Southern Australia. Beu (1988, *Journal Malacological Society* Australia 9: 1-9) noted that most species of *Sassia (Cymatiella)* have subdued sculpture, his new species *S. (C.) ansonae* being the exception. The group has a long fossil record dating from the Eocene in Europe, the Oligocene in America, and the Miocene in Australia. Synonym: *Vernotriton* Iredale, 1936 (*pumilio* Hedley).

Sassia ansonae Beu, 1988

Conical, spire tall, stepped; protoconch solid, very small, wide, of $1^1/_3$ almost smooth, loosely coiled whorls; teleoconch whorls prominently keeled at the periphery, subsutural ramps concave, crossed by weak axial folds; with two prominent spiral cords at the periphery, the upper cord forming the keel, plus about six small spiral threads on the base and canal; axial sculpture of widely spaced costae forming nodules where they cross the spiral cords; varixes low, narrow, at each $^2/_3$ turn of the whorls; anterior canal short, open; anterior fasciole lacking; parietal wall not calloused, columella reflected, adherent, smooth. Translucent, fawn-cream, with narrow red-brown spiral lines.

1.4 cm. Known only from beach drift in the vicinity of Margaret River, WA. This is the most elaborately sculptured of all the Ranellidae, according to Beu. Its few-whorled protoconch indicates that it may have lecithotropic development, which may account for its restricted distribution.

Sassia columnaria (Hedley & May, 1908)

Spire tall, whorls convex with deeply impressed suture; protoconch relatively large; varixes weak, located slightly more than every half whorl; sculptured with strong axial costae crossed by spiral cords; aperture circular, outer lip thin, with about five weak inner nodules, anal notch lacking; anterior canal short, open, inclined to the left. Cream.

1.5 cm. Bass St. to SA. This species is not common but its tall spire, relatively large protoconch and strong sculpture make it easily recognisable.

Sassia eburnea (Reeve, 1844)

Pl. 40; fig. 13 Fusiform, spire tall with deeply impressed suture and convex whorls, anterior canal short and straight, varixes low; sculptured with rather weakly nodulose spiral ribs, 13 on the body whorl, with the shoulder nodules aligned on low axial folds; outer lip with 6 or 7 inner denticles; parietal callus and ridge lacking, columellar callus narrow, bearing 3 strong anterior denticles. White, apex often pink.

2 cm. Bass St. The finer sculpture distinguishes this species from *S. verrucosa* and *S. sexcostata*. Synonym: *lesueuri* Iredale, 1929.

Sassia pumilio (Hedley, 1903)

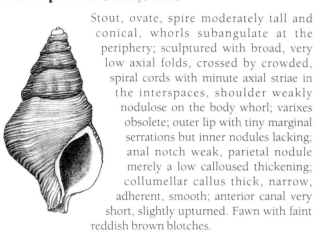

Stout, ovate, spire moderately tall and conical, whorls subangulate at the periphery; sculptured with broad, very low axial folds, crossed by crowded, spiral cords with minute axial striae in the interspaces, shoulder weakly nodulose on the body whorl; varixes obsolete; outer lip with tiny marginal serrations but inner nodules lacking; anal notch weak, parietal nodule merely a low calloused thickening; collumellar callus thick, narrow, adherent, smooth; anterior canal very short, slightly upturned. Fawn with faint reddish brown blotches.

2.5 cm. Southern Qld and northern NSW. Dredged to 50 m.

Sassia sexcostata (Tate, 1888)

Pl. 40; fig. 15 a-b Elongate, spire high, with impressed suture; penultimate and body whorls slightly distorted; sculptured with spiral rows of regular nodules and fine threads in the interspaces, the nodules also aligned on axial folds, 4 spiral rows on the spire whorls, 7 on the body whorl plus several spiral cords on the canal; parietal callus ridge lacking, columella short, narrow, with small anterior denticles; anterior canal short and straight. Brown; aperture white; apex white.

3 cm. Bass St. to SA. Synonym: *gaimardi* Iredale, 1929.

Sassia verrucosa (Reeve, 1844)

Pl. 40; fig. 14 Fusiform; body whorl inflated, slightly distorted; spire tall, with impressed suture and convex whorls, anterior canal short, straight, varixes low; body whorl sculptured with 6 spiral ribs, plus several cords on the canal, axial folds which form weak nodules where they cross the ribs, and minute spiral threads throughout; outer lip erect, with 5 tiny denticles within; parietal callus lacking; columella thin, narrow and short, bearing 3 or 4 sharp anterior plicae. Cream or white beneath a thin brown periostracum.

2.5 cm. NSW to SA and possibly southern WA. Very common. Synonym: *peroniana* Iredale, 1929.

FAMILY **TONNIDAE**

Tun Shells

This family takes its name from the thin and globose shells. In most species the spire is only moderately elevated and the whorls are inflated and sculptured with spiral ribs. There are no varixes, axial sculpture or operculum. The bodies of tuns are large and attractively coloured. The head tentacles are long, the proboscis long and extensible, and there is a very long anterior siphon. Beu (1980, *Records Australian Museum* 33 (5): 248-324) has given a brief summary of the anatomy.

The Tonnidae is a tropical and subtropical group. Most species inhabit relatively shallow water, the exception being *Eudolium*. During the day tuns bury themselves in sandy substrates. At night they emerge to hunt their prey. They are fast-moving and make a spectacular sight as they crawl across the sea bed. Holothurians are the prey which they swallow whole (Morton, 1991, *J. Molluscan Studies* 57: 11-19).

Egg masses consist of wide gelatinous ribbons containing small, transparent eggs. The veligers hatch as planktotrophic larvae and there is a very long development time. Two Atlantic species have been recorded as having a larval life of six to eight months. The initial larval shell is uncalcified but after settlement the animal fills it with calcium carbonate so that it becomes solid in adults.

Marshall has recently revised the genus *Eudolium* (1992, *Nautilus* 106 (1): 24-38) but otherwise there has been no review of the taxonomy of this family in Australia since that of Hedley (1919, *Records Australian Museum* 12 (11): 329-336) and several additions have been made to our fauna since then.

▲ GENUS **EUDOLIUM** Dall, 1889
Dolium crosseanum Monterosato

Spire moderately high; whorls sculptured with strong spiral cords; outer lip reflected; anterior canal produced, tapered; umbilicus lacking; parietal shield weak; protoconch conical but with large whorls.

Cosmopolitan. A deep water genus with origins in the Oligocene. There are three living species, two of them discovered in Australian waters only recently. *Eudolium* was a replacement name for *Doliopsis* Monterosato, 1872 (not Vogt, 1852). Synonyms: *Galeodolium* Sacco, 1891; ?*Tuberculodolium* Sacco, 1891; ?*Simplicodolium* Sacco, 1891.

Eudolium bairdii (Verrill & Smith, 1881)

Pl. 39; fig. 3   Protoconch of about five convex whorls, sculptured with three narrow, spinose spiral threads; teleoconch whorls with numerous, widely spaced, rounded primary spiral cords and microscopic axial striae in the interspaces, the latter tending to disappear on the body whorl; 2-3 ribs on the shoulder are finely nodulose; outer lip reflexed, beaded on its upper side where the spiral cords meet, weakly toothed on the inner side; columella with a thick, spirally plicate basal callus; parietal area with 1-3 elongate denticles. White, major spiral ribs yellowish to reddish brown; protoconch deep reddish brown.

6 cm. Atlantic, Mediterranean, New Zealand and Indo-West Pacific; Rowley Shoals, WA to Port Stephens, NSW. Trawled to 500 m. The remarkable distribution of this species became evident with the revision of the genus by Marshall (1992). Synonyms: *crosseanum* var. *solidior* Dautzenberg & Fischer, 1906; *lineata* Schepman, 1909; *inflatum* Kuroda & Habe, 1952; *kuroharai* Azuma, 1960; *solidor* Piani, 1977.

Eudolium pyriforme (Sowerby, 1914)

Pl. 39; fig. 6   Rather high-spired; protoconch of about five convex whorls; teleoconch whorls sculptured with close, rounded spiral ribs alternating with fine spiral threads, axial striae obsolete; outer lip weakly thickened and reflexed, sometimes with pairs of weak denticles on its inclined inner side. White to buff, maculated and suffused with pale tan which sometimes forms indistinct axial bands; outer lip brown or with a pinkish flush.

8 cm. Indo-West Pacific; Kimberley, WA to Bateman Bay, NSW. Also found in northern New Zealand. Trawled in very deep water.

▲ GENUS **MALEA** Valenciennes, 1833
M. latilabris Valenciennes (= *Cassis ringens* Swainson)

Globose, whorls moderately thick; outer lip thick, reflexed, toothed; anterior canal slightly produced, anal canal notch present; columella calloused, with several plaits; not umbilicate.

Indo-West Pacific. Synonym: *Quimalea* Iredale, 1929 (*pomum* Linnaeus).

Malea pomum (Linnaeus, 1758)

Pl. 39; fig. 12 a-c   Spire with a nucleus of three smooth whorls and weakly incised suture; postnuclear whorls four, sculptured with strong, rounded spiral ribs (about 12 on the body whorl) separated by wide, shallow grooves; outer lip with about 10 strong teeth on its inner margin, those at the anterior end continuing across the lip to form outward-pointing spines. Fawn or cream-orange with large white spots on the ribs; interior yellow, columella and outer lip white or cream.

6 cm. Indo-West Pacific; Abrolhos, WA to southern Qld. WA specimens tend to be very small (3-4 cm). See p. 368.

▲ GENUS **TONNA** Brünnich, 1772
Buccinum galea Linnaeus

Globose to elongate-ovate; whorls thin and inflated; outer lip thin and simple, anterior canal a simple U-shaped notch; usually umbilicate; sculpture of spiral ribs or cords.

Cosmopolitan in tropical and warm temperate seas. Synonyms: *Cadus* Röding, 1798 (*perdix* Linnaeus); *Dolium* Lamarck, 1801 (*galea* Linnaeus); *Cadium* Link, 1807 (*perdix* Linaeus); *Perdix* Montfort, 1810 (*reticulatus* Montfort = *perdix* Linnaeus); *Macgillivrayia* Forbes, 1852 (*pelagica* Forbes); *Foratidolium* Rovereto, 1899 (replacement for *Perdix* Montfort); *Parvitonna* Iredale, 1931 (*perselecta* Iredale = *cumingi* Reeve). Further study is needed to determine the species of this genus in Australia.

Tonna allium (Dillwyn, 1817)

Pl. 39; fig. 2   Whorls with strong, rounded, widely spaced, rounded spiral ribs, about 13 on the body whorl; outer lip slightly reflexed, crenulate, with about 28 inner denticles which tend to be paired; columellar callus thin, reflexed over the umbilicus. Cream or fawn, sometimes a few ribs are brown.

10 cm. Indo-West Pacific; Shark Bay, WA to northern NSW. Another species on the Qld list, *T. ampullacea* Philippi, 1845 is very like *T. allium* but is said to have riblets in the interspaces. Synonym: *costata* Menke, 1828.

Tonna cepa (Röding, 1798)

Pl. 39; fig. 7   Thin; spire with three smooth nuclear whorls and widely and deeply channelled suture; postnuclear whorls (about five) sculptured with wide low spiral ribs (about 17 on the body whorl) separated by narrow grooves; posterior ribs on the body whorl flat, anterior ribs rounded; columella vertical, almost straight, reflected over a deep umbilicus; anterior fasciole broad but low; anterior canal notch wide, shallow, oblique. Light yellow-brown, darker on the ribs than in the grooves, with brown and white blotches at the sutures; deep interior yellow becoming white near the margin.

12 cm. Indo-West Pacific; Abrolhos, WA to southern Qld. This species has been known as *T. canaliculata* (Linnaeus, 1758) but Cernohorsky has shown that is a *nomen dubium*.

Tonna chinensis (Dillwyn, 1817)

Pl. 39; figs. 4, 10   Spire with a nucleus of three smooth whorls and shallow suture; teleoconch with four whorls sculptured with strong rounded spiral ribs (about 17 on body whorl) separated by grooves slightly narrower than the ribs; anterior fasciole wide and raised; columella thin, reflected over a wide umbilicus, anterior canal notch deep, oblique. Light yellow-brown or cream with white dashes and brown spots on the ribs, interior yellow-brown becoming white near the lip.

6 cm. Indo-West Pacific; Cape Leeuwin, WA to NSW. Specimens of this species may be misidentified as immature *T. variegata* but they are consistently smaller for any given number of whorls. Common in shallow water and sandy pockets of intertidal reefs.

Tonna cumingii (Reeve, 1849)

Pl. 39; fig. 11 Globose, spire low; sculptured with flat, close-set spiral cords (about 21 on the body whorl) and threads in some interspaces; outer lip fluted, little thickened, weakly toothed; anterior fasciole thick and twisted; columellar callus small, reflected over a deep umbilicus. Cream or fawn, with diffuse brown spots and axial streaks, sometimes with brown chevron marks or white spots; outer lip dark brown.

11 cm. Indo-West Pacific; Qld. Also reported from New Zealand. Synonym: *perselecta* Iredale, 1931.

Tonna perdix (Linnaeus, 1758)

Pl. 39; fig. 13 Thin; spire high, with incised suture and a nucleus of two smooth whorls; five teleoconch whorls sculptured with about 18 broad, low, rounded spiral ribs; anterior canal notch wide, very shallow and oblique, anterior fasciole broad but low; columellar margin gently curved, columellar callus thin and reflected over a deep umbilicus. Light brown with white lines in the grooves between the ribs and numerous crescent-shaped white markings on the ribs; deep interior yellow-brown becoming white near the lip, columellar callus white.

14 cm. Indo-West Pacific; Fremantle, WA to southern Qld.

Tonna sulcosa (Born, 1778)

Pl. 39; fig. 8 Suture impressed; sculptured with flat spiral cords (about 20 on the body whorl); interspaces on the early whorls axially striate, smooth on the body whorl; outer lip inclined, weakly crenulate, thickened, with 15-17 pairs of denticles along its innter side; columella with a thin callus glaze, strongly twisted. White, with 3-4 brown bands; apex purplish.

12 cm. Indo-West Pacific; Qld. The flatter, more numerous ribs, and prominent banding distinguish this from *T. allium*.

Tonna tessellata (Lamarck, 1816)

Pl. 39; fig. 1 a-b Spire with a nucleus of two smooth whorls; teleoconch whorls with broad rounded spiral ribs (about 14 on the body whorl) interspaces wide and smooth; outer lip slightly reflected, crenulate; outer wall of aperture deeply spirally grooved by impressions of the external ribs; columellar callus reflected, forming a shield over the umbilicus; umbilicus small but deep. Fawn or blue-grey between the ribs, white or pale fawn with large squarish brown blotches, interior brown.

10 cm. Indo-West Pacific; Dampier Archipelago, WA to eastern Qld. The slightly thickened and reflexed outer lip is unusual in the genus but the species fits here better than in *Malea* or *Eudolium*. *T. dolium* (Linnaeus, 1758) is very similar but is larger and thinner, has a twisted columella, lacks a parietal shield, and the outer lip is fluted but not thickened. This latter species is illustrated by a Thai specimen (Pl. 39, fig. 5) because it might be expected to occur in northern Australia although I have found no record of it.

Tonna tetracotula Hedley, 1919

Thin, spire rather high; whorls with strong spiral ribs and small cords in the spaces between the ribs on the shoulders; otherwise very like *T. variegata*. White or pale orange, sometimes with light brown spiral bands.

22 cm. NSW and New Zealand.

Tonna variegata (Lamarck, 1822)

Pl. 39; fig. 9 Thin; apex of three smooth whorls; five teleoconch whorls, all strongly sculptured with broad, rounded, spiral ribs (about 17 on the body whorl) and smaller interstitial cords on the posterior parts of the whorls; columella reflected over a wide umbilicus, anterior fasciole thick. Yellow, cream, or light brown, with at least some ribs white or pale cream and sometimes brown spots.

20 cm. Southern NSW to Shark Bay, WA. Also found in New Zealand. The typical form of this species was described from WA. The form from south-eastern Australia and New Zealand was named *T. cerevisina* Hedley, 1919; it tends to lack interstitial cords on the posterior parts of the whorls and may be merely a geographic variant, or at best a subspecies of *T. variegata*.

SUBORDER **HETEROPODA**

Heteropods

The heteropods and the pteropods are two groups of gastropods of very different origin which have adapted to pelagic life in the mid oceans. The pteropods are opisthobranchs and will receive no further attention here. The heteropods are prosobranchs and include three families within the superfamily Carinarioidea. Planktonic molluscs are not the subject of this book and these beautiful creatures are mentioned here in passing simply for completeness.

Unlike the pteropods, heteropods have separate sexes and simple shells. Because of their pelagic life style most species are widespread throughout the world's oceans. The families are:

ATLANTIDAE

Small snails with fragile, sometimes transparent, keeled, discoidal shells. Common genera in Australian waters are: *Atlanta* Lesueur, 1817, with the species *inflata* Eydoux & Souleyet, 1852, *peronii* Lesueur, 1817; *rosea* Eydoux & Souleyet, 1852 and *turriculata* Orbigny, 1836; *Oxygyrus* Benson, 1837 with the species *keraudrenii* Lesueur, 1817.

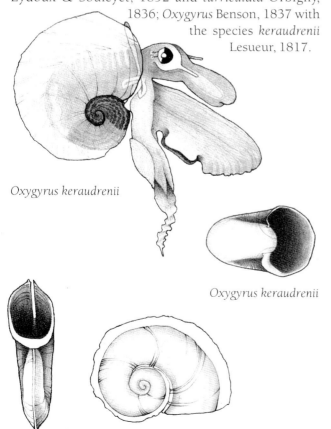

Oxygyrus keraudrenii

Oxygyrus keraudrenii

Atlanta peronii

CARINARIIDAE

Slug-like creatures with the foot modified to form a muscular paddle and a small, cap-shaped shell similar to that of *Capulus*. There are two genera, *Carinaria* Lamarck, 1801 and *Cardiapodia* Orbigny, 1835. *Carinaria australis* Quoy & Gaimard, 1833 is a common species in the ocean off south-eastern Australia. *Carinaria lamarcki* Péron and Lesueur, 1810, is a cosmopolitan species, probably present in the warm waters off northern Australia.

Carinaria australis

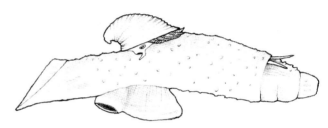

Carinaria lamarcki (redrawn from the original description).

FIROLIDAE

Finned, slug-like creatures without a shell. Genera found in Australian waters are: *Firoloida* Lesueur, 1817, with the species *desmaresti* Lesueur, 1817 and *kowalewskyi* Vassiere, 1904; *Pterotrachea* Forskal, 1775, with the species *mutabilis* Tesch, 1906.

SUPERFAMILY
TRIPHOROIDEA

TWO families belong in this group of small, sponge-eating creepers. Their taxonomy has been problematical for many years. Grouping them together in this way follows the classification of Ponder & Warén (1988).

In this account representatives of each of the Australian genera are illustrated, but readers will surely have trouble identifying specimens. A key to the genera, based on the one published by Laseron (1956), is given in the hope that it may assist.

FAMILY **CERITHIOPSIDAE**

Cerithiopsids

These small creepers have many of the features of the Cerithiidae, where they have sometimes been classified, but their anatomy is very similar to that of triphoras.

Cerithiopsids have long, dextral, many-whorled shells with a small aperture and a straight anterior canal. A posterior canal or sinus is lacking. The protoconch may have few or many whorls but apparently the first whorl is always tilted at an angle. The teleoconch whorls usually bear spiral ribs which may be crossed by axial ribs with nodules at the intersections.

Cerithiopsids feed on the tissues of sponges. Females lay their eggs on the host sponges and, in the few studied species, the larvae hatch as planktonic veligers.

There are very many described Australian species, and undoubtedly many more await discovery, especially in the west where the family has been hardly studied at all. Cotton (1951) reviewed the SA species described to that time, and Laseron gave accounts of the NSW species (1951) and NT and Queensland species (1956). Unfortunately these authors were both chronic "splitters" and based their work solely on shell characters. Many of their generic names may be shown to be synonyms of older names when an anatomical review is done.

References

Cotton, B. (1951). Australian Recent and Tertiary Mollusca, family Cerithiopsidae. *Records SA Museum* **9**: 383-395.

Laseron, C. F. (1951). Revision of the NSW Cerithiopsidae. *Australian Zoologist* **11**: 351-368.

Laseron, C. F. (1956). The family Cerithiopsidae (Mollusca) from the Solanderian and Dampierian zoogeographical provinces. *Australian Journal Marine and Freshwater Research* **7**: 151-182.

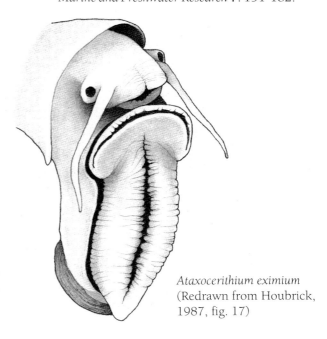

Ataxocerithium eximium
(Redrawn from Houbrick, 1987, fig. 17)

Key to the Australian Genera of the Cerithiopsidae

(Modified from Laseron, 1956)

1   Sculpture both spiral and axial..........2
    Sculpture predominantly spiral **10**
    Sculpture predominantly axial..........**Altispecula**

2   Spire carrot-shaped (inverted) ....**Conciliopsis**
    Spire short, convex, inflated **3**
    Spire elongate, narrow......................**5**

3   3 rows of nodules per whorl..............**Joculator**
    2 rows of nodules per whorl..............**4**

4   Protoconch inflated, 1½ whorls........**Potenatomus**
    Protoconch not inflated,
    multispiral ....**Horologica**

5   Spire turreted **6**
    Spire not turreted **7**

6   Protoconch smooth, slender,
    tapering ....**Clathropsis**
    Protoconch reticulate ....**Zaclys**
    Protoconch glassy ....**Ataxocerithium**

7   Spire subcylindrical, suture
    deep **8**
    Spire, whorls flat, suture shallow **9**

8   Protoconch smooth, slender,
    tapering ....**Synthopsis**
    Protoconch, inflated ....**Specula**

9   Protoconch smooth, slender,
    tapering ....**Tubercliopsis**
    Protoconch axially striate, the
    first whorl flat-topped ....**Socienna**

10 Teleoconch whorls with 4-5
    spirals..**Pilaflexis**
    Teleoconch whorls with 3 spirals **11**

11 Inflated, sides convex ....**Parasella**
    Tall and flat-sided............................**12**

12 Anterior canal extended, reflexed .....**Binda**
    Anterior canal short, wide **13**

13 Protoconch mamillate, rounded ....**Seila**
    Protoconch mamillate but with
    terminal spike ....**Euseila**

▲ GENUS **ATAXOCERITHIUM** Tate, 1894

Cerithium serotinum Adams

Elongate, turreted, thin-shelled; protoconch whorls glassy; varixes lacking, delicately sculptured with both spiral and axial cords; posterior canal lacking, anterior canal short, oblique, tubular; aperture round, outer lip expanded and flaring, columellar lip flanged and elevated, both lips tending to form a high rim encircling the round aperture; anterior end of the outer lip covering the canal.

Indo-West Pacific and Southern Australian Region. These animals are found alive on sponges and probably feed on them. There are about seven available names for living Australian species in this group but a revision is needed. Richard Houbrick (1987, *Nautilus* 101 (4): 155-161) has settled the question of the family placement of this genus by showing that the species are cerithiopsids and not cerithiids.

After a cursory examination of the many species attributed to *Ataxocerithium sensu lato*, he suggested that several genera are probably involved. In addition to the species illustrated and described here, available names used in the Australian context are: *applenum* Iredale, 1936 and *gemmulatum* Woolacott, 1957, both from NSW; *abbreviatum* Brazier, 1877 from Qld.

Ataxocerithium beasleyi Cotton & Godfrey, 1938

Resembles *A. serotinum* in overall form but the whorls are more flat-sided, the body whorl angulate at the periphery and the spiral sculpture is stronger and sometimes dominating the axials. The outer lip is simple, not flaring and the anterior canal is only subtubular. White or uniformly brown.

10 mm. SA to the western Great Australian Bight, WA. Dredged to 300 m. The illustration is drawn from the holotype. Given the many differences in shell morphology it seems unlikely that this species is congeric with *A. serotinum*.

Ataxocerithium eximium Houbrick, 1987

Protoconch of three smooth, inflated whorls, the first pointed; teleoconch sculpture cancellate, prickly, rasp-like, with 5-6 spiral cords crossing numerous axial riblets and pointed beads at the intersections; siphonal canal long, tubular, reflected to the left; columella concave, with a high lip. Cream, spiral cords golden tan-orange with white beads and a golden-tan subsutural band.

4.6 cm. Central NSW. Described from 421 m off Botany Bay. This is the largest species of the genus. Shells were illustrated by Houbrick (1987: figs. 1-12) who also described its anatomy and radula. See also Houbrick's illustration of the animal's head-foot, copied on p. 255 with his kind permission. While provisionally describing this species as a member of *Ataxocerithium*, Houbrick noted significant differences from other species in the genus.

Ataxocerithium scruposum Iredale, 1936

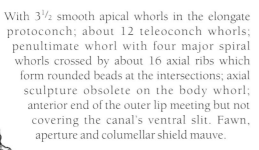

With 3½ smooth apical whorls in the elongate protoconch; about 12 teleoconch whorls; penultimate whorl with four major spiral whorls crossed by about 16 axial ribs which form rounded beads at the intersections; axial sculpture obsolete on the body whorl; anterior end of the outer lip meeting but not covering the canal's ventral slit. Fawn, aperture and columellar shield mauve.

12 mm. NSW. Iredale described two other species at the same time as this one. *A. conturbatum* Iredale, 1936 with more numerous but less pronounced axial ribs and more convex whorls; the type was from Sydney Harbour. *A. applenum* Iredale, 1936, from 128 m off Green Cape, NSW (the same type locality as *A. scruposum*) is shorter and broader, with a protoconch of one tumid whorl, but otherwise very similar to *A. scruposum*. The illustrated specimen was found with many others on sponges at a depth of about 20 m off the "Pinnacles", Sydney.

Ataxocerithium serotinum (Adams, 1855)

a

With about two smooth apical whorls and 13 teleoconch whorls; body whorl with nine rather flat spiral ribs crossed by axial cords becoming obsolete near the lip, earlier whorls with prominent axial ribs becoming broad and nodulose toward the anterior suture, giving the spire a pagoda-like appearance; anterior end of the outer lip meeting the end of the columellar flange with only a chink separating them, or overlapping it. Apex purple or mauve, teleoconch cream, fawn or brown, columella usually pink.

b

18 mm. NSW to Geraldton, WA. Commonly found alive on sponges below tide level. *Cerithium rhodostomum* Adams, 1855 is probably a synonym. In series examined from WA there is considerable variation in form and sculpture, similar to the range of NSW specimens which led Iredale, 1936 to introduce several new names. Fig. (a) represents a specimen from Cape Naturaliste. Figure (b), illustrated from Esperance, is of the form with pronounced axial ribs.

▲ GENUS **ALTISPECULA** Powell, 1930
Cerithiopsis geniculosus Hedley

With the characters of the type species.

Southern Australia. Monotypic.

Altispecula geniculose (Hedley, 1911)

Tall and slender with very convex whorls and deeply impressed suture; protoconch conical, of 2½ smooth, glassy whorls; teleoconch of about 11 whorls, sculptured with widely spaced, prominent, rounded axial lirae, spiral sculpture lacking; aperture ovate to subrectangular, columella narrow; anterior canal short, slightly reflected. White.

1 cm. Described from material collected by the research vessel *Endeavour* in 183 m off Cape Wiles, SA in 1909. The illustrated specimen is drawn from a syntype.

▲ GENUS **CONCILIOPSIS** Laseron, 1956
C. carrota Laseron

With the characters of the type species.

Eastern Australia. There is only one described species.

Conciliopsis carrota Laseron, 1956

Broad at the base and tapering, like an inverted carrot; protoconch tall and narrow, with about five smooth whorls; suture distinct but not deeply impressed; teleoconch with 6-7 whorls, sculptured with three rows of solid, rounded nodules which are joined at their bases, plus a fourth rib on the excavate base; columella short, slightly curved, pointed. Straw-coloured.

3.8 mm. Described from Eclipse I., Qld.

▲ GENUS **JOCULATOR** Hedley, 1909
Cerithiopsis ridicula Watson

Stout, ovate or bulbous; protoconch conical, smooth, usually multispiral; sutures shallow but usually distinct; teleoconch whorls with three spiral rows of rounded tubercles tending to be axially as well as spirally aligned and interconnected by ridges, base bearing an additional spiral; columella broad, straight, truncate; aperture subquadrate, outer lip thin and entire.

Northern and Southern Australia and New Zealand Laseron (1951, 1958) listed many species from eastern and northern Australia; there are two species in the south.

Joculator ridicula (Watson, 1886)

Minute, sides convex; protoconch tiny, of four whorls; teleoconch with a channelled suture; base excavate and bearing three spiral cords. With the characters of the genus; the tiny but bulbous protoconch is usually lost on adult shells. Reddish brown.

3 mm. North Qld. This tiny shell was described from material collected by the Challenger Expedition in coral mud off Wednesday I., Torres St. Its geographic distribution remains unknown.

▲ GENUS **POTENATOMUS** Laseron, 1956
P. caputumere Laseron

Minute, ovate, inflated; protoconch swollen of 1^1/$_2$ smooth whorls; teleoconch whorls about four, sutures indistinct, contracted at the base with a weakly nodulose spiral; whorls with two spiral ribs bearing prominent, rounded tubercles, columella short, broad, anteriorly truncate, lirate; aperture large.

Indian Ocean. There are two described species. Distinguished from *Horologica* only by its very small size and bulbous protoconch.

Potenatomus caputumere Laseron, 1956

Axial sculpture backing. Yellow.

1 mm. Described from about 100 m off North West Cape, WA. This is the smallest of the Australian cerithiopsids. The other species of the genus, *P. secundus* Laseron, 1956, has strong axial lamellae connecting the tubercles; it comes from Christmas I. The figure is drawn from the holotype.

▲ GENUS **HOROLOGICA** Laseron, 1956
H. bicolor Laseron

Small, ovate to bulbous; protoconch, where known, tall, multispiral, smooth; teleoconch whorls 5-9, sculptured with two spiral rows of tubercles which are axially aligned and connected, plus an additional row on the base, upper row usually the larger, rows sometimes divide to form five on the body whorl; suture indistinct; base slightly restricted; aperture subquadrate; columella straight, broad, truncate.

Northern Australia. Laseron (1956) listed ten named species in the genus, which is like *Joculator* but distinguished by two, not three, spiral rows of tubercles.

Horologica westiana (Hedley, 1909)

Protoconch like a tall, narrow spike, with five smooth, convex whorls; first teleoconch whorl with an abrupt increase in size and change in sculpture, consisting of six whorls, encircled by two spiral ribs intersected by axial ribs forming prominent tubercles at the intersections and a strong fenestrate pattern with deep pits in the interspaces.

2.5 mm. Described from the Hope Is., Qld, where Hedley collected numerous specimens at depths from 5-10 fathoms. The figure is drawn from a paratype.

▲ GENUS **CLATHROPSIS** Laseron, 1956
C. impedita Laseron

Elongate and slender, turreted, whorls rounded with deeply impressed suture; protoconch of 3-6 whorls, smooth; teleoconch sculpture of three or more spiral ribs bearing small nodules, crossed by axial bars; base excavate; columella pointed.

Northern and eastern Australia. Laseron (1956) listed four species from northern Australia and four from NSW.

Clathropsis impedita Laseron, 1956

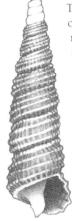

Thin, translucent; protoconch very high, conical of 3^1/$_2$ whorls; teleoconch of seven whorls, relatively long and rounded; base slightly excavate; sculpture delicate, cancellate with four spirals on the body whorl crossed by weak axials forming weak tubercles at the intersections; aperture large, anterior canal short, wide, open. Pale yellow with chocolate bands at the base of the whorls, base golden brown.

3.6 mm. Described from Mast Head I., Qld from specimens collected by Charles Hedley. The figure is drawn from the holotype.

▲ GENUS **ZACLYS** Finlay, 1927
Cerithiopsis sarissa Murdoch

Elongate, slender; protoconch of 4-4^1/$_2$ whorls, inflated and carinate, finely reticulate; teleoconch with a deep suture, spire turreted, sculptured with both spiral and axial riblets; aperture subquadrate; columellar pillar short, straight or slightly twisted.

Southern Australia and New Zealand. The type species is from New Zealand. Cotton (1951) assigned four southern Australian species to this genus but Laseron (1956) reassigned all but *Z. styliferus* to other genera.

Zaclys styliferus Cotton, 1951

Protoconch of four whorls, the third inflated, the fourth carinate; subsutural rib low, weakly nodulose, remaining two ribs more prominent and strongly nodulose; body whorl with a peripheral rib; axial lamellae prominent; anterior canal shallow and narrow, oblique. White.

7 mm. SA to King George Sound, WA; dredged to 300 m.

▲ GENUS **SYNTHOPSIS** Laseron, 1956
S. cylindrica Laseron

Subcylindrical with convex sides and 5-6 adult whorls; protoconch smooth, multispiral, diameter of its last whorl less than that of the first teleoconch whorl; suture impressed; teleoconch whorls with three weakly nodulose spiral ribs and weak axial connectors; base excavate; columella broad, pointed or truncate.

Northern and eastern Australia. Laseron (1956) listed six species from northern Australia and three from NSW, noting: "seems a natural genus, and included a group of species with a distinct facies."

Synthopsis cylindrica Laseron, 1956

Narrowly cylindrical, spire slightly convex with little taper; protoconch slender, narrower than the summit of the teleoconch, with 5½ whorls, terminated by a varix; suture deep; keels connected by axial cords, nodulose at the intersections; deeply channelled between the spirals; straw-coloured.

3.2 mm. Described from Michaelmas Cay, Qld from material collected by Tom Iredale during the Great Barrier Reef Expedition. The illustration is drawn from the holotype.

▲ GENUS **SPECULA** Finlay, 1927
Cerithiopsis styliformis Suter

Spire tall, cylindrical; protoconch of 1½-2 smooth, inflated whorls; teleoconch whorls 8 or more, slightly convex, with three beaded spiral ribs, sutures usually deeply channelled; aperture subquadrangular, anterior canal short, open.

Southern Australia and New Zealand. Cotton (1951) assigned three southern Australian species to this genus. Two are illustrated here. The third is *S. turbonilloides* Tenison Woods, 1879 described from Victoria.

Specula mammilla (May, 1919)

Protoconch mammillate; suture not channelled, sculpture variable, the three spiral ribs weakly or strongly beaded, not with a smooth spiral peripheral rib on the body whorl. White or brown.

4 mm. Tas to SA. Although Cotton (1951) assigned this species to *Specula*, it lacks the deeply channelled sutures of the other two Australian species.

Specula regina Cotton, 1951

Minute; suture channelled, containing a golden thread which becomes a peripheral rib around the base of the body whorl; base flat, columellar pillar short, straight; sculpture of three strongly beaded spiral ribs, the beads connected by axial lamellae so that the general appearance is fenestrate. White except for the golden sutural thread.

3.25 mm long; Gulf of St Vincent, SA to King George Sound, SA. Dredged to 300 m although the type was beach collected. This little species takes its name from the golden thread in the suture. The illustration is drawn from the holotype.

▲ GENUS **TUBERCLIOPSIS** Laseron, 1956
T. capricornia Laseron

Elongate, flat-sided, suture indistinct; protoconch subulate, of 2-3½ smooth whorls; 7-11 adult whorls bearing 3-4 rows of rounded nodules which may be connected axially; base excavate.

Eastern and south-eastern Australia. Laseron (1956) listed five species from Qld, eight from NSW and two from south-eastern Australia. While listing the southern shells in the genus, Laseron noted that the type species (from Heron I., Qld) has a rather different protoconch.

Tubercliopsis dannevigi (Hedley, 1911)

Protoconch of four whorls, terminating with a faint varix; eight teleoconch whorls, each with three tuberculate spirals. Dark red-brown.

5 mm. Bass St. to Cape Wiles, SA. Cotton (1959, p. 363) listed this species under Zaclys but Laseron (1956, p. 155) placed it in *Tubercliopsis*. It appears to represent the shell form of the genus fairly well but the protoconch does not seem typical.

▲ GENUS **SOCIENNA** Finlay, 1927
Cerithiopsis apicicostata May

Tall-spired and slender with flat sides; protoconch large, with three whorls, the first smooth and flat-topped, the others axially striate; teleoconch many-whorled, suture shallow, sculptured with three beaded spiral ribs, and weak axials; base excavate; columella short, twisted; aperture subquadrate, anterior canal wide.

Southern Australian Region.

Socienna apicicostata (May, 1919)

Length more than three times the diameter; spiral sculpture moderately developed, low rounded nodules are connected by weak axials. White.

4 mm. Tas to the Great Australian Bight, WA. Dredged to 300 m. The other species assigned to this genus, *S. trisculpta* (May, 1912), is less elongate and more strongly sculptured.

▲ GENUS **PILAFLEXIS** Laseron, 1951
P. regularis Laseron

Spire tall, flat-sided; suture distinct; protoconch of 2-3 axially striate whorls; teleoconch with 4-5 strong, broadly and irregularly tuberculate spiral ribs, with weak axials in the interspaces; columella strongly bent; anterior canal short.

NSW. There are two described species.

Pilaflexis regularis Laseron, 1951

Protoconch blunt, with three weakly spirally sculptured whorls; teleoconch of 12 whorls, sculptured by 4-5 subequal ribs, the subsutural rib thread-like, and about 20 axials, forming irregular tubercles at the intersections.

8.5 mm. Described from a series dredged at depths of 15-25 m in the vicinity of Sydney. The second species of the genus, *P. oculis* Laseron, 1951 has lens-shaped tubercles. Drawn from one of the syntypes.

▲ GENUS **PARASEILA** Laseron, 1951
P. heronensis Laseron

Stout and inflated, sides convex; suture not distinguishable; protoconch inflated, with 1½ -2 whorls;

teleoconch whorls with three smooth spiral ribs, and only minute axial striae in the interspaces; base rounded, with three spirals; columella short, broad, tapering.

Eastern Australia. Close to *Seila* but more inflated and the sculpture continues onto the base. There are three described species.

Paraseila heronensis Laseron, 1951

Deep red-brown.

5 mm. Described from the Capricorn Group, Qld. Intertidal. A second species described from the Capricorn Group, *P. ampulla* Laseron, 1956 is yellow and less bulbous. Laseron (1951) placed the third species, *P. halligani* Hedley, 1918 from NSW, in *Seila* but in 1956 transferred it to this genus.

▲ GENUS **BINDA** Laseron, 1951
B. tasmantis Laseron

Tall and tapering; protoconch with two inflated whorls; teleoconch whorls flat-sided, suture indistinct, sculptured with three flattened spiral keels, axial sculpture weak; aperture quadrate, base excavate; anterior canal extended and reflexed.

South-eastern Australia and possibly New Zealand. This is a deepwater group.

Binda tasmantis Laseron, 1951

Thin, translucent; teleoconch of 12 whorls, spiral keels undulating rather than nodulose. Yellow-brown with darker base.

8 mm. Described from about 60 m off Crookhaven, NSW. *B. cacuminatus* (Hedley & Petterd, 1906), also from south-eastern Australia, is similar but nodulose.

▲ GENUS **SEILA** Adams, 1861
Triforis dextroversus Adams & Reeve

Elongate, slender, flat-sided; suture indistinct; protoconch mammillate, 1-3 whorls; teleoconch whorls with three spiral ribs and a fourth on the base, fine axial striae in the interspaces; columella short, broad, flat, sometimes twisted; anterior canal wide, short; base excavate, smooth.

Cosmopolitan. There are six or seven described species in south-eastern Australia and two in Qld. Some

Australian authors have used the New Zealand generic name *Notoseila* Finlay, 1927 for them but the grounds for this are dubious and the older name is used here.

Seila crocea (Angas, 1871)

Teleoconch whorls with three strong, smooth axial keels which may be angular, plus a fourth on the periphery of the body whorl. Brown, fawn or yellow.

23 mm. NSW to Cape Naturaliste, WA. This striking shell is one of the largest species in the family. Because of the poor original description of this species, its identity has been difficult to establish. The shell illustrated here matches the concept of the species commonly adopted.

Seila marmorata (Tate, 1893)

Slender, flat-sided; later teleoconch whorls with five spiral ribs, plus another on the base of the body whorl. White, marbled with reddish brown.

8 mm. SA. The colouring and greater number of ribs distinguishes this from *N. crocea*.

▲ GENUS **EUSEILA** Cotton, 1951
E. pileata Cotton

With the characters of the type species.

Southern Australia. There is only one described species.

Euseila pileata Cotton, 1951

Tall-spired, turreted, suture indistinct; protoconch large, mammillate, with a terminal spike and two whorls, the last larger than the first teleoconch whorl; teleoconch whorls with three smooth, flat, spiral ribs, and narrow, incised interspaces, axials lacking; columella short, twisted; anterior canal wide, oblique. Fawn, with obscure axial flames.

13 mm. SA to the Great Australian Bight, WA; dredged to 600 m. The tall, slender shell and spiked protoconch make this shell rather easy to recognise, although coming from such depths it is rarely seen.

FAMILY **TRIPHORIDAE**

Triphoras

Triphorids are small, sometimes minute snails with slender, multiwhorled shells resembling ceriths and cerithiopsids, except that most (subfamilies Triphorinae and Adelacerithiinae) are sinistral (left-handed). Members of the small subfamily Metaxiinae are dextral. The Adelacerithiinae is apparently not represented in the Australian fauna.

Marshall (1983) has estimated that there may be as many as 1000 living species in this family. A single sand sample from Euston Reef, Qld yielded at least 80 species. It is not possible to deal comprehensively here with this huge family; only representative species of the selected genera are illustrated.

Many triphorids have conical many-whorled (multispiral) protoconchs indicating planktotrophic larval life. Others have swollen few-whorled (paucispiral) protoconchs indicating short-lived lecithotrophic planktonic life or direct development. In some genera there are both planktotrophic and lecithotrophic species. Minute sculptural details of the protoconch are useful characters for identification of the species.

The teleoconch whorls of nearly all triphorids have four primary spiral cords, although the fourth is usually at least partly hidden at the suture of the spire whorls. The second spiral cord may begin later on the spire than the others and this is another useful diagnostic character. Most species also have axial costae, or at least axial lamellae, between the spirals and there are usually nodules at the intersections.

The operculum of triphorids is horny and spiral, with a central or eccentric nucleus.

Triphorids are predators of sponges. Most species feed selectively on particular sponge species. The proboscis is long and able to penetrate deep into the soft tissues of the sponge. They have specialised radulae which vary enormously among the species, depending on the structure of the sponge prey. The radula ribbon is always narrow but may have as few as five or as many as 63 teeth per transverse row. In this respect triphorids differ from cerithiopsids. Anatomically triphorids have very distinctive characters, notably the presence of a peculiar glandular pouch that opens into the posterior oesophagus and distinguishes them from cerithiopsids.

Shell characters are usually sufficient to separate the genera. Where this is not so the reader should consult a review by Bruce Marshall (1983, *Records Australian Museum, Supp.* **2**: 1-119). Other important references are by Charles Laseron (1954, *Records Australian Museum* **23**: 139-158; 1958, *Australian Journal Marine and Freshwater Research* **9** (4): 569 - 658).

The following key to the genera may assist the reader reduce the options. However, the Australian triphorid fauna is far from being well known; there are many undescribed species and genera which will not key out, and the generic groupings of described species is still not entirely satisfactory.

Key to the Genera of the Southern Australian Triphorids

(Characters derived from Marshall 1983)

1 Shell dextral..................................**Metaxiinae 2**
   Shell sinistral ....**Triphorinae 3**

2 Teleoconch whorls with strong axial costae, nodulose at the intersections ....**Metaxia**
   Teleoconch whorls with weak axial sculpture ....**Seilarex**

3 Posterior canal a simple notch or a round hole..................................**4**
   Posterior canal tubular, projecting **26**

4 Protoconch whorls nodulose ....**Eutriphora (part) and Viriola (part)**
   Protoconch whorls axially ribbed......**5**
   Protoconch whorls smooth or spirally keeled................................**21**

5 Shell with straight or convex sides **6**
   Shell bottle-shaped (inflated medially, narrow **24**

6 First protoconch whorl with hemispherical granules **7**
   First protoconch whorl with T-shaped granules **19**
   First protoconch whorl with reticulate sculpture **20**

7 Later protoconch whorls with two encircling threads........................... **8**
   Later protoconch whorls with one encircling thread........................... **14**

8 Protoconch axial riblets interrupted by a broad, smooth, spiral, adapical zone..... ....**Teretriphora**
   Protoconch axial riblets extend from suture to suture. **9**

9 Third spiral cord of teleoconch whorls originates from the adapical protoconch spiral ....**Talophora**
   Third spiral cord of teleoconch whorls does not originate from protoconch spirals **10**

10 All three teleoconch spirals commence simultaneously, becoming broad with flattened nodules...**Latitriphora**
   Second teleoconch spiral begins later than first and third.................**11**

11 Suture channelled ....**Aclophora**
   Suture merely impressed **12**

12 Teleoconch spirals strong and nodulose.......................................**Nototriphora Eutriphora (part)**
   Teleoconch spirals weakly nodulose or smooth........................ **13**

13 Axials weak or obsolete, when present consisting of a few broad low costae ....**Euthymella**
   Axials in the interspaces numerous, narrow.......................... **Viriola (part)**

14 Later protoconch whorls with a broad, smooth central spiral zone ....**Hedleytriphora**
   Later protoconch whorls with uninterrupted axial ribs................ **15**

15 Spiral cords smooth, axials narrow and weak...........................**Viriolopsis**
   Spiral cords transversely nodulose, axials lacking.......... ....**Torresophora**
   Spiral cords heavily nodulose, axials strong.................................**16**

16 Operculum of two whorls...............**Aclophoropsis**
   Operculum of four whorls **17**

17 Operculum flat ....**Bouchetriphora**
   Operculum shallowly concave........**18**

18 Anterior inward fold of outer lip
   contacting base of columellar lip.....**Obesula**
   Anterior inward fold of outer lip
   overhanging base of columellar
   lip..**Mesophora**

19 Anterior end of outer lip infolded,
   overhanging base of columellar lip..**Monophorus**
   Anterior end of outer lip infolded,
   but not quite meeting base of
   columellar lip................................**Subulophora**

20 Later protoconch whorls
   encircled by two spirals.................**Tetraphora**
   Later protoconch whorls
   encircled by one spiral ....**Sagenotriphora**

21 Outer lip anterior end infolded
   and overhanging base of the
   columellar lip................................**22**
   Outer lip anterior end infolded
   but not meeting base of the
   columellar lip................................**23**

22 Body whorl periphery rounded,
   base inclined..................................**Cheirodonta**
   Body whorl periphery angulate,
   base excavate ....**Magnosinister**

23 Protoconch conical; fourth
   teleoconch spiral well exposed
   at suture..**Inella**
   Protoconch subcylindrical;
   fourth teleoconch spiral slightly
   exposed at suture ....**Hypotriphora**

24 Protoconch at about two whorls .....**Nanaphora**
   Protoconch multispiral...................**25**

25 Shell strongly inflated medially;
   mature whorls less than 10. ....**Opimaphora**
   Shell weakly inflated medially;
   mature whorls 10 or more... ....**Mastonia**

26 Tubular anterior and posterior
   canals long; microsculpture
   weak or lacking.............................**Mastoniaeforis**
   Tubular anterior and posterior
   canals short or of moderate
   length; teleoconch
   microsculpture well developed.......**Iniforis**

SUBFAMILY **METAXIINAE**

Shells dextral, with only notched anterior canal and no posterior canal. With their dextral shells, the members of this group are easy to confuse with the Cerithiopsidae but they have anatomical and radular characters typical of the Triphoridae.

▲ GENUS **METAXIA** Monterosato, 1884
Cerithium rugulosum Adams

Protoconch of the planktotrophic or lecithotrophic types, of $2^{1}/_{2}$-5 whorls, the first encircled by fine spiral zigzag threads, others with six spiral threads and fine axial riblets; teleoconch whorls convex, sculptured with spiral cords and strong axial costae, intersections nodulose; columella tapering to a point; anterior canal a wide simple notch; periostracum axially lamellate.

Cosmopolitan. There are at least 14 described living species. Two named species are known from Australia but there appear to be several undescribed species in the north.

Metaxia fuscoapicata Thiele, 1930

Planktotrophic; protoconch conical, about $4^{1}/_{2}$ finely sculptured whorls; teleoconch whorl nodules rather weak. White, yellow or reddish brown on the summits of the spirals, sometimes forming axial bands.

6 mm. Fremantle to Carnarvon, WA. Thiele's original material was collected in Shark Bay during the so-called "Hamburg Expedition" to south-western Australia in 1905. The other described Australian species is *M. protolineata* (Laseron, 1951), a lecithotrophic species from south-eastern Australia with a ruggedly sculptured protoconch.

▲ GENUS **SEILAREX** Iredale, 1924
Seila attenuata Hedley = *turritelliformis* Angas

Dextral; slender and attenuate; planktotrophic or lecithotropic, protoconch with three or more whorls, the first $1^{1}/_{2}$ whorls ornamented with about six zigzag spiral threads; teleoconch whorls convex, suture deeply impressed, spiral sculpture of five or more smooth, thin cords, axial sculpture weak; base smooth, excavate; columella tapering to a point; anterior canal a wide simple notch; periostracum smooth.

Northern and southern Australia. There are only two described species. Many authors have placed this genus in the Cerithiopsidae.

Seilarex verconis Cotton, 1951

Planktotrophic, protoconch narrow consisting of six whorls; with about 11 convex teleoconch whorls, five spiral ribs on each plus a prominent peripheral rib on the body whorl.

14 mm. Darwin, NT and down the east coast and across the south coast to at least as far west as Eucla, WA. Probably circum-Australian. The figured specimen is the holotype. The second species of the genus, S. *turritelliformis* (Angas, 1877) = *attenuata* Hedley, 1900, is found in eastern Australia from north Qld to Lakes Entrance, Vic. It is darker coloured and more coarsely sculptured than S. *verconis*.

SUBFAMILY **TRIPHORINAE**

Shells sinistral, with a tubular or subtubular anterior canal, the posterior canal forming a notch, a hole or a tube; columella simple.

▲ GENUS **ISOTRIPHORA** Cotton & Godfrey, 1931
Triforis tasmanica Tenison Woods

Protoconch blunt-tipped, $2^{1}/_{2}$ whorls, with two nodulose spiral cords; teleoconch whorls heavily nodulose; aperture round to ovate; outer lip reflected over the columellar lip, closing the anterior canal in some species; anterior canal short to moderately long, usually subtubular.

Southern Australia. There are six described species and a number of known but undescribed species. The nodulose, not axially ribbed, protoconch is said to distinguish this group from *Eutriphora* but the difference is not entirely consistent.

Isotriphora nivea (Verco, 1909)

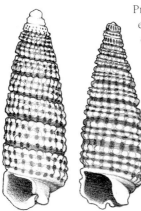

Protoconch whorls angulate; about eight teleoconch whorls with channelled suture and three tuberculate spiral ribs, vaguely connected axially; basal periphery on the body whorl with an additional rib and excavate below.

8 mm. SA to Cape Naturaliste, WA. The holotype, illustrated left, has a smooth-whorled but carinate protoconch, closed anterior canal and deeply channelled suture. In

the WA specimen, illustrated right, from Ellenbrook, (collected and identified as *nivea* by Verco), the second protoconch whorl is axially ribbed and the anterior canal is open. It may represent a different species.

Isotriphora tasmanica (Tenison Woods, 1875)

Conical, spirals and straight axial costae prominent, heavily nodulose at the intersections; fourth spiral partly exposed at the suture; posterior siphonal notch deep but almost entirely closed by the outer lip. Yellow-brown, darker between the nodules of spiral three.

11 mm. Northern NSW to SA; dredged to 200 m. The nearly closed posterior siphonal notch distinguishes this species from others in the genus. *I. echina* Laseron, 1954, described from NSW, is a synonym. Northern and central NSW shells have more numerous axial costae.

▲ GENUS **EUTRIPHORA** Cotton & Godfrey, 1931
Triphora cana Verco

Protoconch either paucispiral and blunt-ending with spiral rows of thick, sometimes elongate nodules, or conical, multispiral with the first whorl minutely granulose and subsequent whorls axially ribbed and with two spiral threads; teleoconch with prominent axials; aperture ovate, outer lip simple or infolded over the base of the columellar lip; anterior canal open or subtubular.

Southern Australia. Marshall (1983) listed three species, with nodulose protoconchs, which he assigned to this genus, viz. *cana* and *tricolor* Laseron, 1954, and *pseudocana* Marshall, 1983. He remarked that these species have a close relationship to *Isotriphora*. A fourth species, *E. armillata* (Verco, 1909), has a conical, multispiral, axially ribbed protoconch and it keys out to *Nototriphora*, but the radula suggests a relationship with *Eutriphora* according to Marshall.

Eutriphora cana (Verco, 1909)

Lecithotrophic; protoconch blunt-ending with $2^{1}/_{4}$-$2^{1}/_{2}$ whorls, all axially ribbed, the first whorl rounded, the second with a prominent central angulation; teleoconch whorls with prominent spiral cords and axial costae, intersections nodular. White apex, teleoconch yellowish or reddish brown.

9 mm. SA to Cape Naturaliste, WA. The figured specimen is a paratype. The teleoconch postion of this species is very similar to that of its congener *E. armillata* which has an overlapping geographic distribution. But the keeled protoconch of *E. cana* is diagnostic.

▲ GENUS **TERETRIPHORA** Finlay, 1927

Triphora huttoni Suter

Protoconch of the planktotrophic or lecithotrophic types; first protoconch whorl with hemispherical granules, subsequent whorls with two median spiral threads and axial riblets which are interrupted by a broad, smooth adapical spiral zone; teleoconch whorls convex, with three spirals commencing simultaneously, either evenly reticulate or with spirals dominant; anterior canal very short; outer lip infolded.

Southern and eastern Australia and New Zealand. There are at least four Australian species. Synonym: *Distophora* Laseron, 1958 (*distorta* Laseron, Qld).

Teretriphora spica (Verco, 1909)

Tall and rather flat-sided; planktotrophic; protoconch with about five convex whorls; teleoconch whorls with strong reticulate sculpture and heavy nodules at the interesections. White, buff, pale, yellow, often with brown axial streaks.

8 mm. Tas to Cap Naturaliste, WA; found from the shallows to at least 200 m. The figured specimen is the holotype. This species is distinctive within the genus because of a broad smooth spiral zone around the protoconch whorls, and its slender form.

▲ GENUS **TALOPHORA** Gründel, 1975

Notosinister subulata Laseron

With the characters of the type species.

Eastern Australia. Apparently there is only the one described species.

Talophora subulata (Laseron, 1958)

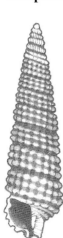

Whorls convex with impressed suture; protoconch multispiral, first whorl with hemispherical granules, subsequent whorls axially ribbed, with two spiral threads; teleoconch whorls reticulate with nodulose intersections, third spiral originates from the adapical spiral of the protoconch whorls; fourth spiral partly exposed at the suture; aperture ovate, anterior end of the outer lip slightly reflected but remains distant from the columellar base; anterior canal open, of moderate length. Pale yellow-brown, first and fourth spirals darker between the nodules.

9.5 mm. North Qld to Sydney, NSW.

▲ GENUS **LATITRIPHORA** Marshall, 1983

Triphora latilirata Verco

Protoconch with about five convex whorls, the first with minute hemispherical granules, subsequent whorls with axial riblets and two spiral threads; three spirals commence simultaneously on the first teleoconch whorl, later teleoconch whorls flat-sided, reticulately sculptured, with nodular intersections; aperture ovate, outer lip infolded; anterior canal long and tubular or short and subtubular.

Indo-West Pacific. Marshall (1983) lists only three species but several undescribed species belong here.

Latitriphora latilirata (Vero, 1909)

Nodules flat-topped and sharp-edged; spiral interspaces very narrow. Buff or white, irregularly patched with pale yellow-brown.

13 mm. SA to Cape Naturaliste, WA. The figured specimen is the holotype. The flat-topped, sharp-edged nodules and narrow interspaces make shells of this species readily distinguishable within the genus.

▲ GENUS **ACLOPHORA** Laseron, 1958

A. robusta Laseron

Protoconch conical, first whorl multispiral with minute hemispherical granules, subsequent whorls with axial riblets and two angulating spiral threads; teleoconch whorls reticulate and nodulose, suture channelled; second spiral beginning later than the others; fourth spiral cord visible at the suture; aperture ovate; outer lip extremity reflexed and overhanging basally; anterior canal short, subtubular; operculum flat, ovate, nucleus subcentral or eccentric.

Indo-West Pacific and Southern Australian regions. There are at least 12 species in northern Australia and one, *A. hedleyi* Marshall, 1983, in the south.

Aclophora robusta Laseron, 1958

Robust with slightly convex sides; protoconch of two whorls, the second spirally keeled; teleoconch of nine whorls with deeply channelled and spirally threaded suture; sculpture of three rows of rounded gemmules; four rows on the body whorl; columellar pillar short, slightly oblique, calloused on its inner margin. Variegated, red-brown, gemmules sometimes white.

8.5 mm. Port Curtis to Moreton Bay, Qld. A common Qld species.

▲ GENUS **NOTOTRIPHORA** Marshall, 1983
Notosinister aupouria Powell

Protoconch multispiral, first whorl with minute hemispherical granules, subsequent whorls axially ribbed and with two spiral threads; teleoconch with reticulate and heavily nodulose sculpture, second spiral commencing later than the others, fourth spiral partly exposed at the suture; whorls flat-sided; columella broad; aperture subcircular; outer lip extremity reflexed, overhanging the base of the columella; anterior canal short, broad, subtubular; operculum multispiral, thick, with a prominent external spire.

Southern Australia and New Zealand. There are four described species in southern Australia. The shells are like those of *Bouchetriphora* but the radulae and operculae are distinctive.

Nototriphora vestita Marshall, 1983

Protoconch narrowly concical, with 4-5 whorls. Teleoconch whorls white with bold oblique yellowish or reddish brown blotches.

7 mm. Tas. to Fremantle, WA. From the shallows to at least 100 m. The figured specimen is the holotype. The narrow protoconch and distinctive colour pattern allow shells of this species to be readily distinguished from the others of the genus.

▲ GENUS **EUTHYMELLA** Thiele, 1929
Euthymia regalis Jousseaume

Protoconch of 4-5 whorls, the first with hemispherical granules, subsequent whorls axially ribbed and with two encircling threads; teleoconch spirals smooth or weakly nodulose, second spiral commencing later than first and third, the third spiral is prominent and angulates the whorls, axial costae few, broad, or obsolete; aperture ovate, sometimes narrow; outer lip crenulate, its anterior end reflexed over the end of the thick columella; anterior canal tubular, short or long, oblique.

Indo-West Pacific. The type species is from New Caledonia. At least four species are in northern Australia.

Euthymella kosugei Marshall, 1983

Teleoconch whorls strongly angulate at spiral three, spiral cords prominent, angular, fourth entirely exposed at the suture, axial sculpture obsolete; base sharply contracted; anterior canal long. Yellow-brown or orange, darker between nodules on the fourth spiral.

3.85 mm. Indo-West Pacific; Qld and south to Coffs Harbour, NSW. For illustrations see Marshall (1983, p. 105; figs. 22 a-c).

▲ GENUS **VIRIOLA** Jousseaume, 1884
V. bayani Jousseaume

Elongate and slender; protoconch of the planktotrophic or lecithotrophic types; first whorl with minute hemispherical granules, subsequent whorls with axial riblets crossed by two median spiral threads, sometimes nodulose; teleoconch whorls with smooth spiral cords and numerous narrow axials, second spiral weak, commences later than the first and third; aperture subquadrate; outer lip strongly reflected, its anterior extremity overhanging the columella; anterior canal subtubular, rather long.

Indo-West Pacific. There are four described species in northern Australia. Synonyms: *Solosinister* Laseron, 1954 (*pagoda* Laseron = *corrugatus* Hinds) and *Orbitriphora* Laseron, 1958 (*iredalei* Laseron = *cancellatus* Hinds).

Viriola truncata Marshall, 1983

Lecithotrophic; protoconch blunt-tipped, of three whorls, sculptured with two rows of prominent nodules; teleoconch whorls with prominent spirals, weaker axial costae, and weak nodules at the intersections. White, axially blotched with brown.

10 mm. Described from Port Hedland, WA. *V. corrugata* (Hinds, 1843) occurs at the same locality but is more slender and its protoconch is multispiral and axially ribbed rather than nodulose. The other species of the genus are *V. cancellata* (Hinds, 1843) and *V. excelsior* (Melvill & Standen, 1899), which are recorded from Qld.

▲ GENUS **HEDLEYTRIPHORA** Marshall, 1983
Triphoris fasciata Tenison Woods

Protoconch of 4-6 whorls, the first with minute hemispherical granules, subsequent whorls with fine axial riblets above and below and a broad smooth median zone encircled by one spiral thread; second teleoconch spiral weak, commencing later than the first and third, later teleoconch whorls with spiral cords and axial coastae and nodules at the intersections; outer lip, infolded anteriorly to cover the base of the thick inner lip; anterior canal subtubular, short.

Southern Australia. Marshall (1983) listed five species.

Hedleytriphora elata (Thiele, 1930)

Slender and rather flat-sided; protoconch narrowly conical, of 4-5 whorls; teleoconch whorls sculptured with prominent spiral cords and axial costae, with nodules at the intersections, four cords on the body whorl plus two on the base; outer lip produced and flaring basally. Yellowish to reddish brown, darker between the nodules, nodules paler.

6 mm. Coffs Harbour, NSW to Quobba, WA.

Hedleytriphora fasciata (Tenison Woods, 1879)

Protoconch of 4-5$^{1}/_{2}$ convex whorls, first whorl with minute granules, subsequent whorls finely, axially ribbed near both upper and lower sutures but smooth at the centre; body whorl weakly angulate, spirals closely spaced, axial costae weak. Yellow-brown, often darker on the base, fourth spiral white or buff and blotched with yellow-brown.

7.3 mm. NSW to Eucla, WA. Very common in shallow water. *H. scitula* (Adams, 1851) is a very similar species occurring in the same areas of SA but is easily distinguished by its almost entirely smooth protoconch.

▲ GENUS **VIRIOLOPSIS** Marshall, 1983
V. occidua Marshall

Protoconch multispiral, sculptured with minute hemispherical granules, subsequent whorls with axial riblets and a central, spiral thread; teleoconch flat-sided; with smooth spiral cords and narrow, close, weak axials, the second spiral commencing later than the first and third; aperture ovate, outer lip infolded to almost cover the base of the thick inner lip; anterior canal short, subtubular.

Indo-West Pacific. Only the type species is yet recorded from Australia but there are at least two others in the Indo-West Pacific region.

Viriolopsis occidua. (Redrawn from Marshall (1983)

Viriolopsis occidua Marshall, 1983

Protoconch with about five whorls; teleoconch spiral cords strong, axial riblets weak and confined to the interspaces. White or buff, spirals alternately white and patched with yellowish or reddish brown.

4.25 mm. Eucla to Exmouth, WA. Common, intertidal to at least 125 m. When naming this species Marshall (1983, p. 50) noted that is closely related to the South African *V. alboguttata* (Tomlin, 1926) and the Indo-West Pacific *V. fallax* (Kay, 1979), which he included in the genus.

▲ GENUS **TORRESOPHORA** Laseron, 1958
T. elongata Laseron

With the characters of the type species.

Qld. Monotypic. Resembles *Viriola* in general form but distinguished by the single spiral on the protoconch and lack of axials on the teleoconch whorls.

Torresophora elongata Laseron, 1958

Elongate, slender; protoconch conical, of four whorls, axially ribbed and weakly angulated by a central spiral thread; suture indistinct; teleoconch of about 20 whorls bearing four spiral rows of transversely elongated, sharp-topped nodules, the rows alternating in size with the fourth row largest, axials lacking; periphery angulate, base flat; aperture round, with a raised rim; outer lip extremity strongly reflected over the columella; anterior canal closed and tubular. Uniformly cream or white.

14 mm. Torres St. to Capricorn Group, Qld. Apparently common in shallow water. Drawn from the holotype which came from Darnley I., Torres St. Laseron noted that the nodules are more rounded in southern specimens.

▲ GENUS **ACLOPHOROPSIS** Marshall, 1983
Triphoris festivus Adams

Protoconch of the planktotrophic or lecitothrophic types, with minute hemispherical granules on the first whorl, subsequent whorls with axial riblets and a central spiral thread; teleoconch whorls evenly reticulate and nodulose, the second spiral commencing later than the others, fourth spiral evident at the suture; outer lip extremity reflexed over the base of the columellar lip; anterior canal short or of moderate length, subtubular; operculum subcircular, thin, of two whorls with subcentral nucleus.

Southern Australia. There are three described species.

Aclophoropsis festiva (Adams, 1851)

Protoconch with 3-4¹/₂ whorls. Colour pattern very distinctive, white, alternatively maculate with white and yellow to dark brown on the sides of the first spiral, base brown.

12 mm. Tas. to Bunbury, WA. Resembles *A. maculosa* (Hedley, 1903) which occurs in NSW and Vic., but that species has a protoconch of the lecithotrophic type (only 2-2¹/₂ whorls) and has a different colour pattern.

▲ GENUS **BOUCHETRIPHORA** Marshall, 1983
Triphoris pallidus Pease

Protoconch multispiral, first whorl with minute hemispherical nodules, subsequent whorls axially ribbed, with a central spiral thread; teleoconch whorls flat-sided, evenly reticulate, intersections nodulose, second spiral commencing later than the first and third; aperture small, ovate; outer lip extremity strongly reflected over the columella; anterior canal short, subtubular; operculum multispiral, flat or with an external spire.

Indo-West Pacific, southern Australia and New Zealand. Three species are described in the Australian fauna.

Bouchetriphora pallida (Pease, 1870)

Easily distinguished by its pure-white shell.

8 mm. Indo-West Pacific, Southern Australia and New Zealand. This is a very common species with an exceptionally wide geographic distribution. Its white shell readily distinguishes it from the brightly coloured species *B. aspergata* (Laseron, 1958) from Qld, and NSW, and *B. marrowi* Marshall, 1983 from NSW. As might be expected with such a wide geographic distribution there are several synonyms. Three relevant in the Australian context are: *leuca* Verco, 1909; *glaciala* Laseron, 1954; *albomicra* Laseron, 1958.

▲ GENUS **OBESULA** Jousseaume, 1898
Mastonia obesula Jousseaume

Protoconch of the planktotrophic or lecithotrophic types, first whorl with hemispherical granules, subsequent whorls axially ribbed and with one central spiral thread, first whorl may be bulbous in lecithotrophic species; teleoconch sculpture evenly reticulate and nodulose, the second spiral commencing later than the others; aperture ovate or subquadrate; end of outer lip reflected to contact the base of the columella; anterior canal short, subtubular; operculum thick, shallowly concave, subcircular, with four whorls and subcentral nucleus.

Eastern and southern Australia. Marshall (1983) listed four living species. The radular and opercular characters are distinctive and consistent but the shells of some species may be confused with those of *Bouchetriphora*.

Obesula mamillata (Verco, 1909)

Lecithotrophic; protoconch with 2-2¹/₂ convex, mamillate whorls. Yellow-brown or pale buff, base reddish brown.

5 mm. NSW to Geographe Bay, WA; intertidal to at least 125 m. The figured specimen is a paratype. Another NSW species, *O. albovittata* (Hedley, 1903), is identical except that the protoconch is of the planktotrophic type. As cases are known where a gastropod species has both planktotrophic and lecithotrophic development, the relationship of these two types of shells should be questioned.

▲ GENUS **MESOPHORA** Laseron, 1958
M. bowenensis Laseron = fusca Dunker

Protoconch multispiral with 3-6 whorls, the first with minute hemispherical granules, subsequent whorls with axial riblets crossed by a central spiral thread; teleoconch sculpture of spiral cords and axial costae, nodulose at the intersections, second spiral begins later than the first and third; aperture ovate; outer lip extremity infolded over the basal end of the columella; anterior canal subtubular; operculum thick, concave, ovate, with four whorls, nucleus subcentral.

Indo-West Pacific. There are about 31 described species in Australia, most of them described by Laseron (1958) from Qld. According to Marshall (1983) this group is closely related to *Mastonia* and *Iniforis*. Synonym: *Coriophora* Laseron, 1958 (*negrita* Laseron).

Mesophora fusca (Dunker, 1860)

Protoconch of about three whorls; teleoconch sculpture of oblique costae and strong spirals forming strong nodules at the intersections. Dark red-brown, nodules often paler.

12 mm. Indo-West Pacific; south to Sydney in eastern Australia. This is a very common intertidal species in the tropics. For an illustration see Marshall (1983, p. 102, figs. 19 i-k).

▲ GENUS **MONOPHORUS** Grillo, 1877

Trochus perversus Linnaeus

Protoconch of the planktotrophic or lecithotrophic types, with 2-5 whorls, the first with T-shaped granules, subsequent whorls with axial riblets crossed by two median spiral threads; teleoconch whorls evenly reticulate with nodular intersections, the second spiral cord commences later than the others, the fourth spiral partly visible at the suture; aperture subquadrate; outer lip extremity strongly reflected to overhang the base of the columellar lip; anterior canal subtubular, of moderate length.

Mediterranean, Southern Australia and New Zealand. The type species is from the Mediterranean. There are 13 described Australian species. Synonyms: *Biforina* Bucquoy, Dautzenberg & Dollfus, 1884 (*perversus* Linnaeus); *Notosinister* Finlay, 1927 (*fascelina* Suter) from New Zealand.

Monophorus angasi (Crosse & Fischer, 1865)

Planktotrophic; protoconch conical, with $3^1/_2$-$4^1/_2$ whorls. Yellowish or reddish brown, darker on the base.

8 mm. NSW to Fremantle, WA. This is one of the most common triphorids along the southern Australian coast and is reasonably representative of the genus.

▲ GENUS **SUBULOPHORA** Laseron, 1958

S. exporrecta Laseron= *rutilans* Hervier

Tall and slender; protoconch of about seven whorls, first with T-shaped granules, subsequent whorls with axial ribs interrupted by a smooth zone below the suture and crossed by two spiral threads; teleoconch whorls reticulate and nodulose; extremity of outer lip reflected to almost meet the columellar lip; anterior canal subtubular, short.

Indo-West Pacific. A close relationship with *Inella* and *Hypotriphora* is indicated. There are at least four species in Qld. See Laseron (1958, pp. 610-613, figs. 125-133).

Subulophora rutilans (left).

Subulophora rutilans (Hervier, 1897)

Flat-sided, suture indistinct; whorls with three spiral ribs bearing prominent equal-sized tubercles. Yellowish.

8 mm. Indo-West Pacific. North Qld. The figured specimen is from New Caledonia. *S. exporrecta* Laseron, 1958 is a synonym.

▲ GENUS **TETRAPHORA** Laseron, 1958

T. mapoonensis Laseron

Sides convex; protoconch with $3^1/_2$-5 whorls, the first reticulately sculptured, subsequent whorls with axial riblets and one spiral thread; teleoconch whorls evenly reticulate with nodular intersections, or with the axials and nodules weak, fourth spiral partly visible at the suture, second spiral commencing later than the first and third; aperture subquadrate; outer lip extremity slightly reflected but distant from the base of the columellar lip; anterior canal open.

Indo-West Pacific. There are five species recorded from Australia. A SA species is figured to represent the genus. Perhaps the most comon species in shallow water is *T. granifera* (Brazier, 1894) a little yellow or reddish brown shell from southern Australia between NSW and Shark Bay, WA. It is quite variable and several of the varieties have been named, e.g. *pocula* and *jacksonensis* by Laseron (1954) from NSW, and *adela* and *albina* by Thiele (1930) from Shark Bay.

Tetraphora mcgilpi (Cotton, 1952)

Stout; protoconch of $3^1/_2$ whorls; teleoconch whorls sculptured with 4 rounded spiral ribs, the first and last at the suture. Ribs white, interspaces light brown.

4 mm. SA. The figured specimen is the holotype. Although the shells lack any axial sculpture, Marshall (1983, p. 33) suggested that this may be a synonym of *T. granifera*.

▲ GENUS **SAGENOTRIPHORA** Marshall, 1983

Triphora ampulla Hedley

Shell stout with convex sides; protoconch with 4-5 convex whorls, the first minutely reticulately sculptured with axial riblets and spiral threads, subsequent whorls with prominent axials and two spiral threads; teleoconch whorls with reticulate sculpture and prominent nodules at the intersections, second spiral commences later than the first and third, fourth spiral partly exposed at the suture;

aperture ovate, anterior end of the outer lip strongly reflexed to overhang the base of the columellar lip; anterior canal subtubular, short.

Indo-West Pacific. There are many species but only one recorded from Australia. Probably related to *Tetraphora*.

Sagenotriphora ampulla (Hedley, 1903)

Bottle-shaped; boldly patterned with dark reddish brown patches on a white or buff background, base red-brown.

5 mm. NSW to SA; also from New Zealand. The figured specimen is a paratype. The unique bottle-shape and maculate colour pattern makes this little shell easily recognisable. It is a common shell in the intertidal zone.

▲ GENUS CHEIRODONTA Marshall, 1983
Cerithium perversum var. *pallescens* Jeffreys

Protoconch multispiral, granular on the first whorl with axial riblets and two angulating median spiral threads, or lecithotropic with a paucispiral protoconch whose whorls bear two spiral cords; teleoconch whorls flat-sided, reticulately sculptured and nodulose at the intersections, the second spiral commences later than the others, fourth spiral partly visible at the suture; outer lip extremity reflexed but not covering the slightly open anterior canal.

Cosmopolitan. The type species is from the Atlantic.

Cheirodonta labiata (Adams, 1851)

Lecithotrophic, protoconch with $2^{1}/_{2}$ whorls sculptured with two smooth angular cords. Dark red-brown, usually with the first spiral a different shade.

5 mm. NSW. The dark colouring, small size and blunt protoconch distinguish this species. *Notosinister conferta* Laseron, 1954 is a synonym. The type localities of both *labiata* and *conferta* are in the vicinity of Sydney.

▲ GENUS MAGNOSINISTER Laseon, 1954
M. hedleyi Laseron

With characters of the type species.

South-eastern Australia. There is only one described species and a close relationship to *Inella* is apparent.

Magnosinister hedleyi Laseron, 1954

Shell massive; protoconch with $2^{1}/_{2}$-3 whorls, the first convex, smooth, remainder with two prominent but rounded spiral cords, plus a suprasutural cord on the last whorl; teleoconch whorls reticulate and nodulose, the fourth spiral well evident at the suture; aperture subquadrate, outer lip infolded at its anterior end to almost contact the thick columellar lip; anterior canal subtubular, of moderate length. Upper whorls white, later whorls buff, spotted and blotched with yellow-brown.

22 mm. NSW in shallow water. The figured specimen is a paratype.

▲ GENUS INELLA Bayle, 1879
Triforis (Ino) gigas Hinds

Protoconch conical, of the planktotrophic or lecithotrophic types, smooth with 1-2 spiral keels, or with a subsutural row of nodules; teleoconch whorls with three spiral cords commencing simultaneously (the first usually weak) and a fourth on the body whorl periphery which is well exposed at the suture on the spire; teleoconch sculpture reticulate and nodulose at the intersections; aperture subquadrate; end of the lip reflected but remaining distant from the columella; anterior canal subtubular, short.

Cosmopolitan. This is a very large genus which probably encompasses several different natural groups. There are at least six described species in southern Australia and an unknown number in the north. *Inella gigas* (Hinds, 1843) is an Indo-Pacific, planktotrophic species recorded from the NT. The generic affinity of the southern species needs to be confirmed. Two southern species are illustrated to represent the genus.

Inella intercalaris Marshall, 1983

Shell tall and slender; lecithotrophic; protoconch concial, blunt-tipped, of nearly three whorls, carinate; teleoconch whorls reticulately sculptured with spiral cords and axial costae, nodulose at the intersections; aperture subquadrate. White, later whorls maculated with pale yellowish brown.

9 mm. St Vincent Gulf, SA to Eucla, WA: dredged to 148 m. This may not be a good choice to represent the genus, as Marshall himself indicated that it might prove to be better located in *Teretriphora*. *I. carinata* Marshall, 1983 is a similar species from the same area.

Inella spina (Verco, 1909)

Very tall and slender, suture very shallow; protoconch of the lecithotrophic type, conical, of 3-4 smooth but angulate whorls; teleoconch whorls flat-sided, with reticulate sculpture of strong spiral cords and much weaker axials, four spiral cords on the body whorls plus 1-2 on the base, cords one and three nodular at the intersections; base contracted, anterior canal short, oblique, subtubular. Early whorls white, later whorls either uniformly pale yellowish brown or maculated with yellowish brown on a white ground.

13 mm. Tas. to SA. The unicarinate, smooth protoconch and very slender form readily distinguish this species from others in the genus. The figure is drawn from the holotype.

▲ GENUS **HYPOTRIPHORA** Cotton & Godfrey, 1931
Triphora subula Verco, 1909

With the characters of the type species.

Southern Australia. Monotypic. Probably closely related to *Inella*; there are only minor shell characters distinguishing between them.

Hypotriphora subula (Verco, 1909)

Very slender; protoconch of the lecithotrophic type, subcylindrical, with about 3½ whorls, smooth but with two strong, rounded spiral cords; teleoconch whorls flat-sided, sculpture reticulate with well-defined spiral cords, axial costae and nodular intersections, fourth spiral slightly exposed at the suture; aperture subquadrate; anterior canal open, almost straight. Uniform brownish yellow.

12 mm. SA; dredged, 18-101m. The shell is tall and slender like *I. spina* but has a subcylindrical protoconch. Drawn from a paratype.

▲ GENUS **NANAPHORA** Laseron, 1958
N. torquesa Laseron = *leucomys* Hervier

Shell small, bottle-shaped, narrow apically, inflated medially; protoconch with many or few whorls, sculptured with axial riblets and two spiral threads; teleoconch whorls heavily nodulose, with microsculpture of granulated spiral threads in the interspaces; aperture circular, outer lip strongly reflected over the columellar lip; anterior canal short, tubular, opening within the aperture, posterior canal round or a U-shaped notch.

Indo-West Pacific. Laseron (1958) listed 10 species from Qld but there is doubt whether they are all congeneric. Although the type species has a protoconch of the lecithotrophic type, others assigned to the genus have a multi-spiral protoconch more like that of the species Laseron assigned to *Opimaphora*. The multi-spiral species of *Nanaphora* will key out to *Opimaphora* in the key to the genera given here.

Nanaphora leucomys (Hervier, 1897)

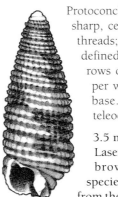

Protoconch mammillate, of two whorls, the first with a tiny tilted apex, inflated sides and prominent axial cords, the second shorter and with a single, central keel. Protoconch white, later teleoconch whorls uniform straw-yellow, suture of the teleoconch indistinguishable, whorls sculptured with two spiral rows of large, rounded gemmules, 14 per whorl, connected axially by low, rounded bars; base with small, smooth spiral keels, gemmules brown.

4 mm. Qld. Laseron noted that the unusual protoconch and sculpture should afford ready recognition of this species. The illustration is drawn from the type of *N. torquesa* Laseron, 1958, which is a synonym.

▲ GENUS **OPIMAPHORA** Laseron, 1958
O. sarcira Laseron

Very like *Nanaphora* but the protoconch is of the planktotrophic type with many whorls; mature whorls less than 10.

Northern Australia. Laseron (1958) described seven species in this genus but the relationship with *Nanophora* is problematical. If the two should prove to be synonymous *Nanaphora* would have page priority. However, both groups differ little in shell features from *Mastonia*, which is a much older name.

Opimaphora sarcira Laseron, 1958

Protoconch conical, with four whorls and a single sharp, central spiral keel crossed by fine axial threads; teleoconch of seven whorls with ill-defined suture; whorls sculptured with two rows of large, close rounded gemmules, 16 per whorl, and an extra three rows on the base. Protoconch white or dark brown; teleoconch cream with brown blotches.

3.5 mm. Cairns to Capricorn Group, Qld. Laseron noted that the colour pattern of brown blotches is unique among the species of the genus. The figure is drawn from the holotype.

▲ GENUS **MASTONIA** Hinds, 1843
Triphoris (Mastonia) rubra Hinds

Bottle-shaped; protoconch slender, multispiral, axially ribbed and with two spiral threads; spire teleoconch whorls with two spiral rows of smooth, rounded, widely separated nodules and conspicuous microsculpture of granulate spiral threads in the interspaces, there is a third row of nodules around the periphery of the body whorl; suture indistinguishable; aperture round with an elevated rim and reflexed outer lip extremity; anterior canal tubular, posterior canal a circular hole but not tubular.

Indo-West Pacific. There are five described species in Qld. The presence of microsculpture and two spirals on the protoconch distinguish this group from *Mesophora*.

Mastonia rubra Hinds, 1843

Protoconch of 4-5 whorls, bicarinate, keels crossed by numerous axial bars; teleoconch whorls numbering about 18, sculptured with two rows of large, rounded gemmules, about 22 per whorl, interspaces finely wrinkled, with a sinuous spiral cord between the rows of gemmules becoming segmented on the later whorls and finally a third row of gemmules on the body whorl. Bright reddish purple to lilac; apex and lower row of gemmules cream.

9 mm. Western Pacific; Qld. This is a very common and widely distributed species found under coral stones in shallow water. The bright colour of the shell fades dramatically after the animal has died.

▲ GENUS **MASTONIAEFORIS** Jousseaume, 1884
M. chaperi Jousseaume

Protoconch with 4-5 whorls, the early whorls finely granulose and the later whorls axially ribbed; teleoconch of 4-5 whorls with two rows of large nodules plus a third row of smaller nodules at the periphery of the body whorl, microsculpture weak or lacking; aperture round, reflexed over the columellar lip; posterior and anterior canals long and tubular in adults.

Indo-West Pacific. Marshall (1983, p.103, figs 20 e, f) illustrated a shell from Euston Reef, Qld which he said was almost indistinguishable from the syntype of the type species. According to Marshall *Epiforis* Laseron, 1958 (*australis* Laseron) and *Contraforis* Laseron, 1958 (*insulana* Laseron) are synonyms.

Mastoniaeforis insulana (Laseron, 1958)

Elongate, subcylindrical; suture distinct; protoconch unknown; teleoconch whorls numbering about 15, sculpture of unequal, wedge-shaped gemmules, the anterior row larger, rising above well defined spiral keels, axial connections absent, body whorl with two additional small rows on the base; aperture round, tubular and slightly extended; anterior canal long, tubular, straight in the line of the spire, posterior canal long, tubular, projecting obliquely, swollen near the base. White.

6.6 mm. Described from Christmas I. The distinctive sculpture separates this species from others in the genus. The figure is drawn from the holotype. *M. australis* (Laseron, 1958) described from Michaelmas Cay, Qld is similar but is bottle-shaped and has an indistinct suture.

▲ GENUS **INIFORIS** Jousseaume, 1884
I. malvaceous Jousseaume

Protoconch multispiral or paucispiral, when the former it is axially ribbed with one spiral thread; teleoconch whorls with two rows of large nodules and granulose interspaces, plus a third row of smaller nodules at the periphery of the body whorl; suture indistinguishable; aperture round, anterior end of the outer lip strongly reflexed over the columellar lip; posterior and anterior canals short and tubular (in adults).

Indo-West Pacific. There appear to be several species of *Iniforis* in northern Australia (*violaceus* Quoy & Gaimard, 1834; *violaceus evanidus* Laseron, 1958; *zonata* Laseron, 1958; *porrecta* Laseron, 1958). They are characterised by short tubular anterior and posterior canals and microsculpture in the interspaces. The relationship of this genus to *Mastoniaeformis* is problematical as the main differences relate to the length of the tubular canals and the presence or absence of microsculpture.

The type species of *Iniforis* is known from New Caledonia and was figured by Marshall (1983, p. 103, Fig. 20 a). For descriptions and illustrations of the known Australian species see Laseron (1958, pp. 579-582, figs. 1-13). For comments on the genus and illustrations of the type and other species, see Laseron (1958, pp. 636-637, figs. 214-219) and Marshall (1983, pp. 44-45, fig. 20a).

JANTHINOIDEA

*J*N this group there are two families which feed on coelenterates. Although their shells do not look alike, their relationship is established by anatomical and radula features as well as their feeding habits.

FAMILY EPITONIIDAE

Wentletraps

The family Epitoniidae is represented in all tropical, subtropical and temperate seas. There are very many species and the taxonomy of the Australian representatives remains difficult. Iredale (1936) introduced many new generic and species names for Australian wentletraps but there has been no thorough review. Only a selection of representative kinds are illustrated here. For convenience the generic arrangement follows that given by Vaught (1989, *A Classification of the Living Mollusca*, American Malacologists Inc., Melbourne, Florida).

The shells of this family are usually turreted, with many convex, strongly ribbed whorls. These ribs are sometimes called varixes but this is not correct. Some species have true varixes as well as axial ribs. There may also be spiral sculpture. The aperture is usually round or oval with a raised, entire rim (peristome) and the outer lip is usually thickened. Sometimes the whorls are only loosely coiled and are united only by axial lamellae. There is a thin, horny, many-whorled operculum.

Most wentletraps live associated with and presumably feed on anemones, although some associate with other coelenterates. For example, *Epitonium ulu* Pilsbry, 1921 lives on the fungiid coral *Fungia actiniformis* (Bosch, 1965). Ian Loch (1982) has reported the same association in north Queensland and another species living associated with dendrophyllid corals of the genera *Dendrophyllia* and *Tubastrea*. Egg masses and the body of the latter species are bright yellow-orange, matching the colour of the host polyps.

Robert Robertson (1963; 1966; 1983) has published several papers on the habits of wentletraps.

References

Bosch, H.F. (1965). A gastropod parasite of solitary corals in Hawaii. *Pacific Science* **19**: 267-268.

Iredale, T. (1936). Australian Molluscan Notes. No. 2. *Records Australian Museum* **19**: 294-306.

Loch, I. (1982). Queensland epitoniids. *Australian Shell News* No. **39**: 3-6.

Robertson, R. (1963). Wentletraps feeding on anenomes and corals. *Proceedings Malacological Society* London **35**: 51-61.

Robertson, R. (1966). Coelenterate-associated prosobranch gastropods. American Malacological Union Annual Reports **1965**: 6-8.

Robertson, R. (1983). Observations on the wentletrap *Epitonium albidum* in the West Indies. *American Malacological Bulletin* **1**: 1-12.

▲ GENUS **ACRILLA** H. & A. Adams, 1860
Scalaria acuminata Sowerby

With the characters of the type species.

Indo-West Pacific. Monotypic.

Acrilla acuminata (Sowerby, 1844)

Tall, acuminate, whorls convex, suture impressed, basal periphery subangulate; umbilicus lacking, aperture round, outer lip simple; with about 11 teleoconch whorls, sculptured with crowded, fine, slightly sinuous axial cords and a basal rib at the periphery; operculum thin, yellow. Brown, columella white.

4 cm. Indo-West Pacific; recorded from Keppel Bay, Qld. This species is easily recognisable and, although there are few locality records, where it occurs it may be quite common.

▲ GENUS **CIRSOTREMA** Mörch, 1852
Scalaria varicosum Lamarck

Elongate and tapering, whorls not separating, umbilicus lacking; base demarcated by a strong spiral rib originating at the suture; sculpture cancellate.

Cosmopolitan. Some of these shells are particularly beautiful. There are several species in Australia. Vaught (1988) lists *Plastiscala* and *Rectacirsa* as possible subgenera and this is followed here, although neither group conforms very well with the typical *Cirsotrema*.

Cirsotrema bavayi de Boury, 1912

Pl. 44; fig. 2　Whorls with convex sides, sculptured with close axial and spiral ribs forming a cancellate patern with small but deep pits in the interspaces and small nodules at the intersections; axial ribs constructed of adhering lamellae crossing the suture and attaching to the preceding whorl; varixes thick, rounded, two per whorl, roughly aligned up the sides, bearing thick rugae where the spiral ribs cross; basal rib consisting of overlapping, flattened nodules situated close to the thick columella, the deep slot between columella and rib crossed by axial lamellae. White.

3.5 cm. Indo-West Pacific; Exmouth Gulf, WA to Yeppoon, Qld. The more slender form, denser sculpture and lamellate structure of the axial ribs distinguish this species from *C. varicosum*.

Cirsotrema kieneri Tapperone-Canefri, 1876

Pl. 44; fig. 3　Tall and slender; suture wide, deep, varixes lacking; axial ribs fine, sharp, numerous, crossed by fine spiral ridges of almost equal size forming a cancellate pattern,

interspaces relatively wide and rectangular, axials not crossing the suture; basal rib thin, distant from the columella, the space between finely sculptured with subequal axial and spiral cords. White.

8 cm. Indo-West Pacific; Dampier Arch., WA to southern Qld.

Cirsotrema varicosum (Lamarck, 1822)

Pl. 44; fig. 1 a-b　Moderately tall and slender, whorls convex, with two strong, irregularly spaced, nodulose varixes on each whorl; suture deep, crossed by thick lamellae; axial ribs high, slightly flattened, connected by spiral ribs forming a cancellate pattern, and with small, rough, sometimes subspinose nodules at the intersections; basal rib moderate, with thin axial lamellae connecting to the columella. White.

6.5 cm. Indo-West Pacific; Abrolhos, WA to southern Qld.

▲ SUBGENUS **PLASTISCALA** Iredale, 1936
Scala morchi Angas

Basal rib relatively weak.

Eastern and southern Australia. *Plastiscala* was introduced as a generic name. Since then Cotton (1959) has added the southern species *invalida* Verco, 1906 and *verconis* Cotton, 1938 to the group.

Cirsotrema mörchi (Angas, 1871)

Small, elongate, whorls weakly convex; protoconch mammillate, of 1 1/2 whorls; teleoconch of 8-9 whorls, the last with about 20 rounded, depressed axial costae overridden by stout spiral cords extending onto the base, basal rib indistinct; aperture oval.

1 cm. NSW. Verco (1906, pp.147-148) recorded this species from the outer edge of the continental shelf off SA, discussed the considerable variation of the shell, and illustrated the operculum (Pl. 4, figs. 102). Iredale (1936) introduced two subspecies names for varieties from the NSW continental shelf, viz. *mörchi bentha* and *mörchi profundior*. Do not confuse this with *Problitora moerchi*.

▲ SUBGENUS **PROPESCALA** Cotton & Godfrey, 1931
Scala translucida Gatliff

Shell translucent; with few teleoconch whorls and a large, paucispiral, mammillate protoconch of 1 1/2-2 whorls; sculpture of well spaced, rounded axial ribs and fine spiral lamellae in the interspaces; imperforate; periphery of the body whorl bounded by a prominent spiral rib.

Southern Australia. There are two described species, notable for large protoconchs and few teleoconch whorls. The prominent basal rib suggests an affinity with *Cirsotrema* rather than separate generic status.

Cirsotremum translucida (Gatliff, 1906)

With four convex teleoconch whorls and distinct suture; protoconch of $1^{1}/_{2}$ bulbous whorls, rapidly descending; about 16 axial ribs on the body whorl and obscure spiral lamellae in the interspaces.

4 mm. Described from Port Phillip, Vic. Cotton (1959, p. 376) also listed the species from Tas, SA and "? SWA". The illustrated specimen is missing the last whorl. Cotton included *Scala valida* Verco, 1906 as a second species in *Propescala*. Verco's species was described from Backstairs Passage, SA. Shells of the two are very similar and may be the same thing. The major difference is that *valida* has stronger spiral lamellae and about eight teleoconch whorls but there is little information on variation in either form.

▲ SUBGENUS **RECTACIRSA** Iredale, 1936
R. fregata Iredale

With characters of the type species.

Eastern Australia. Monotypic. Although the type species has a pronounced basal rib it is not very similar to *Cirsotrema* s.s. and Iredale may have been correct in giving it full generic rank.

Cirsotrema fregatum (Iredale, 1936)

Small, whorls convex, basal rib strong; protoconch tall, slightly tilted, of $3^{1}/_{2}$ smooth whorls, stopped by a varix; six teleoconch whorls, axial ribs elevated, sharply cut, 11 on the last whorl not extending onto the base, interspaces spirally striate; aperture oval, stoutly varicose. White.

4 mm. Described from 468 m off Sydney, NSW.

▲ GENUS **DANNEVIGENA** Iredale, 1936
D. martyr Iredale

With the characters of the type species.

Eastern Australia. Monotypic.

Dannevigena martyr Iredale, 1936

Pl. 44; fig. 4   Tall with straight sides, thick, with 9-10 slightly convex whorls; umbilicus lacking; axial ribs close together, about 16 on the last whorl, high, thick, lamellar, recurved and delicately frilled, interspaces faintly concentrically striate but spiral ribs lacking; aperture subcircular, peristome complete, outer lip lamellately varicose but varixes lacking on the earlier whorls; basal rib formed by connections of the axial ribs below the periphery, with axial lamellae connecting to the fasciole. White.

5 cm. Described from off Gabo I., Bass St. in about 400 m. This genus and species were named in honour of Captain Dannevig, who was lost at sea during a research cruise.

▲ GENUS **EGLISIA** Gray, 1847
Turritella spirata Sowerby

Tall-spired, turreted, whorls convex; sculptured with spiral cords and fine axial lamellae, varixes weak; peripheral rib moderate.

Cosmopolitan in warm seas. There is little published information about this genus. There are several species in the Indo-West Pacific region, one in northern Australia. The type species is from the Caribbean.

Eglisia tricarinata Adams & Reeve, 1850

Pl. 14; figs. 7, 10 a-b   With about 15 teleoconch whorls, deep suture and slightly excavated base; sculpture of three or four angulate spiral cords, the upper and lower cords weak (one or both sometimes lacking) the two central cords strong, sometimes with minute spiral threads in the interspaces; axial lamellae fine, regular and crowded, slightly foliose at the suture; varixes hardly discernible, consisting of thin axial ribs which interrupt the spiral cords; peripheral rib angulate, basal sculpture finely cancellate with axial lamellae crossing almost equal spiral threads; aperture round, columella basally thick and rolled, forming a slightly projecting anterior end. Buff.

4.5 cm. Indo-West Pacific. Rottnest I., WA to southern Qld. Lives in sandy substrates from the shallows to the outer edge of the shelf. The shell may be mistaken for a turritellid at first glance and is illustrated with shells of that family for comparison.

▲ GENUS **EPITONIUM** Röding, 1798
Turbo scalaris Linnaeus

Whorls convex, sometimes not touching; sculptured with strong, lamellar, blade-like axial costae, interspaces smooth or weakly striate; usually umbilicate, outer lip usually in the same plane as the shell axis.

Cosmopolitan. There are very many synonyms of *Epitonium* s.s. Some which have been used in the Australian context are: *Acutiscala* Boury, 1909 (*philippinarum* Sowerby); *Parviscala* Boury, 1887 (*algeriana* Weinkauff); *Scala* Bruguière, 1792 (*scalaris* Linnaeus); *Scalara* Lamarck, 1801; *Solvaclathrus* Iredale, 1936 (*jacobiscala* Iredale, 1936).

Several subgenera have been recognised but their diagnoses are poor and there has been no comprehensive review. It is not possible to assign Australian species to subgenera with any confidence and no attempt is made to do so here. The following list of subgeneric names used in the Australian context is extracted from the classification used by Vaught, 1988.

Foliaeiscala de Boury, 1912 (*dubia* Sowerby) = *Crenuliscala* Iredale, 1936 (*pindasa* Iredale).

Globiscala de Boury, 1909 (*bullata* Sowerby).

Hirtoscala Monterosato, 1890 (*cantrainei* Weinkauff) = *Pudentiscala* Iredale, 1936 (*christyi* Iredale).

Laeviscala de Boury, 1909 (*subauriculata* Souverbie)

Lamelliscala de Boury, 1909 (*fasciata* Sowerby).

Limiscala de Boury, 1909 (*lyra* Sowerby) = *Crenuliscala* Iredale, 1936 (*pindasa* Iredale).

Epitonium acanthopleura (Verco, 1906)

Rather solid, wide, with eight rapidly increasing whorls; protoconch conical, of three smooth whorls; teleoconch whorls sculptured with solid axial ribs, about 20 on the body whorl, ribs bearing prickle-like tubercles, interspaces minutely spirally striate with the striae mounting the sides of the ribs; aperture round, outer lip with a shallow gutter at the base, columella twisted and broadly toothed, inner lip thin and erect. Whorls pale buff, ribs white.

4 mm. Described from 187 m off the Neptune Is., SA. For an illustration see the original description (1906, *Transactions Royal Society SA* **30**: 145-146, Pl. 4, fig 8).

Epitonium alatum (Sowerby, 1844)

Pl. 44; fig. 32 Stout, rather solid, loosely coiled, umbilicus open; teleoconch of about five whorls, sculptured with prominent, erect, sharp-edged, slightly hooked axial ribs, six ribs per whorl, ribs connect across the suture, interspaces smooth; aperture round, outer lip with a wide, flaring varix, hooked at the suture, peristome complete. White.

1.5 cm. Indo-West Pacific; Qld south to Yeppoon.

Epitonium auritum (Sowerby, 1844)

Pl. 44; fig. 14 Tall and acuminate, suture impressed, umbilicus wide; with 11 convex teleoconch whorls, sculptured with low, narrow, rounded axial ribs, about nine ribs per whorl, interspaces smooth; aperture oval, peristome complete, angled posteriorly. Fawn to light brown, glossy, ribs white.

3 cm. Indo-West Pacific; Ningaloo, WA to Cairns, Qld.

Epitonium barissum (Iredale, 1936)

Pl. 44; fig. 24 Thin, whorls quite rounded, narrowly umbilicate; protoconch of three elongate glassy whorls; teleoconch of 10 whorls sculptured with many fine axial costae of varying strength, 35-40 on the last whorl and spiral

cords, fine spiral cords in the interspaces; aperture oval, inner lip reflected a little but just joins the outer lip to complete the peristome. Off-white.

1.8 cm. Port Douglas, Qld to Sydney Harbour, NSW. Described by Iredale as a species of *Foliaeiscala*, together with *F. carchedon* and *F. antisoa*. The spiral cords in the interspaces are an unusual and characteristic feature.

Epitonium bullatum (Sowerby, 1844)

Pl. 44; fig. 9 Stout, thin, bulliform, suture impressed, with 4-5 convex and inflated teleoconch whorls, umbilicus a chink only; aperture oval, peristome incomplete, anterior end slightly produced; sculpture weak, consisting of low, thin to obsolete axial cords and obscure spiral lines. White.

2.5 cm. Indo-West Pacific; Qld and south to Long Reef, NSW. Type of *Globiscala* Boury, 1909. Synonym: *woolacottae* Kerslake, 1958.

Epitonium christyi (Iredale, 1936)

Pl. 44; fig. 27 Moderately stout but thin, teleoconch with about nine inflated whorls, suture very deep; umbilicus narrowly chinked; axial ribs erect, recurved, a little peaked below the suture, extended over the base, 10-12 per whorl, minutely striate in the interspaces; aperture subcircular, varicose, peristome complete. White.

1.7 cm. Described from Sydney Harbour. Iredale placed this species in *Acutiscala* but introduced the subgeneric name *Pudentiscala* for it.

Epitonium coreta (Iredale, 1936)

 Tall, acuminate, with about seven rounded teleoconch whorls, suture deeply impressed, umbilicus lacking; sculptured with strong axial ribs, sharp-edged, slightly reflected, hooked at the shoulders, about eight ribs per whorl, interspaces smooth; aperture round, peristome complete. White.

1 cm. NSW. Specimens seen from Cheyne Bay, southern WA may be this species.

Epitonium delicatulum (Crosse, 1864)

Pl. 44; fig. 15 Tall and slender, suture impressed and oblique, umbilicus lacking; teleoconch with about eight convex whorls; sculpture of thin, sharp axial ribs, about 22 per whorl, which tend to become elevated at the suture and on the base, usually slightly hooked below the suture, spiral sculpture lacking; peristome just complete. White.

2.5 cm. Keppel Bay, Qld to Semaphore, SA.

Epitonium fabia (Iredale, 1936)

Tall and slender, whorls convex, suture shallow, umbilicus lacking; aperture oval, peristome incomplete; with eight teleoconch whorls, sculptured with thin, sharp-edged axial ribs which are continuous from whorl to whorl, about 20 on the body whorl. Off-white.

1 cm. NSW from Port Stephens to Sydney Harbour where the original specimens were dredged.

Epitonium fasciatum (Sowerby, 1844)

Pl. 44; fig. 19 Whorls convex, about six, suture deeply impressed, umbilicus lacking; aperture round, peristome complete; sculptured with thin, erect, sharp-edged ribs, slightly hooked at the suture, about eight ribs per whorl, interspaces smooth. White, glossy; operculum brown.

1.5 cm. Indo-West Pacific; north Qld at least as far south as Cairns. Type species of *Lamelliscala* Boury, 1909.

Epitonium godfreyi Cotton, 1938

Pl. 44; fig. 16 Slender, suture deep, umbilicus closed; with about 10 whorls, the last with eight axial costae with rolled edges which continue over the base, interspaced smooth. Like *E. tacitum* in size and form but there are no spiral threads in the interspaces and no rib formed by coalescence of the axial ribs on the base behind the fasciole. White, glossy.

2.5 cm. SA to Fremantle, WA.

Epitonium helicornum (Iredale, 1936)

Pl. 44; fig. 25 Rather thin, moderately tall, with eight convex teleoconch whorls, umbilicus very small; aperture oval, peristome almost complete, slightly flaring anteriorly; axial ribs low, thin, rounded, about 18 ribs on the last whorl, not continuous from whorl to whorl. Fawn, sometimes with a brown peripheral stripe around the base, ribs white.

1.9 cm. Central NSW. This appears in many collections as *E. tenellum* (Hutton, 1885), a New Zealand species, but Iredale showed that the NSW shell has more ribs and is larger and narrower.

Epitonium imperialis (Sowerby, 1844)

Pl. 44; fig. 7 Thin and fragile, rather stout, whorls convex; suture and umbilicus deep; axial ribs thin, about 30 on the last whorl and extending onto the base, interspaces smooth and glossy; aperture round, peristome complete. Pale fawn to purple-brown, axial ribs white.

4 cm. Indo-West Pacific; Cape Naturaliste, WA to southern Qld. Lives associated with sea anemones buried in sand in the shallow sublittoral zone.

Epitonium jacobiscala Iredale, 1936

Small and glassy, uncoiled; with three apical whorls, adult whorls seven; distantly ribbed, ribs lamellate, 7-8 on the body whorl, interspaces smooth; aperture round, peristome complete, with a rimmed lip.

1.4 cm. Described from Sydney Harbour. Type species of *Solvaclathrus* Iredale, 1936.

Epitonium jukesianum (Forbes, 1852)

Pl. 44; fig. 22 Small, slender, teleoconch of about eight convex whorls, suture deep, umbilicus lacking; axial ribs strong, reflected at the edge, about 10 ribs per whorl, coalescing to form a thickening behind the fasciole; peristome just complete. White to pale purple-brown, ribs white.

1.5 cm. Sydney, NSW to Fremantle, WA. This little shell is quite common across the southern Australian coast. Synonyms: *delicatula* Crosse & Fischer, 1864; *ampacta* Iredale, 1936.

Epitonium lyrum (Sowerby, 1844)

Pl. 44; fig. 28 Stout but thin and fragile, teleoconch of about six inflated whorls, suture deeply impressed, umbilicus narrow, peristome complete; axial ribs low, thin and sharp, curving into the umbilicus, about 40 per whorl, with minute spiral threads in the interspaces. Fawn with two darker reddish brown spiral bands, ribs white.

1.5 cm. Indo-West Pacific; recorded from Port Douglas to Capricorn Group, Qld. This is the type species of *Limiscala* de Boury, 1909.

Epitonium marmoratum (Sowerby, 1844)

Pl. 44; fig. 18 Moderately stout, whorls convex, suture deep with the whorls usually slightly separate, umbilicus lacking, aperture round, peristome complete; sculpture of thin axial ribs, sharp and rolled over at the edges, hooked at the shoulders, fused together where they meet the columella, eight ribs per whorl, interspaces smooth. Apex purple, whorls fawn, marbled with light brown, ribs and peristome white.

2.5 cm. Indo-West Pacific; north Qld at least as far south as Cairns. The colouring of this species is distinctive and quite unusual in the family.

Epitonium neglectum (Adams & Reeve, 1885)

Pl. 44; fig. 31 Stout, teleoconch of 4-5 loosely coiled whorls, joined by corresponding axial ribs, suture deep, umbilicus wide and deep; axial ribs sharp and high, 12-13 per whorl, slightly hooked at the suture, extending into the umbilicus, interspaces minutely striate; aperture oval, peristome complete. Fawn or pale brown, ribs white.

1.5 cm. Qld. See comments on *E. pallasi*.

Epitonium pallasi Kiener, 1838

Pl. 44; fig. 6 a-b   Stout, teleoconch of 5-6 convex, loosely coiled whorls, joined by corresponding axial ribs, suture deep, umbilicus wide and deep; axial ribs sharp and high, extending over the base into the umbilicus, about 10 ribs on the last whorl, interspaces glossy smooth; aperture round, peristome complete, columella slightly reflected basally. Fawn or pale brown with white ribs.

2.5 cm. Indo-West Pacific; Darwin, NT to Port Stephens, NSW. *E. neglecta* (Adams & Reeve, 1885), also from Qld, is similar but the ribs are terminally hooked and more numerous.

Epitonium parspeciosum (Iredale, 1929)

Pl. 44; fig. 33   Stout, with 6-7 inflated teleoconch whorls, suture very deep, umbilicus wide, aperture round, peristome complete; axial ribs strong, edges sharp and slightly rolled over, hooked at the shoulder, curve into the umbilicus, about 10 per whorl. Pale fawn, ribs white.

1.6 cm. Port Stephens to Sydney Harbour, NSW.

Epitonium perplexum Pease, 1860

Pl. 44; fig. 13   Moderately tall, a little loosely coiled with a deep suture; umbilicus lacking; protoconch conical, of about three whorls; teleoconch of 9-12 convex whorls; axial ribs widely spaced, sharp and terminally hooked, about 13 ribs on the last whorl, extending onto the base, with a thin spiral cord between the ribs on the base; aperture subcircular, peristome just complete, columella reflected. White or fawn, sometimes with obscure reddish spiral bands, varixes white.

4 cm. Indo-West Pacific; Rottnest I., WA to Long Bay, NSW. A common shallow water species found in tide pools on rocky shores. In most respects the form and sculpture are typical of *Epitonium* but the presence of a basal rib is not. Because of this, and the closed umbilicus, Iredale (1936) introduced the generic name *Pomiscala* for *perplicata* Iredale which is a synonym. Synonyms: *lamellosa* Bryon, 1915; *perplicata* Iredale, 1929; *reevesbyi* Cotton, 1938; *pyramis* Tinker, 1952.

Epitonium philippinarum (Sowerby, 1844)

Pl. 44; fig. 23 a-b   Elongate, with about 10 convex whorls, umbilicus closed or represented by a minute chink; axial ribs strong, appressed to the preceding whorl, usually slightly elevated at the suture, edges rolled, about nine ribs per whorl; aperture oval, peristome complete. Shiny white or purplish brown with white ribs.

2.7 cm. Indo-West Pacific; eastern Australia as far south as Sydney Harbour. Iredale (1936) described a shell from Sydney Harbour as *Acutiscala minora* (fig. 22b) noting its similarity to *philippinarum*, type of *Acutiscala*, but he gave no diagnostic characters, except that the Sydney shells were white and the other "coloured". Qld material includes both coloured and white specimens in the same samples. Iredale's name is here regarded as a synonym.

Epitonium pindasum (Iredale, 1936)

Broadly awl-shaped, umbilicus closed by the basal rib; teleoconch of 7-8 whorls, the penultimate with about nine strong axial costae and smaller irregular ribs between, costae slightly peaked at the suture; aperture oval, peristome complete. White.

1.4 cm. Described from Sydney Harbour, NSW. Only the holotype is known. For an illustration see the original description (1936, *Records Australian Museum* **19**: p. 301, Pl. 22, fig. 18). Iredale described this as a species of *Foliaceiscala* but introduced the subgeneric name *Crenuliscala* for it.

Epitonium platypleura (Verco, 1906)

Moderately solid and elongate, with eight convex whorls; protoconch of two smooth, convex whorls; axial ribs low, doubly flanged so that a free edge projects slightly on either side, edges minutely cut, subangular below the suture, about 15 ribs on the body whorl, obscure spiral striae present in the interspaces; aperture with an oblique gutter at the base. White.

5 mm. Described from 40 m in Backstairs Passage, SA. Verco noted that this species differs from *E. jukesianum* by its more rapid increase of its whorls and more solid ribs. For an illustration see the original description (1906, *Transactions Royal Society SA* **30**: 145, Pl. 4, fig. 6).

Epitonium pyramidalis (Sowerby, 1844)

Pl. 44; fig. 12   Large and acuminate; about nine whorls convex and separate, umbilicus very small; aperture oval, peristome complete, lip thickened and sometimes forming a shallow spout anteriorly; axial ribs strong, sharp, reflected, not always connecting from whorl to whorl, slightly hooked at the shoulders. White.

5 cm. Indo-West Pacific; Qld south at least as far as Innisfail. Resembles a large *E. perplexum* but lacks the characteristic basal rib of that species.

Epitonium replicatum (Sowerby, 1844)

Pl. 44; fig. 20   Whorls convex, slightly separated, numbering seven; umbilicus minute; aperture round, peristome thick and complete; axial ribs erect and lamellate, sharp-edged, hooked at the shoulders, about 11 ribs per whorl, interspaces smooth. White.

1.8 cm. Indo-West Pacific; Qld south to the Whitsunday Group. With fewer axial ribs than *E. alatum* but with more and stronger ribs than *E. christyi*.

Epitonium scalare (Linnaeus, 1758)

Pl. 44; fig. 17 a-b   Stout with broadly convex whorls and a very deep umbilicus; suture so deep that the whorls are separated, connected by the conjoined axial ribs; about eight lamellar axial ribs on the last whorl, edges slightly rolled over, extending onto the base; aperture round, peristome complete. White to pale grey or fawn, ribs white; operculum black.

5 cm. Indo-West Pacific; Qld south to Hervey Bay. A very beautiful shell occasionally taken by trawlers.

Epitonium sexcostum Jousseaume, 1912

Pl. 44; fig. 26   Stout and strong, with five convex teleoconch whorls which are separate, and a small umbilicus; aperture round, peristome complete, outer lip with an expanded varix; axial ribs strong, erect and slightly reflected, hooked at the shoulders, six per whorl, interspaces smooth.

1.5 cm. Indo-West Pacific; Qld south as far as Moreton Bay. Very common in Keppel Bay.

Epitonium stigmaticum (Pilsbry, 1911)

Pl. 44; fig. 34   Tall and slender, with nine convex teleoconch whorls which are barely connected except by the axial ribs which cross the suture; umbilicus small; aperture round, peristome complete, outer lip expanded anteriorly; axial ribs of moderate strength, slightly hooked at the suture, about nine ribs per whorl. Fawn, ribs and peristome white, glossy.

3 cm. Indo-West Pacific; Qld south as far as Yeppoon.

Epitonium tacitum (Iredale, 1936)

Pl. 44; fig. 11   Elongate, moderately thin, suture very deep, umbilicus closed; whorls 11, the last with about eight axial costae which are continuous from whorl to whorl, have a rolled-back edge, and fuse into a basal spiral rib at the fasciole; interspaces finely striate; aperture oval, lips just meeting to form a complete free peristome. White, glossy when taken live.

2.5 cm. Sydney, NSW to southern WA. Iredale described this as a species of *Laeviscala*, noting that previous authors (e.g. Hedley in his Qld list) had recorded it as *Scala aculeata* Sowerby, 1844, which is a different Indo-West Pacific species with smooth interspaces. See note on *godfreyi* Cotton, 1938.

▲ GENUS **FILISCALA** Boury, 1911
Scala martini Wood

With the characters of the type species.

Indo-West Pacific. Monotypic.

Filiscala martini (Wood, 1828)

Pl. 44; fig. 21 a-c   Small, spire tall, turreted, whorls convex, suture impressed, base rounded, umbilicus narrow; teleoconch of seven whorls, sculptured with widely spaced, thin, rounded axial cords and slightly closer together spiral cords of equal size; aperture oval, peristome thickened and reflected but incomplete. White.

1 cm. Indo-West Pacific; recorded from Innisfail, Qld.

▲ GENUS **MAZESCALA** Iredale, 1936
M. thrasys Iredale

Whorls tightly coiled, umbilicus lacking; apical whorls thin and glassy, teleoconch whorls numerous, subshouldered, sculptured with many erect axial lamellae.

Eastern Australia. In addition to the type species Iredale included the NSW species *M. heloris* Iredale, 1936 and *M. bellicosa* (Hedley, 1907) in the genus, while Cotton later added *M. friabilis* (Sowerby, 1844) from SA and southern WA.

Mazescala thrasys Iredale, 1936

Small, thin, elongately awl-shaped, suture deep, mouth subcircular, peristome complete, umbilicus closed; protoconch subturbinate, of $3^{1}/_{2}$ whorls; about 10 teleoconch whorls which are slightly shouldered, rounded, the last with about 20 thin, erect axial lamellae which are peaked above the periphery. White.

1.5 cm. Described from off Green Cape, NSW, in 130 m.

▲ GENUS **MURDOCHELLA** Finlay, 1927
Scala laevifoliata Murdoch & Suter

Minute, elongate, sculpture of axial wrinkles, apex stout, outer lip thin.

New Zealand and eastern Australia. Iredale expressed doubt whether this genus belongs in the Epitoniidae but there seems no better place for it at the moment.

Murdochella macrina Iredale, 1936

Apex large, with an incurved tip, axially ribbed and spirally striate, ends with a varix; teleoconch of 7-8 convex whorls, sculptured with spiral cords, four on the last whorl, the last forming a basal ridge, cords overriden by axial wrinkles; base flat and a little excavate, columella almost straight, peristome incomplete. Off-white.

5 mm. Described from off Narrabeen, NSW in 150 m. No specimen of this tiny shell was available for illustration. See original description (1936, *Records Australian Museum* **19**: Pl. 22, fig. 19).

▲ GENUS **NARVALISCALA** Iredale, 1936
N. dorysa Iredale

With the characters of the type species.

South-eastern Australia. This genus shows similarities to *Cirsotrema* and may be a synonym.

Narvaliscala dorysa Iredale, 1936

Pl. 44; fig. 8 Elongate, acuminate, strongly varixed, aperture circular, peristome incomplete, umbilicus lacking; teleoconch of 15 whorls, the last with about 20 low, rounded axial costae, overridden by concentric threads; base flattened, spirally corded, granulose. White.

2.7 cm. Dredged off Gabo I. Cotton & Godfrey (1938) named a similar SA species *N. flindersi*.

▲ GENUS **OPALIA** H. & A. Adams, 1853
Scalaria australis Lamarck

Elongate, whorls not separating, umbilicus lacking; sculptured with nonlamellar axial ribs, spiral sculpture variable; outer lip thickened by the last rib, its plane approximately that of the shell axis; base with or without a spiral peripheral rib.

Cosmopolitan.

▲ SUBGENUS **GRANULISCALA** de Boury, 1909
Scalaria granulosa = granosa Quoy & Gaimard

Peripheral basal rib weak or lacking, axial sculpture becoming obsolete on the later whorls, interspaces minutely striate and granulose.

Southern Australia. Vaught (1989) treated this as a synonym of the subgenus *Dentiscala* Boury, 1886.

Opalia granosa (Quoy & Gaimard, 1834)

Pl. 44; fig. 30 a-b Moderately elongate, whorls convex, sometimes varixed; axial ribs low, rounded, about 10 but usually obsolete on the last whorl; aperture thickly varixed, peristome incomplete; basal rib weak. White.

4 cm. Bass St. to Fremantle, WA. *O. ballinensis* (Smith, 1891) is a similar species from northern NSW.

▲ SUBGENUS **NODISCALA** de Boury, 1889
Scalaria bicarinata Sowerby

Base without a peripheral rib, small, solid, few-whorled, suture crenulate, spiral ribs present.

Cosmopolitan.

Opalia apostolorum (Iredale, 1936)

Whorls rather flattened, about six; protoconch stout, incurved, smooth, of less than one whorl; axial ribs rounded, flattened at the suture, resting on the base of the preceding whorl, about 12 ribs on the body whorl, base bicarinate.

7 mm. Described from Sydney Harbour, NSW. *N. subcrassa* (Cotton & Godfrey, 1938) from SA may be the same thing.

▲ SUBGENUS **OPALIA** s. s.

Base with a prominent peripheral rib, suture not crenulate, spiral sculpture weak.

Cosmopolitan.

Opalia australis (Lamarck, 1822)

Pl.44; fig. 29 a-b Slender, suture impressed; about eight axial ribs on the last whorl ending at a thick basal spiral rib; aperture oval, peristome incomplete. White.

4 cm. Port Jackson, NSW to Fremantle, WA. *Opalia consors* Crosse & Fischer, 1864 is another species recorded from WA but its status is uncertain.

▲ GENUS **PROBLITORA** Iredale, 1931
Amauropsis moerchi Adams & Angas

Shell thin, globose to turbinate, smooth, body whorl inflated with a wide, round aperture and thin outer lip; periostracum hairy.

Eastern and southern Australia. There are two described species but little is known of their distribution or biology. There have been diverse views about the taxonomic postion of this genus. The type species was originally introduced as a naticid. In introducing this generic name Iredale put it in the Littorinidae. Anders Warén (pers. comm.) has suggested that it belongs in the Epitoniidae and that view is adopted here.

Problitora globula (Angas, 1880)

Spire height about $1/3$ of total shell length, suture canaliculate, aperture pyriform, columella slightly curved. Light brown.

1.8 cm. Bass St. to SA. Dredged in the shallow sublittoral zone.

Problitora moerchi (Adams & Angas, 1864)

Spire acuminate, height almost half total shell length, suture impressed; aperture wide, columella straight with a narrow parietal callus, narrowly umbilicate; operculum with $1 1/2$ whorls. Light brown.

1.1 cm. NSW. Inhabits muddy areas in the intertidal zone.

▲ GENUS **VARICISCALA** de Boury, 1909
Scalaria raricostata Lamarck

With the characters of the type species.

Indo-West Pacific. There are two described species.

Variciscala raricostata (Lamarck, 1822)

Pl.44; fig. 10 Stout, solid, suture deeply impressed, narrowly umbilicate, irregularly varicate; with five inflated teleoconch whorls, sculptured with numerous, regularly spaced, low, subequal axial and spiral cords; aperture oval, outer lip heavily varixed, the varix weakly nodulose and expanded below the umbilical area, peristome complete. White.

2 cm. Indo-West Pacific; recorded from Cairns, Qld. Iredale (1930) described a Qld shell as *Cirsotrema kelea* but in a subsequent paper (1936) he noted the lack of a basal cord, the "essential" character of *Cirsotrema*, and moved it to *Variciscala*, noting a close relationship with *V. raricostata*.

FAMILY **JANTHINIDAE**

Violet Sea Snails

The janthinids are pelagic molluscs which live at the surface of the sea suspended by a float made of air held within mucous bubbles. They live associated with the colonial, pelagic coelenterate *Velella,* upon which they feed. Because of this oceanic habitat they are rarely collected alive but are often cast ashore on beaches during storms. Janthinids are protandric hermaphrodites, i.e. each individual begins life as a male, some later changing to become females. There are two genera in Australian waters.

▲ GENUS **JANTHINA** Röding, 1798
janthina Linnaeus

Pelagic snails with fragile, globosely turbinate shells of violet colour.

Cosmopolitan. The name has been spelt *Ianthina* but the above is now currently accepted. The beautiful violet shells of this genus are so coloured for protection against predators. Three species may be found on beaches in southern Australia. The group was revised by D. Laursen (1953. The genus *Janthina*. A monograph. *Dana Reports* 6 (38): 1-40). Synonyms *Iodina* Mörch, 1860 (*exigua* Lamarck); *Violetta* Iredale, 1929 (*globosa* Swainson).

Janthina exigua Lamarck, 1816

Pl. 44; fig. 37 a-b Spire tall, suture impressed, whorls moderately rounded, body whorl with rounded periphery; outer lip with a deep V-shaped central sinus; columella sinuous, forming a short anterior spout at its junction with the outer lip; surface sculptured with conspicuous chevron-shaped growth striae reflecting the form of the lip sinus. Violet, paler near the sutures.

2 cm. Cosmopolitan; Southern Australia. This is the smallest of the Australian species, easily recognised by its tall spire and deep lip sinus. Synonyms: *globosa* Swainson, 1826; *capreolata* Montrouzier, 1859.

Janthina janthina (Linnaeus, 1758)

Pl. 44; fig. 36 a-b Spire domed, with a shallow subsutural sulcus, periphery subangulate, columellar column and the outer lip meet at a moderate angle; outer lip with a broad, shallow central sulcus; surface smooth. Violet, paler near the suture.

3 cm. Cosmopolitan; Southern Australia. Synonyms: *fragilis* Lamarck, 1801; *striolata* and *contorta* Carpenter, 1857; *carpenteri* Mörch, 1860.

Janthina pallida (Thompson, 1840)

Pl. 44; fig. 38 a-b Spire low, suture shallowly channelled, whorls inflated with rounded periphery; outer lip flaring and with a central sinus of moderate depth, columellar column arched, meeting the outer lip in a tight curve and forming an anterior spout. Violet, with a narrow pale subsutural band.

2.5 cm. Cosmopolitan; southern Australia. The central position of the lip sinus and anterior spout distinguish this species from *J. janthina*.

▲ GENUS **RECLUZIA** Petit, 1853
R. rollandiana Petit

Shells like those of *Janthina* but thicker and white beneath a thin brown periostracum.

Cosmopolitan. These animals are cast ashore much less frequently than those of *Janthina* and specimens are very rare in collections. One species is recorded from beaches of eastern Australia.

Recluzia hargravesi Cox, 1870

Pl. 44; fig. 35 a-b Spire tall, suture impressed and narrowly channelled, teleoconch with about five convex whorls; columella sinuous, thin but slightly raised and reflected, anterior end produced and forming a broad spatulate spout.

3 cm. Distribution uncertain, recorded in Australia from Caloundra, Qld and northern NSW.

EULIMOIDEA

THE members of this remarkable group live in varying degrees of parasitic relationship with echinoderms. Some are free-living, at least temporarily, others are permanently attached ectoparasites, while a few are internal parasites. Earlier authors recognised two or several families in the superfamily, e.g. Eulimidae (= Melanellidae), supposed to include elongate species mostly parasitic on holothurians, and Stiliferidae supposed to include more globose species parasitic on echinoids and starfish. Most recent studies agree that the genera intergrade and that all should be classed in a single family.

FAMILY **EULIMIDAE**

Eulimids

The shells of eulimids tend to be rather featureless so that classification of the group is difficult. They may be slender and pointed or stout and turbinate. Often there are axial scars representing the positions of earlier outer lips. There is no siphonal canal. Species with strong shells have an operculum which may have pegs, folds or other reinforcements.

Eulimid anatomy exhibits several unusual features associated with the parasitic life style. The foot is usually small and there may be a large fold (pseudopallium) reflected over the shell. The proboscis is often not retractable. In most genera there is no radula, food intake being accomplished by a suctorial, pump-like arrangement in the mouth. In those genera where a radula is present it is ptenoglossate. The stomach and intestine are either very small or lacking. Sexes may be either separate or hermaphroditic and most species have planktotrophic larvae. Egg capsules contain a few to many eggs and are usually laid attached to the host, the snail's own shell, or to hard objects on the sea-floor.

Many eulimids appear to be "host specific"; that is, they parasitise only particular echinoderm host species, or a few related host species which have similar biology. Often the snails spend part of their time living in the sediments and part attached to the host. It is believed that they feed by sucking out fluids from the coelomic cavity of the host, which is probably the reason echinoderms are vulnerable to these parasites as they have peculiar coelomic cavities unique in the animal kingdom. The eulimids make a hole through the echinoderm test (shell) through which the proboscis is inserted. The method of making the holes is unknown although their shape suggests that a dissolving secretion is involved. After the parasite has fed and moved on the holes are repaired by the victim.

There are very many species in the family. A revision of NSW eulimids was published by Charles Laseron (1955). Anders Warén published a revision of the genera (1984) and a series of subsequent generic studies which are referred to below in the appropriate sections. The taxonomic account given here is a summary of those works with respect to the Australian fauna. Because of the inadequate knowledge of the group only representative species of each of the genera are illustrated.

References

Laseron, C. (1955). Revision of the NSW Eulimoid shells. *Australian Zoologist,* **12**: 83-107.

Warén, A. (1984). A generic revision of the family Eulimidae. *Journal Molluscan Studies Supp.* **13**: 1-96.

▲ GENUS **APICALIA** Adams, 1862
A. gibba Adams

Shell very solid, rissoid, porcelaineous and highly polished; whorls rounded and inflated, suture indented; scars of the previous outer lips conspicuous; operculum thin, transparent. Proboscis completely retractile, acrembolic; head well defined with eyes and tentacles, foot functional, anterior pedal gland conspicuous behind the right tentacle.

Indo-West Pacific. When known the host is a starfish. The genus was revised by Warén (1981, *Zoologica Scripta* 10: 133-154). Synonym: *Chryseulima* Laseron, 1955 (*brazieri* Angas).

In addition to the illustrated species the following are recorded from Australia: *angulata* Warén, 1981, "on a starfish", Heron I., Qld; *cicatricosa* Warén, 1981, host not known, Eagle I., Qld; *echinasteri* Warén, 1981, on *Echinaster luzonicus*, Cape Leeuwin, WA to Heron I., Qld.

Apicalia brazieri (Angas, 1877)

Perfectly smooth and lustrous, sometimes with a slight bend in the spire; protoconch of about two whorls, teleoconch of six convex whorls. White.

8 mm (female). Exmouth, WA to Westernport, Vic. On the starfish *Coscinasterias calamaria*. Synonym: *immaculata* Pritchard & Gatliff, 1900.

▲ GENUS **ASTEROLAMIA** Warén, 1980
A. hians Warén

Shells small, broadly conical, with a wide, angulated aperture and strong scars from earlier outer lips.

Central Indo-West Pacific. Ectoparasites on starfish. There are two described species, one from Australia.

Asterolamia hians Warén, 1980

Short, irregularly conical, thin, transparent, body whorl relatively large; protoconch of two whorls, teleoconch of about four convex whorls, sculptured with numerous growth lines and deep old lip scars; aperture distinctly angulated anteriorly, columella slightly reflected, outer lip slightly concave in lateral view.

5 mm (female; males about half size). Described from Townsville, Qld. Attached to the starfish *Astropecten indicus*. Drawn from the holotype.

▲ GENUS **CURVEULIMA** Laseron, 1955
C. cornuta Laseron

Small, more or less conical but with a bent or twisted spire; vitreous, transparent, white to colourless; aperture short and exsert, outer lip thin, sinuate below the suture, rounded when viwed laterally; inner lip slightly reflected; operculate.

Australia. The type species is a parasite of comatulid crinoids. At least one species (*commensalis*) is said to parasitise echinoids.

Laseron (1955, *Australian Zoologist* 12: 88-91) described a number of NSW species which he assigned to his new genus and provided a key to their identification, viz.: *abrupta, cornuta, lata, litoris, manifesta, obtusa, subobtusa*. Several SA and Tas species were referred to this group by Cotton (1959), viz.: *commensalis* and *indiscreta* both of Tate, 1898; *petterdi* Beddome, 1882; *triggi* and *edwardsi* both of Cotton & Godfrey, 1932.

Warén (1981) tentatively placed *subobtusa* and *lata* (which he regarded as synonymous) in *Parvioris*, and other species listed here may also need to be reassigned, given the apparent diversity of hosts.

Curveulima cornuta Laseron, 1955

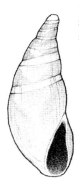

Spire strongly curved, with eight whorls, suture weak, body whorl about half the total shell length, evenly rounded at the base; aperture long, lunate, not produced anteriorly, outer lip arcuate, with a narrow sinus below the suture, making a sharp angle with the nearly straight inner lip which is only slightly reflected.

4.5 mm. Quobba, WA to Cronulla, NSW. Parasitic on comatulid crinoids. The illustrated specimen is one of the syntypes from Manly, NSW.

▲ GENUS **ECHINEULIMA** Lutzen & Nielsen, 1975
Stylifer mittrei Petit

Solid, rather inflated to globose, highly polished, with one to several strong scars on the outer lip; larval shell 300-500 microns high, of at least three whorls.

Indo-West Pacific. Ectoparasitic on sea urchins. The genus was revised by Warén (1980, *Zoologica Scripta* 9: 187-210) with a key to the species.

In addition to the illustrated species the following are recorded from Australia: *mittrei* Petit, 1851 (on the ventral side of diadematid sea urchins, widely distributed in the Indo-West Pacific with a record from Wheeler Reef, Townsville, Qld); *thanuumi* Pilsbry, 1921 (on *Echinothrix* sp., known only from the type material from Townsville, Qld).

Echineulima ponderi Warén, 1980

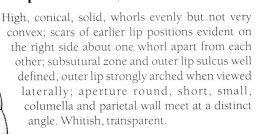

High, conical, solid, whorls evenly but not very convex; scars of earlier lip positions evident on the right side about one whorl apart from each other; subsutural zone and outer lip sulcus well defined, outer lip strongly arched when viewed laterally; aperture round, short, small, columella and parietal wall meet at a distinct angle. Whitish, transparent.

6 mm (female). North Qld. Attach to the lantern area of the sea urchin *Parasalenia gratiosa*. Males are about half the size of females. The illustrated specimen came from Broadhurst Reef off Townsville.

▲ GENUS **EULIMA** Risso, 1826

Designation of type species pending decision of the ICZN.

Elongate, slender, apex evenly tapered, sometimes bent to one side, surface highly polished and vitreous; protoconch dome-shaped, teleoconch with many, slightly convex whorls and shallow suture, body whorl long; aperture long, narrow, sharply angled posteriorly.

Cosmopolitan. This group probably contains several different lines and is in need of revision. In *E. glabra* there is a retractile proboscis and a radula with many teeth. According to Warén (1983, *Zoologica Scripta* 12 (4): 273-294) some species parasitise ophiuroids but the hosts are not known for most of the species. The slender form and elongate aperture distinguish this genus from *Melanella*. Synonyms: *Cuspeulima* Laseron, 1955 (*acutissima* Sowerby, 1866); *Leiostraca* H. & A. Adams, 1853 (*subulatus* Donovan, 1804 = *glabra* Da Costa, 1778).

In addition to the illustrated species chosen to represent this large group, the following from Australia are tentatively included in the genus: *australasica* Melvill & Standen (Indo-West Pacific with records from Torres St., Qld); *amabilis* Brazier, 1876 (Torres St., Qld); *augur broadbentae* Cotton & Godfrey, 1932 (SA); *cumingii* Adams, in Reeve, 1866 (Indo-West Pacific with records from North West Cape, WA); *grandis* Adams, 1851 (Indo-West Pacific with records from Broome, WA to Keppel Bay, Qld); *joshuana* Gatliff & Gabriel, 1910 (Vic. & Tas. to southern WA); *major* Sowerby, 1834 = *arcuata* Sowerby, 1866 and *flexiosa* Adams, 1851 (Indo-West Pacific with records from North West Cape, WA to Cairns, Qld); *metcalfei* Adams, 1853 (Indo-West Pacific, north Qld); *piperita* Hedley, 1909 (Qld south to Caloundra); *roegerae* Cotton & Godfrey, 1932 (SA); *sobrina* Laseron, 1955 (NSW to Eucla, WA).

Eulima acutissima (Sowerby, 1866)

Tall, very slender, acuminate, sides straight, with about 11 whorls; aperture narrow and long, outer lip rather straight except for a slight subsutural sinus. Colourless and semitransparent.

1.5 cm. NSW. The largest of the Australian members of this genus. The figured specimen came from Twofold Bay. Synonym: *lesbia* Angas, 1871.

▲ GENUS **EULITOMA** Laseron, 1955
E. nitens Laseron

With the characters of the type species.

South-eastern Australia. In addition to the type species Laseron named another, *E. castanea,* which he assigned to this genus, but the distinction between the two is dubious.

Eulitoma nitens Laseron, 1955

Elongate, flat-sided, with seven teleoconch whorls, body whorl long; protoconch blunt, dome-shaped; suture oblique; aperture pyriform, rather short, laterally expanded and rounded, outer lip with a wide subsutural sinus, inner lip almost straight, slightly reflected; vitreous and highly polished. White.

Holotype: 4.2 mm. Recorded from Sydney to Kangaroo I., SA. The figured shell is drawn from the holotype.

▲ GENUS **FUSCEULIMA** Laseron, 1955
F. jacksonensis Laseron

Rather short and stout, flat-sided, few-whorled, suture slightly impressed; protoconch blunt and dome-shaped; body whorl long, aperture short and wide, outer lip with a wide, shallow subsutural sinus; vitreous and transparent.

South-eastern Australia. Ectoparasites on sea urchins. In addition to the type the following Australian species may belong to this genus: *brunnea* Tate, 1887 (on the sea urchin *Heliocidaris erythrogramma*, Bass St., Vic. to SA); *flava* Laseron, 1955 (NSW); *perexigua* Tate & May, 1900 (Vic., Tas. & SA.); *sucina* Laseron, 1955 (NSW).

Fusceulima jacksonensis Laseron, 1955

With shell characters typical of the genus; with four adult whorls. Brown with a lighter subsutural band.

2.5 mm. NSW to Tas. Laseron's species *flava* and *sucina,* which were named at the same time as this, differ mainly in colouration and may be synonyms.

▲ GENUS **HEBEULIMA** Laseron, 1955
Leiostraca inusta Hedley

Elongate, cylindrical, many-whorled, early whorls sometimes distorted, outer lip almost straight.

Eastern and southern Australia. When introducing this generic name Laseron included the following Australian species as well as the two illustrated here: *columnaria* May, 1915 (Tas.); *crassiceps* Laseron, 1955 (NSW); *fricata* Hedley, 1907 (NSW to SA); *kilcundae* Gatliff & Gabriel, 1914 (Bass St.); *perexigua* Tate & May, 1900 (Vic. and Tas.); *tumere* Laseron, 1955 (NSW).

Hebeulima inusta (Hedley, 1906)

Thin, almost transluscent, smooth except for weak growth lines, suture ragged; protoconch blunt; teleoconch cylindrical, of about six slightly convex whorls; aperture oval, peristome slightly thickened and reflected. White, base and aperture cinnamon-brown.

2.8 mm. NSW. Drawn from a specimen collected in beach sand at Port Stephens.

▲ GENUS **HEMILIOSTRACA** Pilsbry, 1918
Leiostraca distorta Pease = *peasei* Tryon

Shells smooth and polished, tall-spired, strongly bent; columella flattened; patterned with brown bands.

Indo-West Pacific. Hosts unknown except for *H. sloani,* which is an ectoparasite on an ophiuroid. According to Warén (1984 *loc. cit.*) the anatomy resembles that of *Eulima* and there are many undescribed species of the genus in tropical waters.

Hemiliostraca peasei (Tryon, 1886)

Whorls slightly convex, dorso-ventrally compressed, spire bent, body whorl half the shell length, suture very shallow; aperture elongate, acuminate posteriorly, U-shaped anteriorly, columella slightly thickened. White, faintly banded.

5 mm. Indo-West Pacific; North West Cape, WA to north Qld. *Leiostraca distorta* Pease, 1861, is a preoccupied earlier name.

▲ GENUS **HYPERMASTUS** Pilsbry, 1899
H. coxi Pilsbry

Shell small to large, colourless, cylindrical or slender and conical with slowly increasing diameter of the flat or very slightly convex whorls; aperture rather small, columellar-parietal wall almost straight, slightly arched or with a weak angle at the transition; outer lip profile straight or curved with the median part protruding; proboscis fully retractile with a small collar.

Indo-West Pacific. The genus was revised by Warén & Crossland (1991, *Records Australian Museum* 43: 85-112). There are 24 described species; those with known hosts are parasitic exclusively on sand dollars (clypeasteroid sea urchins).

Only the type species has been chosen to represent the genus. Others in Australia are: *colmani* Warén & Crossland, 1991 (Urangan, Qld on the large red sand dollar *Peronella leseuri*); *georgiregis* Cotton & Godfrey, 1932 (King George Sound, WA); *mucronata* Sowerby, 1866 (NSW); *placentae* Warén & Crossland, 1991 (north Qld on the sand dollar *Arachoides placenta* on intertidal sand flats); *williamsi* Cotton & Godfrey, 1932 (St. Vincent Gulf, SA).

Hypermastus coxi Pilsbry, 1899

Cylindrical, sides slightly convex; protoconch small, mamillate, of about $1^1/2$ whorls; teleoconch of about five whorls; aperture ovate, outer lip almost straight in lateral view, columella almost straight, inclined;

4.5 mm. Port Stephens, NSW to Vic. at depths to 128 m. Resembles *H. mucronata* (Sowerby, 1866) which is more elongate and slender and found in the same region. The small mamillate protoconch is distinctive within the genus. Few specimens exist in collections and the host has not been recorded.

▲ GENUS **LEUTZENIA** Warén, 1980
L. asthenosomae Warén

Shells large, inflated, fragile, with about eight convex teleoconch whorls. Animal with big epipodial flaps covering the base of the shell; snout occupying a central position in the aperture; foot, operculum and tentacles present. Sexes separate.

Western Pacific. Parasitic on sea urchins. Warén included the Japanese species *toki* Habe, 1974 in the genus (1980, *Zoologica Scripta* 9: 283 -306).

Leutzenia asthenosomae Warén, 1980

Ovately conical, very thin and fragile, semitransparent; teleoconch of eight whorls with a shallow but distinct suture, indistinct growth lines and a prominent subsutural zone. Greyish.

1 cm (female). Qld to Port Stephens, NSW. Parasitic on *Asthenosoma*. For an illustration see original description (1980, *loc. cit.*, fig. 120).

▲ GENUS **MEGADENUS** Rosén, 1910
M. holothuricola Rosén

Shells globose, fragile, at least the lower part covered by a very large pseudopallium.

Pantropical. Internal parasites living in the cloacal cavity of holothuria. There are several Indo-West Pacific species. There is an undescribed species parasitic on *Holothuria atra* in Qld and for that reason the genus is mentioned here.

▲ GENUS **MELANELLA** Bowdich, 1832
M. dufresnei Bowdich

Elongately conical, rather flat-sided, with many whorls and an oily, glossy surface; apex evenly tapered, often bent; aperture short, inner lip slightly reflected; operculate. Sexes separate; sexually dimorphic, males being much smaller than females.

Pantropical. Mostly parasites on holothurians and starfish. The relationship to *Eulima* remains to be clarified. Synonyms: *Balcis* Leach, 1847 (*alba* Da Costa); *Eulimaustra* Laseron, 1955 (*proxima* Sowerby).

In addition to the illustrated species the following Australian species may be attributed to this genus: *acicula* Gould, 1849 (on *Stichopus chloronotus*, Indo-West Pacific, north Qld); *aciculata* Pease, 1861 (on *Holothuria atra* and other holothurians, Indo-West Pacific, north Qld); *anomala* Laseron, 1955 (on a holothurian, NSW); *attenuata* Sowerby, 1866 (Indo-West Pacific, Shark Bay, WA to north Qld); *cuspidata* Adams, 1851 (Indo-West Pacific, north Qld); *oblonga* Boettger (Indo-West Pacific, north Qld); *oxytata* Watson, 1883 (Indo-West Pacific; Shark Bay, WA to north Qld); *proxima* Sowerby, 1866 (NSW); *spina* Boettger (Indo-West Pacific; north Qld).

Melanella dufresnei Bowdich, 1832

Strong, bent, whorls slightly convex; columella narrowly thickened. White. Parasitic on holothuria.

2.5 cm. Indo-West Pacific; north Qld. The illustrated specimen came from Lizard I.

▲ GENUS **MICROSTILIFER** Warén, 1980
Stylifer auricula Hedley

With the characters of the type species.

Indo-West Pacific. Monotypic.

Microstilifer auricula (Hedley, 1907)

Shell very small, smooth, globosely turbinate, spire subulate; protoconch of one whorl, teleoconch of six whorls, the last globose; sculpture of weak growth lines; aperture semilunate, outer lip expanded, columella broadly reflected, umbilicus deep and broad.

2.2 mm. Described from Hope I. and Mast Head I., Qld; Warén (1980) reported the species also from New

Caledonia and Madagascar. Host unknown. For an illustration see Warén (1980, *Zoologica Scripta* 9: figs. 132-133).

▲ GENUS **MUCRONALIA** Adams, 1860
M. bicincta Adams

Shell form variable but usually tall-spired, mucronate, glossy smooth, with convex whorls; outer lip curved, parietal wall with a well-developed callus; often with yellowish spiral bands.

Pantropical. There are many species referred to this genus (listed by Warén 1980, *loc. cit.* 187-210). Hosts are known for only a few of them but those known are ophiuroids. The shells resemble those of *Hypermastus* in form but the outer lip is more curved and the parietal callus is stronger. Also, *Hypermastus* shells lack colour pattern. As well as the illustrated species, *M. xanthias* (Watson, 1886) has been described from Australia (off Cape York).

Mucronalia trilineata Warén, 1980

Tall-spired and slender, suture distinct but flat; protoconch of about three whorls, distinctly mucronate, cylindrical; teleoconch whorls evenly convex, with an abrupt increase in breadth after the third whorl, a prominent lip scar present close to the existing outer lip and others at about one whorl intervals; aperture squarish, outer lip retracted at the suture, evenly curved below; operculum long and slender. Greyish, semitransparent, with brownish spiral lines and brown axial lines near the lip and the old lip scars.

4.6 mm. Described from Broadhurst Reef, Qld on the ophiuroid *Ophioarachnella septem-spinosa*. Also recorded by Warén from further south at Noosa Heads. Drawn from the holotype. The odd change of form after the third whorl from slender and cylindrical to broad and tapering is an unusual feature of this species. It is further characterised by its prominent colour bands.

▲ GENUS **PARAMEGADENUS** Humphreys & Lutzen, 1972
Megadenus arrhynchus Ivanov

Elongate but with tumid, convex whorls and impressed suture; protoconch blunt; operculum with distinct growth zones.

The type species is from the North Pacific; Warén, 1980 (*loc. cit.*) described two Australian species. They are parasites on starfish. The type species (North Pacific) makes a depression on the dorsal side of the host and covers the walls of the depression with a pseudopallium originating from the snout.

Paramegadenus scutellicola Warén, 1980

Thin, transparent, cylindrical; protoconch of two short whorls, teleoconch of about seven whorls, previous lip scars weak; aperture rather small, columella meeting the parietal wall at a distinct angle.

6.3 mm. Holothuria Bank, NT to Capricorn Group, Qld. Lives attached to the tube feet of the starfish *Stellaster*. Probably a protandric hermaphrodite. The thin shells and short apertures distinguish this species. Also, the females have a distinctively cylindrical form. The illustrated specimen was collected at North Keppel I.

The second Australian species, *P. incerta* Warén, 1980, was described from Wide Bay, Qld where it was found on a starfish. It is known only from the holotype and its generic placement here is uncertain. The body whorl of the shell is inflated, rather like that of a *Stilifer*, but in that genus there is no operculum.

▲ GENUS **PARVIORIS** Warén, 1981
Eulima fulvescens Adams

Shell tall, slender with flat whorls and small aperture; outer lip usually straight when seen from the side; several distinct scars of previous outer lips visible; operculum with a half-moon-shaped fold serving for muscle attachment. Radula lacking; penis with two distinct lobes.

Indo-West Pacific and Mediterranean. Ectoparasites on starfish. There are many species in this group and they have rather similar shells which are difficult to tell apart. See Warén, 1981, *Zoologica Scripta* 10: 133-154.

In addition to the illustrated species the following are recorded from Australia: *equestris* Koehler & Vaney, 1912 (on *Stellaster equestris,* Indo-West Pacific, Murray I., Qld); *fulvescens* Adams, 1866 (on *Astropecten polyacanthus* and *Archaster typicus,* Indo-West Pacific, "N. Australia"); *noumeae* Warén, 1981 (on *Astropecten* sp., New Caledonia and central Qld); *styliferoides* Melvill & Standen, 1901 (on *Nepanthia brevis,* Indo-West Pacific, north Qld); *subobtusa* Laseron, 1955 (host unknown, NSW).

Parvioris australiensis Warén, 1981

Shell solid, smooth, slightly concave or straight conical, with a single scar from the outer lip about one whorl from the aperture; with about three protoconch whorls and 6-7 rather convex teleoconch whorls; suture not very deep but distinct. Yellowish, transparent.

5.3 mm. The unique type material was collected on *Archaster* sp. on an intertidal sandflat at Pasco I., WA. For illustrations see original description (1981, *loc. cit.* figs. 77-81, 111).

▲ GENUS **PEASISTILIFER** Warén, 1980
Mucronalia nitidula Pease

Shells solid, rather broad, conical, with convex whorls and short aperture; operculum with a peg.

Indo-West Pacific. Ectoparasites on holothurians. The genus was described in *Zoologica Scripta* **9**: 283-306. There are several species. The opercular peg is unique in the family and this appears to be the diagnostic feature of the genus. *Chryseulima solitaria* Laseron, 1950, described from NSW, may also belong here.

Peasistilifer nitidula (Pease, 1860)

Moderately elongate, polished, suture impressed, protoconch mucronate, of about four whorls, slightly oblique; teleoconch of 4-6 convex whorls; aperture ovate, rounded anteriorly. Opaque, white, sometimes with a brown spiral band.

4 mm (females; males about ²/₃ the size). Indo-West Pacific; Qld. Parasitic on *Holothuria atra*. The illustrated specimen came from Heron I., Qld. These parasites feed on blood cells obtained by inserting the proboscis into the spaces on the body wall. Hoskin & Cheng (1969, *Proceedings Symposium Mollusca, Marine Biological Association, India* **3**: 780-798) described the anatomy of this species. Synonyms: *gracilis* Pease, 1867; *holothuricola* Adam, 1934; *rosea* Pease, 1860.

▲ GENUS **PELSENEERIA** Köhler & Vaney, 1908
P. profunda Köhler & Vaney

Shell thin, globose, mucronate, colourless or light brown; snout very large; simultaneous hermaphrodites.

Pantropical. Parasitic in echinoids. There are about seventeen described species, but only one recorded from Australian waters. Synonyms: *Turtonia* Rosén, 1910 (*stylifera* Broderip); *Parastilifer* Ivanov, 1952 (*sibogae* Schepman); *Rosenia* Schepman, 1914 (new name for *Turtonia*); *Stylina* Fleming, 1828 not Lamarck, 1816 (*stilifer* Turton); *Venustilifer* Powell, 1939 (*bountyensis* Powell).

Pelseneeria brunneus (Tate, 1888)

Globose, but with an acuminate spire; sculptured throughout with minute spiral and axial striae. Dark brown.

5mm. NSW to SA. Parasites on the sea urchins *Heliocidaris erythrogramma* and *H. tuberculata*.

▲ GENUS **PICTOBALCIS** Laseron, 1955
Eulima articulata Sowerby

With the characters of the type species.

Eastern Australia. Monotypic. The massive, spotted, axially sculptured shell is said to differentiate the genus.

Pictobalcis articulata (Sowerby, 1834)

Solid, elongate, with many flat-sided whorls and distinct axial striae; aperture short; porcelaineous, not highly polished. Yellowish with brown spots or patches.

2.1 cm. Central Qld to central NSW. Host unknown. The figured shell was beach collected at Long Reef, NSW.

▲ GENUS **PULICICOCHLEA** Ponder & Gooding, 1978
P. calamaris Ponder & Gooding

Shells thin, glossy, uniformly dark brown; with an elongate aperture and sinuate outer lip; apertural side of the body whorl tending to be concave; protoconch squat, rounded, of few whorls; spire short to tall, tending to become asymmetrically immersed in the last whorl. Simultaneous hermaphrodites.

Indo-West Pacific. Ectoparasites on diadematid sea urchins. The genus and its four known species were described, with habitat and anatomical information, by Ponder & Gooding (1978, *Pacific Science* **32** (2): 157-181). One minute species is recorded from Australia.

Pulicicochlea fusca Ponder & Gooding, 1978

Protoconch of two whorls, the last enveloped by the penultimate teleoconch whorl; teleoconch of less than two whorls, the last very large, asymmetrical; aperture large, expanded, outer lip with a deep sinus posteriorly, anterior end broadly spatulate; columella thin, simple. Brown.

1.1 mm. Central Indo-West Pacific; north Qld. Parasitic on *Diadema setosum* and *D. savignyi*. For an illustration see original description (1978, *loc. cit.*, figs. 8, 9).

▲ GENUS **PYRAMIDELLOIDES** Nevill, 1884
Rissoina miranda Adams

Shell solid, tall and slender, teleoconch strongly sculptured; protoconch tall, smooth, of about three whorls.

Gill lacking; supraoesophageal ganglion single, undivided; osphradium simple, short, rounded.

Indo-West Pacific. Ectoparasites on ophiuroids. This genus has often been placed in the Rissoidae but the anatomy and parasitic life style show that it is a eulimid (Warén, 1983, *Zoologica Scripta* **12**: 273-294). The strongly sculptured teleoconch whorls of the shells are unusual in the family. The taxonomy is confused but the species illustrated here may represent the genus.

Pyramidelloides miranda (Adams, 1861)

Spire straight or slightly bent; sculptured with four strong spiral ribs, the upper one bearing heavy tubercles, and conspicuous axial striae in the interspaces; aperture narrowly ovate, outer lip thickened.

3.5mm. Indo-West Pacific; Port Hedland, WA to north Qld. A very common species which lives on *Macrophiothrix*. Warén (1983, *loc. cit.*) noted that there are two sizes in most samples, and assumed that the smaller and more slender specimens were males. It is quite variable and there are many synonyms including *pacifica*, *cylindrica* and *turris*, all of Laseron, 1956. The figured shell came from Lizard I., Qld.

▲ GENUS **SABINELLA** Monterosato, 1890
Eulima piriformis Monterosato (not Brugnone) = *pachya* Dautzenberg & Fischer

Shell pointed, conical, with a wide and angular aperture and several strong scars of earlier outer lips; possibly protandric hermaphrodites.

Cosmopolitan. Parasites of echinoids, some forming galls in the base of the spines. Synonym: *Eulimoda* Laseron, 1955 (*munita* Hedley).

Sabinella munita (Hedley, 1903)

Thin, transluscent, pyramidal, base subangulate, spire tall and acuminate; with 11 flat-sided whorls, impressed suture and irregular, but tressed varixes, surface sculptured with shallow, sometimes punctate grooves; aperture large, oblique, subquadrate, outer lip thin, inner lip spreading a sheet of callus over the parietal wall, columella broad, rather straight. White.

8 mm. NSW to SA. Parasitic on the sea urchin *Goniocidaris tubaria*. The figured specimen was dredged in Twofold Bay, NSW.

▲ GENUS **STICTEULIMA** Laseron, 1955
S. cameroni Laseron

Usually tall-spired, slender, with a high protoconch of 2-3 flat whorls; spotted or colourless.

Eastern Australia. Warén (1984) reported that there are many tropical species but noted that the type species "deviates from most" he included in the genus by its rather large size and distorted shell. Hosts unknown. When introducing this genus Laseron noted its affinity to *Curveulima*. Synonym: *Lentigobalcis* Habe, 1961 (*lentignosa* Adams).

Sticteulima cameroni Laseron, 1955

Subporcelaineous, transluscent, moderately tall, acuminate, with a bent spire; body whorl long, extended anteriorly, moderately inflated; aperture long, outer lip curved, with a strong subsutural sinus, inner lip strongly reflected and with a layer of callus on the parietal wall. White, irregularly covered with chestnut dots.

5.5 mm. North Qld to northern NSW. The figured shell came from the Swain Reefs, Qld.

▲ GENUS **STILAPEX** Iredale, 1925
S. lactarius Iredale

Shell thin, vitreous; whorls very inflated, spire tall and acuminate, lacking scars from the previous outer lips; protoconch of less than $2^{1}/_{2}$ whorls; aperture fairly wide, outer lip sinuous, columella a little thickened and reflected, umbilicus lacking. Proboscis large, pseudopallium, head and tentacles well developed, tentacles bearing black eyes; foot large and probably functional; operculum thin, paucispiral.

Indo-West Pacific and Southern Australia. Ectoparasites on ophiuroids. A replacement name for the preoccupied *Lambertia* Souverbie, 1869 (*montrouzieri* Souverbie). The genus was revised by Warén (1981, *Zoologica Scripta* **10**: 133-154).

In addition to the type species, the following Indo-West Pacific species have been recorded from Australia: *ophiuraphila* Habe, 1976 (on *Ophiactis savignyi*; Darwin, NT to Torres St., Qld); *montrouzieri* Souverbie, 1869 (on *Macrophiothrix* cf. *longipeda*, Murray I., Qld); *parva* Schepman, 1909 (on *Ophiothrix crassispina*, NSW to SA); *thielei* Sturany, 1903 (on *Macrophiothrix longipeda*, NT to Keppel Is., Qld).

Stilapex lactarius Iredale, 1925.

Protoconch of two slender whorls, teleoconch of about six slightly shouldered whorls, the last very tumid.

8 mm. NSW to Tas.; also recorded from Japan. Warén recorded a male specimen from Tas. attached to the base of an arm of *Ophiothrix aristulata*. The illustrated shell came from 125 m in Bass St.

▲ GENUS **STILIFER** Broderip, 1832
S. astericola Broderip

Shell thin, fragile, inflated, especially the body whorl, aperture markedly oblique, with sculpture of distinct incremental lines; larval shell mucronate.

Indo-West Pacific. Cyst or gall-forming parasites on starfishes, adults becoming embedded in the host test but with the tip of the spire often projecting from the small opening. There is evidence that in *S. astericola* the males remain as ectoparasites. These animals are consecutive hermaphrodites, males being much smaller than females. The genus has been revised by Warén (1980, *Zoologica Scripta* **9**: 187-210) giving a key to the species.

Australian species: *concavus* Warén, 1980 (described from an undetermined starfish, Murray I. to Cape Moreton, Qld); *inflatus* Warén, 1980 (in *Linckia laevigata* at a depth of 15 m, described from Lizard I., Qld); *linckiae* Sarasin & Sarasin, 1887 (in species of *Linckia,* Indo-West Pacific, south to the Capricorn Group, Qld); *utinomi* Habe, 1951 (in species of *Linckia, Ophidiaster* and *Tamaria*, Japan and south to the Capricorn Group, Qld); *variabilis* Boettger, 1893 (host not known, central Indo-West Pacific, Exmouth, WA to Groote I., NT).

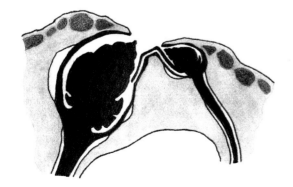

Schematic drawing of *Stilifer* sp. in their echinoderm host. Large specimen female; small specimen male. (Redrawn from Warén, 1980, p. 193, fig. 22)

Stilifer inflatus Warén, 1980

Broad, solid, semitransparent; spire concave, body whorl large and inflated; aperture more than half the shell length. Without spots.

7.4 mm. North Qld. Parisitic on the starfish *Linkia laevigata*. The illustrated specimen is the holotype collected at 15 m, North Barrier Reef I., Qld.

▲ GENUS **SCALENOSTOMA** Deshayes, 1863
S. carinata Deshayes

Tall-spired shells, often with the apex bent or distorted, columella well formed and reflected.

Cosmopolitan. This genus has been classified as an epitoniid and an aclidid but in a recent revision Warén (1980, *Zoologica Scripta* **9**: 187-210) argues for a place for it in the Eulimidae. Unlike others in the family, however, the species appear to be parasites of corals. There are three described species, one from Australia. Synonyms: *Hyperlia* Pilsbry, 1918 (*apiculata* Souverbie); *Stilimella* Laseron, 1955 (*lodderae* Petterd).

Scalenostoma lodderae Petterd, 1884

Sexually dimorphic with females broader than males; whorls convex, suture impressed; columellar lip reflected, sometimes slightly spouted anteriorly. Polished, white.

1.2 cm. NSW to Tas. Host not known. According to Warén, *robusta* Petterd was based on a tumid female specimen while *lodderae* was described from a more slender male specimen. Synonyms: *robusta* Petterd, 1884; *crotaphis* Watson, 1886; *petterdi* Tate & May, 1901.

▲ GENUS **THYCA** H. & A. Adams, 1854
Pileopsis astericola Adams & Reeve

Adult shells limpet-shaped, resembling capulids but with a typical eulimid larval shell.

Indo-West Pacific. Ectoparasites on starfish. It has been customary to classify these starfish ectoparasites in the ciliary feeding family Capulidae, but Warén (1980) has listed a number of differences showing that they are better placed here. The genus was revised by Warén (1991, *Zoologica Scripta* **9**: 187-210). Illustrations may be found in that publication. Synonyms: *Kiramodulus*

Kuroda, 1949 (*lacteus* Kuroda); *Bessomia* Berry, 1959 (*callista* Berry); *Granulithyca* Habe, 1976 (*nardoafrianti*).

The following species are reported from Australia: *cf. astericola* Adams & Reeve, 1850 (central Indo-West Pacific, central Qld); *crystallina* Gould, 1846 (on *Linckia,* Indo-West Pacific, north Qld) = *pellucida* Kukenthal,1897; *Thyca stellasteris* Köhler & Vaney, 1912 (WA on *Stellaster equestris*).

▲ GENUS **TROCHOSTILIFER** Warén, 1980
T. domus Warén

Shell trochiform with distinct spiral sculpture; males dwarf.

Indo-West Pacific. Parasitic on cidaroid sea urchins, sometimes forming galls in the test or spines of the host. The trochiform shells are quite distinctive within the family. There are five species, one from Australia.

Trochostilifer striatus (Hedley, 1905)

Conical, whorls slightly convex, basal periphery subangulate, suture impressed; protoconch domed, of about 1½ whorls; teleoconch of five whorls, sculptured with numerous strong spiral cords; aperture wide, slightly flaring anteriorly, columella a little reflected. White.

5.5 mm. Recorded from Japan and Qld south to Whitsunday Group. Attached to the sea urchin *Prionocidaris baculosa* and possibly *Goniocidaris* sp. The illustrated specimen was parasitising a sea urchin (*Phyllacanthus annulifera*) found off Gladstone, Qld. Synonym: *prionocidaricola* Habe, 1974.

▲ GENUS **VITREOBALCIS** Warén, 1980
Apicalia holdsworthi Adams, 1874

Shells conical with convex whorls and sloping, evenly rounded base; aperture short, ovate. Proboscis with a large, retractile, rounded collar which protects the soft parts and the base of the shell.

Indo-West Pacific. Ectoparasites on sea urchins. Warén noted that the shells of this genus resemble those of *Echineulima*, which have large snouts but lack the proboscis collar and live permanently attached to their sea urchin hosts, and *Apicalia*, which have a rounded aperture and parasitise starfish.

Vitreobalcis laevis Warén, 1980

Vitreous, rather thin; protoconch tall, of about three whorls; teleoconch of 5-6 convex and perfectly smooth whorls, lacking scars from earlier lips; outer lip retracted at the suture, slightly flaring, distinctly protruding at its centre when viewed from the side. White.

4.2 mm. Known only from the holotype collected on the apodan holothurian *Synaptula purpurea,* off Townsville, Qld. Although Warén provisionally assigned this to his new genus *Vitreobalcis* he noted that "it probably does not belong here". The host is certainly different from that of the type species.

INDEX TO SHELL PLATES

Plate 1

ACMAEIDAE, PATELLIDAE

Plate 2

NERITIDAE, NERITOPSIDAE

1 *Nerita costata* (p. 40)
      a-b Yeppoon, Qld.

2 *Nerita grossa* (p. 40)
      a-b Cocos Keeling.

3 *Nerita reticulata* (p. 41)
      a-c St. George Basin, WA.

4 *Nerita squamulata* (p. 41)
      a   Ningaloo Marine Park, WA.
      b   Long I., WA.
      c-e Port Essington, NT.

5 *Nerita atramentosa* (p. 40)
      a-b Point Peron, WA.

6 *Nerita chamaeleon* (p. 40)
      a-e Woody I., Qld.

7 *Nerita undata* (p. 41)
      a-b One Arm Pont, WA.
      c   Cape Leveque, WA.

8 *Nerita plicata* (p. 40)
      a-b Ningaloo Marine Park, WA.

9 *Nerita balteata* (p. 40)
      a-b St. George Basin, WA.

10 *Nerita polita* (p. 40)
      a-c Qld.

11 *Nerita planospira* (p. 40)
      a-b North Qld.

12 *Nerita albicilla* (p. 40)
      a-c Ningaloo Marine Park, WA.

13 *Neritopsis radula* (p. 38)
      a-b Ashmore Reef, NT.

14 *Neritina zigzag* (p. 41)
      a-b Marike, PNG.

15 *Neritina cf. zigzag* (p. 41)
      a-d west of Darwin, NT.

16 *Neritina cf. variegata* (p. 42)
      a-b Kalumburu, WA.

17 *Neritina oualaniensis* (p. 42)
      a-i Cairns, Qld.

18 *Neritina violacea* (p. 41)
      a-b Marike, PNG.

19 *Neritina violacea* (p. 41)
      a-b Ross Creek, Yeppoon, Qld.
      c-d Keppel Bay, Qld.

Approx. x 1

Plate 3
PLEUROTOMARIIDAE(BELOW)

1A

1B

1   *Perotrochus westralis* (p. 45)
         a-b   North West Shelf,
             off Port Hedland, WA.

Approx. x .75

Plate 4
HALIOTIDAE(OPPOSITE)

1   *Haliotis squamata* (p. 50)
         Ningaloo Marine Park, WA
         (see also Pl. 5 figs. 10 a-b).

2   *Haliotis elegans* (p. 48)
         a-b   Rottnest I., WA.

3   *Haliotis asinina* (p. 48)
         a-b   Ningaloo Marine Park, WA.
         (see also p. 300).

4   *Haliotis roei* (p. 49)
         a   Marmion Marine Park, WA.
         b   Point Brown, SA.

5   *Haliotis ovina* (p. 49)
         a-b   Barrow I., WA.

6   *Haliotis emmae* (p. 49)
         Flinders, Vic.

7   *Haliotis scalaris* (p. 50)
         a-b   Marmion Marine Park, WA.
         (see also illustration on p. 47)

8   *Haliotis cyclobates* (p. 48)
         a-b   Oyster Harbour, SA.

9   *Haliotis conicopora* (p. 48)
         Garden I., WA.

10   *Haliotis laevigata* (p. 49)
         Two Peoples Bay, WA.

Approx. x .6

Plate 5

HALIOTIDAE

1   *Haliotis brazieri* (p. 48)
         a-b Sydney, NSW.

2   *Haliotis varia* (p. 50)
         a   One Arm Point, WA.
         b   Norfolk I.

3   *Haliotis hargravesi* (p. 49)
              Jervis Bay, SA.

4   *Haliotis melculus* (p. 49)
              Point Cartwright, Qld.

5   *Haliotis ethologus* (p. 49)
              Point Cartwright, Qld.

6   *Haliotis planata* (p. 49)
              Cartier I., NT.

7   *Haliotis coccoradiata* (p. 48)
         a   Botany Bay, NSW.
         b   Sydney, NSW.

8   *Haliotis semiplicata* (p. 50)
         a   Marmion Marine Park, WA.
         b   Cottesloe, WA.
         c   Rottnest I., WA.

9   *Haliotis crebrisculpta* (p. 48)
         a   Wilson I., Qld.
         b   Ningaloo Marine Park, WA.

10   *Haliotis squamata* (p. 50)
         a-b Ningaloo Marine Park, WA
              (see also Pl. 4, fig. 1).

Haliotis asinina has a large and attractive animal. Ningaloo Marine Park, WA. (Photo: Barry Wilson)

Approx. x 1.1

Plate 6

FISSURELLIDAE

1 *Macroschisma munita* (p. 54)
    a   Marmion Marine Park, WA.
    b   North West Cape, WA.

2 *Macroschisma producta* (p. 54)
    Augusta, WA.

3 *Macroschisma hiatula* (p. 54)
    Lord Howe I.

4 *Amblychilepas javanicensis* (p. 53)
    Yanchep, WA.

5 *Amblychilepas nigrita* (p. 53)
    Bunker Bay, WA.

6 *Amblychilepas oblonga* (p. 53)
    Mundrabilla, WA.

7 *Macroschisma bakiei* (p. 54)
    a   Point Peron, WA.
    b   Penguin Island, WA.

8 *Emarginula incisura* (p. 55)
    a   80 Mile Beach, WA.
    b   Yeppoon, Qld.

9 *Emarginula patula* (p. 56)
    Windy Harbour, WA.

10 *Emarginula dilecta* (p. 55)
    Two Peoples Bay, WA.

11 *Emarginula candida* (p. 55)
    Two Peoples Bay, WA.

12 *Emarginula gabensis* (p. 55)
    Gabo I., Vic. (HOLOTYPE)

13 *Amblychilepas crucis* (p. 53)
    Bass Strait, Tas.

14 *Amblychilepas compressa* (p. 53)
    Rottnest I., WA.

15 *Cosmetalepas concatenatus* (p. 53)
    Garden I., WA.

16 *Diodora singaporensis* (p. 52)
    a-b Exmouth Gulf, WA.
    c   Noosa, Qld.

17 *Diodora ticaonica* (p. 53)
    a, c Port Gregory, WA.
    b   Port Hedland, WA.

18 *Diodora ruppelli* (p. 52)
    a-b Oman.

19 *Diodora galeata* (p. 52)
    Green I., Qld.

20 *Diodora granifera* (p. 52)
    Boat Harbour, Qld.

21 *Montfortula rugosa* (p. 57)
    a-b Yamba, NSW.

22 *Montfortula variegata* (p. 57)
    a-b Dampier Arch., WA.

23 *Montfortula pulchra* (p. 57)
    One Arm Point, WA.

24 *Diodora occidua* (p. 52)
    Petit Point, Shark Bay, WA.

25 *Diodora jukesii* (p. 52)
    a   Yeppoon, Qld.
    b   Cockatoo I., WA.
    c   Dampier Arch., WA.

26 *Diodora sp.*
    Stradbroke I., Qld.

27 *Diodora lineata* (p. 52)
    Sydney, NSW.

28 *Diodora lincolnensis* (p. 52)
    Esperance, WA.

29 *Scutus antipodes* (p. 59)
    Marmion Marine Park, WA.

30 *Scutus unguis* (p. 59)
    Port Hedland, WA.

31 *Tugali parmophoidea* (p. 59)
    a   North West Reef, NSW.
    b   Port Stephens, NSW.

32 *Tugali cicatricosa* (p. 59)
    a   Dunsborough, WA.
    b   Duke of Orleans Bay, WA.

33 *Hemitoma excentrica* (p. 57)
    a-c Monte Bello I., WA.

34 *Hemitoma subemarginata* (p. 56)
    a   Albany, WA.
    b   SA.
    c   Windy Harbour, WA.

Approx. x 1.1

1A 1B 2 3 4 5 6 7A 7B

8A 8B 9 10 11 12 13 14 15

16A 17A 18A 21A 21B

16C

16B 17B 17C 18B 19 20 22A 22B 23

24 25A 27 28

26 25B 25C

29 30 31A 32A 33A

31B 32B 33C

33B

34A 34B 34C

Plate 7

TROCHIDAE (OPPOSITE)

1   *Monodonta labio* (p. 78)
       a-b   Exmouth Gulf, WA.

2   *Austrocochlea rudis* (p. 76)
       Two Peoples Bay, WA.

3   *Austrocochlea constricta* (p. 75)
       a   North Sydney, NSW.
       b   Flinders, Vic.

4   *Austrocochlea concamerata* (p. 75)
       a-b   Two Peoples Bay, WA.

5   *Austrocochlea zeus* (p. 76)
       Ningaloo Marine Park, WA.

6   *Austrocochlea odontis* (p. 75)
       Flinders, Vic.

7   *Austrocochlea adelaidae* (p. 75)
       Port Phillip Bay, Vic.

8   *Austrocochlea crinita* (p. 75)
       a-b   Frenchman Bay, WA.

9   *Granata imbricata* (p. 67)
       a-b   Two Peoples Bay, WA.

10   *Trochus cf. stellatus* (p. 90)
       Dampier Arch., WA.

11   *Trochus histrio* (p. 90)
       a-b   One Arm Point, WA.

12   *Tectus triserialis* (p. 90)
       Broadhurst Reef, Qld.

13   *Tectus fenestratus* (p. 89)
       a-b   Port Hedland, WA.

14   *Trochus hanleyanus* (p. 90)
       a-b   Koolan I., WA.
       c   Dampier Arch., WA.

15   *Trochus maculatus* (p. 90)
       a   Scott Reef, WA.
       b   Cocos Keeling.

16   *Trochus niloticus* (p. 90)
       One Arm Point, WA.

17   *Tectus pyramis* (p. 89)
       a-b   Shark Bay, WA.

Plate 7a (BELOW)

1A

1B

1   *Angaria delphinus* (p. 96)
       a-b   One Arm Point, WA

Approx. x .75

Approx. x 1

Plate 8

TROCHIDAE

Approx. x 1.5

Plate 9
TROCHIDAE

1   *Thalotia chlorostoma* (p. 83)
        a   Horrocks Beach, WA.
        b   Geraldton, WA.
        c   Geographe Bay, WA.
        d   Marmion Marine Park, WA.
        e   East Wallaby I., Abrolhos, WA.

2   *Thalotia conica* (p. 83)
           Marmion Marine Park, WA.

3   *Phasianotrochus eximius* (p. 82)
        a   Ellensbrook, WA.
        b   Cape Naturaliste, WA.
        c   Largs Bay, SA.
        d   Long Reef, NSW.

4   *Phasianotrochus bellulus* (p. 82)
        a   Cockburn Sound., WA.
        b   Cape Peron, WA.

5   *Phasianotrochus apicinus* (p. 82)
           Recherche Arch., WA.

6   *Phasianotrochus irisodontes* (p. 82)
        a-b   Fremantle., WA.

7   *Notogibbula preissiana* (p. 79)
           Nannarup, WA.

8   *Notogibbula lehmanni* (p. 79)
        a   Cockburn Sd., WA.
        b   Busselton, WA.

9   *Bankivia fasciata* (p. 91)
        a-j   Bateman Bay, WA.

10   *Cantharidus ramburi* (p. 80)
           Flinders, Vic.

11   *Cantharidus pulcherrimus* (p. 80)
           Flinders, Vic.

12   *Cantharidus lehmanni* (p. 80)
        a-b   Two Peoples Bay, WA.

13   *Jujubinus polychromus* (p. 81)
        a-b   Ningaloo Marine Park, WA.

14   *Jujubinus gilberti* (p. 80)
        a   Dampier Arch., WA.
        b   Admiralty Gulf, WA.

15   *Cantharidus sp.*
           North Sydney.

16   *Jujubinus lepidus* (p. 81)
        a   Two Peoples Bay, WA
           (large protoconch form).
        b   Frenchman Bay, WA
           (small protoconch form).

17   *Calthalotia baudini* (p. 81)
           Shark Bay, WA.

18   *Calthalotia mundula* (p. 82)
        a-b   Shark Bay, WA.
        c   Shark Bay, WA (HOLOTYPE).
        d   Roebuck Bay, WA.

19   *Calthalotia comtessi* (p. 81)
           Sydney, NSW (HOLOTYPE).

20   *Calthalotia arruensis* (p. 81)
        a   Stradbroke I., Qld.
        b   Bowen, Qld.

21   *Calthalotia strigata* (p. 82)
        a   Swan Pt., WA (HOLOTYPE).
        b   Exmouth Gulf, WA.
        c   One Arm Point, WA.

1A

1B

1C

1D

1E

2

3A

3B

3C

3D

4A

4B

5

6A

6B

7

8A

8B

9A

9B

9C

9D

9E

9F

9G

9H

9I

9J

10

11

12A

12B

13A

13B

14A

14B

15

16A

16B

17

18A

18B

18C

18D

19

20A

20B

21A

21B

21C

Plate 10
TROCHIDAE

1   *Hybochelus cancellatus* (p. 69)
     Ningaloo Marine Park, WA.

2   *Tallorbis roseolus* (p. 69)
     Capricorn Group, Qld
     (HOLOTYPE).

3   *Herpetopoma atrata* (p. 68)
     Cape Borda, WA.

4   *Herpetopoma aspersa* (p. 68)
     Marmion Marine Park, WA.

5   *Chrysostoma paradoxum* (p. 77)
     a-b   North Qld.

6   *Calliostoma allporti* (p. 62)
     Westernport, Vic. (HOLOTYPE).

7   *Calliostoma hedleyi* (p. 62)
     Westernport, Vic.

8   *Calliostoma legrandi* (p. 63)
     Westernport, Vic.

9   *Calliostoma comptum* (p. 62)
     Eden, NSW.

10   *Astele allanae* (p. 66)
     a-b   Capricorn Group., Qld.
     (SYNTYPES).

11   *Calliostoma incertum* (p. 64)
     Westernport, Vic.

12   *Astele scitulum* (p. 65)
     Sydney NSW.

13   *Astele multigranum* (p. 65)
     a-b   Dunsborough, WA.

14   *Calliostoma monile* (p. 65)
     Broome, WA.

15   *Calliostoma rubiginosum* (p. 63)
     Leighton, WA.

16   *Calliostoma similarae* (p. 63)
     a   Carnarvon, WA.
     b   Shark Bay, WA.

17   *Astele bularra* (p. 65)
     Cape Moreton, Qld.

18   *Astele ciliare* (p. 65)
     a-b   Esperance, WA.

19   *Astele speciosum* (p. 65)
     North Stradbroke I., Qld.

20   *Astele subcarinatum* (p. 65)
     a-b   Great Australian Bight, WA.

21   *Calliostoma armillatum* (p. 63)
     a-b   Perkins Bay, Tas.

Approx. x 1.2

Plate 11

TROCHIDAE, TURBINIDAE

1   *Phasianella australis* (p. 102)
    a-b   Albany, WA.

2   *Phasianella ventricosa* (p. 102)
    a-b   Duke of Orleans Bay, WA.

3   *Phasianella solida* (p. 102)
    a   Cheyne Beach, W.A.
    b   Garden I., WA.

4   *Phasianella variegata* (p. 102)
    a,e   Cheyne Beach, WA.
    c   Garden I., WA.
    b, d Ningaloo Marine Park, WA.

5   *Microtis rubra* (p. 73)
          Fremantle, WA.

6   *Stomatia phymotis* (p. 74)
    a   Ningaloo Marine Park, WA.
    b   Monte Bello Is., WA.
    c   Cape Borda, WA.

7   *Stomatella impertusa* (p. 74)
    a   Port Lincoln, SA.
    b   North Sydney, NSW.
    c   Mandurah, WA.

8   *Pseudostomatella papyracea* (p. 74)
    a-b   Ningaloo Marine Park, WA.

9   *Pseudostomatella maculata* (p. 73)
    a-b   Port Douglas, Qld.

10 *Pseudostomatella decolorata* (p. 73)
    a   Barrow I., WA.
    b   Pumpkin I., Qld.

11 *Pseudostomatella sp.* (p. 74)
    a   Ningaloo Marine Park, WA.
    b   Cape Borda, WA.

12 *Microtis tuberculata* (p. 73)
    a   Barrow I., WA.
    b   Monte Bello Is., WA.

13 *Ethalia guamensis* (p. 93)
    a   North Qld.
    b-c   Arafura Sea.

14 *Umbonium vestiarium* (p. 95)
    a-f   Indonesia.

15 *Ethalia pulchella* (p. 93)
    a-b   Dampier Arch., WA.

16 *Ethminolia vitiliginea* (p. 93)
    a-b   Dunsborough, WA.

17 *Talopena vernicosa* (p. 95)
    a-b   Abrolhos, WA.

18 *Monilea morti* (p. 94)
          Sydney, NSW (HOLOTYPE).

19 *Monilea callifera* (p. 94)
    a   Bowen, Qld.
    b-c   Cockburn Sound., WA.

20 *Monilea belcheri* (p. 94)
    a-b   North Qld.

21 *Angaria tyria* (p. 96)
    a   Cockburn Sound, WA.
    b   Shark Bay, WA.

22 *Angaria delphinus* (p. 96)
          Scott Reef, WA.
          (See also pl. 7a, fig. 1 a-b).

Approx. x 1

1A 1B 2A 2B

4A 3A 3B 4E

4B 4C 4D 5

6A 6C 7A 7B 7C

6B 8B 9B 10B 11B 12B

8A 9A 10A 11A 12A

13B 14D 14B 14F 15B 16B 17B

13A 14A 14C 14E 15A 16A 17A

13C

18 19A 19B 19C 20A 20B

21A 21B 22

Plate 12
TURBINIDAE

Approx. x .7

Plate 13

TURBINIDAE

Approx. x 1.1

Plate 14

TURRITELLIDAE, MODULIDAE, PLANAXIDAE, EPITONIIDAE

1   *Archimediella maculata* (p. 135)
        a-b   Saudi Arabia.
        c   East Aden.

2   *Colpospira congelata* (p. 138)
        off Onslow, WA; 50 m.

3   *Gazameda iredalei* (p. 139)
        Yorke Peninsula, SA.

4   *Colpospira runcinata* (p. 137)
        southcoast, WA; 180-300 m.

5   *Gazameda gunnii* (p. 139)
        a-b   Botany Bay, NSW; 100 m

6   *Colpospira atkinsoni* (p. 136)
        Babel I., Bass St., Tas.; 119 m.

7   *Eglisia tricarinata* (Epitoniidae; p. 275)
        off Troughton I., WA; 63 m.

8   *Colpospira guilleaumei* (p. 138)
        Twofold Bay, NSW; 9-27 m.
        HOLOTYPE.

9   *Haustator cingulifera* (p. 139)
        a-b   Dampier Arch., WA; 42 m.

10   *Eglisia tricarinata* (Epitoniidae; p. 275)
        a   off Onslow, WA; 52 m.
        b   off Rottnest I., WA; 144 m.

11   *Archimediella fastigiata* (p. 135)
        a-b   Ceram, Indonesia; 58 m.
        c   Port Samson, WA; intertidal.

12   *Gazameda declivis* (p. 139)
        Rottnest I., WA; 150-160 m.

13   *Turritella terebra* (p. 140)
        Darwin, NT.

14   *Maoricolpus roseus* (p. 140)
        Bruny I., Tas.; 11 m.

15   *Modulus tectum* (p. 129)
        a-b   Ningaloo Marine Park, WA.

16   *Hinea braziliana* (p. 130)
        a-b   Maroubra Beach, Sydney, NSW.

17   *Planaxis sulcatus* (p. 131)
        a-b   Ningaloo Marine Park, WA.

18   *Fissilabia decollata* (p. 130)
        a-b   Murray I., Torres St., Qld.

      Approx. x 1

Plate 15

VERMETIDAE, SILIQUARIIDAE, POTAMIDIDAE, DIASTOMATIDAE, CAMPANILIDAE

1   *Serpulorbis sipho* (p. 171)
       Cockburn Sound, WA.

2   *Siliquaria weldii* (p. 141)
       southern WA.

3   *Siliquaria cf. anguina* (p. 141)
       Shark Bay, WA.

4   *Siliquaria anguina* (p. 141)
       Aru, Indonesia.

5   *Siliquaria ponderosa* (p. 141)
       Ningaloo Marine Park, WA.

6   *Petaloconchus sp.* (p. 171)
       Shark Bay, WA.

7   *Siliquaria sp.* (p. 141)
       Marmion Marine Park, WA.

8   *Serpulorbis cf. sipho* (p. 171)
       Shark Bay, WA.

9   *Campanile symbolicum* (p. 112)
       Shoalwater Bay, WA.

10   *Diastoma melanioides* (p. 126)
       a-b Cheyne Beach, WA.

11   *Velacumantus australis* (p. 114)
       Swan Estuary, WA.

12   *Cerithidea largillierti* (p. 133)
       Townsville, Qld.

13   *Zeacumantus diemenensis* (p. 114)
       a-b Oyster Harbour, WA.

14   *Cerithidea cingulata* (p. 133)
       a-c St. George Basin, WA.

15   *Cerithidea obtusa* (p. 133)
       Townsville, Qld.

16   *Cerithidea anticipata* (p. 132)
       Prince Regent River, WA.

17   *Cerithidea reidi* (p. 133)
       a-b Broome, WA.

18   *Terebralia palustris* (p. 134)
       Kimberley, WA.

19   *Terebralia semistriata* (p. 134)
       a-b Barrow I., WA.

20   *Pyrazus ebeninus* (p. 113)
       NSW.

21   *Telescopium telescopium* (p. 133)
       Qld.

Approx. x 1.3

Plate 16

CERITHIIDAE

Approx. x 1

1

2A

2B

3A

3B

4A

4B

5A

6A

6B

7A

7B

8A

8B

5B

9

12A

10

11A

11B

12B

12C

13A

13B

14

15

16

17

Plate 17

CERITHIIDAE

Approx. x 1.5

1A 1B 2 3 4 5
6A 6B 7A 7B 7C 8A 8B 8C 8D
9A 9B 10 11A 11B 12A 12B
13A 13B 13C 14 15 16A 16B
17A 17B 17C 17D 18
19 20 21A 21B 22A 22B

Plate 18

LITTORINIDAE

1   *Littoraria coccinea* (p. 144)
        a-b Murray I., Torres St., Qld.

2   *Littoraria undulata* (p. 144)
        a-b Yeppoon, Qld.
        c-d Ningaloo Marine Park, WA.

3   *Littorina unifasciata* (p. 146)
        Marmion Marine Park, WA.

4   *Littorina praetermissa* (p. 146)
        a-b Port Fairy, Vic.

5   *Nodilittorina pyramidalis* (p. 147)
        a-b Ningaloo Marine Park, WA.

6   *Littoraria articulata* (p. 146)
        a-b Yeppoon, Qld.

7   *Littoraria cingulata pristissini* (p. 145)
        a-d Shark Bay, WA (PARATYPES).

8   *Littoraria cingulata cingulata* (p. 145)
        a-b Broome, WA.

9   *Nodilittorina millegrana* (p. 147)
        a-b Ningaloo Marine Park, WA.

10   *Nodilittorina australis* (p. 147)
        a-b Marmion Marine Park, WA.

11   *Nodilittorina nodosa* (p. 147)
        a-b Ningaloo Marine Park, WA.

12   *Littoraria filosa* (p. 145)
        a-d Darwin, NT.

13   *Littoraria philippiana* (p. 145)
        a-b Orpheus I., Qld.

14   *Littoraria pallescens* (p. 145)
        a-c Gove Peninsula., NT.
        d   Barrow I., WA.

15   *Littoraria sulculosa* (p. 146)
        a   Broome, WA.
        b   Hinchinbrook I., Qld.

16   *Littoraria scabra* (p. 146)
        Cape York, Qld.

17   *Littoraria luteola* (p. 145)
        a-b Port Jackson, NSW.

18   *Littoraria intermedia* (p. 145)
        Magnetic I., Qld.

19   *Bembicium nanum* (p. 143)
        a-c Keppel I., Qld.
        d-e Port Phillip Bay, Vic.

20   *Tectarius rusticus* (p. 148)
        Koolan I., WA.

21   *Bembicium auratum* (p. 143)
        a-b Yeppoon, Qld.

22   *Bembicium flavescens* (p. 143)
        a-b Lord Howe I.

23   *Bembicium vittatum* (p. 144)
        a-b Streaky Bay, SA.

24   *Bembicium melanostomum* (p. 143)
        a-b Westernport, Vic.

Approx. x 1.4

1A

1B

2A

2B

2C

2D

3

4A

4B

5A

5B

6A

6B

7A

7B

7C

7D

8A

8B

9A

9B

10A

10B

11A

11B

12A

12B

12C

12D

13A

13B

14A

14B

14C

14D

15A

15B

16

17A

17B

18

19A

19B

19C

19D

19E

20

21A

21B

22A

22B

23A

23B

24A

24B

Plate 19
STROMBIDAE

1   *Lambis lambis* (p. 154)
      a   Cocos Keeling
           (white form).
      b   Hook Reef, Qld.
           (typical female form).

2   *Lambis scorpius* (p. 154)
      a-b North Qld.

3   *Lambis millepeda* (p. 154)
      East Java.

4   *Lambis crocata* (p. 154)
      Rowley Shoals, WA.

5   *Lambis chiragra* (p. 155)
      a-b Scott Reef, WA.

6   *Lambis truncata* (p. 154)
      Scott Reef, WA.

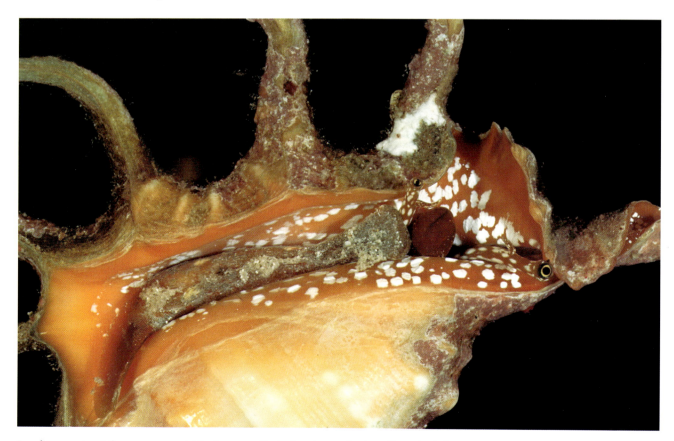

Lambis scorpius. Like most strombids the outer lip is deeply notched to allow the left eye-stalk to protrude while the shell remains aperture downwards. (Photo: Barry Wilson)

Approx. x .5

Plate 20

STROMBIDAE

Strombus bulla dredged in the Arafura Sea. The remarkable eyes of strombs give an impression that they are capable of more acuity than is probably the case. (Photo: Barry Wilson)

Approx. x .6

Plate 21
STROMBIDAE

Approx. x 1.2

1A
1B
2A
2B
3A
3B
4A
4B
5A
5B
5C
6
7
8A
8B
9
10A
10B
11A
11B
12A
12B
13A
13B
14A
14B
15
16A
16B
16C
17
18A
18B

Plate 22

HIPPONICIDAE, CAPULIDAE, CALYPTRAEIDAE, VANIKORIDAE, PHENACOLEPADIDAE

1   *Lippistes gabrieli* (p. 166)
         a-b   Westernport, Vic.

2   *Lippistes helicoides* (p. 166)
         loc. unknown, northern Aust.

3   *Cheilea equestris* (p. 163)
         a-b   Ningaloo Marine Park, WA.

4   *Cheilea flindersi* (p. 164)
         Daly Head, SA (HOLOTYPE).

5   *Cheilea occidua* (p. 164)
         Ellensbrook, WA (HOLOTYPE).

6   *Capulus sycophanta* (p. 165)
         a   SA (probably erroneus).
         b   Keppel Bay, Qld; on *Amusium*.
         c   Little Broadhurst Reef, Qld.

7   *Capulus danieli* (p. 165)
         Ningaloo Marine Park, WA.

8   *Capulus liberatus* (p. 165)
         Thomas Cay, Qld.

9   *Capulus sp.*
         North West Cape, WA.

10   *Capulus violaceus* (p. 166)
         off Sydney, NSW; 503 m.

11   *Hipponix australis* (p. 161)
         a-c   Two Peoples Bay, WA.

12   *Calyptraea calyptraeformis* (p. 163)
         a-b   Sandringham, Vic.

13   *Crepidula aculeata* (p. 164)
         a-c   Shark Bay, WA.

14   *Zeacrypta immersa* (p. 164)
         a-b   Westernport, Vic.

15   *Vanikoro gueriniana* (p. 162)
         Middle Harbour, Sydney, NSW.

16   *Vanikoro cancellata* (p. 162)
         a-b   Lindeman I., Qld.

17   *Cinnalepeta cinnamomea* (p. 43)
         a-b   Lake Macquarie, NSW.

18   *Phenacolepas crenulata* (p. 43)
         a-b   North Keppel I., Qld.

Plate 23

XENOPHORIDAE

1   *Xenophora exuta* (p. 169)
         a-b Joseph Bonaparte Gulf,
               WA; 100 m.

2   *Xenophora solarioides* (p. 168)
         Keppel Bay, Qld; 45 m.

3   *Xenophora indica* (p. 169)
         a-b Keppel Bay, Qld; 30 m.

4   *Xenophora peroniana* (p. 168)
         a-b Wollongong, NSW; 73 m.

5   *Xenophora flindersi* (p. 168)
         Esperance, WA; 9 m.

6   *Xenophora cerea* (p. 167)
         Michaelmas Cay, Qld; 46 m.

7   *Xenophora gigantea* (p. 169)
         a-b off Scott Reef, WA; 400 m.

8   *Xenophora pallidula* (p. 168)
         a-b off Scott Reef, WA; 400 m.
         c   off Clarence R., NSW; 457 m.

9   *Xenophora mekranensis konoi* (p. 168)
         a   Bunker Group, Qld; 220 m.
         b   off Lynx & Leopard Reefs, Qld;
               175 m.

10   *Xenophora chinensis* (p. 168)
         West of Dirk Hartog I., WA;
         133 m.

Xenophora cerea. Piru Bay, Ceram, Indonesia. (Photo: Barry Wilson)

Approx. x .6

Plate 24

OVULIDAE, TRIVIIDAE, LAMELLARIIDAE

Approx. x 2

339

Plate 25
OVULIDAE (BELOW)

| 1 | *Ovula ovum* (p. 202) | | 2 | *Ovula costellata* (p. 202) | Approx. x .6 |
| | a-b Ningaloo Marine Park, WA. | | | a-b Southern Qld. | |

Left - A lamellariid from Cape Naturaliste, WA.

Opposite page - *Volva volva* living on fronds of a soft-coral trawled off Townsville.

Photos: Barry Wilson

Plate 26

CYPRAEIDAE

Cypraea (Zoila) friendii complex (pp. 193 - 195)

1   *Cypraea (Zoila) friendii jeaniana*
         off Quobba, WA (*sherylae* form).

2   *Cypraea (Zoila) friendii jeaniana*
         off Quobba, WA (*sherylae* form).

3   *Cypraea (Zoila) friendii jeaniana*
         off Quobba,WA (*sherylae* form
         dwarf).

4   *Cypraea (Zoila) friendii jeaniana*
         off Quobba, WA (*sherylae* form).

5   *Cypraea (Zoila) friendii jeaniana*
         West of Dorre I., WA; 110 m.

6   *Cypraea (Zoila) friendii jeaniana*
         West of False Entrance, Shark Bay
         WA; 55 m.

7   *Cypraea (Zoila) friendii friendii*
         Eagle Bay, Cape Naturaliste, WA;
         9 m.

8   *Cypraea (Zoila) friendii friendii* (dwarf)
         Quindalup, WA; 3 m.

9   *Cypraea (Zoila) friendii friendii*
         Marmion Marine Park, WA; 5 m.

10   *Cypraea (Zoila) friendii friendii*
         Parmelia Bank, Fremantle, WA; 5 m.

11   *Cypraea (Zoila) friendii friendii*
         off Mandurah, WA; trawled, 60 m.

12   *Cypraea (Zoila) friendii friendii*
         Geographe Bay, WA; 18 m.

13   *Cypraea (Zoila) friendii vercoi*
         Esperance, WA; 5 m.

14   *Cypraea (Zoila) friendii vercoi*
         King George Sound, WA; 25 m.

15   *Cypraea (Zoila) friendii vercoi*
         Hopetoun, WA.

16   *Cypraea (Zoila) friendii vercoi*
         Eucla, WA.

17   *Cypraea (Zoila) friendii thersites*
         Lusby Rocks, St. Joseph Banks
         Group, SA.

18   *Cypraea (Zoila) friendii thersites*
         Reevesby I., SA.

Cypraea (Zoila) friendi friendii.
(Photo: Barry Wilson)

Plate 27

CYPRAEIDAE

Cypraea (Zoila) venusta (pp. 197 - 198) and *Cypraea (Zoila) friendii* (pp. 193 - 195) complexes - pale forms

1	*Cypraea (Zoila) venusta* off Beagle Is., WA; 160 m.
2	*Cypraea (Zoila) venusta* off Abrolhos Is., WA.
3	*Cypraea (Zoila) venusta* Hopetoun, WA; 25 m.
4	*Cypraea (Zoila) venusta* Cervantes, WA; 18 m.
5	*Cypraea (Zoila) venusta* off Mandurah, WA; 12 m.
6	*Cypraea (Zoila) venusta* Draper I., WA; 57 m.
7	*Cypraea (Zoila) venusta* Bremer Bay, WA; 30 m.
8	*Cypraea (Zoila) venusta* Esperance, WA; 35 m.
9	*Cypraea (Zoila) venusta* Recherche Arch., WA.
10	*Cypraea (Zoila) venusta* Esperance, WA.
11	*Cypraea (Zoila) venusta* Esperance, WA.
12	*Cypraea (Zoila) friendii jeaniana* North West Shelf edge, off Onslow ("*aurata*" form).

13 *Cypraea (Zoila) friendii jeaniana*
west of Point Maud, WA; 110 m
("*aurata*" form).

14 *Cypraea (Zoila) friendii jeaniana*
west of Point Maud, WA; 110 m.
("*aurata*" form).

15 *Cypraea (Zoila) friendii jeaniana*
trawler operating from Broome,
WA
("*aurata*" form).

16 *Cypraea (Zoila) friendii vercoi*
King George Sound, WA; 28 m
(intermediate form).

17 *Cypraea (Zoila) friendii vercoi*
east of Esperance, WA; trawled
outer shelf
("*contraria*" form).

18 *Cypraea (Zoila) friendii vercoi*
50 miles east of Esperance, WA;
145-180 m
("*contraria*" form).

19 *Cypraea (Zoila) friendii vercoi*
off Eucla, WA; trawled outer shelf
("*contraria*" form).

Approx. x .5

Plate 28

CYPRAEIDAE

1   *Cypraea (Zoila) venusta* (p. 198)
         Cape Naturaliste, WA; 12 m
         (*"episema"*) form.

2   *Cypraea (Zoila) venusta* (p. 198)
         Dunsborough, WA; 15 m
         (*"episema"*) form.

3   *Cypraea (Zoila) venusta* (p. 198)
         Geographe Bay, WA; 12 m
         (*"episema"*) form.

4   *Cypraea (Zoila) venusta* (p. 198)
         Sorrento Beach, WA; 6 m
         (*"sorrentensis"*) form.

5   *Cypraea (Zoila) venusta* (p. 198)
         Canal Rocks, Yallingup, WA; 20 m
         (*"episema"*) form.

6   *Cypraea (Zoila) venusta* (p. 198)
         off Mandurah, WA; 12 m.

7   *Cypraea (Zoila) decipiens* (p. 193)
         Port Hedland, WA.

8   *Cypraea (Zoila) decipiens* (p. 193)
         Broome, WA.

9   *Cypraea (Zoila) decipiens* (p. 193)
         Port Hedland, WA.

10   *Cypraea (Zoila) decipiens* (p. 193)
         Port Hedland, WA.

11   *Cypraea (Zoila) decipiens* (p. 193)
         Broome, WA.

12   *Cypraea (Zoila) eludens* (p. 193)
         Exmouth Gulf , WA; 25-35 m.

13   *Cypraea (Zoila) eludens* (p. 193)
         Exmouth Gulf , WA; 25-35 m.

14   *Cypraea (Zoila) eludens* (p. 193)
         Exmouth Gulf , WA; 25-35 m.

15   *Cypraea (Zoila) eludens* (p. 193)
         Exmouth Gulf , WA; 25-35 m.

16   *Cypraea (Zoila) rosselli* (pp. 197-198)
         Esperance, WA.

17   *Cypraea (Zoila) rosselli* (pp. 197-198)
         Esperance, WA.

18   *Cypraea (Zoila) rosselli* (pp. 197-198)
         Rottnest I., WA.

19   *Cypraea (Zoila) rosselli* (pp. 197-198)
         Rottnest I., WA.

Approx. x .6

Plate 28a (BELOW)

Cypraea (Zoila) mariellae (p. 197)
    1 a-b   HOLOTYPE. Locality uncertain -
          believed to be off the Kimberley
          coast, WA.

Cypraea (Zoila) perlae (p. 197)   Approx. x .6
    2 a-b   North West Shelf; exact locality
          unknown.

Plate 29

CYPRAEIDAE

Cypraea (Zoila) marginata complex (pp. 196 - 197)

1   *Cypraea (Zoila) marginata ketyana*
         northern WA - exact locality
         unknown.

2   *Cypraea (Zoila) marginata ketyana*
         north end of Bernier I., WA; 7 m.

3   *Cypraea (Zoila) marginata*
         (intermediate form)
         Jurien Bay, WA; 55 m.

4   *Cypraea (Zoila) marginata*
         (intermediate form)
         Geraldton, WA; 30 m.

5   *Cypraea (Zoila) marginata consueta*
         west of Mandurah, WA;
         trawled.

6   *Cypraea (Zoila) marginata*
         (intermediate form)
         a-b Abrolhos, WA; 30 m.

7   *Cypraea (Zoila) marginata*
         (intermediate form)
         Cervantes, WA; 55 m.

8   *Cypraea (Zoila) marginata consueta*
         Mewstone Rocks, Fremantle, WA.

9   *Cypraea (Zoila) marginata consueta*
         Rottnest I., WA.

10   *Cypraea (Zoila) marginata consueta*
         Carnac I., Fremantle, WA;
         5-10 m.

11   *Cypraea (Zoila) marginata consueta*
         Carnac I., Fremantle, WA;
         5-10 m.

12   *Cypraea (Zoila) marginata consueta*
         Carnac I., Fremantle, WA;
         5-10 m.

13   *Cypraea (Zoila) marginata albanyensis*
         Albany, WA; 5-10 m.

14   *Cypraea (Zoila) marginata consueta*
         Carnac I., Fremantle, WA;
         5-10 m.

15   *Cypraea (Zoila) marginata consueta*
         Carnac I., Fremantle, WA;
         5-10 m.

16   *Cypraea (Zoila) marginata albanyensis*
         Esperance, WA; 36 m.

17   *Cypraea (Zoila) marginata albanyensis*
         Esperance, WA.

18   *Cypraea (Zoila) marginata marginata*
         St. Francis I., SA; 55 m.

19   *Cypraea (Zoila) marginata marginata*
         Port Lincoln, SA.

20   *Cypraea (Zoila) marginata ? consueta*
         Rottnest I., WA; 32 m.

21   *Cypraea (Zoila) marginata marginata*
         Grindle I., Port Lincoln, SA.

Plate 30

CYPRAEIDAE

Subgenus *Umbilia*

1   *Cypraea (Umbilia) capricornica* (p. 192)
        Swain Reefs, Qld; 240 m.

2   *Cypraea (Umbilia) capricornica* (p. 192)
        S.E. of Hixson Cay, Swain Reefs,
        Qld; 225 m.

3   *Cypraea (Umbilia) capricornica* (p. 192)
        off Lady Musgrave I., Qld; 250 m.

4   *Cypraea (Umbilia) capricornica* (p. 192)
        off Hixson Cay, Swain Reefs, Qld;
        180 m.

5   *Cypraea (Umbilia) capricornica* (p. 192)
        Swain Reefs, Qld.

6   *Cypraea (Umbilia) capricornica* (p. 192)
        off Beacon Reef, Swain Reefs, Qld;
        245 m.

7   *Cypraea (Umbilia) hesitata* (p. 192)
        trawled off Coffs Harbour, NSW;
        173 m ("*beddomei*" form).

8   *Cypraea (Umbilia) capricornica* (p. 192)
        off Lady Musgrave I., Qld; 216 m.

9   *Cypraea (Umbilia) hesitata* (p. 192)
        off Montague I., NSW
        ("*howelli*" form).

10 *Cypraea (Umbilia) hesitata* (p. 192)
        off Lakes Entrance, Vic.

11 *Cypraea (Umbilia) hesitata* (p. 192)
        off Lakes Entrance, Vic.

12 *Cypraea (Umbilia) armeniaca* (p. 191)
        off Albany, WA.

13 *Cypraea (Umbilia) armeniaca* (p. 191)
        off Esperance, WA.

14 *Cypraea (Umbilia) armeniaca* (p. 191)
        off Esperance, WA.

15 *Cypraea (Umbilia) armeniaca* (p. 191)
        off Albany, WA; 150 m.

16 *Cypraea (Umbilia) armeniaca* (p. 191)
        Western Great Australian Bight,
        WA; 160 m.

Approx. x .6

Plate 31

CYPRAEIDAE

Cypraea eglantina. Dampier Archipelago, WA. Approx. natural size. (Photo: Barry Wilson)

Cypraea tigris. Cairns Reef, Qld. Approx. x .3. In daytime this species is often found in tide pools. (Photo: Barry Wilson)

Approx. x .6

Plate 32

CYPRAEIDAE

1 *Cypraea chinensis* (p. 181)
    a, c North West Cape, WA.
    b   Madang, PNG.

2 *Cypraea helvola* (p. 180)
    a-b Ashmore Reef, NT.

3 *Cypraea teres* (p. 177)
    a   Ashmore Reef, NT.
    b   Cocos Keeling.

4 *Cypraea alisonae* (p. 177)
    Kendrew I., WA.

5 *Cypraea annulus* (p. 184)
    a-b North West Cape, WA.

6 *Cypraea moneta* (p. 184)
    a-b South Muiron I., WA.

7 *Cypraea isabella* (p. 182)
    a   Ningaloo Marine Park, WA.
    b-d Ashmore Reef, NT.

8 *Cypraea stolida* (p. 177)
    a   Orpheus I., Qld.
    b   Delambre I., WA.
    c   Keppel Bay, Qld.

9 *Cypraea brevidentata* (p. 176)
    Langford Reef, Qld.

10 *Cypraea flaveola* (p. 180)
    a-b Stradbroke I., Qld.

11 *Cypraea coxeni* (p. 177)
    a-b Malaita, Solomon Is.

12 *Cypraea cernica* (p. 179)
    a   Garden I., WA.
    b   Abrolhos, WA.

13 *Cypraea miliaris* (p. 180)
    a-b Townsville, Qld; "*eburnea*" form.
    c-d Exmouth Gulf, WA; typical form.

14 *Cypraea walkeri* (p. 182)
    a-b Bramston Reef, Qld.

15 *Cypraea pyriformis* (p. 182)
    a, c Cairns, Qld.
    b, d Lee Point, NT.
    e, f One Arm Point, WA (*smithi*).

16 *Cypraea subviridis* (p. 182)
    a   off Stradbroke I., Qld; 165 m.
    b   Rottnest I., WA.
    c   Dampier, WA.
    d, e Humpy I., Qld.

17 *Cypraea xanthodon* (p. 182)
    a   Keppel Bay, Qld.
    b   Humpy I., Qld.

Cypraea walkeri on sponge trawled off Townsville, Qld.
(Photo: Barry Wilson)

Approx. x 1.1

1A   1B   1C   2A   2B   3A   3B   4

5A   5B   6A   6B   7A   7B   7C   7D

8A   8B   8C   9   10A   10B   11A   11B

12A   12B   13A   13B   13C   13D

14A   14B   15A   15B   15C   15D   15E   15F

16A   16B   16C   16D   16E   17A   17B

Plate 33
CYPRAEIDAE

1   *Cypraea felina* (p. 181)
          a-b   Southport, Qld.
          c-d   Heron I., Qld.

2   *Cypraea gracilis* (p. 188)
          a-b   One Arm Point, WA.
          c-d   Barrow I., WA.

3   *Cypraea poraria* (p. 180)
          a   Cocos Keeling.
          b   South Muiron I., WA.

4   *Cypraea asellus* (p. 187)
          a-b   Seringapatam Reef, WA.

5   *Cypraea hammondae* (p. 188)
          a-b   Moreton Bay, Qld.
          c   Port Hedland, WA.

6   *Cypraea minoridens* (p. 189)
          a-b   Christmas I.

7   *Cypraea fimbriata* (p. 188)
          Clerke Reef, Rowley Shoals, WA.

8   *Cypraea microdon* (p. 189)
          a-b   Siasi Is., Philippines.

9   *Cypraea punctata* (p. 185)
          Keppel Bay, Qld.

10   *Cypraea contaminata* (p. 188)
          a-b   Cartier I., NT.

11   *Cypraea hirundo* (p. 176)
          a-c   Delambre I., WA.

12   *Cypraea kieneri* (p. 176)
          a, c Mackay, Qld.
          b   Green I., Qld.

13   *Cypraea ursellus* (p. 177)
          a-b   Mermaid Reef, Rowley Shoals,
          WA.

14   *Cypraea pallidula* (p. 177)
          a   Ashmore Reef, NT.
          b   Langford Reef, Qld.
          c   Ningaloo Marine Park, WA.

15   *Cypraea quadrimaculata* (p. 177)
          a-b   Cape Leveque, WA.

16   *Cypraea clandestina* (p. 188)
          a   Broome, WA.
          b-c   Rosemary I., WA.

17   *Cypraea lutea* (p. 189)
          a-c   Port Hedland, WA.

18   *Cypraea ziczac* (p. 189)
          a   Broome, WA.
          b-c   Rabaul, PNG.

19   *Cypraea saulae* (p. 189)
          Broome, WA.

Approx. x 1.8

1A   1B   1C   1D   2A   2B

3A   3B   4A   4B   2C   2D

6A   7   8B   10A

5B   5C   6B   8A   9   10B

5A   11B   12A   12C   13A   13B

11A   11C   12B   16A

14B   14C   15B   16B

14A   17B   15A   16C

17A   17C   18B   18A   18C   19

Plate 34

CYPRAEIDAE

1   *Cypraea caurica* (p. 180)
         a-c Ashmore Reef, NT.

2   *Cypraea caputserpentis* (p. 179)
         a-b Ningaloo Marine Park, WA.
         c   Cape Naturaliste, WA.

3   *Cypraea cylindrica* (p. 181)
         a-b One Arm Point, WA.

4   *Cypraea erosa* (p. 180)
         a   Ashmore Reef, NT.
         b   Ningaloo Marine Park, WA.

5   *Cypraea errones* (p. 181)
         a   One Arm Point, WA.
         b   Ningaloo Marine Park, WA.

6   *Cypraea ovum* (p. 181)
         a-b Aru, Indonesia.

7   *Cypraea carneola* (p. 183)
         a-b Ningaloo Marine Park, WA.
         c   Stradbroke I., Qld.

8   *Cypraea lynx* (p. 183)
         a   Broome, WA.
         b   South Molle I., Qld.
         c   Ningaloo Marine Park, WA.

9   *Cypraea vitellus* (p. 183)
         a-c Abrolhos, WA.

10   *Cypraea hungerfordi* (p. 181)
         off Cape Moreton, Qld; 150 m.

11   *Cypraea hirasei* (p. 190)
         off Cape Moreton, Qld;
         HOLOTYPE of *queenslandica*.

12   *Cypraea langfordi* (p. 190)
         a   off Lynx & Leopard Reefs, Qld;
              166 m.
         b   Swain Reefs, Qld; 230 m.
         c   Southern Qld.

13   *Cypraea porteri* (p. 183)
         a-b off Hixon Cay, Swain Reefs, Qld;
              150-240 m.

Cypraea cylindrica; Port Hedland, WA.
(Photo: Barry Wilson)

Approx. x .8

1A 1B 1C 2A 2B 2C

3A 3B 4A 4B 5A 5B 6A

6B 7A 7B 7C 8A 8B 8C

9A 9B 9C 10 11

12A 12B 12C 13A 13B

Plate 35
CYPRAEIDAE

1   *Cypraea limacina* (p. 191)
        a   Dampier Arch., WA.
        b   Langford Reef, Qld.

2   *Cypraea nucleus* (p. 185)
        a   Cartier I., NT.
        b   Ashmore Reef., NT.

3   *Cypraea staphylaea* (p. 191)
        a   Ashmore Reef, NT.
        b   Langford Reef, Qld.

4   *Cypraea childreni* (p. 190)
        a-b Rowley Shoals, WA.

5   *Cypraea mariae* (p. 190)
        a-b Scott Reef, WA.

6   *Cypraea beckii* (p. 179)
        a   Mactan, Philippines.
        b   Rowley Shoals, WA.

7   *Cypraea bistrinotata* (p. 190)
        a-b Rowley Shoals, WA.

8   *Cypraea cicercula* (p. 190)
        a   Dawson Reef, Qld.
        b   Orpheus I., Qld.

9   *Cypraea globulus* (p. 190)
        a, c-d Cocos Keeling.
        b   Christmas I.

10   *Cypraea humphreysii* (p. 189)
        a-b Central Qld.

11   *Cypraea saulae* (p. 189)
        a   Princess Charlotte Bay, Qld.
        b   Langford Reef, Qld.

12   *Cypraea piperita* (p. 187)
        a-b Thorny Passage, SA.
        c   Albany, WA.
        d   Kangaroo I., SA.

        e-f Cape Naturaliste, WA.
        g   Augusta, WA.

13   *Cypraea comptoni* (p. 186)
        a   Anxious Bay, SA.
        c   Eyre Peninsula, SA.
        b, d-f Thorny I., SA.

14   *Cypraea pulicaria* (p. 187)
        a-d Yallingup, WA.

15   *Cypraea reevei* (p. 176)
        a   Frenchman Bay, WA.
        b-c Cape Naturaliste, WA.
        d   Carnac I., WA.

16   *Cypraea angustata* (p. 186)
        a   Kangaroo I., SA.
        b   Thorny Passage, SA.
        c   Port McDonnell, SA.

17   *Cypraea comptoni* (p. 186)
        Port McDonnell, SA;
        "*subcarnea*" form.

18   *Cypraea declivis* (p. 186)
        a   Port McDonnell, SA.
        b-c Port Arthur, Tas.

19   *Cypraea cribraria* (p. 178)
        a   Dampier Arch., WA.
        b   Port Gregory, WA;
            ("*exmouthensis*" form).
        c   Exmouth Gulf, WA.
        d   Rottnest I., WA;
            ("*exmouthensis*" form).

20   *Cypraea fallax* (p. 178)
        a   South coast, WA.
        b   Canal Rocks, WA.
        c   Denmark, WA.

Approx. x 1

Plate 36

NATICIDAE

Approx. x .9

Plate 37
CASSIDAE

1   *Semicassis labiata* (p. 233)
         a-b Iluka, NSW.

2   *Phalium bandatum* (p. 233)
             Nickol Bay, WA.

3   *Phalium areola* (p. 232)
         a-b Shelly Beach, Qld.

4   *Semicassis thomsoni* (p. 234)
         a-b Off Lady Musgrove, Qld.

5   *Semicassis angasi* (p. 233)
         a   Tin Can Bay, Qld.
         b   Off Onslow, WA.

6   *Semicassis whitworthi* (p. 234)
         a-b Jurien Bay, WA.

7   *Semicassis bisulcata* (p. 233)
         a   Shark Bay, WA.
         b-c North West Shelf, WA.

8   *Semicassis cf. pyrum* (p. 234)
             off Port Hedland, WA; 270 m.

9   *Semicassis paucirugis* (p. 234)
         a-b off Rottnest I., WA.

10   *Semicassis semigranosum* (p. 235)
         a-b SA.

11   *Semicassis pyrum* (p. 234)
         a   Port Phillip Bay, Vic.
         b   Esperance, WA; "*nivea*" form.
         c   Cape Howe; "*stadiale*" form.
         d   SW Rottnest I., WA; 460 m.
         e   Port Phillip Bay, Vic.

12   *Cypraecassis rufa* (p. 230)
             Locality uncertain, possibly
             Torres Strait.

13   *Phalium glaucum* (p. 233)
         a-b ?North Qld.

Semicassis angasi; Cairns Reef, Queensland.
(Photo: Barry Wilson)

Approx. x .6

Plate 38

CASSIDAE

1   *Casmaria ponderosa* (p. 231)
      a   Cartier I., NT.
      b   Great Barrier Reef, Qld.
      c   Scott Reef, WA.

2   *Casmaria erinacea* (p. 231)
      a   Seringapatam Atoll, WA.
      b   Orpheus I., Qld.
      c   Scott Reef, WA.

3   *Semicassis adcocki* (p. 234)
      a   Port Lincoln, SA.
      b   Eucla, WA.

4   *Semicassis sinuosum* (p. 235)
      a   Bunbury, WA.
      b   Recherche Arch., WA;
         "*angustatum*" form.

5   *Galeodea maccamleyi* (p. 231)
      a-b   Lady Musgrove I., Qld; 220 m.

6   *Galeodea echinophorella* (p. 231)
      a-b   near Rowley Shoals, WA; 460 m.

7   *Galeodea nipponica* (p. 231)
      a-b   near Rowley Shoals, WA; 460 m.

8   *Cassis fimbriata* (p. 230)
      a-b   Esperance, WA.
      c   Busselton, WA.

9   *Cassis nana* (p. 230)
      a-b   Swain Reefs, Qld; 190 m.

10   *Echinophoria wyvillei* (p. 232)
       Rowley Shoals, WA; 460 m.

11   *Echinophoria carnosa* (p. 232)
      a   near Rowley Shoals, WA; 460 m.
      b   Bonaparte Gulf, WA.

12   *Echinophoria kurodai* (p. 232)
      a-b   North West Shelf, WA.

Casmaria erinacea; Arafura Sea
(Photo: Barry Wilson)

Approx. x .7

Plate 39

TONNIDAE

Malea pomum; Ningaloo Marine Park, WA.
(Photo: Barry Wilson)

1A　　　　1B　　　　2　　　　3A　　　　3B

4　　　　5　　　　6　　　　7

8　　　　9　　　　10

11　　　　12A　　　12B　　　12C　　　13

Plate 40
FICIDAE, RANELLIDAE

1   *Ficus eospila* (p. 235)
        a-b Cape Borda, WA.
        c   Esperance, WA.

2   *Ficus subintermedia* (p. 236)
        North Qld.

3   *Ficus filosa* (p. 236)
        North West Shelf, WA.

4   *Biplex pulchellum* (p. 242)
        a-b Gladstone, Qld.

5   *Biplex pulchrum* (p. 242)
        a-b Onslow, WA.

6   *Gyrineum gyrinum* (p. 241)
        Palm I., Qld.

7   *Gyrineum bituberculare* (p. 241)
        Torres St, Qld.

8   *Gyrineum lacunatum* (p. 241)
        a   North Qld.
        b-c Darwin, NT.

9   *Sassia mimetica* (p. 250)
        a-b eastern SA; 75-155 m.

10   *Sassia parkinsonia* (p. 249)
        a-b Port Hacking, NSW.

11   *Sassia kampyla* (p. 249)
        a-b off NSW.

12   *Sassia petulans* (p. 250)
        Pirate Bay, Tas. (HOLOTYPE).

13   *Sassia eburnea* (p. 251)
        Port Arthur, Tas.

14   *Sassia verrucosa* (p. 251)
        Lorne, Vic.

15   *Sassia sexcostata* (p. 251)
        a-b Semaphore, SA.

16   *Sassia bassi* (p. 249)
        Stanley, Tas.

17   *Sassia subdistorta* (p. 250)
        a   Portland, Vic.
        b   Port Arthur, Tas.
        c   Esperance, WA.

18   *Sassia garrardi* (p. 250)
        Botany Bay, NSW; (PARATYPE).

19   *Sassia apenninica remensa* (p. 248)
        Botany Bay, NSW; (PARATYPE).

20   *Distorsio habei* (p. 239)
        a-b North West Shelf, WA.

21   *Distorsio anus* (p. 239)
        a-b Scott Reef, WA.

22   *Distorsio reticulata* (p. 239)
        a   central Qld.
        b   northern NSW.

Ficus eospila; Dampier Archipelago, WA. (Photo: Barry Wilson)

Approx. x 1

1A 1B 1C 2 3
4A 4B 5A 5B 9B
6 7 8A 8B 8C 9A
10A 10B 11A 11B 12 13 14 15A 15B
16 17A 17B 17C 18 19 20B
20A 21A 21B 22A 22B

Plate 41
RANELLIDAE

1   *Cymatium lotorium* (p. 244)
        Christmas I.

2   *Cymatium aquatile* (p. 245)
        North Qld.

3   *Cymatium succinctum* (p. 244)
        Keppel I., Qld.

4   *Cymatium pileare* (p. 246)
        Dampier Arch., WA.

5   *Cymatium parthenopeum* (p. 246)
        Sydney, NSW.

6   *Cymatium muricinum* (p. 244)
        North West Cape, WA.

7   *Cymatium cutaceum* (p. 244)
        Brunswick Heads, NSW.

8   *Argobuccinum pustulosum* (p. 241)
        Bicheno, Tas.

9   *Cymatium nicobaricum* (p. 245)
        Ningaloo Marine Park, WA.

10   *Fusitriton magellanicus* (p. 241)
        Broken Bay, NSW; 500 m.

11   *Ranella australasia* (p. 242)
        a   Port Arthur, Tas.
        b   Albany, WA.

12   *Cabestana tabulata* (p. 243)
        Esperance, WA.

13   *Cabestana spengleri* (p. 243)
        Long Reef, NSW.

14   *Charonia tritonis* (p. 243)
        Abrolhos, WA.

15   *Charonia lampas* (p. 243)
        a   Byron Bay, NSW; typical form.
        b   Margaret River, WA.

Cymatium parthenopeum; Marmion Marine Park, WA. Most ranellids have colourful animals. (Photo: Barry Wilson).

Approx. x .6

Plate 42

RANELLIDAE

1   *Cymatium exaratum* (p. 245)
    a   Marmion Marine Park, WA.
    b   Cockburn Sound, WA.

2   *Cymatium pfeifferianum* (p. 248)
    Bonaparte Gulf, WA.

3   *Cymatium vespaceum* (p. 246)
    a-b Cape Borda, WA.

4   *Cymatium labiosum* (p. 248)
    a-b Dampier Arch., WA.

5   *Cymatium cynocephalum* (p. 246)
    Shark Bay, WA.

6   *Cymatium caudatum* (p. 246)
    off Carnarvon, WA.

7   *Cymatium sarcostomum* (p. 247)
    a-b Rosemary I., WA.

8   *Cymatium pyrum* (p. 247)
    North Qld.

9   *Cymatium gutturnium* (p. 247)
    a-b North Qld.

10   *Cymatium sinense* (p. 247)
    southern Qld.

11   *Cymatium mundum* (p. 245)
    Torres St., Qld.

12   *Cymatium caudatum* (p. 246)
    Port Hedland, WA.

13   *Cymatium oblitum* (p. 247)
    North West Cape, WA.

14   *Cymatium occidentale* (p. 247)
    a   Solitary Is., NSW.
    b   Marmion Marine Park, WA.

15   *Cymatium rubeculum* (p. 248)
    a-b Dampier Arch., WA.

16   *Cymatium hepaticum* (p. 247)
    a   Scott Reef, WA.
    b   Rowley Shoals, WA.

Plate 43

BURSIDAE

Approx. x .6

Plate 44

JANTHINIDAE, EPITONIIDAE

1	*Cirsotrema varicosum* (p. 274) a-b Lodestone Reef, Qld.		21	*Filiscala martini* (p. 279) a Cockburn Sound, WA. b-c Qld
2	*Cirsotrema bavayi* (p. 274) Hope I., Qld.		22	*Epitonium jukesianum* (p. 277) Garden I., WA.
3	*Cirsotrema kieneri* (p. 274) Broome, WA.		23	*Epitonium philippinarum* (p. 278) Port Douglas, Qld.
4	*Dannevigena martyr* (p. 275) Jervis Bay, NSW; 137 m.		24	*Epitonium barissum* (p. 276) Sydney Harbour, NSW.
5	*Cirsotrema sp.* off Port Hedland, WA; 400 m.		25	*Epitonium helicornum* (p. 277) Port Hacking, NSW.
6	*Epitonium pallasi* (p. 278) a Port Stephens, NSW. b Yeppoon, Qld.		26	*Epitonium sexcostum* (p. 279) Keppel Bay, Qld.
7	*Epitonium imperialis* (p. 277) Garden I., WA.		27	*Epitonium chrystyi* (p. 276) Port Hacking, NSW.
8	*Narvaliscala dorysa* (p. 280) Nowra, NSW, 360 m.		28	*Epitonium lyrum* (p. 277) North West I., Qld.
9	*Epitonium bullatum* (p. 276) Long Reef, NSW.		29	*Opalia australis* (p. 280) a-b Marmion Marine Park, WA.
10	*Variciscala raricostata* (p. 281) Green I., Qld.		30	*Opalia granosa* (p. 280) a Long Reef, NSW; ("*ballinensis*"). b Fitzgerald Inlet., WA.
11	*Epitonium tacitum* (p. 279) Sydney Harbour (HOLOTYPE).		31	*Epitonium neglectum* (p. 277) Keppel Bay, Qld.
12	*Epitonium pyramidalis* (p. 278) Kurrimine Beach, Qld.		32	*Epitonium alatum* (p. 276) Yeppoon, Qld.
13	*Epitonium perplexum* (p. 278) Rottnest I., WA.		33	*Epitonium parspeciosum* (p. 278) Port Stephens, NSW.
14	*Epitonium auritum* (p. 276) Cape Borda, WA.		34	*Epitonium stigmaticum* (p. 279) Keppel Bay, Qld.
15	*Epitonium delicatulum* (p. 276) Westernport, Vic.		35	*Recluzia hargravesi* (p. 281) a-b Caloundra, Qld.
16	*Epitonium godfreyi* (p. 277) Semaphore, Adelaide, SA.		36	*Janthina janthina* (p. 281) a-b Bremer Bay, WA.
17	*Epitonium scalare* (p. 278) a-b Yeppoon, Qld.		37	*Janthina exigua* (p. 281) a-b Bermagui, NSW.
18	*Epitonium marmoratum* (p. 277) Green I., Qld.		38	*Janthina pallida* (p. 281) a Kilcare, NSW. b Two Peoples Bay, WA.
19	*Epitonium fasciatum* (p. 277) Upolu Cay., Qld.			
20	*Epitonium replicatum* (p. 278) Kurrimine Beach., Qld.			

Approx. x 1.2

GLOSSARY

aculeate prickly

acuminate tapered to a point

anterior canal anterior notch or trough-like or tubular extension of the shell supporting the anterior siphon (= siphonal canal)

aperture opening or entrance of the shell (= mouth)

apex tip of the spire

apical whorls those whorls near the apex

apophysis peg on underside of operculum

auriform ear-shaped

axial parallel or nearly so to the shell axis (= longitudinal)

axis imaginary line through the apex, about which the whorls are coiled

benthic living on the sea bed

biconic resembling two cones placed base to base

body whorl last and usually the largest whorl of the coiled shell

callus calcareous thickening (= callosity)

canaliculate channelled or grooved

cancellate ornamented with intersecting spiral and axial ridges

carinate keeled

cingulus ring

circumtropical found throughout the world's tropical zone

clathrate latticed, crossing to form a network

clavate club-shaped

columella pillar along the axis of a coiled shell, formed by the inner walls of the whorls and often forming a thickened inner apertural lip

columellar lip inner edge of the aperture comprising the visible part of the columella

concentric refers to spiral sculpture

congeneric of the same genus

conspecific of the same species

convolute with the last whorl completely concealing the earlier ones

cord fine round-topped spiral or axial ridge

coronate with tubercles or nodules around the shoulders of the whorls

costa rib (plural costae)

costate ribbed

crenate with the edge notched (= crenulate)

cusp point or pointed projection

decollate apex broken off

decussate with a pattern of intersecting striae

dentate toothed (= denticulate)

dentition tooth structure - in gastropods referring to the elements of the radula or nodules along the shell aperture

dextral coiled with a right-hand spiral, i.e. clockwise when viewed from the apex

dimorphic of two types

direct development refers to larval development which takes place entirely within a protective capsule - without a planktonic phase

dorsum back of the body whorl

ectoparasite a parasite living on the outside of the host body

edentulous without teeth

endemic peculiar to a region

entire uninterrupted

epipodial of the foot; usually referring to tentacles formed around the sides of the foot

exsert protruding

fasciate banded or striped

fasciole spiral band formed by successive edges of a canal or notch (either anterior or posterior)

flexuous winding, crooked

foliaceous with thin leaf-like lamellae

foot the muscular ventral part of the gastropod body used for locomotion

fossula shallow linear depression on the anterior part of the inner lip

frondose leafy

funicle ridge spiralling into the umbilicus

fusiform spindle-shaped, swollen at the centre and tapering almost equally toward the ends

genus group of genetically related species possessing certain characters in common (plural *genera*)

gibbous humpbacked

glabrous bald, smooth

granose surface with minute grains (= granulose)

growth line more or less axial line representing former resting position of the outer lip (= incremental stria)

height distance from the apex to the anterior tip of the shell (sometimes also called length)

hermaphrodite with functional male and female organs in the same individuals

holotype the single specimen designated as the "type" by the original author of a species name at the time of the original description

homonym one of two or more identical but independently proposed names for different species, genera or other taxa

hyaline glassy

imbricated with laminae overlapping each other like roof tiles

imperforate without an umbilicus

impressed indented

inner lip edge of the aperture near the shell axis, extending from the suture to the anterior end of the columella (= labium or columellar lip)

interspaces grooves or spaces between ribs or cords (= interstices)

intertidal living between high and low tide levels

intervarical between the varixes

lamella thin plate (plural lamellae)

lanceolate elongate, lance-like

larviparous retaining larvae within the protection of the mother's body

lecithotrophic refers to planktonic larval development where little yolk is supplied in the egg and development time in the plankton is short

lira prominent elevated ridge (plural lirae)

littoral the zone of the shore between low and high tide levels (= intertidal zone)

maculate patterned with blotches, i.e. with maculations

mammillate rounded or dome-shaped protuberance like a teat

mantle the thin skin-like organ which covers the body mass of molluscs and contains the shell-secreting glands (=pallium)

mantle cavity the space beneath the rim of the mantle which contains the gills and the openings of the alimentary, excretory and reproductive organs

monotypic a genus with only one species

multispiral with many whorls

nacreous with a layer of nacre or "mother of pearl"

nebulous hazy, clouded

nodulose with conspicuous nodules (= nodose)

nominate subspecies (or subgenus) having the same name and the same type as the species (or genus) which has been subdivided

nucleus earliest formed part of a shell or operculum

obconical with the form of an inverted cone

obtuse blunt or rounded

operculum horny or calcareous structure borne by the foot and serving for the closure of the aperture

outer lip outer margin of the aperture (= labrum)

oviparous bringing forth eggs which hatch after leaving the mother's body

ovoviviparous retaining eggs within the protection of the mother's body until hatching

pallial relating to the mollusc pallium or mantle

papillate teat-like (= papillose)

paratype a specimen other than the holotype which was seen and designated by the author at the time of the original description of a species or subspecies

parietal area that part of the body whorl forming the inner side of the aperture between the suture and the columella

paucispiral with few whorls

pelagic inhabiting the water column of the open ocean

penultimate next to the last-formed; refers to the last whorl prior to the body whorl

periostracum outer layers of horny material covering the calcareous shell (= epidermis)

periphery greatest circumference of a whorl

peristome edge of the aperture; said to be complete when the inner and outer lips are entire

pillar structure calloused part of the body whorl beside the columella

plait spiral fold or ridge on the columellar lip (= plication)

planispiral coiled in a single plane

planktonic drifting or weakly swimming in the ocean

planktotrophic refers to planktonic larval development where the larvae feeds itself on micro-organisms and development time in the plankton is long

polytypic with many forms

porcellaineous surface and texture as in porcellain

posterior canal notch or trough-like or tubular extension of the posterior apertural margin supporting the posterior siphon (= anal canal, excurrent siphonal canal)

preoccupied a name which is unavailable because it had been used before for another taxon

presutural above (before) the suture

proboscis an extensible snout, usually with the mouth at its end

produced elongate, extended

protandry sex reversal - the individual begins life as a male but later changes to female

protoconch embryonic shell, present in the adult as the apical or nuclear whorls and often demarcated from the teleoconch whorls by a change of sculpture

punctate minutely pitted

pyriform pear-shaped

rachiglossate type of radula with three teeth per row, a central tooth and a lateral on each side

recurved with the distal end (e.g. anterior canal) bent away from the shell axis

reflected turned outward and backward at the margin

reticulate forming a network of intersecting lines

rhipidoglossate type of radula with many marginal teeth

rib round-topped elevated ridge of moderate width and prominence

rostrate beaked

rugose rough

scabrous scaly

selenizone spiral band of crescentic growth lines generated by a notch or slit in some archae-gastropods

sensu lato (s.l.) in the broad sense

sensu stricto (s.s.) in the strict sense

septum a thin plate (plural septae)

serrate notched or toothed at the edge like a saw

sinistral coiled in a left-handed spiral, i.e. anti-clockwise when viewed from the apex

sinuate wavy (= sinuous)

sinus a deep indentation or cavity

spatula spoon-shaped muscle scar in limpets

spatulate spoon-shaped

species group of actually or potentially interbreeding natural populations which are reproductively isolated (unable to interbreed) from other such groups

spinose spiny

spire the coiled part of the shell consisting of all the whorls except the last

squamose with scales (= squamiferous)

stria narrow and shallow incised groove (plural *striae*)

striate marked with striae

sublittoral below low-tide level

subspecies a geographically defined group of populations comprising individuals which possess characteristics distinguishing them from other such subdivisions of the species

substrate the material of the sea bed (= substratum)

subsutural below the suture

subulate slowly tapering to a sharp point (as in *Terebra*)

sulcate grooved or furrowed

sulcus groove or furrow (plural *sulci*)

suture continuous line on the shell surface where successive whorls adjoin

synonym each of two or more different names for the same species or other taxon

synonymy a list of synonyms

syntype one of several type specimens where none has been designated as holotype

taenioglossate type of radula with five teeth per row, i.e. a lateral and two marginals on each side

taxon a taxonomic group that is distinct from other groups and has been distinguished by a name (plural *taxa*)

taxonomy the theory and practice of classifying organisms

teleplanic refers to species with very long-lived planktonic larvae

tessellate with a colour pattern consisting of regular patches

trochiform with a conical shell like that of *Trochus*

truncate cut off at the end, blunt

tubercle protuberance or knob

tumid swollen

turbinate turban-shaped

turbiniform with the form of a turban, as in the genus *Turbo*

turreted with the spire whorls like a succession of turrets emerging one above the other

type see holotype

type locality locality at which the type specimen was collected

type species the species which was designated type of a genus or subgenus

umbilicus cavity or hollow around which the axis of shells form when the inner walls of successive whorls do not meet

varicose with varixes (= varixed)

varix elevated ridge formed by a thickened and reflected former outer lip (plural *varixes*)

ventricose swollen

ABBREVIATIONS

Arch. Archipelago

auct. of various authors

Aus. Australia

cf compare with

cm centimetres

CSIRO Commonwealth Scientific and Industrial
Research Organisation

et al. and others

Gp. Group

I. Island

ibid. in the same place

ICZN International Commission for Zoological
Nomenclature

Is. Islands

loc. cit. location previously cited

m metres

mm millimetres

NSW New South Wales

NZ New Zealand

NT Northern Territory

Pen Peninsula

PNG Papua New Guinea

Pt. Point

Qld Queensland

SA South Australia

St. Strait

Tas. Tasmania

var. a variety of

Vic. Victoria

viz "which see"; thus

WA Western Australia

INDEX

KEY TO INDEX　　family name — HALIOTIDAE;　common name — abalone;　species or subspecies — *aboriginea*;　genus — **Margovula**;
subgenus — **Acutospira**;　subgenus synonym — *Actinoleuca*　[= is a synonym of;　subsp. is a subspecies of]

INDEX　387

KEY TO INDEX   family name — HALIOTIDAE;   common name — abalone;   species or subspecies — *aboriginea*;   genus — **Margovula**;   subgenus — **Acutospira**;   subgenus synonym — *Actinoleuca*   [= is a synonym of;   subsp. is a subspecies of]

388   AUSTRALIAN MARINE SHELLS

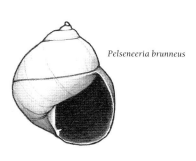

Pelseneeria brunneus

C

KEY TO INDEX   family name — HALIOTIDAE;   common name — abalone;   species or subspecies — *aboriginea*;   genus — **Margovula**;
    subgenus — **Acutospira**;   subgenus synonym — *Actinoleuca*   [= is a synonym of;   subsp. is a subspecies of]

INDEX   **389**

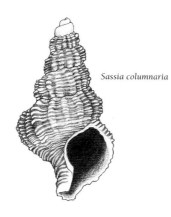

Sassia columnaria

KEY TO INDEX   family name — HALIOTIDAE; common name — abalone; species or subspecies — *aboriginea*; genus — **Margovula**;
    subgenus — **Acutospira**; subgenus synonym — *Actinoleuca* [= is a synonym of; subsp. is a subspecies of]

INDEX **391**

D

E

Phenacovolva dancei

KEY TO INDEX   family name — HALIOTIDAE; common name — abalone; species or subspecies — *aboriginea*; genus — **Margovula**;
subgenus — **Acutospira**; subgenus synonym — *Actinoleuca*   [= is a synonym of; subsp. is a subspecies of]

392   AUSTRALIAN MARINE SHELLS

F

KEY TO INDEX   family name — HALIOTIDAE; common name — abalone; species or subspecies — *aboriginea*; genus — **Margovula**;
        subgenus — **Acutospira**; subgenus synonym — *Actinoleuca* [= is a synonym of; subsp. is a subspecies of]

INDEX 393

Tricolia gabiniana

KEY TO INDEX   family name — HALIOTIDAE; common name — abalone; species or subspecies — *aboriginea*; genus — **Margovula**;
    subgenus — **Acutospira**; subgenus synonym — *Actinoleuca*   [= is a synonym of; subsp. is a subspecies of]

394   AUSTRALIAN MARINE SHELLS

H

I

KEY TO INDEX   family name — HALIOTIDAE;   common name — abalone;   species or subspecies — *aboriginea*;   genus — **Margovula**;
     subgenus — **Acutospira**;   subgenus synonym — *Actinoleuca*   [= is a synonym of;   subsp. is a subspecies of]

INDEX   395

KEY TO INDEX   family name — HALIOTIDAE;   common name — abalone;   species or subspecies — *aboriginea*;   genus — **Margovula**;
subgenus — **Acutospira**;   subgenus synonym — *Actinoleuca*   [= is a synonym of;   subsp. is a subspecies of]

M

Plesiotrochus monachus

N

KEY TO INDEX   family name — HALIOTIDAE; common name — abalone; species or subspecies — *aboriginea*; genus — **Margovula**;
   subgenus — **Acutospira**; subgenus synonym — *Actinoleuca*   [= is a synonym of; subsp. is a subspecies of]

398   AUSTRALIAN MARINE SHELLS

O

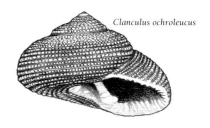

Clanculus ochroleucus

KEY TO INDEX   family name — HALIOTIDAE; common name — abalone; species or subspecies — *aboriginea*; genus — **Margovula**;
subgenus — **Acutospira**; subgenus synonym — *Actinoleuca* [= is a synonym of; subsp. is a subspecies of]

INDEX 399

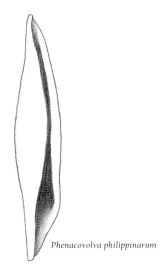

Phenacovolva philippinarum

KEY TO INDEX family name — HALIOTIDAE; common name — abalone; species or subspecies — *aboriginea*; genus — **Margovula**;
  subgenus — **Acutospira**; subgenus synonym — *Actinoleuca* [= is a synonym of; subsp. is a subspecies of]

INDEX 401

KEY TO INDEX   family name — HALIOTIDAE; common name — abalone; species or subspecies — *aboriginea*; genus — **Margovula**;
subgenus — **Acutospira**; subgenus synonym — *Actinoleuca*   [= is a synonym of; subsp. is a subspecies of]

S

Royella sinon

KEY TO INDEX   family name — HALIOTIDAE; common name — abalone; species or subspecies — aboriginea; genus — **Margovula**;
    subgenus — **Acutospira**; subgenus synonym — Actinoleuca   [= is a synonym of; subsp. is a subspecies of]

INDEX **403**

T

Larinopsis turbinatus

KEY TO INDEX   family name — HALIOTIDAE; common name — abalone; species or subspecies — *aboriginea*; genus — **Margovula**;
subgenus — **Acutospira**; subgenus synonym — *Actinoleuca*   [= is a synonym of; subsp. is a subspecies of]

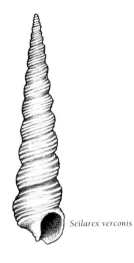

Seilarex verconis

W

KEY TO INDEX   family name — HALIOTIDAE; common name — abalone; species or subspecies — *aboriginea*; genus — **Margovula**;
   subgenus — **Acutospira**; subgenus synonym — *Actinoleuca*   [= is a synonym of; subsp. is a subspecies of]

X

Y

Z

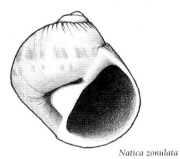

Natica zonulata

KEY TO INDEX   family name — HALIOTIDAE; common name — abalone; species or subspecies — *aboriginea*; genus — **Margovula**;
subgenus — **Acutospira**; subgenus synonym — *Actinoleuca*   [= is a synonym of; subsp. is a subspecies of]

INDEX 407

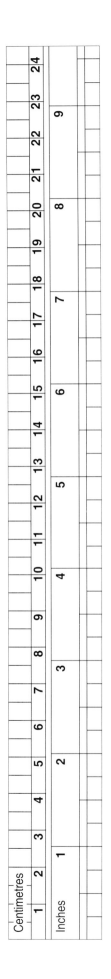

METRIC CONVERTOR

Inches	Millimetres
1/8	3.1750
1/4	6.3500
3/8	9.5250
1/2	12.7000
3/4	19.0500
5/8	15.8750
7/8	22.2250
1	25.400
2	50.800
3	76.200
4	101.600
5	127.000
6	152.400
7	177.800
8	203.200
9	228.600
10	254.000
11	279.400
12	304.800

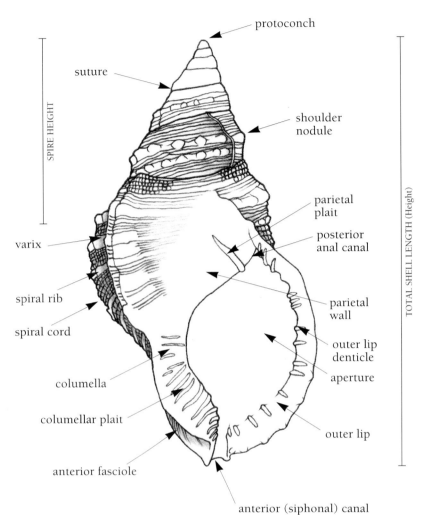

CONVERSION FACTORS

Quantity	Imperial Unit	Metric Unit	Imperial to Metric	Metric to Imperial
LENGTH	inch (in)	millimetre (mm) or centimetre (cm)	1 in = 25.4 mm	1 cm = 0.394 in
	foot (ft)	centimetre or metre (m)	1 ft = 30.5 cm	1 m = 3.28 ft
	yard (yd)	metre (m)	1 yd = 0.914 m	1 m = 1.09 yd
	mile	kilometre (km)	1 mile = 1.61 km	1 km = 0.621 mile
TEMPERATURE	degree Farenheit (°F)	degree Celsius (°C)	$°C = {}^5/_9(°F - 32)$	$°F = \frac{9 \times °C}{5} = 32$